T0331015

DOUBLY FED INDUCTION MACHINE

DOUBLY FED INDUCTION MACHINE

MODELING AND CONTROL
FOR WIND ENERGY GENERATION

Gonzalo Abad
Jesús López
Miguel A. Rodríguez
Luis Marroyo
Grzegorz Iwanski

Mohamed E. El-Hawary, *Series Editor*

IEEE PRESS

A JOHN WILEY & SONS, INC., PUBLICATION

Library of Congress Cataloging-in-Publication Data:

Doubly fed induction machine : modeling and control for wind energy generation /
G. Abad... [et al.].
    p. cm.
  Includes bibliographical references.
  ISBN 978-0-470-76865-5 (hardback)
  1. Induction generators–Mathematical models. 2. Induction generators–Automatic control. 3. Wind turbines–Equipment and supplies. I. Abad, G. (Gonzalo), 1976-
  TK2451.D68 2011
  621.31'6–dc22

                                                                        2011006741

Printed in the United States of America

ePDF ISBN: 978-1-118-10494-1
ePub ISBN: 978-1-118-10495-8
oBook ISBN: 978-1-118-10496-5

10 9 8 7 6 5 4 3 2 1

■ CONTENTS

The IEEE Press Series on Power Engineering

Over the last years, there has been a strong penetration of renewal energy resources into the power supply network. Wind energy generation has played and will continue to play a very important role in this area for the coming years.

Doubly fed induction machine (DFIM) based wind turbines have undoubtedly arisen as one of the leading technologies for wind turbine manufacturers, demonstrating that it is a cost effective, efficient, and reliable solution. This machine, a key element of the wind turbine, is also known in the literature as the wound rotor induction machine (WRIM). It presents many similarities with the widely used and popular squirrel cage induction machine (SCIM). However, despite the parallelism of both machines, the DFIM requires its own specific study for an adequate understanding.

Although there have been a significant number of excellent textbooks on the subject of induction machine modeling and control, books containing a significant portion of material related to the DFIM are less common. Therefore, today this book seems to be the unique and comprehensive reference, exclusively dedicated to the DFIM modeling and control and applied to wind energy generation.

This book provides the reader with basic and advanced knowledge about DFIM based wind turbines, including market overview and tendencies, discussing realistic and practical problems with numerical and graphical illustrative examples, as well as providing guidance to help understand the new concepts.

The technical level of the book increases progressively along the chapters, covering first basic background knowledge, and later addressing advanced study of the DFIM. The book can be adopted as a textbook by nonexpert readers, undergraduate or postgraduate students, to whom the first chapters will help lay the groundwork for further reading. In addition, a more experienced audience, such as researchers or professionals involved in covered topics, would also benefit from the reading of this book, allowing them to obtain a high level of understanding and expertise, of DFIM based wind turbines.

It must be mentioned that, by means of this book, the reader not only will be able to learn from wind turbine technology or from the DFIM itself, but also enhance his/her knowledge on AC drives in general, since many aspects of this book present universal character and may be applied to different AC machines that operate on different applications.

On the other hand, it is the belief of the authors that what makes this DFIM based wind turbine technology cost effective (i.e., its reduced size converter requirement due to the double supply nature of the machine), makes its study challenging for new readers. The combination and coordination of the converter supply and grid supply, compared to single supplied machines such as asynchronous or synchronous machines, lead us to a more enriching environment in terms of conceptual understanding.

In addition, the direct grid supply can be a disadvantage, when the machine must operate under a faulty or distorted grid voltage scenario; especially if its disconnection must be avoided, fulfilling the generation grid code requirements. This mainly occurs because the stator windings of the machine are directly affected by those perturbations. In order to move forward with these problematic but realistic and unavoidable situations, additional active hardware protections or increased size of supplying converter are commonly adopted, accompanied by special control adaptations. Because of this, the work focuses on voltage disturbance analysis for DFIM throughout the book.

It is clear that this work is not intended to be a defense of DFIM based wind turbines, as the best technological solution to the existing alternative ones. Instead, the objective of this book is to serve as a detailed and complete reference, of the well established wind turbine concept.

No matter what the future holds, DFIM based wind turbines have gained an undoubtedly distinguished place that will always be recognized in the history of wind energy generation.

Finally, we would like to first express our sincere gratitude to Professor M. P. Kazmierkowski, for encouraging us to write this book. We wish to also thank everyone who has contributed to the writing of this book. During the last ten years, there have been a significant number of students, researchers, industry, and university colleagues who have influenced us, simply with technical discussions, or with direct and more concise contributions. Thanks to your daily and continuous support, this project has become a reality.

To conclude, we would like also to acknowledge IEEE Press and John Wiley & Sons, for their patience and allowance of the edition of the book.

GONZALO ABAD
JESÚS LÓPEZ
MIGUEL A. RODRÍGUEZ
LUIS MARROYO
GRZEGORZ IWANSKI

CHAPTER 1

Introduction to A Wind Energy Generation System

1.1 INTRODUCTION

The aim of this chapter is to provide the basic concepts to understand a wind energy generation system and the way it must be operated to be connected to the utility grid.

It covers general background on wind turbine knowledge, not only related to the electrical system, but also to the mechanical and aerodynamics characteristics of wind turbines.

In Section 1.2 the components and basic concepts of a fixed speed wind turbine (FSWT) are explained, as an introduction to a modern wind turbine concept; also, energy extraction from the wind and power–torque coefficients are also introduced.

In Section 1.3 a simple model for the aerodynamic, mechanical, and pitch systems is developed together with a control system for a variable speed wind turbine (VSWT). This section explains the different configurations for the gearbox, generator, and power electronics converter, used in a VSWT.

Section 1.4 describes the main components of a wind energy generation system (WEGS), starting with a VSWT based on a doubly fed induction motor (DFIM); then a wind farm electrical layout is described and finally the overall control strategy for the wind farm and the wind turbine.

In Section 1.5, the grid integration concepts are presented since the rising integration of wind power in the utility grid demands more constraining connection requirements.

Since the low voltage ride through (LVRT) is the most demanding in terms of control strategy, Section 1.6 deals with the LVRT operation description. The origin, classification, and description of voltage dips are given in order to understand specifications for the LVRT. The section finishes by describing a grid model suitable to validate the LVRT response of wind turbines.

Section 1.7 provides a survey of solutions given by different wind turbine manufacturers. And finally a 2.4 MW VSWT is numerically analyzed.

To conclude, the next chapters are overviewed in Section 1.8.

Doubly Fed Induction Machine: Modeling and Control for Wind Energy Generation,
First Edition. By G. Abad, J. López, M. A. Rodríguez, L. Marroyo, and G. Iwanski.
© 2011 the Institute of Electrical and Electronic Engineers, Inc. Published 2011 by John Wiley & Sons, Inc.

1.2 BASIC CONCEPTS OF A FIXED SPEED WIND TURBINE (FSWT)

1.2.1 Basic Wind Turbine Description

The basic components of a wind turbine are described by means of a fixed speed wind turbine, based on a squirrel cage (asynchronous machine) and stall–pitch power control. This technology, developed in the late 1970s by pioneers in Denmark, was widely used during the 1980s and 1990s, and was the base of wind energy expansion in countries like Spain, Denmark, and Germany during the 1990s.

The main manufacturers developing this technology have been Vestas, Bonus (Siemens), Neg-Micon and Nordtank, in Denmark, Nordex and Repower in Germany, Ecotècnia (Alstom), Izar-Bonus and Made in Spain, and Zond (Enron-GE) in the United States. At present, many other small manufacturers and new players such as Sulzon in India or GoldWind in China are in the market.

The first fixed speed wind turbines were designed and constructed under the concept of reusing many electrical and mechanical components existing in the market (electrical generators, gearboxes, transformers) looking for lower prices and robustness (as the pioneers did when they manufactured the first 25 kW turbines in their garages in Denmark). Those models were very simple and robust (most of them are still working, and there is a very active secondhand market).

To achieve the utility scale of 600,750, and 1000 kW, development of wind turbines took only ten years, and around two-thirds of the world's wind turbines installed in the 1980s and 1990s were fixed speed models.

Before we describe the FSWT, let's have a look at the main concepts related to this technology:

- The fixed speed is related to the fact that an asynchronous machine coupled to a fixed frequency electrical network rotates at a quasifixed mechanical speed independent of the wind speed.
- The stall and pitch control will be explained later in the chapter, but is related to the way the wind turbine limits or controls the power extracted from the wind.

Figure 1.1 shows the main components of a fixed speed wind turbine.

The nacelle contains the key components of the wind turbine, including the gearbox and the electrical generator. Service personnel may enter the nacelle from the tower of the turbine.

To the left of the nacelle we have the wind turbine rotor, that is, the rotor blades and the hub. The rotor blades capture the wind and transfer its power to the rotor hub. On a 600 kW wind turbine, each rotor blade measures about 20 meters in length and is designed much like the wing of an aeroplane.

The movable blade tips on the outer 2–3 meters of the blades function as air brakes, usually called tip brakes. The blade tip is fixed on a carbon fiber shaft, mounted on a bearing inside the main body of the blade. On the end of the shaft inside the main blade, a construction is fixed, which rotates the blade tip when subjected to an outward movement. The shaft also has a fixture for a steel wire, running the length of the blade from the shaft to the hub, enclosed inside a hollow tube.

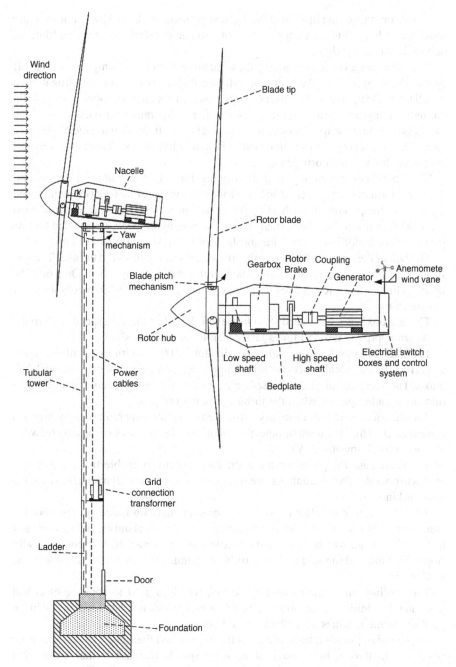

Figure 1.1 Main components of a fixed speed wind turbine.

During operation, the tip is held fast against the main blade by a hydraulic cylinder inside the hub, pulling with a force of about 1 ton on the steel wire running from the hub to the blade tip shaft.

When it becomes necessary to stop the wind turbine, the restraining power is cut off by the release of oil from the hydraulic cylinder, thereby permitting centrifugal force to pull the blade tip outwards. The mechanism on the tip shaft then rotates the blade tip through 90 degrees, into the braking position. The hydraulic oil outflow from the hydraulic cylinder escapes through a rather small hole, thus allowing the blade tip to turn slowly for a couple of seconds before it is fully in position. This thereby avoids excessive shock loads during braking.

The tip brakes effectively stop the driving force of the blades. They cannot, however, normally completely stop blade rotation, and therefore for every wind speed there is a corresponding freewheeling rotational speed. The freewheeling rotational speed is much lower than the normal operational rotational speed, so the wind turbine is in a secure condition, even if the mechanical brake should possibly fail.

The hub of the rotor is attached to the low speed shaft of the wind turbine. The low speed shaft of the wind turbine connects the rotor hub to the gearbox. On a 600 kW wind turbine, the rotor rotates relatively slowly, about 19–30 revolutions per minute (rpm).

The gearbox has a low speed shaft to the left. It makes the high speed shaft to the right turn approximately 50 times faster than the low speed shaft.

The high speed shaft rotates with approximately 1500 revolutions per minute (rpm) and drives the electrical generator. It is equipped with an emergency mechanical disk brake. The mechanical brake is used in case of failure of the aerodynamic brake (movable blade tips), or when the turbine is being serviced.

The electrical generator is usually a so-called induction generator or asynchronous generator. On a modern wind turbine, the maximum electric power is usually between 500 and 1500 kilowatts (kW).

The shaft contains pipes for the hydraulics system to enable the aerodynamic brakes to operate. The hydraulics system is used to reset the aerodynamic brakes of the wind turbine.

The electronic controller contains a computer that continuously monitors the condition of the wind turbine and controls the yaw mechanism. In case of any malfunction (e.g., overheating of the gearbox or the generator), it automatically stops the wind turbine and calls the turbine operator's computer via a telephone modem link.

The cooling unit contains an electric fan, which is used to cool the electrical generator. In addition, it contains an oil cooling unit, which is used to cool the oil in the gearbox. Some turbines have water-cooled generators.

The tower of the wind turbine carries the nacelle and the rotor. Generally, it is an advantage to have a high tower, since wind speeds increase farther away from the ground.

Towers may be either tubular towers (such as the one in Figure 1.1) or lattice towers. Tubular towers are safer for the personnel who have to maintain the turbines, as they may use an inside ladder to get to the top of the turbine. The advantage of lattice towers is primarily that they are cheaper. A typical 600 kW turbine will have a tower of 40–60 meters (the height of a 13–20 story building).

Wind turbines, by their nature, are very tall slender structures. The foundation is a conventional engineering structure that is designed mainly to transfer the vertical load (dead weight). However, in the case of wind turbines, due to the high wind and environmental loads experienced, there is a significant horizontal load that needs to be accounted for.

The yaw mechanism uses electrical motors to turn the nacelle with the rotor against the wind. The yaw mechanism is operated by the electronic controller, which senses the wind direction using the wind vane. Normally, the turbine will yaw only a few degrees at a time, when the wind changes its direction. The anemometer and the wind vane are used to measure the speed and the direction of the wind.

The electronic signals from the anemometer are used by the wind turbine's electronic controller to start the wind turbine when the wind speed reaches approximately 5 meters per second (m/s). The computer stops the wind turbine automatically if the wind speed exceeds 25 meters per second in order to protect the turbine and its surroundings.

The wind vane signals are used by the wind turbine's electronic controller to turn the wind turbine against the wind, using the yaw mechanism.

The wind turbine output voltages were in the low voltage range—380, 400, 440 V—for the first wind turbine models (20–500 kW) in order to be connected directly to the low voltage three-phase distribution grid, but the increasing power demand and the integration in wind farms has increased this voltage to 690 V. When the wind turbine must be connected to the medium voltage distribution grid, a transformer is included (inside the tower or in a shelter outside).

1.2.2 Power Control of Wind Turbines

Wind turbines are designed to produce electrical energy as cheaply as possible. Wind turbines are therefore generally designed so that they yield maximum output at wind speeds around 15 meters per second. Its does not pay to design turbines that maximize their output at stronger winds, because such strong winds are rare.

In the case of stronger winds, it is necessary to waste part of the excess energy of the wind in order to avoid damaging the wind turbine. All wind turbines are therefore designed with some sort of power control.

There are two different ways of doing this safely on modern wind turbines—pitch and stall control, and a mix of both active stall.

1.2.2.1 Pitch Controlled Wind Turbines   On a pitch controlled wind turbine, the turbine's electronic controller checks the power output of the turbine several times per second. When the power output becomes too high, it sends an order to the blade pitch mechanism, which immediately pitches (turns) the rotor blades slightly out of the wind. Conversely, the blades are turned back into the wind whenever the wind drops again.

The rotor blades thus have to be able to turn around their longitudinal axis (to pitch) as shown in Figure 1.1.

During normal operation, the blades will pitch a fraction of a degree at a time—and the rotor will be turning at the same time.

Designing a pitch controlled wind turbine requires some clever engineering to make sure that the rotor blades pitch exactly the amount required. On a pitch controlled wind turbine, the computer will generally pitch the blades a few degrees every time the wind changes in order to keep the rotor blades at the optimum angle and maximize output for all wind speeds.

The pitch mechanism is usually operated using hydraulics or electrical drives.

1.2.2.2 *Stall Controlled Wind Turbines*   (Passive) stall controlled wind turbines have the rotor blades bolted onto the hub at a fixed angle.

Stalling works by increasing the angle at which the relative wind strikes the blades (angle of attack), and it reduces the induced drag (drag associated with lift). Stalling is simple because it can be made to happen passively (it increases automatically when the winds speed up), but it increases the cross section of the blade face-on to the wind, and thus the ordinary drag. A fully stalled turbine blade, when stopped, has the flat side of the blade facing directly into the wind.

If you look closely at a rotor blade for a stall controlled wind turbine, you will notice that the blade is twisted slightly as you move along its longitudinal axis. This is partly done in order to ensure that the rotor blade stalls gradually rather than abruptly when the wind speed reaches its critical value.

The basic advantage of stall control is that one avoids moving parts in the rotor itself, and a complex control system. On the other hand, stall control represents a very complex aerodynamic design problem, and related design challenges in the structural dynamics of the whole wind turbine, for example, to avoid stall-induced vibrations.

1.2.2.3 *Active Stall Controlled Wind Turbines*   An increasing number of larger wind turbines (1 MW and up) are being developed with an active stall power control mechanism.

Technically, the active stall machines resemble pitch controlled machines, since they have pitchable blades. In order to get a reasonably large torque (turning force) at low wind speeds, the machines will usually be programmed to pitch their blades much like a pitch controlled machine at low wind speeds. (Often they use only a few fixed steps depending on the wind speed.)

When the machine reaches its rated power, however, you will notice an important difference from the pitch controlled machines: If the generator is about to be overloaded, the machine will pitch its blades in the opposite direction from what a pitch controlled machine does. In other words, it will increase the angle of attack of the rotor blades in order to make the blades go into a deeper stall, thus wasting the excess energy in the wind.

One of the advantages of active stall is that one can control the power output more accurately than with passive stall, so as to avoid overshooting the rated power of the machine at the beginning of a gust of wind. Another advantage is that the machine can be run almost exactly at rated power at all high wind speeds. A normal passive stall controlled wind turbine will usually have a drop in the electrical power output for higher wind speeds, as the rotor blades go into deeper stall.

The pitch mechanism is usually operated using hydraulics or electric stepper motors.

As with pitch control, it is largely an economic question whether it is worthwhile to pay for the added complexity of the machine, when the blade pitch mechanism is added.

1.2.3 Wind Turbine Aerodynamics

The actuator disk theory explains in a very simply way the process of extracting the kinetic energy in the wind, based on energy balances and the application of Bernoulli's equation. The rotor wind capturing energy is viewed as a porous disk, which causes a decrease in momentum of the airflow, resulting in a pressure jump in the faces of the disk and a deflection of downstream flows (Figure 1.2).

The theory of momentum is used to study the behavior of the wind turbine and to make certain assumptions. The assumptions are that the air is incompressible, the fluid motion is steady, and the studied variables have the same value on a given section of the stream tube of air.

The power contained in the form of kinetic energy in the wind crossing at a speed V_v, surface A_1, is expressed by

$$P_v = \tfrac{1}{2}\rho A_1 V_v^3 \qquad (1.1)$$

where ρ is the air density.

The wind turbine can recover only a part of that power:

$$P_t = \tfrac{1}{2}\rho \pi R^2 V_v^3 C_p \qquad (1.2)$$

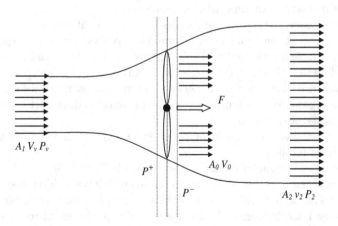

Figure 1.2   Schematic of fluid flow through a disk-shaped actuator.

where R is the radius of the wind turbine and C_p is the power coefficient, a dimensionless parameter that expresses the effectiveness of the wind turbine in the transformation of kinetic energy of the wind into mechanical energy.

For a given wind turbine, this coefficient is a function of wind speed, the speed of rotation of the wind turbine, and the pitch angle.

C_p is often given as a function of the tip speed ratio, λ, defined by

$$\lambda = \frac{R\Omega_t}{V_v} \tag{1.3}$$

where R is the length of the blades (radius of the turbine rotor) and Ω_t is the angular speed of the rotor.

The theoretical maximum value of C_p is given by the Betz limit:

$$C_{p\_theo\_max} = 0.593 = 59.3\%$$

The rotor torque is obtained from the power received and the speed of rotation of the turbine:

$$T_t = \frac{P_t}{\Omega_t} = \frac{\rho\pi R^2 V_v^3}{2\Omega_t} C_p = \frac{\rho\pi R^3 V_v^2}{2\lambda} C_p = \frac{\rho\pi R^3 V_v^2}{2} C_t \tag{1.4}$$

where C_t is the coefficient of torque. The coefficients of power and torque are related by the equation

$$C_p(\lambda) = \lambda \cdot C_t(\lambda) \tag{1.5}$$

Using the resulting model of the theory of momentum requires knowledge of the expressions for $C_p(\lambda)$ and $C_t(\lambda)$. These expressions depend mainly on the geometric characteristics of the blades. These are tailored to the particular site characteristics, the desired nominal power and control type (pitch or stall), and operation (variable or fixed speed) of the windmill.

The calculus of these curves can only be done by means of aeroelastic software such as Bladed or by experimental measurements.

From these curves, it is interesting to derive an analytical expression. This task is much easier than obtaining the curves themselves. Without analytical expression, it would save in table form a number of points on the curves and calculate the coefficient corresponding to a given λ (pitch angle) by means of a double interpolation.

The analytical expression for $C_p(\lambda)$ or $C_t(\lambda)$ may be obtained, for example, by polynomial regression. One typical expression that models these coefficients will be described in the next section.

Figure 1.3 shows an example of $C_p(\lambda)$ and $C_t(\lambda)$ curves for a 200 kW pitch regulated wind turbine.

The power and torque of the turbine are shown in Figure 1.4.

The wind speed V_v of precedent equations is not real; it is a fictitious homogeneous wind. It's a wind, expressed as a point of the area swept by the wind turbine, but the wind must be traceable torque T_t near the field that produced the true wind speed incident on the entire area swept by the rotor.

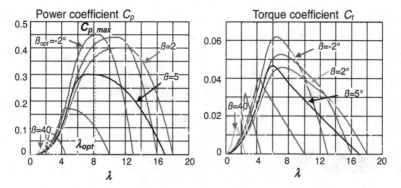

Figure 1.3 Curves of coefficients of power and torque of a 200 kW pitch regulated wind turbine, for different pitch angles β.

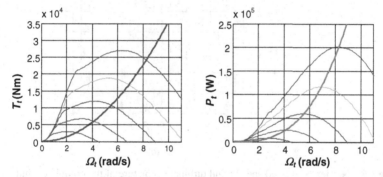

Figure 1.4 Curves of power and torque of a 200 kW pitch regulated wind turbine.

The generation of this fictitious wind can be really complicated depending on the phenomenon to be analyzed, for example, for flicker studies.

1.2.4 Example of a Commercial Wind Turbine

The Nordex N60 (1.3 MW nominal power) is a typical example of a fixed speed wind turbine based on the concepts explained previously. The main characteristics of the turbine are:

- The diameter of the turbine is 60 meters and has a stall power regulation.
- The rotor rotates at 12.8 and 19.2 fixed speeds.
- The gearbox is a three-stage design with a ratio of 78.3 for a 50 Hz wind turbine, with the first stage as a high torque planetary stage and the second and third stages as spur stages.
- The generator is a water-cooled squirrel cage asynchronous type. It is connected to the gearbox by a flexible coupling and it can turn at two speeds

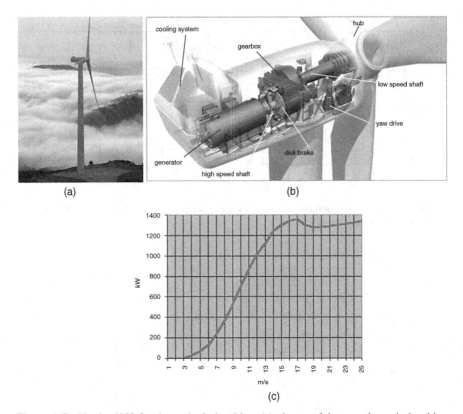

Figure 1.5 Nordex N60 fixed speed wind turbine: (a) picture of the complete wind turbine, (b) nacelle components, and (c) Power curve. (*Source:* Nordex).

(1000 and 1500 rpm), changing the number of pairs of poles of the machine (3 and 2).

- The generator is provided with a thyristor based soft-starter.
- The primary brake system is the aerodynamic blade tip brake. The secondary mechanical brake is a disk brake.

Figure 1.5 shows a picture of the Nordex N60, the main components located in the nacelle, and their power curve.

1.3 VARIABLE SPEED WIND TURBINES (VSWTs)

Figure 1.6 shows the nacelle layout of a Nordex N80 (2.5 MW nominal power), 2.5 MW variable speed wind turbine.

One must appreciate the big differences between the fixed speed and the variable speed wind turbines; it is a technological evolution from the first one. An increase

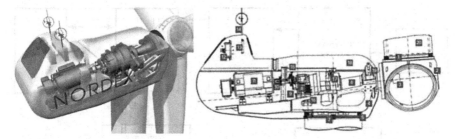

Figure 1.6 Nordex variable speed pitch regulated wind turbine. (*Source:* Nordex).

in size equals an increase in mechanical efforts, and the variable speed and the power control provide the tools to do this without risks. The major differences between them are:

- Power control is by means of pitchable blades.
- Doubly fed induction generator and power converters provide variable speed.

The main components of the nacelle and rotor are: (1) pitch bearing, (2) rotor hub, (3) pitch drive, (4) framework, (5) yaw adjustment bearing, (6) main rotor shaft, (7) yaw brakes, (8) gearbox, (9) holding brake, (10) coupling to generator, (11) generator, (12) cooler for the generator, (13) cooler for the gearbox, (14) wind sensors, (15) on-board crane, (16) yaw drive mechanism, (17) support of the gearbox, (18) nacelle fiberglass housing, (19) rotor bearing, and (20) stem of the rotor blade.

The following subsections will explain the basic models and control for the wind turbine. In Section 1.7 a more detailed description is given of commercial wind turbines.

1.3.1 Modeling of Variable Speed Wind Turbine

The proposed wind turbine model is composed of the following systems:

- Aerodynamic model, evaluates the turbine torque T_t as a function of wind speed V_v and the turbine angular speed Ω_t
- Pitch system, evaluates the pitch angle dynamics as a function of pitch reference β_{ref}
- Mechanical system, evaluates the generator and turbine angular speed (Ω_t and ω_m) as a function of turbine torque and generator torque T_{em}
- Electrical machine and power converters transform the generator torque into a grid current as a function of voltage grid
- Control system, evaluates the generator torque, pith angle and reactive power references as a function of wind speed and grid voltage

Figure 1.7 shows the interaction between the different subsystems.

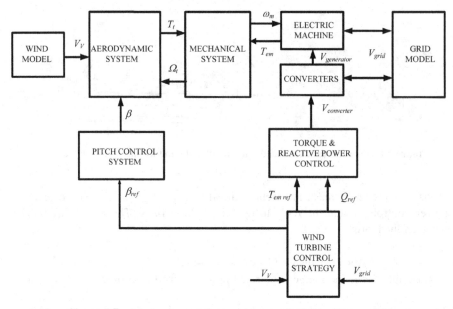

Figure 1.7 Block scheme of a variable speed wind turbine model.

1.3.1.1 Aerodynamic Model

The aerodynamic model represents the power extraction of the rotor, calculating the mechanical torque as a function of the air flow on the blades. The wind speed can be considered as the averaged incident wind speed on the swept area by the blades with the aim of evaluating the average torque in the low speed axle.

The torque generated by the rotor has been defined by the following expression:

$$T_t = \tfrac{1}{2}\rho \pi R^3 V_v^2 C_t \tag{1.6}$$

As mentioned in a previous section, the most straightforward way to represent the torque and power coefficient C_p is by means of analytical expressions as a function of tip step ratio (λ) and the pitch angle (β). One expression commonly used, and easy to adapt to different turbines, is

$$C_p = k_1 \left(\frac{k_2}{\lambda_i} - k_3\beta - k_4\beta^{k_5} - k_6 \right) \left(e^{k_7/\lambda_i} \right) \tag{1.7}$$

$$\lambda_i = \frac{1}{\lambda + k_8} \tag{1.8}$$

with the tip step ratio,

$$\lambda = \frac{R\Omega_t}{V_v} \tag{1.9}$$

1.3.1.2 *Mechanical System* The mechanical representation of the entire wind turbine is complex. The mechanical elements of a wind turbine and the forces suffered or transmitted through its components are very numerous.

It is therefore necessary to choose the dynamics to represent and the typical values of their characteristic parameters. The first is the resonant frequency of the power train. The power transmission train is constituted by the blades linked to the hub, coupled to the slow shaft, which is linked to the gearbox, which multiplies the rotational speed of the fast shaft connected to the generator.

For the purpose of this simulation model, representing the fundamental resonance frequency of the drive train is sufficient and a two mass model, as illustrated in Figure 1.8, can then model the drive train. The second resonance frequency is much higher and its magnitude is lower.

All the magnitudes are considered in the fast shaft. Inertia J_t concerns the turbine side masses, while J_m concerns those of the electrical machine. These inertias do not always represent exactly the turbine and the electrical machine. If the fundamental resonance frequency comes from the blades, part of the turbine inertia is then considered in J_m.

The stiffness and damping coefficients, K_{tm} and D_{tm}, define the flexible coupling between the two inertias. As for the inertias, these coefficients are not always directly linked to the fast shaft but to the fundamental resonance, which may be located somewhere else.

D_t and D_m are the friction coefficients and they represent the mechanical losses by friction in the rotational movement.

The turbine rotational speed and driving torque are expressed in the fast shaft by

$$\Omega_{t\_ar} = N\Omega_t \tag{1.10}$$

$$T_{t\_ar} = \frac{T_t}{N} \tag{1.11}$$

where N is the gearbox ratio.

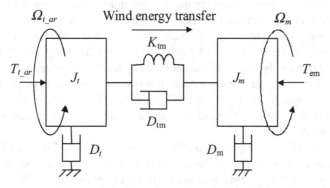

Figure 1.8 Two mass mechanical model.

Next,

$$J_t \frac{d\Omega_{t\_ar}}{dt} = T_{t\_ar} - D_t \Omega_{t\_ar} - T_{em}$$

$$J_m \frac{d\Omega_m}{dt} = T_{em} - D_m \Omega_m + T_{em} \qquad (1.12)$$

$$\frac{dT_{em}}{dt} = K_{tm}(\Omega_{t\_ar} - \Omega_m) + D_{tm}\left(\frac{d\Omega_{t\_ar}}{dt} - \frac{d\Omega_m}{dt}\right)$$

The model can be simplified by neglecting the damping coefficients (D_t, D_m, and D_{tm}), resulting in a model with two inertias (J_t and J_m) and the stiffness (K_{tm}). The resulting transfer function relating the generator torque and speed presents a pole at ω_{01} pulsation and a zero ω_{02} pulsation:

$$\omega_{01} = \sqrt{K_{tm}\frac{J_t + J_m}{J_t J_m}} \qquad (1.13)$$

$$\omega_{02} = \sqrt{\frac{K_{tm}}{J_t}} \qquad (1.14)$$

The pole has a frequency in the range between 1 and 2 hertz for a multimegawatt wind turbine.

1.3.1.3 Pitch System The controller is designed for rotating all the blades at the same angle or each of them independently. This independent regulation gives more degrees of freedom to the control system. This particular operation would reduce the stresses in the blades. The independent regulation of blades is an important innovation that will bring more intelligence into the control system of wind turbines.

In studying a dynamic control system, a blade pitch involves many torques and forces. The representation of this torques requires modeling the structural dynamics of the blade, the behavior of the air around the blades, or the inclusion of friction in the bearings. Moreover, regulation of the speed of rotation around the longitudinal axis of the blades has a bandwidth much greater than that of the control of the angle itself.

Given these last two observations, the most standard approach is to represent the loop control, the rate of change of pitch angle, and a linear system of first order containing the main dynamics of the actuator (hydraulic or electric).

In fact, when modeling the pitch control, it is very important to model the rate of change of this angle. Indeed, given the effort sustained by the blades, the variation of the pitch must be limited. It is limited to about 10°/s during normal operation and 20°/s for emergencies.

Regulation of the blade angle is modeled as shown in Figure 1.9, by a PI controller that generates a reference rate of change of pitch; this reference is limited and a first-order system gives the dynamic behavior of speed control of pitch variation. The pitch angle itself is then obtained by integrating the variation of the angle.

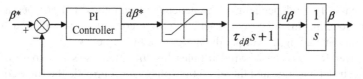

Figure 1.9   Pitch system and control model.

1.3.2 Control of a Variable Speed Wind Turbine

Control of a variable speed wind turbine is needed to calculate the generator torque and pitch angle references in order to fulfill several requirements:

- Extract the maximum energy from the wind.
- Keep the turbine in safe operating mode (power, speed, and torque under limits).
- Minimize mechanical loads in the drive train.

Design of this strategy is a very complicated task strongly related with the aerodynamic and mechanical design of the turbine, and indeed only known by the manufacturers. In this section only the aspects related to the energy extraction and speed–power control will be treated.

Figure 1.10 shows a general control scheme for the VSWT, where the two degrees of freedom are the generator torque and the pitch angle.

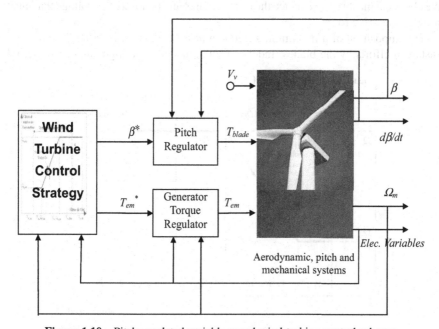

Figure 1.10   Pitch regulated variable speed wind turbine control schema.

This control is independent of the generator technology and can be simulated without modeling the electrical machine, power converters, and their associated controls just including the torque dynamics as a first-order system.

Moreover, for DFIG based wind turbines this limitation also serves to limit the slip of the electrical machine and therefore the voltage must provide the rotor converter.

The following subsections describe the wind turbine control strategy and the control objectives.

1.3.2.1 *Turbine Speed Control Regions* The wind turbine control strategy most commonly used is illustrated in Figure 1.11 and consists of four operation zones:

1. Limit the minimum speed of operation.
2. Follow the curve of maximum power extraction from variable speed operation with partial load.
3. Limit the maximum speed at partial load operation.
4. Limit the maximum operating speed at rated power output.

Figure 1.11 shows the wind turbine speed as a function of the wind speed.

The minimum speed limit is explained by the fact that we must prevent the turbine from rotating at speeds corresponding to the resonant frequency of the tower. This resonance frequency is about 0.5 Hz and a rotational speed too small can excite it.

Moreover, for DFIM based turbines this limitation also serves to limit the sliding of the electrical machine, and hence the rotor voltage, and therefore the voltage that must provide the drive rotor.

The imposition of a maximum speed can also be explained by the limitation of sustained efforts by the blades. Indeed, a rotation speed too high can cause inertial

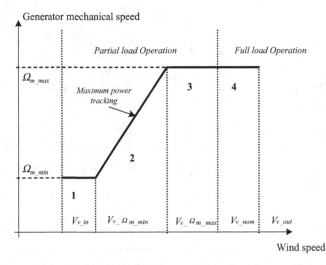

Figure 1.11   Wind turbine control strategy based on four speed regions.

loads unbearable by the blades and the turbine shaft. Also, the linear speed of the tip of the blade must be limited.

For DFIM based turbines, this limitation responds to the desire to limit the slip but also the maximum power that passes through the rotor and therefore by the rotor converter and network. With this strategy, the power to operate the converters will be around 25% of the rated power of the electric generator.

Therefore, the wind generator starts to run at the wind speed connection (cut-in wind speed) with a rotating speed Ω_{t_min}.

When the wind speed becomes more important, it reaches the maximum aerodynamic performance operating in Zone 2. As wind speed increases, the rotation speed also increases until the maximum rotation speed Ω_{t_max}. The wind generator then operates in Zone 3. When wind speed reaches its nominal value, the generator works at the rated mechanical power and the energy captured for higher wind speeds should be regulated at this nominal value.

Zone 4 corresponds to operation at full load. Here, the mechanical power can be limited either by varying the pitch or by torque control. Typically, the electromagnetic torque is maintained at nominal value and adjusts the pitch angle to keep the turbine at maximum speed and rated power.

Figure 1.12 shows the torque and power in different operation modes.

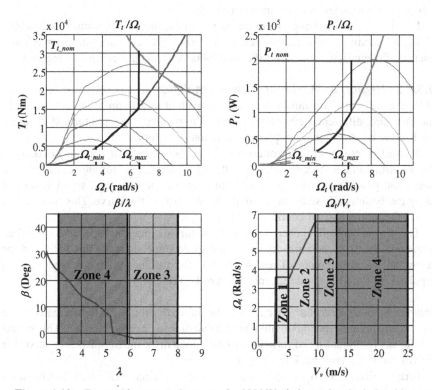

Figure 1.12 Curves of power and torque of a 200 kW pitch regulated wind turbine.

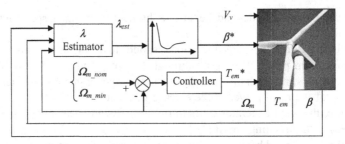

Figure 1.13 Control schemes for regions 1 and 3.

1.3.2.2 Regions 1 and 3: Minimum and Maximum Speed Control The
main objective is to maintain a constant speed of rotation of the turbine at its minimum value in Zone 1 and its nominal value in Zone 3.

Regarding energy efficiency, maximization is not as high a priority as in Zone 2, where the speed of the turbine may evolve to maintain a specific speed λ_{opt} corresponding to the maximum power coefficient C_{p_max}. Here, the generator operates at constant speed.

The specific speed λ varies with wind speed. Depending on the shape of the curves of power coefficient parameterized by the pitch angle, it might be interesting to vary this angle to optimize aerodynamic performance.

It is therefore interesting to plot the curve representing the optimum blade angle, giving it a maximum power coefficient for a given λ. The reference pitch of maximum energy efficiency is λ, a given specific speed obtained from this curve. See Figure 1.13.

1.3.2.3 Region 2: Maximum Power Tracking In this operation region, the
objective of the speed control is to follow the path of maximum power extraction. In the literature, different methods are proposed to regulate the wind turbine at partial load following the maximum power extraction trajectory.

Two different types of controllers have been considered; one consists of taking as the electromagnetic torque reference the electromagnetic torque related to the maximum power curve of Figure 1.12 for each turbine rotational speed value and using the dynamically stable nature of the VSWT around this curve. This controller is called the indirect speed controller (ISC).

The second controller generates the optimal turbine rotational speed (this is linked to the optimal tip speed ratio) for each wind speed value, and uses this as the turbine rotational speed reference. Then, it controls the turbine rotational speed with a regulator. It is called the direct speed controller (DSC).

Indirect Speed Controller It can easily be shown that the WT is dynamically stable around any point of the maximum power curve of Zone 2 of Figure 1.12. This means that for any rotational speed variation around a point in the maximum power curve, the VSWT naturally goes back to its operating point.

Imagine that the VSWT is operating at point *a* of the curve in Figure 1.14a, the wind speed and the electromagnetic torque being fixed. If the turbine rotational speed is

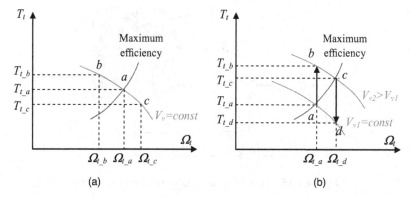

Figure 1.14 Stability study around a point of the maximum power curve.

reduced to Ω_{t_b}, the operating point passes to point b, and the turbine torque is then T_{t_b}. The electromagnetic torque is fixed to its preceding value corresponding to T_{t_a}, so T_{t_b} is higher than T_{em}, and the turbine rotational speed increases until it is again stabilized around the Ω_{t_a} value.

Considering this stability property, the aerodynamic torque T_t can be kept in the maximum power curve in response to wind variations, if the electromagnetic torque T_{em} is controlled in a way to follow this curve. Actually, imagine that the VSWT is operating at point a of the curve in Figure 1.14b.

When the wind speed value increases from V_{v1} to V_{v2}, the operating point becomes b, and the turbine torque becomes T_{t_b}. The controller provides the electromagnetic torque corresponding to the maximum power curve (point c), which is smaller than T_{t_b}. This makes the turbine rotational speed increase until it reaches the equilibrium point c.

When the turbine is working on the maximum power point,

$$\lambda_{opt} = \frac{R\Omega_t}{V_v}, \quad C_p = C_{p\_max}, \quad \text{and} \quad C_t = C_{t\_opt}$$

The aerodynamic torque extracted by the turbine is then given by

$$T_t = \frac{1}{2}\rho\pi R^3 \frac{R^2\Omega_t^2}{\lambda_{opt}^2}\frac{C_{p,\max}}{\lambda_{opt}} \tag{1.15}$$

That is,

$$T_t = \frac{1}{2}\rho\pi \frac{R^5}{\lambda_{opt}^3}C_{p,\max}\Omega_t^2 = k_{opt\_t}\Omega_t^2 \tag{1.16}$$

where

$$k_{opt\_t} = \frac{1}{2}\rho\pi \frac{R^5}{\lambda_{opt}^3}C_{p,\max}$$

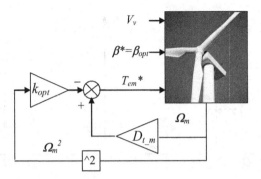

Figure 1.15 Indirect speed control $(D_{tm} = D_t + D_m)$.

It results in an optimal torque evolving as a quadratic function of the wind turbine speed.

Moreover, from Equation (1.12) written in steady state,

$$0 = \frac{T_t}{N} - D_t\Omega_t N - K_{tm}(\Omega_{t\_ar} - \Omega_m)$$
$$0 = T_{em} - D_m\Omega_m - K_{tm}(\Omega_m - \Omega_{t\_ar}) \tag{1.17}$$

where $\Omega_m = N\Omega_t$.

$$T_{em} = -\frac{T_t}{N} + (D_t + D_m)\Omega_m \tag{1.18}$$

Replacing T_t in Equation (1.18) by the expression (1.16), we have

$$T_{em} = -k_{opt}\Omega_m^2 + (D_t + D_m)\Omega_m \tag{1.19}$$

where

$$k_{opt} = \frac{1}{2}\rho\pi\frac{R^5}{\lambda_{opt}^3 N^3}C_{p\_max} \tag{1.20}$$

This last expression leads to the controller illustrated in Figure 1.15.

As seen in Equation (1.19), the behavior of the rotational speed Ω_t depends on the dynamics of the mechanical coupling.

With the ISC method, the behavior of the electromagnetic torque T_{em} and that of Ω_t is the same, since the relation between Ω_t and T_{em} has no dynamics. The electromagnetic torque is not used to increase the Ω_t dynamics as it could be if it were the output of a regulator. Thus, the main disadvantage of the ISC is that the mechanical coupling dynamics is not cancelled out, leading to a fixed soft response of the system.

Direct Speed Controller   The DSC tracks the maximum power curve more closely with faster dynamics.

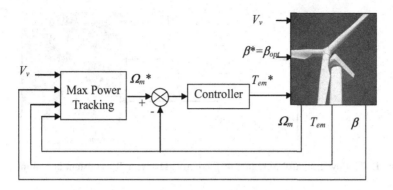

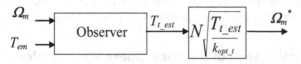

Figure 1.16 Direct speed control.

Knowing the definition of the tip speed ratio λ, the optimal VSWT rotational speed Ω_{t_opt} could be found from the wind speed (V_v). Unfortunately, V_v cannot be measured because it is a fictitious wind speed; it does not exist.

The rotational speed optimal value can nevertheless be obtained from an estimation of the aerodynamic torque. An observer based on Equation (1.12) and using magnitudes such as the electromagnetic torque T_{em} and the turbine rotational speed Ω_t, directly linked to measured signals, can easily be designed to estimate the turbine aerodynamic torque T_{t_est}.

Thus, from Equation (1.16), in the optimal operating point,

$$\Omega_m^* = N\sqrt{\frac{T_{t\_est}}{k_{opt\_t}}} \qquad (1.21)$$

Once the rotational speed reference is generated, a regulator controls Ω_t using the electromagnetic torque value T_{em}. The diagram of the DSC is illustrated in Figure 1.16.

1.3.2.4 Region 4: Power Control The most common control structure for controlling the wind turbine in this region is illustrated in Figure 1.17. Here the electromagnetic torque is held constant at its nominal value. Most of the electrical power generated is that of the stator, that is, the electromagnetic torque produced by the electrical stator pulsation ω_s; this structure leads to proper regulation of electric power. The flicker emission is therefore low with this configuration. The electromagnetic torque does not, however, contribute to regulation of the speed of

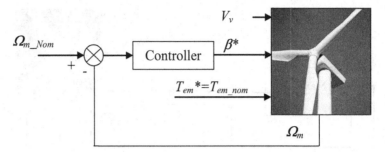

Figure 1.17   Power control in region 4: β controls Ω_m with T_{em} constant at nominal value T_{em_nom}.

rotation. Another disadvantage is that, since T_{em} is constant, the mechanical coupling at low fundamental resonance and flexibility of this coupling cannot be dumped.

1.3.3   Electrical System of a Variable Speed Wind Turbine

Until the mid-1990s, most of the installed wind turbines were fixed speed ones, based on squirrel cage induction machines directly connected to the grid, and the generation was always done at constant speed.

Today, most of the installed wind turbines are variable speed ones, based on a doubly fed induction generator (DFIG), sharing the market with the wound rotor synchronous generators (WRSGs) and the new arrivals, based on the permanent magnet synchronous generators (PMSGs). All of these generator choices allow variable speed generation.

In this section, the evolution of the variable speed generation systems is briefly described. Looking at the generator used in the generation system of the wind turbine, the variable speed wind turbine basic topologies can be classified into three different categories.

1.3.3.1 Doubly Fed Induction Generator Solutions   The doubly fed induction generator has been used for years for variable speed drives. The stator is connected directly to the grid and the rotor is fed by a bidirectional converter that is also connected to the grid (Figure 1.18).

Using vector control techniques, the bidirectional converter assures energy generation at nominal grid frequency and nominal grid voltage independently of the rotor speed. The converter's main aim is to compensate for the difference between the speed of the rotor and the synchronous speed with the slip control.

The main characteristics may be summarized as follows:

- Limited operating speed range (-30% to $+20\%$)
- Small scale power electronic converter (reduced power losses and price)
- Complete control of active power and reactive power exchanged with the grid

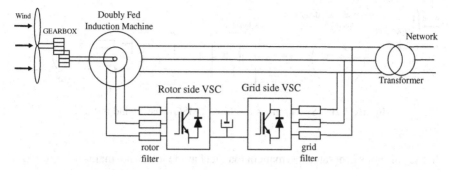

Figure 1.18   Doubly fed induction machine based wind turbine.

- Need for slip-rings
- Need for gearbox (normally a three-stage one)

1.3.3.2 Full Converter Geared Solutions

The full converter with gearbox configuration is used with a permanent magnet synchronous generator (PMSG) and squirrel cage induction generator (SCIG). Using vector control techniques again, a bidirectional converter assures energy generation at nominal grid frequency and nominal grid voltage independently of the rotor speed.

The SCIG uses a three-stage gearbox to connect the low speed shaft to the high speed shaft. Although today the PMSG machine also uses a two-stage gearbox, the objective is to decrease the gearbox from two stages to one, since the nominal speed of the machine is medium.

The induction generator–squirrel cage rotor has the following main characteristics (Figure 1.19):

- Full operating speed range
- No brushes on the generator (reduced maintenance)
- Full scale power electronic converter
- Complete control of active power and reactive power exchanged with the grid
- Need for gear (normally three-stage gear)

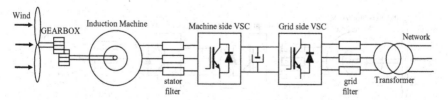

Figure 1.19   Induction machine (SCIG) based wind turbine.

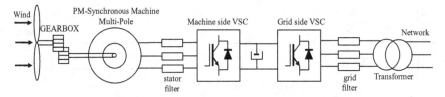

Figure 1.20 Synchronous machine (PMSG) based wind turbine.

The synchronous generator–permanent magnet has the following main characteristics (Figure 1.20):

- Full operating speed range
- No brushes on the generator (reduced maintenance)
- Full scale power electronic converter
- Complete control of active power and reactive power exchanged with the grid
- Possibility to avoid gear
- Multipole generator
- Permanent magnets needed in large quantities
- Need for gear (normally one- or two-stage gear)

1.3.3.3 Full Converter Direct Drive Solutions   Two solutions are proposed in the market:

- Multipole permanent magnet generator (MPMG)
- Multipole wound rotor synchronous generator (WRSG)

The multipole permanent magnet generator allows connecting the axis of the machine directly to the rotor of the wind turbine. Using vector control techniques, a bidirectional converter assures energy generation at nominal grid frequency and nominal grid voltage independently of the rotor speed.

The biggest disadvantage of this technique is the size of the bidirectional converter, which must be of the same power level as the alternator. Also, the harmonic distortion generated by the converter must be eliminated by a nominal power filter system. The advantage of this technique is the elimination of the mechanical converter (gearbox

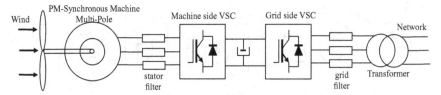

Figure 1.21 Synchronous machine direct drive based wind turbine.

coupling) because the machine can operate at low speed. Another disadvantage is that the multipole machine requires an elevated number of poles, with the size of the machine being bigger than the generators with the gearbox coupling. See Figure 1.21.

1.4 WIND ENERGY GENERATION SYSTEM BASED ON DFIM VSWT

1.4.1 Electrical Configuration of a VSWT Based on the DFIM

The configuration adopted in this book (Figure 1.22) connects the stator directly to the grid, and the rotor is fed by a reversible voltage source converter, as first proposed Peña and co-workers.

The stator windings are supplied at constant frequency and constant three-phase amplitude, since it is directly connected to the grid.

The rotor windings are supplied by a power electronics converter able to feed the DFIM with a variable voltage and frequency three-phase voltages.

This configuration is especially attractive as it allows the power electronic converter to deal with approximately 30% of the generated power, reducing considerably the cost and the efficiency compared with full converter based topologies.

The following subsections briefly describe the main components of the electrical system.

1.4.1.1 Generator   The doubly fed induction machine (DFIM), doubly fed induction generator (DFIG), or wound rotor induction generator (WRIG) are common terms used to describe an electrical machine with the following characteristics:

- A cylindrical stator that has in the internal face a set of slots (typically 36–48), in which are located the three phase windings, creating a magnetic field in the air gap with two or three pairs of poles.

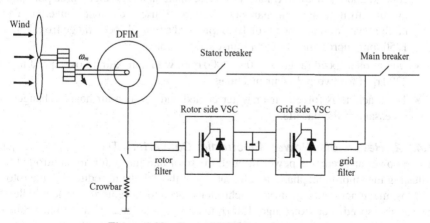

Figure 1.22   General DFIM supply system.

- A cylindrical rotor that has in the external face a set of slots, in which are located the three phase windings, creating a magnetic field in the air gap of the same pair of poles as the stator.
- The magnetic field created by both the stator and rotor windings must turn at the same speed but phase shift to some degrees as a function of the torque created by the machine.
- As the rotor is a rotating part of the machine, to feed it, it's necessary to have three slip rings. The slip ring assembly requires maintenance and compromises the system reliability, cost, and efficiency.

Figure 1.23 shows the different components of the machine.

From the point of view of a variable speed wind turbine (VSWT), several characteristics are required:

- The stator windings are designed for low voltage levels (400, 690, 900 V) in the majority of manufacturers with the exception of Acciona, which uses a medium voltage winding (12 kV) with the aim of reducing the size of the input transformer.
- The rotor windings are designed for medium voltage windings in order to fit the nominal voltage of the converter with the rotor voltage at maximum speed (slip). For example, for a machine with nominal stator and rotor line-to-line voltage of 690 V, with a maximum slip of 33%, the maximum rotor voltage will be 0.33 of the rotor nominal voltage, that is, 228 V. If the rotor winding is sized to 2090.9 volts (690/0.33), the maximum voltage will be 690 V, that is, the maximum available voltage for a back-to-back converter connected to a 690 grid. Note that in older machines the rotor voltage was 420 V in order to reduce the voltage level of the power converter.
- The number of pole pairs is currently selected as two. This implies synchronous speeds of 1500 rpm for a 50 Hz grid frequency, and a typical speed range from 1000 to 2000 rpm approximately. Several manufacturers select three pole pairs in order to minimize mechanical efforts or to use a cheaper converter. This implies synchronous speeds of 1000 rpm, and a typical speed range from 750 to 1250 rpm approximately for a four-quadrant converter.
- Operational speed range is 900 to 2000 rpm, with a maximum overspeed up to 2200 rpm for two pole pair machines.
- The machine is forced air or water cooled and a water–air heat exchanger is necessary in the nacelle.

1.4.1.2 Reversible Power Electronic Converter
The generator torque active power and reactive power through the rotor and the stator are controlled by adjusting the amplitude, phase, and frequency of the voltage introduced in the rotor.

Most manufacturers adjust the synchronous speed to be centered in the middle of the variable speed operation range (1500 rpm for two pole generators in wind turbines with a variable speed range from 1000 to 2000 rpm), which means that the machine,

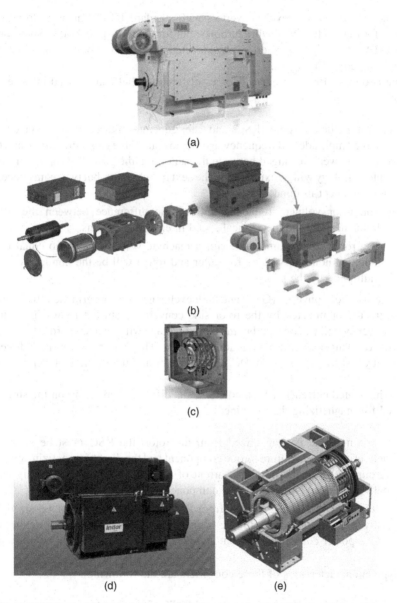

Figure 1.23 (a) Picture of the machine; (b) CAD representation of DFIG components; (c) slip rings (source ABB); (d) picture of the machine; and (e) CAD representation. (*Source*: Indar-Ingeteam.)

working at subsynchronous and hypersynchronous speeds with positive and negative torques, needs to be fed by a four-quadrant power electronic converter.

The standard power electronic converter used in this application is a back-to-back converter composed of two three-phase inverters sharing the DC bus. At present, most manufacturers uses two-level converters with standard IGBTs in order to reduce the cost for the 1.5 to 3 MW wind turbines; but for the most powerful offshore ones (3 to 6 MW), three-level converters are expected to be the best option. Both options will be studied in Chapter 2.

The two converters have two degrees of freedom that can be used in different ways:

- The rotor side converter (RSC) and filter generates a three-phase voltage with variable amplitude and frequency in order to control the generator torque and the reactive power exchanged between the stator and the grid. The rotor converter control strategy will be explained in the next few chapters, but two main concepts should be kept in mind:
  o The rotor voltage frequency will be the difference between the stator frequency and the mechanical speed in electrical radians.
  o The rotor voltage amplitude (for a machine with a turns ratio equal to 1, identical nominal voltage for stator and rotor) will be the nominal voltage multiplied by the slip.
- The grid side converter (GSC) and filter exchanges with the grid the active power extracted or injected by the rotor side converter from the rotor. The output frequency will be constant but the output voltage will change in order to modify the exchanged active and reactive power. The active power is indirectly controlled by means of the DC bus controller and the reactive power.

The sizing (rated current) of both converters is different depending on the strategy selected for magnetizing the machine:

1. If the machine is magnetized from the rotor, the RSC must be sized for delivering the quadrature torque component and the direct magnetizing current (around 30% of the nominal current of the machine). The GSC only must deliver the active power current component.
2. If the machine is magnetized from the stator, the RSC must be sized for delivering the quadrature torque component. The GSC only must deliver the active power current component and the reactive power current component.

The typical characteristics of these converters are the following:

- Vector control or direct torque control (DTC) for generator and grid converter control; active and reactive power control; and grid code support
- Two-level, three-phase converter with IGTBs, at switching frequency of 2.5–5 kHz
- LCL filter for the GSC, and dv/dt filter for the RSC
- Nominal power: 500 to 2500 kVAs
- Nominal voltage 690 V, $+10\%$ to -15%.

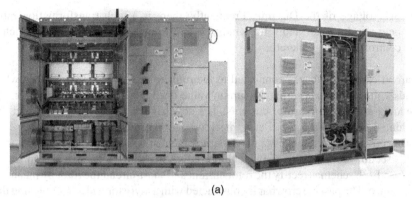

(a)

(b)

Figure 1.24 Back-to-back IGBT based 2L-VSC for 690 V DFIM based wind turbines. (*Source*: Ingeteam (a) and ABB (b).)

- Nominal DC bus voltage 1000 V
- Very low total harmonic distortion (THD < 3%)
- Air or water cooled, in order to reduce the size of the cabinets

Figure 1.24 shows two examples of converters.

1.4.1.3 Crowbar Protections When a voltage dip occurs in the network, current transients in the stator windings (due to the stator's direct connection to the grid) and grid side converter are produced. Hence, these two behaviors are completely different:

- The grid side converter doesn't lose current control in most cases.
- Stator disturbance is transmitted to the rotor, causing uncontrollable currents that can produce damage to the rotor converter due to the overcurrents and the

overvoltage of the DC link. Frequently, there is a high transformation ratio between the stator and rotor windings; thus, the rotor converter has restricted control over the generator.

A circuit called crowbar is connected to the rotor to protect the RSC. The crowbar avoids voltage bus exceed his maximum value once the RSC loses current control providing a path for the rotor currents. The crowbar short-circuit the rotor and the machine operates as a squirrel cage machine, see Figure 1.22.

The crowbar power converter may be implemented with several power structures. In this section we will analyze two configurations that allow a passive and an active crowbar. Both schemas rectify the rotor current and short-circuit the rotor by means of a resistance. The passive crowbar is constructed with a thyristor and allows closing the circuit but does not allow it to open until the crowbar current is extinguished. The active crowbar is constructed with an IGBT and allows opening the circuit in forced commutation.

The control system of the crowbar may be materialized in many ways depending on the power converter structure and the desired performances. After a voltage dip, the rotor current regulators lose control and an energy flow from the stator to the rotor charges the bus capacitor. To avoid the bus voltage from reaching the converter limits, it is necessary to break this energy flow, and the simplest method is to short-circuit the rotor when the bus voltage reaches a limiting value.

With a passive control, the crowbar act as a protection system; the time necessary to open the stator breaker is approximately 100 milliseconds, causing at the end the disconnection of the wind turbine. When the control objective is to keep the wind turbine connected to the grid during fault, it is necessary to control the bus voltage. The simplest technique consists of comparing the bus voltage with its maximum and normal operation reference values and, depending on that comparison, keeping the crowbar circuit open or closed. This technique is called active crowbar control. The bus capacitor load dynamics is determined by the rotor–bus energy flow, and the discharge is determined by the capacity of the grid side inverter (bus to grid energy flow).

1.4.1.4 Transformer Generally, the output voltage of the wind turbine is not designed to fit with low voltage or medium voltage networks, and a transformer is necessary to adapt the turbine voltage to the coupling point.

There are several ways to connect the stator and the back-to-back converter to the grid (Figure 1.25):

- If the stator voltage and the back-to-back converter are the same (commonly 690 V AC), one transformer rated at full power, with a medium voltage primary of several kilovolts (10–30 kV) and a low voltage secondary, is used.
- If the stator voltage is in the range of medium voltage, one transformer rated at the rotor power is used. In this case, the stator voltage is directly connected to a medium voltage distribution grid or to the medium voltage grid of a wind farm.

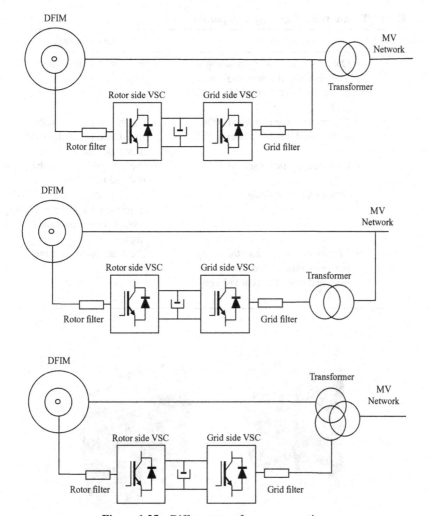

Figure 1.25   Different transformer connections.

- If the stator and back-to-back converter voltages are both in the low voltage range, but different, two secondary windings are used.

All three ransformer configurations have advantages, and the main advantages and disadvantages are summarized in Table 1.1.

The most common technologies used for these transformers are cast resin when they are located in the wind turbine: the nacelle or tower.

Figure 1.26 shows two examples of transformers in liquid filled and resin encapsulated technological solutions.

TABLE 1.1   Transformer Topology Comparison

Option	Advantages	Disadvantages
a	• One single secondary winding • The primary MV winding can be adapted to different MV grids	• Full power transformer
b	• The transformer is rated at 30% of the wind turbine power • Less transformer loses • Stator electrical design	• The transformer leakage impedance limits the short-circuit current • The MV grid must fit the stator MV voltage • Security concerns, as the MV range means more capacitating courses for maintenance employees
c	• The stator voltage and the power electronics can be designed with different voltage levels	• The transformer is more expensive

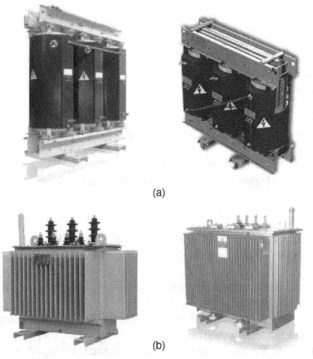

(a)

(b)

Figure 1.26   Picture of (a) a dry transformer (cast resin) and (b) a liquid filled (oil) transformer. (*Source:* ABB.)

1.4.2 Electrical Configuration of a Wind Farm

The first commercial wind turbines were installed as single units on farms or in small villages in Denmark and Germany and were connected directly to the low voltage distribution grid. Due to the financial support of the government, Danish farmers find in wind energy a new income that improves the always complicated economic situation of agriculture.

The increased power of turbines, and their connection to the medium voltage distribution grids, made it necessary to include a transformer in the turbine to connect it to the medium voltage distribution grids (rated 10–33 kV). In the first wind turbines, this transformer and the breaker and protection system were located in shelters near the wind turbine and were shared by several wind turbines. Once the size of the towers became adequate, the transformers were located in the tower. At present, the transformer is located in the nacelle in the biggest wind turbines.

Once wind technology acquires the necessary maturity and governments adopt a green policy for renewable energy generators, utility scale wind energy generation systems will appear. At present, most of the wind turbines are installed in groups currently called "wind farms" or "wind parks"; for example, in Spain the maximum power of a farm is 50 MW, while in the United States you can find wind farms of 200 MW.

These kinds of wind farms are directly connected to the transmission or sub-transmission grids by means of an electrical substation especially constructed for the wind farm.

Electrical collection from each of the wind turbines at the point of interconnection with the wind farm can be done in many different ways but the most standard one is shown in Figure 1.27.

A wind farm has the following components:

- The substation:
  o The input breaker in the high voltage side of the transformer.
  o The coupling transformer to the transmission grid. The primary voltage will vary with each country's standards from 66 to 220 kV. The secondary voltage will vary from 10 to 33 kV.
  o One output breaker on the medium voltage side of the transformer.
  o Static VAR compensation (capacitor banks and inductors), connected by means of medium voltage breakers.
  o Dynamic VAR compensation based on SVC or STATCOM, this last connected by means of a coupling transformer.
  o Inductor or transformer to limit fault currents.
- Medium voltage feeders and circuits for each turbine cluster.
- Wind turbines. In each wind turbine there is a medium voltage breaker, a transformer, and two sectionalizers to open the input and output circuits.

Inside the substation building are located the electrical cabinets for the protective relays and the measuring, communication, and Scada system.

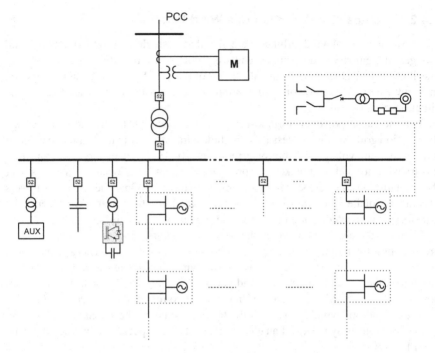

Figure 1.27 Electrical layout of a wind farm connected to the transmission grid.

From the point of view of wind turbine grid integration, the electrical layout of the wind farm is important because of the way the grid outages affect the turbine behavior.

In the following subsections we analyze how the transformer connection types can influence the types of dips that can affect the wind turbine.

1.4.3 WEGS Control Structure

Modern wind farms must be operated as conventional power plants due to the political and institutional agreement for clean energy following the Kyoto Protocol.

As will be explained in the next section, grid codes—the rules that transmission system operators (TSOs) and distribution system operators (DSOs) impose, are more and more restricted, mainly due to the increased penetration of this technology in the generation mix.

One good example of this is Spain. With 19.5 GW of wind power installed in 2009 and a valley consumption of 22 GW, countermeasures must be taken in order to guarantee the stability of the power system when wind power is a 30–40% of the generation mix.

In the last few years, operation and control of wind farms has been one of the most important tasks for the Spanish TSOs, due to the high increment of the wind energy capacity connected to the grid and, specifically, to the high voltage transmission network. This new scenario has modified the traditional criteria for operating the

Spanish power system with the same confidence levels of reliability and security that existed in the past, when this high amount of variable energy did not exist in the generation mix.

This situation has led to the recent publication of technical documents (grid codes) for regulating all aspects of the integration of wind farms in the power systems. At present, in Spain, it is mandatory for all wind farms with a power capacity higher than 10 MW to be connected to a dispatching center with some specific requirements for communication and measurements that must be fulfilled.

Likewise, Spanish TSOs have proposed that when several wind farms evacuate energy to a common point of the transmission network, only one mediating body (usually the owner with the most power connected) must be between the TSO and the rest of the wind farm owners.

The functions of this new figure, named the connection point manager (CPM), are mainly to be in charge of the operation and management of all the power evacuated at that point and to be the only speaker (intermediary) with the TSO.

At present, modern variable speed wind turbines (VSWTs) are capable of exchanging reactive power with the grid and reducing active power by using pitch control systems and dynamic torque control on the electrical generator.

This scenario is where the wind farm controller plays an important role in controlling the active and reactive power injected by the wind farm into the grid according to the references and control mode signals received from the TSO or the CPM. This provides wind farms with the possibility to participate actively in the control tasks of the grid in the same way as conventional power plants do.

Figure 1.28 shows a proposal of the overall control strategy necessary to operate, in a secure way, a grid with increased penetration of renewable energy systems.

At the top, the control center of the TSO or the DSO communicates with the wind farm controller and receives information about the generation state and reactive power capacity of the generators, and sends the references of active–reactive power

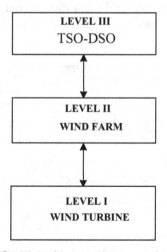

Figure 1.28 Hierarchic control structure for a WEGS.

requirements to operate the grid in an efficient way and maintain the necessary security level. The WEGS must have the same behavior as conventional power plants.

1.4.3.1 *Wind Farm Control System*
The references from the TSO-DSO arrive at the wind farm control and define the requirements for each wind turbine and for the substation reactive power static compensation (capacitor banks or inductors); this is sent by a communication channel in real time.

An overall diagram of this control level is illustrated in Figure 1.29. The centralized control objective must be that the wind farm behaves as a single unit (like a conventional power plant).

The inputs will be the system operator references, measurements from the point of common coupling (PCC), and the state and available power of each wind turbine.

In order to implement a centralized control it is necessary to have an effective communication between the wind farm centralized control (WFCC) and each of the wind turbines. Thus, while each of the wind turbines report to the WFCC the active power and reactive power that they can deliver at any moment, the WFCC should provide each of the wind turbines with references of active and reactive power.

The control objective of the WFCC is the regulation of active and reactive powers injected at the point of common coupling. Figure 1.30 shows a commonly used control structure and flow of signals between different subsystems.

The control system must perform the following tasks for active power:

1. Evaluate the operation mode for active power control: receive a reference or deliver all available power, automatic frequency control, delta or balance control.
2. Limit the generation deviation if necessary (power gradient limiter).
3. Regulate the active power in the PCC.
4. Dispatch the wind turbine active power reference ($P_{ref}^{WT_i}$) as a function of the wind farm power reference (P_{ref}^{WF}). The dispatch function can be done in many different ways, for example, as a function of the available power ($P_{disp}^{WT_i}$) of each

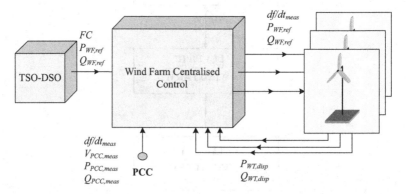

Figure 1.29 Signal flow between grid operator, wind farm control, and wind turbine.

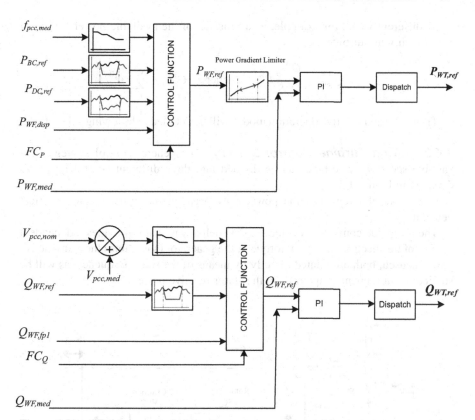

Figure 1.30 Wind farm control: (a) active power control and (b) reactive power control.

wind turbine:

$$P_{ref}^{WT_i} = \frac{P_{disp}^{WT_i}}{P_{disp}^{WF}} P_{ref}^{WF}, \quad \text{where} \quad P_{disp}^{WF} = \sum_{i=1}^{n} P_{disp}^{WT_i}$$

An other option is to send the wind farm reference (per unit) to all the wind turbines and each one will translate the reference to the available wind turbine power.

The control system must perform the following tasks for reactive power:

1. Evaluate the operation mode for reactive power control: receive a reference or deliver all available power, automatic voltage control, reactive control.
2. Regulate the reactive power in the PCC.
3. Dispatch the wind turbine power reference ($Q_{ref}^{WT_i}$) as a function of the wind farm power reference (Q_{ref}^{WF}). The dispatch function can be done in many

different ways, for example, as a function of the available power $(Q_{disp}^{WT_i})$ of each wind turbine:

$$Q_{ref}^{WT_i} = \frac{Q_{disp}^{WT_i}}{Q_{disp}^{WF}} P_{ref}^{WF}, \quad \text{where} \quad Q_{disp}^{WF} = \sum_{i=1}^{n} Q_{disp}^{WT_i}$$

The active and reactive operation modes will be detailed in following subsections.

1.4.3.2 *Wind Turbine Control System* The general control strategy of a variable speed wind turbine can be divided into three different control levels, as depicted in Figure 1.31.

Control level I regulates the power flow between the grid and the electrical generator.

The rotor side converter is controlled in such a way that it provides independent control of the electromechanical torque of the generator (sometimes the stator active power is used, both are related directly by means of the stator frequency, as will be studied in subsequent chapters) and the stator reactive power.

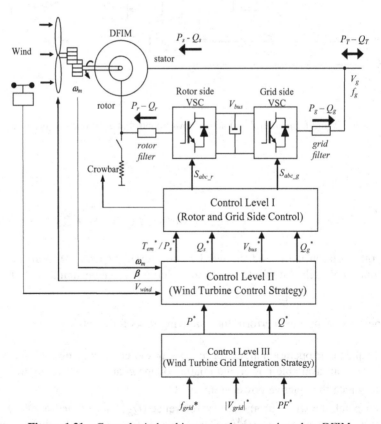

Figure 1.31 General wind turbine control strategy based on DFIM.

In Chapter 7 the vector control techniques for the DFIG will be studied in detail, and in Chapter 8 direct power control techniques will be analyzed.

The grid side converter provides decoupled control of active and reactive power flowing between the converter and the grid. The active power is exchanged between the rotor and the DC bus and the control calculates the active power exchanged with the grid to keep the DC bus voltage constant. In Chapter 2, vector control and direct power control techniques for this converter will be detailed.

The crowbar converter protects the rotor side converter when a voltage dip occurs. In Chapter 9 the most significant hardware solution will be explained.

Control level II is responsible for controlling wind energy conversion into mechanical energy, that is, the amount of energy extracted from the wind by the wind turbine rotor. This control level calculates the references for control level I. Two main operating modes are commonly used:

- Extract the maximum power from the wind, coordinating torque (stator active power) and pitch angle (β) references, always keeping the wind turbine under the speed limits, as explained in preceding subsections. This is the normal operation mode, in order to maximize the return on investment.

- Respond to active and reactive power references from the higher control level. This is necessary to have reserve power in the wind turbine. This is a future operation mode, necessary in grid scenarios with large dependence on wind power.

Control level III is dedicated to the wind turbine–grid integration. This control level performs the same functionalities as the wind farm control:

- Provide ancillary services: voltage (V_{grid}) and frequency (f_{grid}) control (droop characteristics), or inertial response.
- Respond to active and reactive power references from the grid operator or wind farm centralized control.

1.5 GRID CODE REQUIREMENTS

The grid codes of most countries generally aim to achieve the same thing. Electricity networks are constructed and operated to serve a huge and diverse customer demographic.

Electricity transmission and distribution systems serve, by way of example, the following types of user:

- Large high-consumption industrial factories
- High-sensitivity loads requiring high quality and reliable uninterrupted supplies
- Communications systems (e.g., national)
- Farms
- Shops and offices

- Domestic dwellings
- Large power stations

In simple terms, it is vital that electricity supplies remain "on." To do this, the system operator not only balances the system with suitable levels of generation to meet demand, but also requires larger capacity users of the system, including both generation and load, to actively participate in ensuring system security.

To achieve this, the following technical requirements are possibly the most crucial and appear common across most European countries:

- Frequency and voltage tolerance
- Fault ride through
- Reactive power and voltage control capability
- Operating margin and frequency regulation
- Power ramping

And future technical requirements may include:

- Inertial response
- Power system stabilizer
- Wind farm control

1.5.1 Frequency and Voltage Operating Range

The electrical behavior of the network, in terms of frequency and voltage, due to its dynamic nature is continuously changing. Generally, these changes occur in very small quantities. It is a requirement that users of the transmission system are able to continue operating in a normal manner over a specified range of frequency and voltage conditions. With respect to frequency and for a 50 Hz system, this would be in the range of 49 to 51 Hz. With respect to voltage, this range could be ±10% of the nominal voltage.

However, at times the ranges could be wider, although it would normally be expected that the user would continue operating under an extreme condition for a defined period of time, for example, 47 Hz for 15 seconds or +20% of the nominal voltage for 1 hour. Beyond these extremes, the user would normally be required to disconnect from the system.

Table 1.2 shows a common range of conditions within which a user would be required to operate.

TABLE 1.2 Example of Possible Frequency and Voltage Tolerance Requirements

	Normal Continuous Operation	Required	Very Short Term
Frequency	49–51 Hz	47.5–49 Hz and 51–52 Hz	47–47.5 Hz
Voltage	±5%	±10%	

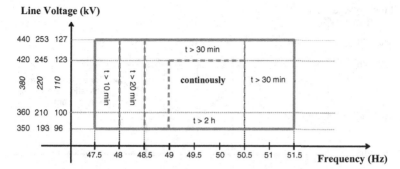

Figure 1.32 EON Netz GMBH grid code related to frequency.

As an example, the EON grid code related to the operating times as a function of frequency is illustrated in Figure 1.32.

1.5.2 Reactive Power and Voltage Control Capability

1.5.2.1 Power Factor Control To minimize losses and thus maintain high levels of efficiency, it is preferable that networks operate with voltage and current in-phase—that is, the power factor is unity.

However, users of electrical systems often tend to have inductive loads or generation facilities that operate such that voltage and current are out-of-phase. In addition, power system components including lines and transformers, for example, produce or consume large levels of reactive power. From the network user's perspective (looking from the installation toward the network) an inductive load/generation facility is said to have a leading power factor because the current leads the voltage. Put another way, the user is consuming reactive power.

As the behavior of the network is continuously changing, users are required to have the ability to adjust their reactive power production or consumption, in order that reactive power production and consumption are balanced over the entire network. In most cases, it is the generators who provide this control ability. The range, for example, could be from 0.95 leading to 0.95 lagging.

An example is the proposed curve of power factor as a function of grid voltage in the EON grid code (see Figure 1.33).

1.5.2.2 Voltage Support A generating station may be required to operate over a range of power factors to provide or consume reactive power as discussed and shown above. Alternatively, the installation may be required to operate in voltage control mode, that is, to adjust its reactive power production or consumption in order to control voltage on the local network.

If the network voltage decreases to a level below a predefined range, an installation may be required to supply reactive power to the network to raise the voltage. Conversely, if the network voltage increases to a level above a predefined upper

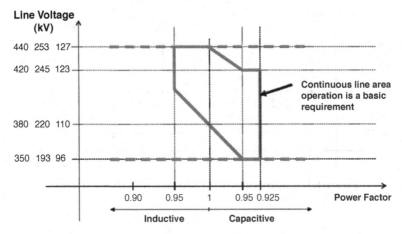

Figure 1.33   Reactive power control as a function of grid voltage (EON Netz GmbH).

limit, then the installation would be required to consume reactive power to bring the voltage back within acceptable limits.

1.5.2.3 Reactive Power Wind Turbine Operation Modes   In order to accomplish the functionalities mentioned, the following operation modes (Figure 1.34) for the reactive power control of the wind turbine are typically defined:

- Reactive power control—the wind turbine is required to produce or absorb a constant specific amount of reactive power.
- Automatic voltage control—the voltage in the wind turbine point of common coupling (PCC) is controlled. This implies that the wind farm can be ordered to produce or absorb an amount of reactive power.

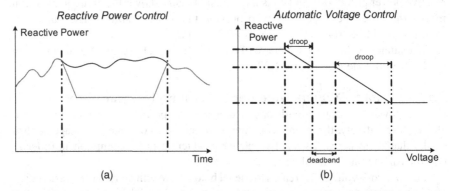

Figure 1.34   Reactive power operation modes: (a) reactive power Control and (b) automatic voltage control.

1.5.3 Power Control

Some network operators impose limits on power output. This could be during normal continuous operation and/or during ramping up to an increased output or ramping down to a decreased output.

This requirement might be necessary in the first case to limit output because of limitations in the capabilities of other generators or the transmission or distribution networks. In the latter case, this might be necessary so that network control systems and other generators have time to respond to a new operating state.

When a generator comes on-line it is providing the network with an increased amount of power which not only affects frequency, but also requires existing operating installations to adjust their operational characteristics to adapt to the "new" operating state.

With respect to wind power, for decreasing wind speed conditions there are limitations in the capabilities of wind turbines. If the wind speed is falling, a wind turbine may not be able to maintain its output or fully control the rate of decrease of output. However, wind speed profiles can be predicted and so if a controlled ramp down or clearly defined power reduction rate is required, a wind turbine's output can be reduced early, thus reducing the maximum rate of change of output power.

An example is the proposed curve of power (as a percentage of the momentary possible power production) as a function of grid frequency in the ESB National Grid code (see Figure 1.35).

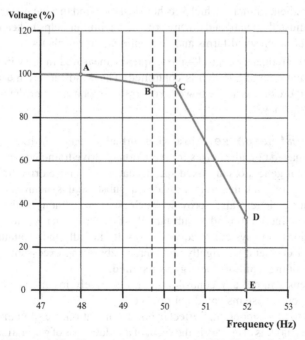

Figure 1.35   Power versus frequency in ESB National Grid

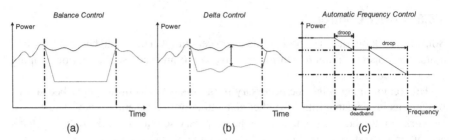

Figure 1.36 Active power operation modes: (a) balance control, (b) delta control, and (c) automatic frequency control.

1.5.3.1 Active Power Wind Turbine Operation Modes
In order to accomplish the functionalities mentioned, the following operation modes (Figure 1.36) for the power control of the wind turbine are typically defined:

- Balance control—whereby wind turbine production can be adjusted downward or upward, in steps, at constant levels.
- Delta control—whereby the wind turbine is ordered to operate with a certain constant reserve capacity in relation to its momentary possible power production capacity. This operation mode allows an installation to operate at a margin below its rated power output–maximum power extraction so that it may respond to significant changes in frequency by increasing or decreasing its output.
- Power gradient limiter—which sets how fast the wind turbine power production can be adjusted upward and downward. Such a limiter helps to keep production balance between wind farms and conventional power plants.
- Automatic frequency control—the frequency measured in the wind farm point of common coupling (PCC) is controlled. The wind turbine must be able to produce more or less active power in order to compensate for a deviant behavior in the frequency.

1.5.3.2 Inertial Response
Most wind turbine concepts utilize variable rotor speed, as this has major advantages for reduction of drive train and structural loads. All conventional generators are fixed speed; that is, the entire drive train rotates at synchronous speed and therefore provides a substantial synchronously rotating inertia. Rotating loads also provide such inertia, although the generators dominate. This inertia provides substantial short-term energy storage, so that small deviations in system frequency result in all the spinning inertias accelerating or decelerating slightly and thereby absorbing excess energy from the system or providing additional energy as required.

This happens without any control system, effectively instantaneously. Without this, modern power systems could not operate.

In addition to this "smoothing" effect in normal operation, the spinning inertia also provides large amounts of energy in the event of a sudden loss of generation: the rate of decrease of system frequency in the first second or so after such an event is entirely governed by the amount of spinning inertia in the system.

Variable speed wind turbines have less synchronously connected inertia, and in the case of the FC concept, none at all. As wind turbines displace conventional generation, there will be less spinning inertia, and therefore the system will become harder to control and more vulnerable to sudden loss of generation.

It is feasible that future grid codes will require all or some generators to provide an inertia effect.

This can in principle be provided by variable speed wind turbines, but this requires a control function and cannot occur without intervention. The control function will sense frequency changes and use this to adjust generator torque demand, in order to increase or decrease output power.

The effect is similar to the frequency-regulation function discussed earlier, but is implemented by generator torque control rather than pitch control. A more complex implementation could also include pitch control.

It is possible that wind turbines would not need to provide this function for small-scale frequency deviations, as conventional generation capacity may still be sufficient. Instead, the requirement could be limited to responses to large-scale deviations associated with a sudden loss of generation.

Initial studies show that, in principle, variable speed wind turbines can provide a greater inertia effect than conventional synchronous machines, because generator torque can be increased at will, extracting relatively large amounts of energy from the spinning wind turbine rotor. This decelerates the wind turbine rotor rapidly, and so may not be sustained for very long before aerodynamic torque is reduced. High generator torque also results in high loads on the drive train, which may add significant cost.

It is concluded that an inertia effect is available, in principle, but may have implications for wind turbine design and cost. It is not clear if some of the FSWT concepts may provide the necessary control.

1.5.4 Power System Stabilizer Function

Power system stabilizer (PSS) functions can be provided by conventional generators. In essence, the output power of the generator is modulated in response to frequency deviations, in order to damp out resonances between generators. These resonances are most likely to occur between two groups of large generators separated by a relatively weak interconnection.

Again, because of the tight control of generator torque provided by the DFIG concept, and possibly also FSWT, this function should also be able to be provided if required.

However, it should be pointed out that because variable speed wind turbines have very little synchronously connected inertia, the risk of such resonances actually reduces as wind penetration increases. Thus, there is an argument that PSS functions should be provided only by conventional generation.

1.5.5 Low Voltage Ride Through (LVRT)

The LVRT fulfillment for wind turbines has become a major requirement from the TSO-DSOs all around the world.

The first wind turbines based on squirrel cage asynchronous generators were very sensitive to grid outages. The protections were tuned in such a way that the wind turbine disconnected with even minor disturbances.

This caused two major problems for the TSO-DSO:

1. The protections were unable to detect faults in lines near wind farms, due to loss of short-circuit current from the wind farms.
2. The loss of wind power generation (reconnection of a fixed speed wind farm takes several minutes) necessitates fast response generation plants (such as hydro) or an increase in the fast reserve power.

So, as mentioned earlier, the first requirement of the TSO is to "keep connected."

But wind turbine behavior during a grid fault is very different depending on the technology (fixed speed or variable speed, and full converter technologies or doubly fed) and also different from conventional power plants based on synchronous generators.

Thus, the TSOs decided to standardize the pattern of current versus voltage during faults, that is, the current that the generator consumes during the fault.

In Chapter 9, the crowbar control strategy and design will be studied in detail. The control strategy must allow the wind turbine:

- To remain connected to the power system and not consume active power during the fault
- To be provided with reactive power during the fault to assist voltage recovery
- To return to normal operation conditions after the fault

Voltage recovery is a complementary requirement; wind generators try to minimize the impact of wind parks in the power system during a short circuit and fault clearing. Voltage recovery is performed by reactive current injection of the wind generator.

1.6 VOLTAGE DIPS AND LVRT

If a defined fault occurs on the transmission system, it is a normal requirement of the transmission system operator that a generating station remain in operation and connected to the system—thus, it "rides through" the fault.

The definition of the fault is derived from the response time of the network protection systems to clear the fault. A normal duration to clear a fault is in the range of hundreds of milliseconds; hence, the requirement of the user will be to ride through a fault that has a significant drop in voltage of some hundreds of milliseconds in duration.

Because this issue is very important for the grid integration of wind energy systems, the following subsections will be oriented to describe the basic concepts:

- The electric power system (EPS) and the origin of the dips in the EPS
- The definition, classification, and transmission in the EPS of dips
- The procedure to validate the simulated LVRT requirements for wind turbines in Spain.

1.6.1 Electric Power System

1.6.1.1 Description   The electric power system (EPS) is the set of infrastructures responsible for the generation, transport, and distribution of electrical energy and can be considered as some of the biggest infrastructures in the world.

Electric power transmission is the bulk transfer of electrical energy from generating power plants to substations located near population centers. This is distinct from the local wiring between high voltage substations and customers, which is typically referred to as electricity distribution. Transmission lines, when interconnected with each other, become high voltage transmission networks.

Figure 1.37 shows a schematic diagram of an EPS, with traditional distribution from generation to a residential user.

Historically, transmission and distribution lines were owned by the same company, but over the last decade or so many countries have introduced market reforms that have led to the separation of the electricity transmission business from the distribution business. Thus, in many countries, there are one or two transmission system operators (TSOs), several distribution system operators (DSOs), and trading companies.

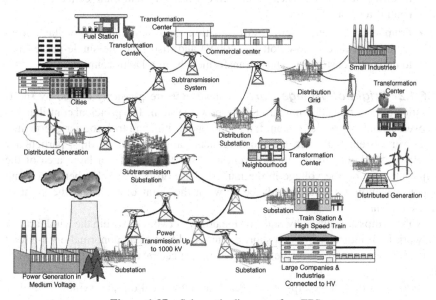

Figure 1.37   Schematic diagram of an EPS.

Wind power generators are commonly connected at all the voltage levels of the EPS, from the most powerful wind farms (from 25 to 250 MW) that are connected to transmission or subtransmission lines (from 66 to 745 kV) to the lower power wind turbines (50 to 500 kW) connected directly to the low voltage (380, 400, 440 V) distribution lines.

Therefore, the connection rules (grid codes) for each voltage level are elaborated by the different companies that operate the networks.

Another important point is the connection of different transformers and ground (related to the protective systems for lines) at different levels; for example, Figure 1.38 shows a diagram of the Spanish transmission and distribution network.

The transmission system operator is responsible for operating the high voltage transmission system. In Spain the TSO operates the 220 and 400 kV transmission system. The 220 and 400 kV systems are connected by means of autotransformers. Big power generation plants such as nuclear, coal, and hydro are connected at this level.

The distribution system operators are responsible for operating the distribution lines and also the subtransmission system necessary to connect the consumer centers to the transmission system. Typical voltages in this subtransmission system are 69 and 132 kV in Spain. The subtransmission system and the transmission system are connected by means of star–star transformers with neutral connected to earth.

Once the lines arrive at the consumer centers, it's necessary to step down the voltage to the proper level for domestic customers. This is done in two steps:

- A high voltage line arrives at the distribution substations where the DSO decreases the voltage levels to 10, 20, 30 kilovolts and distributes the power in different feeders. These transformers have a Delta-Star connection with neutral to ground.
- From each feeder several transformation centers reduce the voltage from medium voltage to low voltage (400 V) and distribute the single-phase lines to domestic customers. These transformers have a Delta connection.

1.6.1.2 *Origin of Voltage Dips*
Voltage dips are primarily caused by short-duration overcurrents flowing through the power system. The principal contributions to overcurrents are power system faults, motor starting, and transformer energizing.

Power system faults are the most frequent cause of voltage dips, particularly single-phase short circuits. In the event of a short circuit, for a large area of the adjacent network, the voltage in the faulted phase drops to a value between 0 and 1 p.u., depending on the impedance between the point of fault and the point of measurement.

Voltage dips are caused by faults on the utility network or within the wind farm. A network fault indicates either a short-circuit condition or an abnormal open-circuit condition. The nature of voltage dips can be influenced by the symmetry of a network fault.

Two types of voltage dips are depicted: asymmetrical dip and symmetrical dip.

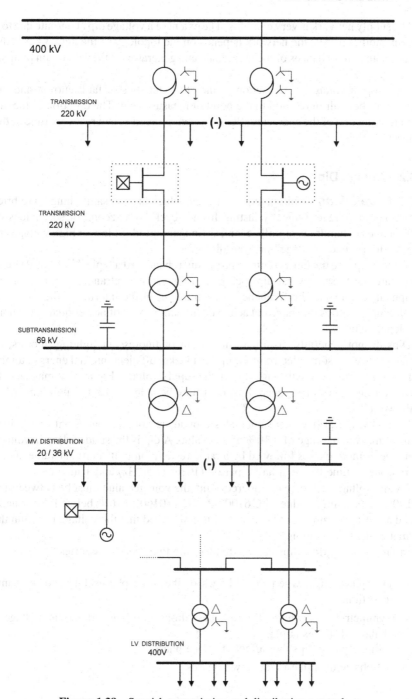

Figure 1.38 Spanish transmission and distribution network.

The supply network is very complex. The extent of a voltage dip at one site due to a fault in another part of the network depends on the topology of the network and the relative source impedances of the fault, load, and generators at their common point of coupling.

The drop in voltage is a function of the characteristics of fault current and the position of the fault in relation to the point of measurement. The duration of the dip event is a function of the characteristics of system protection and recovery time of the connected loads.

1.6.2 Voltage Dips

1.6.2.1 Definition

Voltage dips—or sags, which are the same thing—are brief reductions in voltage, typically lasting from a cycle to a second or so, or tens of milliseconds to hundreds of milliseconds. (Longer periods of low or high voltage are referred to as "undervoltage" or overvoltage.")

Voltage dips are the most common power disturbance. At a typical industrial site, it is not unusual to see several dips per year at the service entrance, and far more at equipment terminals. The frequency is even higher in the interior of the site or in developing countries that have not achieved the same levels of power quality as more developed nations.

Dips do not generally disturb incandescent or fluorescent lighting, motors, or heaters. However, some electronic equipment lacks sufficient internal energy storage and therefore cannot ride through dips in the supply voltage. Equipment may be able to ride through very brief, deep dips, or it may be able to ride through longer but shallower dips.

Normally, the grid voltage fluctuates around its nominal value with variations within a maximum range of $\pm10\%$ of that value. A dip is the sudden drop in voltage from one or more phases followed by a rapid restoration to its nominal value after a short space of time between half a period (10 ms at 50 Hz) and 1 minute.

For the voltage fall to be considered a dip, the voltage value must be between 1% and 90% of its nominal value (IEC 61000-2-1, EN 50160). A drop below 1% is usually called a short interruption. Above 90% it is considered that the voltage is within the normal range of operation.

In three-phase grids, dips can be divided into two broad categories:

- Three-phase dips, when the voltages of the three phases fall into the same proportion.
- Asymmetric dips, where all three phase drops are not equal and the voltage is unbalanced: for example,
  o Single-phase dips that affect only one phase.
  o Biphasic dips that involve two phases.

The voltage dips are normally characterized by the depth and duration, as can be seen in Figure 1.39.

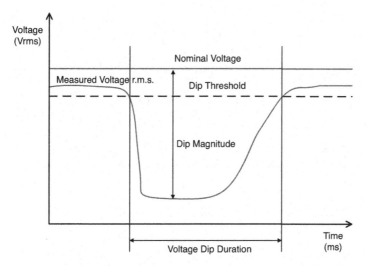

Figure 1.39   Parameters of a dip.

The depth measures the voltage drop in relative terms. It is measured at the deepest point of the valley. The voltage is usually measured by the rms value calculated for each half-period ($RMS_{1/2}$); hence, the minimum duration for a voltage dip is set for half a period. The duration is defined as the length of time that voltage is less than 90% of the nominal value.

1.6.2.2 Classification   Bollen and co-workers propose a more intuitive approach to the characterization of three-phase voltage dips. The ABC classification method distinguishes between seven dip types (A to G) by analyzing the possible types of short circuits and the dip propagation through transformers.

Table 1.3 summarizes these dip classes as a function of fault type, location and connection of measuring instruments in the AC grid.

Table 1.4 shows voltage phasors for different dip classes, in which the positive-sequence, negative-sequence, and zero-sequence impedances are considered to be equal.

If the amplitude depth is defined with the variable p, the fault voltages are defined in the equations for each fault type.

TABLE 1.3   Dip Classes of Several Faults Measured at Different Locations

Fault Type	Dip Class (measured between phase and neutral)	Dip Class (measured between phase and neutral after a Δy or YΔ transformer)	Dip Class (measured between phases)	Dip Class (measured between phases after a Δy or YΔ transformer)
Three-phase	A	A	A	A
Single-phase	B	C	D	C
Phase-to-phase	C	D	C	D
Two-phase-to-ground	E	F	F	G

TABLE 1.4 Voltage Dip Types

Type	Phasor Diagram	Phasor Amplitudes	Type	Phasor Diagram	Phasor amplitudes
A		$\underline{V_a} = 1-p$ $\underline{V_b} = 1-p$ $\underline{V_c} = 1-p$	E		$\underline{V_a} = 1$ $\underline{V_b} = a^2 \cdot (1-p)$ $\underline{V_c} = a \cdot (1-p)$
B		$\underline{V_a} = 1-p$ $\underline{V_b} = a^2$ $\underline{V_c} = a$	F		$\underline{V_a} = 1-p$ $\underline{V_b} = \dfrac{p-1}{2} - j \cdot \dfrac{3-p}{\sqrt{12}}$ $\underline{V_c} = \dfrac{p-1}{2} + j \cdot \dfrac{3-p}{\sqrt{12}}$
C		$\underline{V_a} = 1$ $\underline{V_b} = a^2 + j \cdot \dfrac{\sqrt{3}}{2} \cdot p$ $\underline{V_c} = a - j \cdot \dfrac{\sqrt{3}}{2} \cdot p$	G		$\underline{V_a} = 1 - \dfrac{p}{3}$ $\underline{V_b} = \dfrac{p-3}{6} - j \cdot \dfrac{\sqrt{3}}{2} \cdot (1-p)$ $\underline{V_c} = \dfrac{p-3}{6} + j \cdot \dfrac{\sqrt{3}}{2} \cdot (1-p)$
D		$\underline{V_a} = 1-p$ $\underline{V_b} = a^2 + \tfrac{1}{2}p$ $\underline{V_c} = a + \tfrac{1}{2}p$			

1.6.2.3 Transformer Effect The ABC classification method was introduced to describe the propagation of voltage dips through transformers. This method of voltage dip classification is the oldest and the most commonly used, possibly due to its simplicity.

Three groups of transformer connections must be analyzed:

- Wye–wye with neutral point connected
- Delta–delta and wye–wye without connecting neutral point
- Wye–delta and delta–wye

The first group of transformers allow circulating common mode currents. The second group of transformers do not allow circulating common mode currents. This means that the common mode component is eliminated.

The third group of transformers, apart from common mode component elimination, introduce a phase angle between primary and secondary voltages. Note that positive and negative sequences are phase shifted by the opposite signed angle.

Due to these reasons, the type of voltage dip in the primary side can be changed in the secondary side. A Type A voltage dip does not change with a transformer because it only has a positive sequence.

Figure 1.40 represents the transformation of different types of voltage dip according to the ABC classification method.

1.6.2.4 Transformer Effect in a Wind Farm The wind turbine can suffer grid outages at the point where it's connected to the high, medium, or low voltage network. In low voltage networks the turbine is directly connected to the low voltage distribution grid without a transformer. In medium voltage grids (distribution), the turbine is connected by means of the turbine step-up transformer.

Turbines connected to high voltage grids (transmission and subtransmission) are usually found in wind farms with the electrical layout explained in Section 1.4.2. In this case a typical example of the electrical circuit between the point of common coupling of the wind farm to the wind turbine is represented in Figure 1.41.

In order to study the effect of transmission network short circuits on the wind turbine, the following elements are modeled:

- The high voltage equivalent grid, represented by the equivalent Thèvenin circuit
- The substation transformer with a short-circuit impedance
- The impedance of the feeder from the substation to the wind turbine
- The coupling transformer with a short-circuit impedance
- The internal electrical circuit of the turbine

The short circuits can be located in high voltage (HV) or medium voltage (MV) grids and will affect the wind turbine in different ways but due to the transformer

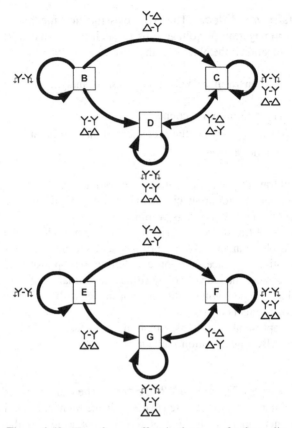

Figure 1.40   Transformer effect in the type of voltage dip.

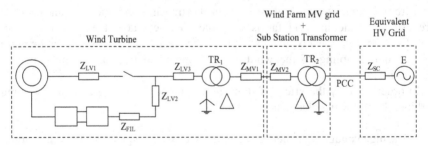

Figure 1.41   Transformer effect in the type of voltage dip for high voltage grid.

configuration only three-phase and two-phase voltage dips will occur at the low voltage connection point of the turbine.

The electrical configuration of wind farms is different in each country and the same goes for how the transmission and distribution systems are operated; each example must be considered separately.

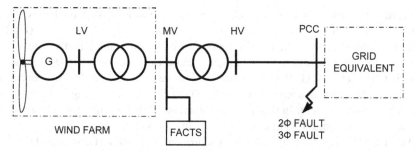

Figure 1.42 Single-line schematic of the electrical system layout.

1.6.3 Spanish Verification Procedure

The purpose of the procedure is to measure and assess the response of wind farms in the event of voltage dips. This procedure ensures uniformity of tests and simulations, precision of measurements, and assessment of the response of wind farms in the event of voltage dips. The requirements for the response to dips are specified in the electrical system Operational Procedure 12.3.

The following subsections will describe the most relevant points of the procedure related to the simulation of the behavior of the wind turbine to verify a response to voltage dips.

1.6.3.1 Topology of Electrical System In order to carry out wind farm simulations, an electrical configuration as described in Figure 1.42 will be used, which must contain at least the following elements:

- Wind farm devices and their specifications.
- Medium to high voltage step-up transformer
- Evacuation line (AT - PCR)
- Grid equivalent

All groups of the WTGs connected through the electrical power circuits will be connected at the medium voltage (MV) point. If additional loads or other wind farms are connected between the high voltage (HV) point and Evacuation line (HV - PCC) the grid connection point (GCP), also called point of common coupling (PCC), those additional loads or wind farms will be taken out of the simulation, in order to avoid any modification of the equipment that connects to the GCP (transformers and lines).

1.6.3.2 Equivalent Electrical Grid The rest of the electrical grid that does not belong to the wind farm being studied must be modeled so that the fault clearance at the grid connection point reproduces the usual voltage profile in the Spanish electrical system—a sudden increase upon clearing of the fault and a slower recovery afterwards. This profile will be considered fixed and independent of the geographic location of the wind farm to be studied.

In order to simulate the equivalent electrical grid, a dynamic system has been chosen consisting of one node in which the equivalent dynamic model of the UCTE (UCTE node) is modeled, along with another node in which an equivalent model reflects the dynamic characteristics due to the hypothetical closest electrical grid (RED node) and a third that represents the GCP (GCP node). These nodes are separated by impedances of predetermined values in such a way as to reproduce the typical voltage profile of the Spanish electrical system. In this way, it is guaranteed that all of the wind farms are tested, by simulation, for short circuits with equal characteristics.

The UCTE equivalent includes a synchronous generator (Generator 1) of an apparent power that reflects a realistic value for the interconnected apparent power and therefore the inertia of the UCTE system. This generator is modeled in 20 kV bars with a step-up transformer. The demand of the UCTE system is modeled as a load in the equivalent node of the system.

In order to consider the dynamic side of the equivalent of the closest grid, a synchronous generator has been included (Generator 2) and a demand. Generator 2 is modeled in 20 kV bars with a step-up transformer and the demand is modelled as a load in 20 kV bars connected to the RED node through a transformer.

The properties of the equivalent electrical grid must include at least the elements represented in Figure 1.43.

1.6.3.3 Evaluation of Response to Voltage Dips
The final part of the simulation procedure consists of the strict evaluation of the wind farm's response to voltage dips. Once the electrical system, its associated dynamic elements, and the starting conditions before simulation have been defined, a fault can be applied to the grid connection point.

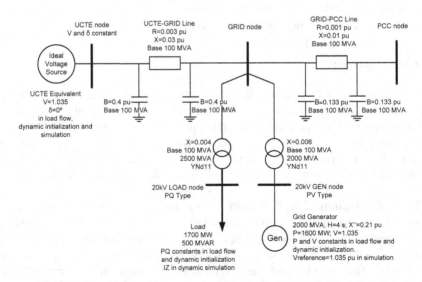

Figure 1.43   Model of the equivalent electrical grid (single-line scheme).

Once the initial conditions are adjusted at the WTG terminals, the active generated power corresponding to the full or partial load and zero reactive power will be considered the initial conditions prior to the simulation. Once the simulations have been carried out, the following requirements must be achieved for each test category:

1. *Continuity of Supply.* During the simulation, it must be shown that the wind farm withstands the specified dips in the test procedure without disconnection. To carry out these simulations, it is necessary that the simulation model includes internal protections, which determine the triggering of the WTG in case of voltage dips and return the resulting disconnection signal.

If the entire wind farm (without aggregation) is simulated, the simulated installation guarantees a continuity of supply if the number of machines remaining connected during the dip is such that the loss of generated active power does not exceed 5% of the power previous to the fault. If an equivalent wind farm (with aggregation) is used, the triggering of the WTG will determine the continuity of supply for the complete wind farm.

2. *Voltage and Current Levels at the WTG Terminals.* After checking the voltage level during the no load test, the voltage and current values in each phase for the four categories described above must be measured and recorded during the load tests (WTG connected during the short circuit).

3. *Exchanges of Active and Reactive Power as Described in OP 12.3.* In the wind farm simulation, it must be shown that neither a specific beginning or clearance point for the voltage dip nor a power factor for the WTGs, which are especially favorable for compliance with the requirements set in OP 12.3, has been chosen.

1.7 VSWT BASED ON DFIM MANUFACTURERS

This section discusses manufacturers that propose wind turbines based on the DFIM, and then develops a complete model for a 2.4 MW wind turbine along with the basic analysis of the aerodynamic, mechanical, and electrical magnitudes.

1.7.1 Industrial Solutions: Wind Turbine Manufacturers

The main players in the market have product solutions based on the DFIM. Table 1.5 shows some of their characteristics for the 50 Hz model wind turbines. The aim of this table is to give readers an idea of the main parameters of the turbines.

In order to understand the table, it's necessary to explain some concepts.

- Turbine speed: minimal–nominal–maximum: The turbine operates in the maximum power tracking region between the minimal and the nominal speed. The maximum speed limit is due to the difficulty in regulating the nominal speed during gusts and lessening the mechanical stress on the turbine when the turbine is operating at nominal power.

TABLE 1.5 Main Characteristics of Commercial Wind Turbines

Company	Model/Power	Turbine Speed Minimal–Nominal–Maximal	Rotor Diameter	Gearbox Ratio Type	Generator Voltage Stator/Rotor	Generator Speed Pole Number Minimal–Nominal–Maximal
Vestas	V80–2 MW	10.8–16.7 rpm 19.1 rpm	80	100.33 Three-stage planetary/helical	690 V – 480 V/—	2 1083/1672 1916
Repower	MM82–2.0 MW	8.5 – 17.1 rpm 17.1 rpm (+12.5%)	82	Approximately 105.4 Planetary/spur wheel	690 V/—	2 900–1800 rpm
	5M–5.075 kW	7.7 – 12.1 rpm 12.1 rpm (+15.0%)	126	Approximately 97 Two helical planetary/one spur wheel	950 V/660 V	3 750–1170 rpm 1170 rpm (+15.0%)
Gamesa	G52–850 kW	14.6 – 30.8 rpm —	52	1:61.74 One planetary/Two parallel axis	690 V/—	2 1000–1950
	G87–2.0 MW	9.0 – 19.0 rpm —	87	1:100.5 One planetary/Two parallel axis	690 V/—	2 900–1680 rpm 1900 rpm
Acciona	AW 1500 1.5 MW	20.2 rpm	70	1:59 Four-stage planetary	12 kV/—	3 770–1320 rpm 1200
	AW 3000 3 MW	14.2 rpm	100	1:77	12 kV/—	3 770–1320 rpm 1.100 rpm

Ecotecnia	ECO 80 2 MW	— 9.7–19.89 rpm	80	1:100.6	690 V	2 1000–1950 rpm 1800 rpm		
	ECO 100 3 MW	— 7.5–14.25 rpm.	100	1:126.319	1.000 V/—	2 1000–1950 rpm 1800 rpm		
Nordex	S70 1.5 MW	— 10.6–19 rpm	70	1:74	690 V	1000–1800 rpm 1800 rpm (+10.0%)		
	N80 2.5 MW	9.6–18 rpm	80	1:68.7	660 V	3 740–1300 rpm		
Mitsubishi	MWT92 2.4 MW	9–15 rpm 16.9 rpm	92	1:76.9 Three-stage, one planetary and two parallel	690 V	3 690–1153.5 1300 rpm		

- The rotor diameter is always a little bigger than the length of the blades, due to the hub diameter and because sometimes the blades have extensions. The tip maximum speed in meters per second is the turbine maximum speed in radians per second; multiplied for the rotor ratio, this speed is around 300 kilometers per hour in most turbines.
- Generator speed: minimal–nominal–maximal: The generator speed must be calculated by multiplying the turbine speed by the gearbox ratio. The minimal, nominal, and maximal values must correspond with the turbine ones.
- Generator pole number: The generator synchronous speed is 1500 rpm for a two pole pair generator, and 1000 rpm for a three pole pair generator connected to a 50 Hz grid.
- The typical stator voltage is 690 V or less than 1000 V, due to the fact that security rules for low voltage implies AC voltages upto 1000 volts and DC voltages upto 1500 V. Medium voltage equipments, for example, in power electronics, will require more strict security rules and ratified formation.
- The typical ratio between the rotor and stator voltages is in the range of 2.6 to 2.8, resulting in open rotor voltages of 1794–1932 V. As explained earlier, the rotor voltage is a function of the slip, which is limited to 500/1500 rpm in most wind turbines, resulting in a maximum rotor voltage in the range of 598–644 V rms line to line.

Before starting the next section, it's necessary to mention that all the information shown in this section comes from the manufacturer's web pages and wind turbines brochures.

1.7.1.1 *Alstom-Ecotècnia*

Alstom-Ecotècnia was a manufacturer of FSWTs until the 1990s when it started to develop model ECO 74, the first VSWT based on the DFIG that the company commercialized. At present the company's product range is based on two platforms—the 2 and the 3 MW.

Figure 1.44 shows a picture, a three-dimensional CAD drawing, and the brochure description of the main components of an ECO 100 wind turbine nacelle.

The main characteristics of the nacelle design are:

- A modular design that permits on-site testing and assembly of components (rotor hub, frame, housing).
- Housing that is made of three independent elements. Lateral housings provide extra space to install the power transformer, the inverter, and control cabinets. Placing the power transformer in the nacelle reduces the power lost during transmission from the generator to the transformer.

Alstom wind turbines are based on a unique mechanical design concept: the ALSTOM PURE TORQUE™ concept. The hub is supported directly by a cast frame on two bearings, whereas the gearbox is fully separated from the supporting

Figure 1.44 Picture and main components of the nacelle for an ECO 100 wind turbine. (*Source*: Alstom.)

structure, as shown Figure 1.45. As a consequence the deflection loads (red arrows) are transmitted directly to the tower whereas only torque is transmitted through the shaft to the gearbox.

Figure 1.46 shows the power curve.

Figure 1.45 Pure torque system to transmit the mechanical efforts between the rotor and the tower. (*Source*: Alstom.)

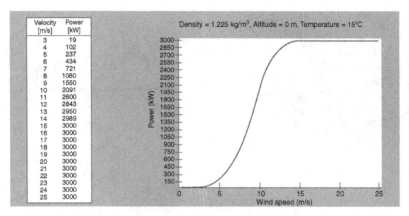

Figure 1.46   Power curve of the ECO 100 wind turbine. (*Source*: Ecotècnia.)

1.7.1.2 *Gamesa*

Gamesa is a Spanish manufacturer with a wide range of products, and one of the first to propose a VSWT based on DFIG. At present it offers three platforms—G5X, G8X, and G9X based on DFIG.

Figure 1.47 shows a picture, the brochure description of the main components in the nacelle, and the power curve for a G87 2 MW wind turbine.

The main characteristics of the nacelle design are:

- Drive train with the main shaft supported by two spherical bearings that transmit the side loads directly to the frame by means of the bearing housing. This prevents the gearbox from receiving additional loads, thus reducing malfunctions and facilitating its service.
- All the components of the drive train (low speed axel, gearbox, disk brake, and generator) are located in series with the transformer after them to equilibrate the weight.

Figure 1.48 shows the nacelle assembly procedure in the Gamesa facilities. The nacelle is transported assembled in one single truck without the hub and cone.

1.7.1.3 *Acciona*

The AW 3000 is the most powerful wind turbine manufactured by Acciona. Figure 1.49 shows a wind turbine picture and a drawing with the main components of the nacelle.

Some interesting characteristics of the nacelle mechanical design are:

- Robust double frame that reduces the stress on the drive train.
- Yaw system that uses a gear ring integrated into the tower and six geared motors integrated into the nacelle.

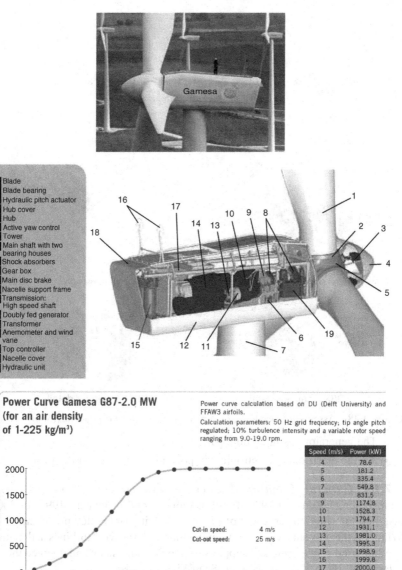

1	Blade
2	Blade bearing
3	Hydraulic pitch actuator
4	Hub cover
5	Hub
6	Active yaw control
7	Tower
8	Main shaft with two bearing houses
9	Shock absorbers
10	Gear box
11	Main disc brake
12	Nacelle support frame
13	Transmission: High speed shaft
14	Doubly fed generator
15	Transformer
16	Anemometer and wind vane
17	Top controller
18	Nacelle cover
19	Hydraulic unit

Power Curve Gamesa G87-2.0 MW (for an air density of 1-225 kg/m³)

Power curve calculation based on DU (Delft University) and FFAW3 airfoils.

Calculation parameters: 50 Hz grid frequency; tip angle pitch regulated; 10% turbulence intensity and a variable rotor speed ranging from 9.0-19.0 rpm.

Cut-in speed: 4 m/s
Cut-out speed: 25 m/s

Speed (m/s)	Power (kW)
4	78.6
5	181.2
6	335.4
7	549.8
8	831.5
9	1174.8
10	1528.3
11	1794.7
12	1931.1
13	1981.0
14	1995.3
15	1998.9
16	1999.8
17	2000.0
18	2000.0
19-25	2000.0

Figure 1.47 Picture, main components of the nacelle for a G87 wind turbine, and power curve. (*Source*: Gamesa.)

From the electrical point of view:

- The doubly fed generator generates at medium voltage (12 kV of stator voltage), which reduces losses and avoids the need for a transformer in many cases. The transformer is rated for the rotor power.

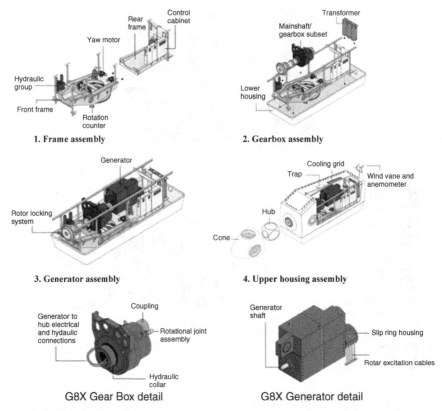

Figure 1.48 Assembly procedure of main components in a G87 nacelle. (*Source*: Gamesa.)

- The generator has three pole pairs.
- All the electric and electronic components are located in the tower.

1.7.1.4 General Electric The entity was created as a developer (not manufacturer), Zond, in 1980. Enron acquired Zond in January 1997. In 2002 GE acquired the wind power assets of Enron during its bankruptcy proceedings. Enron Wind was the only surviving U.S. manufacturer of large wind turbines at the time, and GE increased engineering and supplies for the Wind Division and doubled the annual sales to $1.2B in 2003. It acquired ScanWind in 2009.

The GE 1.5 MW is the most widely used wind turbine in its class. Developed by Zond in collaboration with the U.S. DOE (Department of Energy), it was initially commercialized by Enron in 1996 and improved by GE from 2002 on. Some interesting data as of March 2009 are the following:

- 12,000 + turbines are in operation worldwide.
- 19 countries employ them.
- 170 + million operating hours have been logged.
- 100,000 + GWh have been produced.

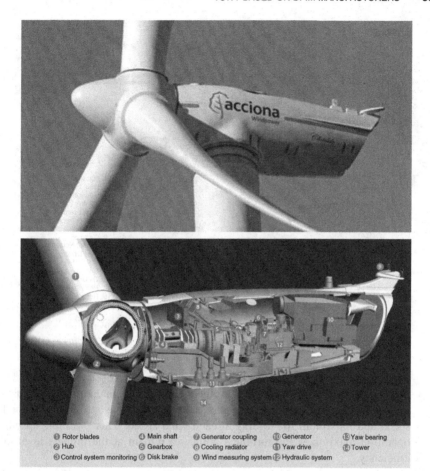

Figure 1.49 | Rotor blades | Main shaft | Generator coupling | Generator | Yaw bearing |
| Hub | Gearbox | Cooling radiator | Yaw drive | Tower |
| Control system monitoring | Disk brake | Wind measuring system | Hydraulic system | |

Figure 1.49 Main components of an AW 3000 nacelle. (*Source*: Acciona.)

Figure 1.50 shows a picture and the brochure description of the main components of the nacelle.

Some of the performance characteristics of the turbine are:

- Variable speed control: GE technology features unique variable speed control technology to maximize energy capture from the wind and minimize turbine drive-train loads.

- Unique wind volt-amp-reactive ("WindVAR") technology: This control provides support to and control of local grid voltage, improving transmission efficiencies and providing the utility grid with reactive power (VARs), increasing grid stability.

- Low voltage ride-through technology: For the first time, wind turbines can remain online and feed reactive power to the electric grid right through major system disturbances.

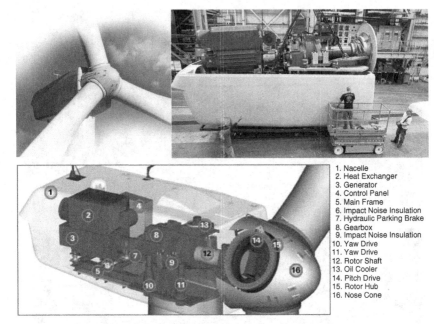

1. Nacelle
2. Heat Exchanger
3. Generator
4. Control Panel
5. Main Frame
6. Impact Noise Insulation
7. Hydraulic Parking Brake
8. Gearbox
9. Impact Noise Insulation
10. Yaw Drive
11. Yaw Drive
12. Rotor Shaft
13. Oil Cooler
14. Pitch Drive
15. Rotor Hub
16. Nose Cone

Figure 1.50 Pictures and main components of a 1.5 MW nacelle. (*Source:* GE.)

- Advanced electronics: The wind turbine's control system continually adjusts the wind turbine's blade pitch angle to enable it to achieve optimum rotational speed and maximum lift-to-drag at each wind speed. This "variable speed" operation maximizes the turbine's ability to remain at the highest level of efficiency.
- Variable speed operation enables the loads from the gust to be absorbed and converted to electric power. Generator torque is controlled through the frequency converter. This control strategy allows the turbine rotor to over-speed operation in strong, gusty winds, thereby reducing torque loads in the drive train.
- Active damping: The variable speed system also provides active damping of the entire wind turbine system, resulting in considerably less tower oscillation when compared to constant speed wind turbines. Active damping of the machine also limits peak torque, providing greater drive-train reliability, reduced maintenance cost, and longer turbine life.

Figure 1.51 shows the power curve for a 1.5 MW wind turbine.

1.7.1.5 Vestas Traditionally, Vestas has been using two technologies in most of its models with power ranges between 600 and 2000 kW: the Opti Slip (speed control by rotor resistance control) and the Active Stall (power control combining pitch

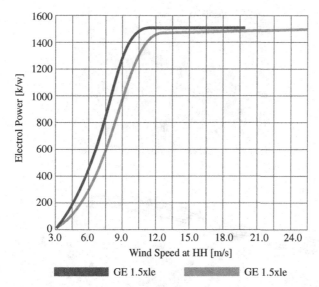

Figure 1.51   Power curve for a 1.5 MW wind turbine. (*Source:* GE.)

control and stall). These technologies provide a speed variation of 10%, enhancing power quality and reducing mechanical stress during wind gusts.

The V80, V90, and V120 wind turbines have a wide range of variable speed and are based on DFIG. Figure 1.52 shows a wind turbine picture and a drawing with the main components of the nacelle.

An interesting characteristic of the mechanical design is the OptiTip® pitch regulation system. This system features microprocessors that rotate the blades around their longitudinal axes, thus ensuring continuous adjustment to maintain optimal blade angles in relation to the prevailing wind. At the same time, sound levels are maintained within the limits stipulated by local regulations.

From the electrical point of view, the OptiSpeed® system:

- Allows the turbine rotor speed to vary between 9 and 19 rpm, depending on conditions.
- Reduces wear and tear on the gearbox, blades, and tower on account of the lower peak loading. Moreover, as turbine sound is a function of wind speed, the lower rotation speeds made it possible to reduce sound levels.
- Helps the turbine deliver better quality power to the grid, with rapid synchronization, reduced harmonic distortion, and less flicker.

Figure 1.53 shows the power curve for different noise levels.

1.7.1.6 Repower   Repower is a German manufacturer, at present the property of Sulzon, that specializes in multimegawatt wind turbines, such as the MM82–2 MW and the 5M, with 6M development just about complete.

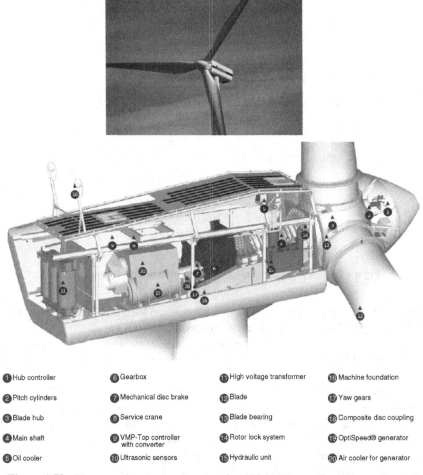

① Hub controller	⑥ Gearbox	⑪ High voltage transformer	⑯ Machine foundation
② Pitch cylinders	⑦ Mechanical disc brake	⑫ Blade	⑰ Yaw gears
③ Blade hub	⑧ Service crane	⑬ Blade bearing	⑱ Composite disc coupling
④ Main shaft	⑨ VMP-Top controller with converter	⑭ Rotor lock system	⑲ OptiSpeed® generator
⑤ Oil cooler	⑩ Ultrasonic sensors	⑮ Hydraulic unit	⑳ Air cooler for generator

Figure 1.52   Picture and main components of a V80 2 MW nacelle. (*Source:* Vestas.)

Figure 1.54 shows a picture, the brochure description of the main components of the nacelle, and the power curve for a Repower MM82–2 MW wind turbine.

The main characteristics of the design are:

- Rotor bearing and shaft: High performance spherical roller bearing with adjusted bearing housing and permanent lubrication for prolonged service life. Rotor shaft forged from heat-treated steel and optimized for power flow.
- Holding brake: Secure holding of rotor due to generously dimensioned disk brake. Soft-brake function reduces stress on the gearbox.
- Generator and converter: Yield-optimized variable speed range. Low conversion loss and high total efficiency as converter output is limited to a maximum of 20%

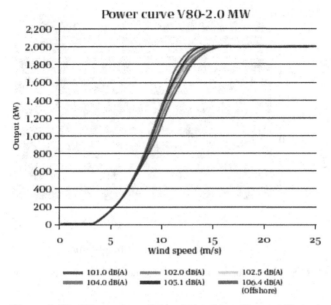

Figure 1.53 Power curve of the wind turbine. (*Source:* Vestas.)

of the overall output. Fully enclosed generator with air/air heat exchanger. Optimized temperature level in generator, even at high outside temperatures.

• All the components of the drive train (low speed axle, gearbox, disk brake, and generator) are located in series over the main frame.

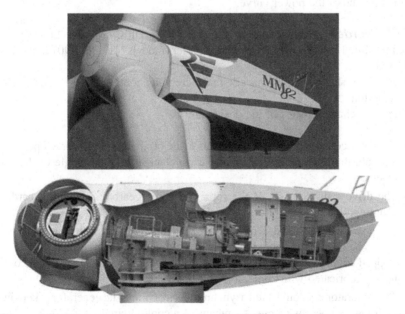

Figure 1.54 Picture and main components of a MM82 nacelle. (*Source:* Repower.)

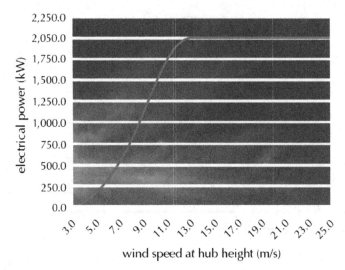

Figure 1.55 Power curve of the MM82 wind turbine. (*Source:* Repower.)

- Pitch system: Virtually maintenance-free electronic system. High quality, generously dimensioned blade bearing with permanent track lubrication. Maximum reliability via redundant blade angle detection by means of two separate measuring systems. Fail-safe design with separate control and regulation systems for each rotor blade.

Figure 1.55 shows the power curve.

1.7.1.7 Nordex Nordex started with the VSWT DFIG based technology with model S70 rated at 1.5 MW and actually also offers models N80 and N90 rated at 2500 and 2300 kW.

Figure 1.56 shows the N80 nacelle in different versions, a picture, and the CAD representation.

The main characteristics of the design are:

- The rotor consists of three rotor blades made of fiberglass-reinforced polyester, the hub, the pitch bearings, and drives to change the pitch angle of the rotor blades.
- The drive train consists of the rotor shaft, the gearbox, an elastic cardanic coupling, and the generator.
- The gearbox is designed as a two-stage planetary gearbox with a one-stage spur gear. The gearbox is cooled by means of an oil–water–air cooling circuit with stepped cooling capacity. The bearings and tooth engagements are kept continuously lubricated with cooled oil.
- The generator is a double-fed asynchronous machine. The generator is kept in its optimum temperature range by means of a cooling circuit.

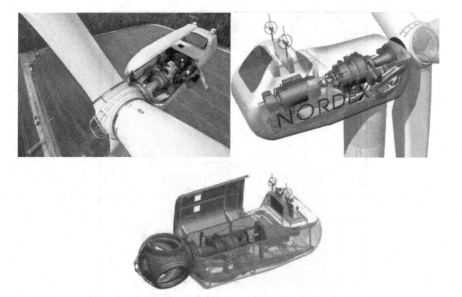

Figure 1.56   Nordex N90 variable speed pitch regulated wind turbine. (*Source:* Nordex).

- The gearbox, generator, and converter have cooling systems that are independent from each other. The cooling system for the generator and converter is based on a water circuit. This ensures optimum operating conditions in all types of weather.
- The three redundant and independently controlled rotor blades can be set at full right angles to the rotation direction for aerodynamic braking. In addition, the hydraulic disk brake provides support in the event of an emergency stop.
- The hydraulic system provides the oil pressure for the operation of different components: the yaw brakes, rotor brake, and nacelle roof.
- The nacelle consists of the cast machine frame and the nacelle housing. The nacelle housing is made of high-quality fiberglass-reinforced polyester (GRP). The roof of the nacelle is opened hydraulically.
- The wind direction is continuously monitored by two redundant wind direction sensors on the nacelle. If the permissible deviation is exceeded, the yaw angle of the nacelle is actively adjusted by means of two geared motors.
- The wind turbine has two anemometers. One anemometer is used for controlling the turbine, the second for monitoring the first. All operational data can be monitored and checked on a control screen located in the switch cabinet.

Figure 1.57 shows the power curve of the turbine.

1.7.1.8 Mitsubishi   Mitsubishi started with the VSWT DFIG based technology with model MWT 92 rated at 2.4 MW and actually also offers models MWT 95, 100, and 102 rated at 2400 kW and MWT 92 rated at 2300 kW.

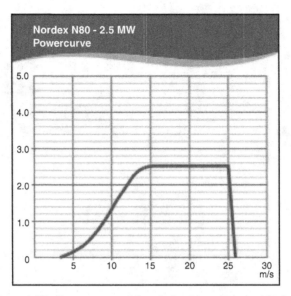

Figure 1.57 Power curve of the wind turbine. (*Source:* Nordex.)

Figure 1.58 shows some pictures of the MWT92 wind turbine and the components' distribution in the nacelle.

The main characteristics of the turbine design are:

- All the components of the drive train (low speed axle, gearbox, disk brake, and generator) are located in series in a lateral of the nacelle.
- The transformer is in the other lateral to equilibrate the weight.
- The power electronic converters are located at the bottom of the nacelle, and the water to air heat exchanger after the cabinets.
- Individual blade pitch control for aerodynamic load reduction.
- "Smart yaw" technology to minimize extreme wind load.

Figure 1.59 shows the power curve of the turbine.

1.7.2 Modeling a 2.4 MW Wind Turbine

Notice that the information that manufacturers provide in their brochures is very restrained. The aim of this section is to provide some basic numbers and ideas for modeling a wind turbine.

A typical wind turbine has a nominal power between 1.5 and 3 MW as shown in preceding sections. In this section a 2.4 MW wind turbine has been selected for modeling. The most significant models on market are the GE 2.5, the Nordex N80, and the Mitsubishi MWT 92.

Wind turbine sizing is a complex task that requires knowledge of multiple disciplines, starting with the aerodynamic behavior and finishing with the electrical machine.

Schematic Diagram

① Blade	⑤ Oil Cooler	⑨ Nacelle Bed Plate	⑬ Control Panel
② Hub	⑥ Generator	⑩ Yaw Bearing	⑭ Inverter
③ Blade Bearing	⑦ Service Winch	⑪ Yaw Gear	⑮ Cooler for Inverter
④ Main Bearing	⑧ Exhaust Duct for Generator	⑫ Transformer	⑯ Tower

Figure 1.58 Mitshibishi MWT 92 wind turbine. (*Source:* Mitsubishi.)

From data extracted from manufacturers' brochures and other specialized references, it's possible to propose some parameters that represent the energetic behavior of the wind turbine and its main mechanical and electrical dynamics.

The two main aspects to take into account are:

- The aerodynamic behavior of the rotor. In our case the analytic expression defined in Section 1.3 must be parameterized.
- The wind turbine control strategy parameters must be selected; that is, the maximum and minimum turbine speed for a nominal wind speed and a minimum wind speed.

Once the energetic behavior of the turbine is defined, it's possible to proceed to the sizing of the power electronic converters.

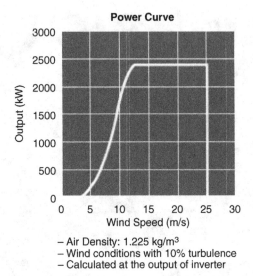

Figure 1.59   Power curve of the wind turbine. (*Source:* Mitsubishi.)

1.7.2.1 *Model of the Aerodynamic System*   From manufacturers' brochures it's possible to get the basic data for a 2.4 MW turbine:

- The radius of the blade is in the range of 40–45 meters, depending on the wind class of the turbine.
- The nominal power is extracted for a wind speed between 11 and 13 m/s.
- The rotor speed (low speed axle) is in the range of 8.5 to 20 rpm.
- The gearbox ratio is around 100 for a two-pole generator and 50 Hz grid.

Another important parameter is the tip blade maximum speed, which is around 80–90 m/s (325 km/h), directly dependent on the turbine radius and maximum speed.

The aerodynamic behavior of the turbine rotor is more complicated because it is one of the differential factors between manufacturers; only Nordex in their N80–90 brochure gives some information about the power coefficient of their turbine. In this sense the only available information is the typical power versus wind speed curve.

Table 1.6 summarizes the selected parameters for turbine sizing.

It's possible to adjust the coefficients of Equation (1.7) for the 2 MW wind turbine. The next expression shows the numerical result:

$$C_p = 0.46\left(\frac{151}{\lambda_i} - 0.58\beta - 0.002\beta^{2.14} - 13.2\right)\left(e^{-18.4/\lambda_i}\right)$$
$$\lambda_i = \frac{1}{\lambda + 0.02\beta} - \frac{0.003}{\beta^3 + 1} \tag{1.22}$$

TABLE 1.6   Turbine Parameters

Parameter	Value	Unit
Radius	42	m
Nominal wind speed	12.5	m/s
Variable speed ratio (minimum–maximum turbine speed)	9–18	rpm
Optimum tip speed ratio λ_{opt}	7.2	—
Maximum power coefficient C_{p_max}	0.44	—
Air density ρ	1.1225	kg/m^3

Figure 1.60 shows the plot of power coefficient as a function of tip speed ratio for Equation (1.7); notice that the maximum power coefficient is 0.44, and the optimum tip speed ratio is 7.2.

The power curves as a function of wind speed are shown in Figure 1.61, along with the typical speed limits (9 and 18 rpm) and the nominal power (2.4 MW).

1.7.2.2 Gearbox and Mechanical Model   The gearbox ratio is a function of the machine nominal and maximum speed and the turbine maximum speed, so a ratio N of 100 has been chosen for a two-pole generator.

Table 1.7 shows the selected two mass mechanical system parameters translated to the high speed axle.

The corresponding resonance frequency is 2 Hz.

1.7.2.3 Generator Characteristics   Table 1.8 shows the electrical generator's main characteristics.

Table 1.9 shows the electrical generator's equivalent schema parameters.

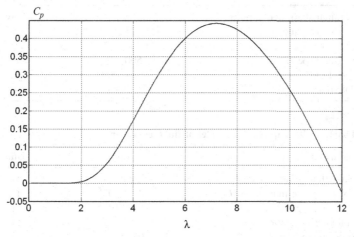

Figure 1.60 Power coefficient versus tip speed ratio.

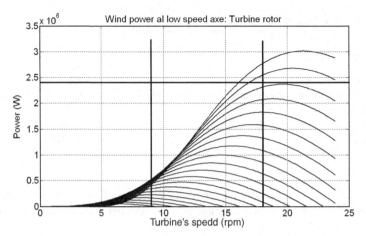

Figure 1.61 Power versus tip speed ratio for wind speeds between 3.5 and 13 m/s in steps of 0.5 m/s.

TABLE 1.7 Mechanical System Parameters

Parameter	Value	Unit
Low speed axle inertia $J_{t,a}$	800	kg·m²
Low speed axle friction $D_{t,a}$	0.1	Nm·s/rad
Coupling stiffness K_{tm}	12500	Nm/rad
Coupling damping D_{tm}	130	Nm·s/rad
High speed axle inertia J_m	90	kg·m²
High speed axle friction D_m	0.1	Nm·s/rad

TABLE 1.8 Main Characteristics of the Generator

Parameter	Value	Unit
Nominal stator active power	2.0	MW
Nominal torque	12732	Nm
Stator voltage	690	V
Nominal speed	1500	rpm
Speed range	900–2000	rpm
Pole pairs	2	—

TABLE 1.9 Equivalent Model of the Generator

Parameter	Value	Unit
Magnetizing inductance L_m	2.5×10^{-3}	H
Rotor leakage inductance L_{or}	87×10^{-6}	H
Stator leakage inductance L_{os}	87×10^{-6}	H
Rotor resistance R_r	0.026	Ω
Stator resistance R_s	0.029	Ω

1.7.2.4 Power Curves as a Function of the Wind Turbine Control Strategy
In order to select the wind turbine control strategy, it is very important to keep three parameters in mind:

1. The optimum tip speed ratio of 7.2
2. The rotor radius of 42
3. The speed range of the turbine from 900 to 1800 rpm

The next range of wind speed for the variable speed turbine is 5.5 m/s for 900 rpm and 11 m/s for 1800 rpm (maximum power tracking region). The nominal power of the turbine is reached around 12.5 m/s and 1800 rpm.

Figure 1.62 represents the evolution of generator speed as a function of wind speed (averaged wind speed in the turbine swept area).

Notice that the incident wind on the turbine is not constant and varies widely. Figure 1.62 shows that the average behavior of the turbine, during real operation, must be defined as an envelope as a function of the control dynamics (maximum and minimum values of turbine speed for each ideal operating point).

For that reason the turbine is designed to work in an overspeed range, for example, 2050 rpm.

Figure 1.63 shows the mechanical power of the turbine. In a real wind turbine, the transition between variable speed–partial load operation is much smoother because:

1. The real turbines currently operate with a slight speed slope (e.g., between 1850 and 1900 rpm).
2. At nominal speed the turbine operates in a stall mode in region 3 (from wind speed of 11m/s until the nominal power is achieved).

Figure 1.64 shows the resulting torque of the turbine, resulting in a nominal torque for the turbine of 12,700 Nm.

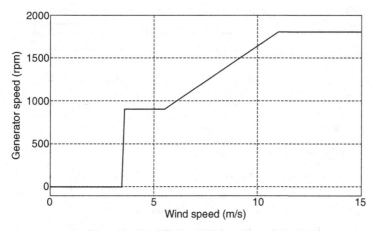

Figure 1.62 Wind turbine speed control.

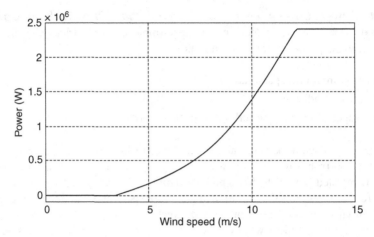

Figure 1.63 Wind turbine power versus wind speed.

Once the wind turbine control strategy has been defined, the power extraction as a function of the wind speed is represented in Figure 1.65.

The different control regions of the turbine are as follows:

1. Constant low speed region for wind speed in the range of 3.5–5.5 m/s.
2. Maximum power tracking region for wind speed in the range of 5.5–11 m/s.
3. Constant nominal speed region for wind speed in the range of 11–12 m/s.

For increasing wind speeds, the pitch control will work in order to diminish the aerodynamic performance of the turbine and limit the power extraction.

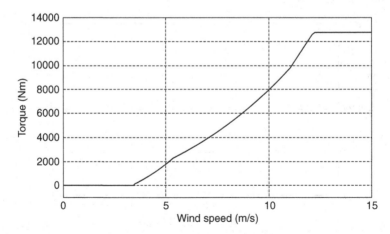

Figure 1.64 Wind turbine torque versus wind speed.

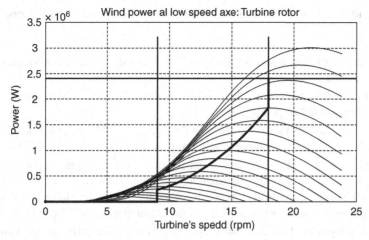

Figure 1.65 Wind turbine control strategy as a function of different wind speeds.

1.7.3 Steady State Generator and Power Converter Sizing

From the wind turbine power and speed, it's possible to calculate all the main electrical magnitudes of the wind turbine generator and its associated converters.

The starting point is the characteristic power as a function of the mechanical speed. To simplify Table 1.10, only some points of Figure 1.65 have been

TABLE 1.10 Operating Points for the Generator

Point	Generator Speed (rpm)	Electric Power Output (kW)	Power Factor	Line Voltage (V_{RMS})
1	900	212	1.0	690
2	1013	326	1.0	690
3	1216	563	1.0	690
4	1397	852	1.0	690
5	1509	1075	1.0	690
6	1600	1280	1.0	690
7	1712	1580	1.0	690
8	1800	1831	1.0	690
9	1800	2400	1.0	690
10	1800	2400	1.0	621
11	1800	2400	1.0	759
12	1800	2400	0.9 inductive	690
13	1800	2400	0.95 capacitive	690
14	1944	2592	0.95 capacitive	690
15	1944	2592	0.9 inductive	690

taken, and some others have been added in order to consider special operating points:

- Normal operation at nominal stator voltage and unity power factor is represented in points 1 to 9.
- Operation at rated power with voltage variation of ±10% from the nominal one are represented in points 10 and 11.
- Operation at rated power and nominal voltage with 0.9 power factor inductive and 0.95 capacitive is represented in points 12 and 13.
- Operation at rated voltage and overspeed of 8% is represented in points 14 and 15. This is a typical situation when the turbine is delivering the nominal power and the speed is controlled by means of a pitch control that has a low dynamic response, and it is normal to overpass the nominal speed (overspeed operation).
- Operating point 14 is a transient operating point; the control system maintains nominal torque while the generator speed rises from 1800 to 1944 rpm.
- The last point defines the operating capacity of the system overload control. The power factor under overload conditions must be inductive 0.9 or better.

The resulting operating table for the generator and converters is illustrated in Table 1.10.

As mentioned in preceding subsections, the machine can be magnetized from the stator or from the rotor; both solutions produce different sizing for the rotor and grid side converter. The next few paragraphs will show the results for the first case. All variables that can be seen in the following figures are represented as in-phase RMS values.

Figure 1.66 shows the evolution of electromagnetic torque and the generator slip. In the first graph you can see the generator electromagnetic torque, calculated from the mechanical generator and speed in Table 1.10. The torque increases as the wind speed increases and thus the speed of the turbine until it arrives at this nominal value.

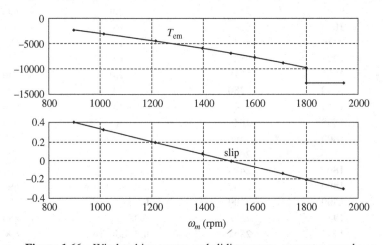

Figure 1.66 Wind turbine torque and sliding versus generator speed.

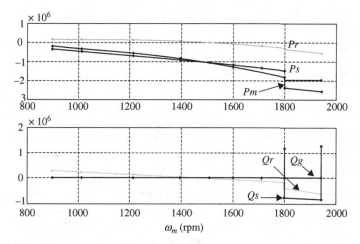

Figure 1.67   Active and reactive power for stator, rotor, and grid side converter.

In the second graph the motor slip can be seen; as the speed increases, a change from the hyper-to subsynchronous operating mode appears at synchronous speed (1500 rpm). If the motor mechanical speed is less than the synchronous speed, the engine is running in subsynchronous operating mode, while if it is greater it is running in hypersynchronous operating mode. In this particular case, the slip rises 40% at minimum speed and 30% at overload speed.

Figure 1.67 represents the mechanical power and the active and reactive power delivered by the rotor (RSC), the stator, and the grid side converter.

In the first graph we can see that the machine is running as a generator, the mechanical power and the stator active power are always negative; in subsynchronous operation, the rotor power is positive (the rotor is taking power from the grid). However, in the hypersynchronous mode, the stator and rotor are delivering active power to the grid (both take negative values).

In the second graph we can see the following:

- The power factor is unitary for the grid side converter (Q_g is zero).
- The stator reactive power (Q_s) is equal to zero while the RSC provides the reactive power necessary to magnetize the machine; the first few dots occur while the speed is lower than 1800 rpm. It is also seen that the rotor of the generator consumes reactive power in subsynchronous operation (as an inductor), while in hyper synchronous mode power returns (as a capacitor).
- However, when the power factor is different from one, the stator must provide the necessary reactive power to reach the required power factor (0.95 or 0.9). For capacitive ones the reactive power of the rotor must increase while for inductive ones it must decrease.

Figure 1.68 shows the rotor voltage and current; that is, the amplitude of the rotor side converter (RSC) voltage and current.

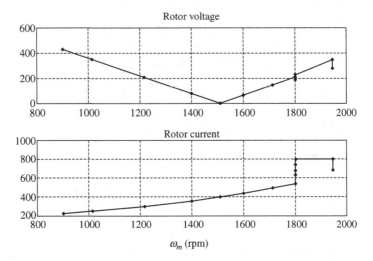

Figure 1.68 Rotor voltage and current.

In the first graph we can see that the magnitude of the rotor voltage is minimal near the synchronism speed, while at subsynchronous and hypersynchronous operation, the voltage increases. This voltage is proportional to the slip, but it's modified by the rotor resistance voltage. For a 690 V grid (line-to-line voltage), the phase voltage is 400 V; it can be appreciated that the rotor voltage is slightly lower at overspeed operation (around 350 V). For this example the relation between the rotor and stator voltages has been selected as 2.6.

In the second graph you can see that as the electromagnetic torque increases, the rotor current increases because the electromagnetic torque is controlled by the quadrature component of the rotor current.

Figure 1.69 plots the stator voltage and current.

The stator voltage is established in the initial table, but the plotted value corresponds to the calculated value once the magnetizing levels of the machine are derived. That's the reason for the slight difference from the 400 V indicated.

The stator current increases as the machine torque grows. The reason is that this current is directly related to the electromagnetic torque due to the fact that the machine is magnetized from the rotor side.

Figure 1.70 plots the grid side converter voltage and current.

When the power factor is unitary, the voltage of the grid side inverter is nearly equal to the grid voltage; only a low voltage difference is introduced by the filter.

The current evolution is proportional to the active power of the grid side converter as the power factor is one:

- In subsynchronous operation the rotor power is positive and decreasing to zero.
- At synchronous speed the rotor power is zero.
- In hypersynchronous operation the rotor power is increasing from zero to their maximum value at maximum speed. Note than maximum rotor current is higher than GSC one, due to the fact that generator is magnetized from rotor.

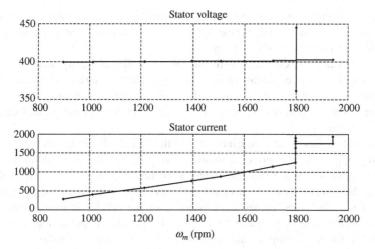

Figure 1.69   Stator voltage and current.

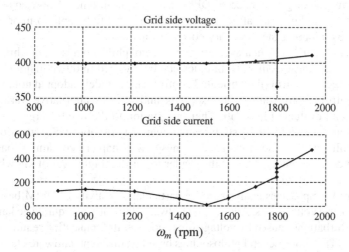

Figure 1.70   Grid side converter voltage and current.

1.8 INTRODUCTION TO THE NEXT CHAPTERS

Once this introductory chapter has immersed the reader in the most basic and general knowledge about doubly fed induction machine based wind turbines, the following chapters cover topics related to more technical electronic aspects.

Chapter 2 presents the back-to-back converter employed to supply the rotor of the wind turbine's generator, that is, the DFIM. It is possible to make the wind turbine operate at variable speed, delivering energy from the wind to the grid and meeting the grid code requirements. Thus, this chapter includes detailed and basic knowledge related to converter modeling and control, permitting one to understand the function of this element in the wind turbine system, as well as the supply capacities and limitations of these kind of converters.

Chapter 3 deals exclusively with the DFIM model at steady state operation. By inclusion of mathematical models and equivalent electric circuits, the basic operation of this machine is analyzed. Then an exhaustive evaluation of the machine is carried out, revealing the performance characteristics of this machine at steady state.

Chapter 4 studies the dynamic modeling of the DFIM. Thanks to this more general modeling approach, it is possible to study the machine in a more complete way. Dynamic behavior as well as the steady state regime can be analyzed in detail, enabling the reader to reach a reasonably high level of understanding of the machine.

Chapter 5 discusses the testing procedure of the DFIM. It includes practical information to characterize machines on the basis of off-line experimental tests, when they are performed during a realistic wind energy generation scenario.

Chapter 6 goes further with modeling, analyzing the dynamic behavior of the DFIM supplied by a disturbed voltage. Grid voltage dips can strongly deteriorate the performance of grid connected DFIM based wind turbines. Knowledge of how these types of turbines are affected is crucial, in order to be able to improve their behavior, by adopting the necessary corrective actions.

Chapters 7 and 8 present a wide range of control philosophies for machine DFIM wind turbines. These control techniques can be grouped into two major concepts, known as vector control (field oriented control, also a widely adopted notation) and direct control. Both control philosophies and several alternative variants are analyzed and evaluated in detail. Their adaptation to the wind energy generation environment is considered, by studying their capacity to solve faulty situations such as grid voltage dips and imbalances. These two chapters containing basic and advanced control concepts allow the reader to reach a high level of knowledge of DFIM based wind turbines.

Chapter 9 emphasizes analysis of the particular sensitivity of DFIM based wind turbines to grid voltage dips. Additional hardware elements are required, to handle the strong perturbations caused by voltage dips and meet the connection requirements of grid codes. Therefore, several philosophies based on different hardware elements are presented and discussed in this chapter.

Chapter 10 deals mainly with some particular needs of DFIM based wind turbines, when they operate with a reduced size of back-to-back converter (cost-effective solution). The specific start-up procedure of these kinds of wind turbines is analyzed, accompanied by several illustrative examples.

Chapter 11 ends the comprehensive study of DFIM based wind turbines, extending the analysis to a slightly different scenario: stand-alone operation. Thus, this concept of wind turbines not connected to the grid, but supplying energy directly to one or several loads, can be studied with most of the knowledge acquired in the preceding chapters. However, there is also new and important modeling and control aspects covered in this chapter, necessary to understand the stand-alone generation system and its special traits.

Finally, Chapter 12 discusses the future challenges and technological tendencies of wind turbines in general, focusing not only on DFIM based ones, but also looking at different and innovative wind turbine concepts and their supporting technology.

BIBLIOGRAPHY

1. J. L. R. Amenedo, et al. *Sistemas eólicos de producción de energía eléctrica*. Rueda, 2003 (in Spanish).
2. I. Munteanu, I. U. Bratcu, N. A. Cutululis, and E. Ceanga, *Optimal Control of Wind energy Systems*. Springer, 2008.
3. M. Stiebler, *Wind Energy Systems for Electric Power Generation*. Springer, 2008.
4. E. Hau, *Wind Turbines: Fundamentals, Technologies, Applications, Economics*. Springer, 2000.
5. F Bianchi, Wind turbine control systems: principles, modelling & gain scheduling design, Springer, 2006.
6. M. P. Kazmierkowski, R. Krishnan, and F. Blaagberg, *Control in Power Electronics Selected Problems*. Academic Press, 2002.
7. F. Blaagberg and Z. Chen, *Power Electronics for Modern Wind Turbines*. Morgan & Claypool Publishers, 2006.
8. I. Boldea, *Variable Speed Generators*. CRC Press/Taylor & Francis, 2006.
9. I. Boldea, *Synchronous Generators*. CRC Press/Taylor & Francis, 2005.
10. T. Ackerman, *Wind Power in Power Systems*. Wiley, 2005.
11. H. Camblong,"Minimización del impacto de las perturbaciones de origen eólico sobre la producción de electricidad en aerogeneradores de velocidad variable." Ph.D. Thesis from University of Mondragon, 2003 (in French).
12. Math Bollen et al. "Voltage Dip Immunity of Equipment and Installations", CIGRE Brochure 412, 2010.
13. J.A. Martinez-Velasco et al. "Voltage dip Evaluation and Prediction Tools", CIGRE Brochure 372, 2009.
14. M. Santos,"Aportaciones al control centralizado de un parque eólico." Ph.D. Thesis from University of Mondragon, 2007 (in Spanish).
15. Catalogs from wind turbine manufacturers on the Internet.

CHAPTER 2

Back-to-Back Power Electronic Converter

2.1 INTRODUCTION

In Chapter 1, a wide perspective of the wind turbine technology in general and of the doubly fed induction machine (DFIM) based wind turbines in particular is provided. It has already been clearly established that the main subject of this book is the study of DFIM based wind turbines. This broadly employed wind turbine is supplied by the rotor with a back-to-back converter, also known in the literature as a reversible or bidirectional converter.

Prior to addressing the analysis of the DFIM and its control, this chapter first covers the most relevant knowledge about the back-to-back converter. Thus, once the most important background of the back-to-back converter is understood, the book looks ahead to the DFIM and its control, focusing always on the wind energy generation application.

Therefore, in this chapter, the basic concepts related to the back-to-back converter are analyzed, paying particular attention to the grid side rather than to the rotor side. Thus, not only the converter itself is studied but also its closely associated elements such as the filters, DC link, and the control (only grid side). In this way, the chapter is structured as follows.

First, the classic and commonly used two-level voltage source converter (VSC) is described, its model is examined, and presenting different pulse generation possibilities for its controlled switches, together with a brief description of the filters employed to mitigate undesired but unavoidable effects of voltage source converters (current ripples, dv/dt, etc.). These aspects are studied for both the rotor and grid side converter.

Second, study of the converter is enhanced by extending the exposition to newer emerging topologies, such as multilevel converters. These newer topologies, among other advantages, basically allow one to increase the power and voltage handled by the converter, thus enabling an increase in the overall wind turbine power to the

Doubly Fed Induction Machine: Modeling and Control for Wind Energy Generation,
First Edition. By G. Abad, J. López, M. A. Rodríguez, L. Marroyo, and G. Iwanski.

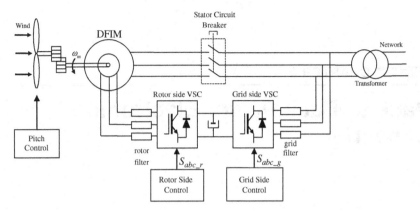

Figure 2.1   System configuration of the DFIM based wind turbine.

range of multimegawatts. Special focus is paid to the most extended multilevel topology—the three-level neutral point clamped (3L-NPC) VSC. However, some other promising multilevel topologies suitable for the DFIM based wind turbines are also presented briefly.

Finally, control of the grid side converter is examined, by studying steady state and dynamic models of the grid side system formed by the grid, the grid side filter, and the grid side converter itself first. Then, the vector control strategy for the grid side converter is studied. This control strategy enables one to fulfill the two main objectives of the grid side converter: control of the bus voltage of the DC link and control of the active and reactive powers exchanged bidirectionally between the rotor of the machine and the grid. Control of the rotor side converter, with which control of the DFIM is performed, is described in subsequent chapters in detail.

In this way, thanks to the knowledge acquired by the reader in this chapter, it is possible to understand the supply advantages and limitations for the DFIM due to the usage of the back-to-back converter. Once the reader understands the DFIM, taking into account the restrictions imposed by the converter, by means of Chapters 3–6, it will be possible to combine all the knowledge to address DFIM control for wind energy generation applications, by means of Chapters 7–11.

Consequently, the simplified schema of a DFIM based wind turbine already presented in the previous chapter, again repeated as Figure 2.1, shows how the back-to-back converter under study is connected. The back-to-back converter is connected by the rotor side filter to the rotor of the DFIM; it is connected to the grid by the grid side filter.

2.2 BACK-TO-BACK CONVERTER BASED ON TWO-LEVEL VSC TOPOLOGY

This section describes the most important aspects of a back-to-back converter based on the widely used two-level converter. Both side converters are described,

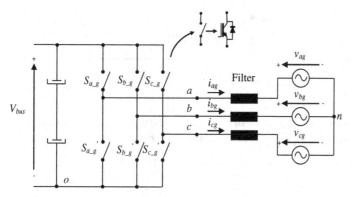

Figure 2.2   Simplified converter, filter, and grid model.

developing models considering ideal switches; in addition, several pulse generation strategies, called modulations, are also studied.

2.2.1 Grid Side System

The grid side system is composed by the grid side converter, the grid side filter, and the grid voltage. Figure 2.2 illustrates a simplified model of the grid side system. We can distinguish the following elements.

- The grid side converter is modeled with ideal bidirectional switches. It converts voltage and currents from DC to AC, while the exchange of power can be in both directions from AC to DC (rectifier mode) and from DC to AC (inverter mode). The ideal switch normally is created by a controlled semiconductor with a diode in antiparallel to allow the flow of current in both directions. In this exposition, the controlled semiconductor used is an insulated gate bipolar transistor (IGBT). It must be remarked again that this chapter treats the controlled switches ideally, not considering real characteristics such as switching time or voltage drops.
- The grid side filter is normally composed of at least three inductances (L), which are the link between each output phase of the converter and the grid voltage. When a high filter requirement is needed, each inductance can be accompanied by one capacitor (LC) or even by one capacitor and one more inductance (LCL).
- The grid voltage is normally supplied through a transformer. This AC voltage is supposed to be balanced and sinusoidal under normal operation conditions. The effect of the transformer or grid impedances is neglected in this chapter.

The following subsections explain each element in more detail.

2.2.1.1 *Converter Model*   The two-level converter is modeled with ideal switches that allow the flow of current in both directions. The command of the

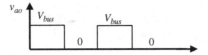

Figure 2.3 Two voltage levels of v_{ao}.

switches is made by means of S_{a_g}, S_{b_g}, and S_{c_g} signals. Note that under ideal conditions, the following order of commands holds:

$$S'_{a\_g} = \overline{S_{a\_g}} \tag{2.1}$$

$$S'_{b\_g} = \overline{S_{b\_g}} \tag{2.2}$$

$$S'_{c\_g} = \overline{S_{c\_g}} \tag{2.3}$$

What this means is that in a leg of the converter, it is not possible to have conduction in both switches. Different output voltages can be distinguished in a converter, for instance, the voltages referenced to the zero point of the DC bus:

$$v_{jo} = V_{bus}S_{j\_g}$$
$$\text{with} \quad S_{j\_g} \in \{0, 1\} \quad \text{and} \quad j = a, b, c \tag{2.4}$$

Thus, by different combinations of S_{a_g}, S_{b_g}, and S_{c_g} it is possible to create AC output voltages with a fundamental component of different amplitude and frequency. As depicted in Figure 2.3, this converter provides two possible voltage levels at each phase.

On the other hand, for modeling purposes, it is very useful to know the converter output voltages referred to the neutral point of the grid three-phase system (n). As Figure 2.4 illustrates, the following voltage relationships are true:

$$v_{jn} = v_{jo} - v_{no} \quad \text{with} \quad j = a, b, c \tag{2.5}$$

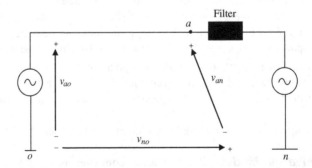

Figure 2.4 Simplified equivalent single-phase grid circuit (a phase).

The voltage between the neutral point (n) and the negative point of the DC bus (o) is needed, so assuming a three-phase grid system that holds, we have

$$v_{an} + v_{bn} + v_{cn} = 0 \tag{2.6}$$

Substituting expression (2.5) into the last expression yields

$$v_{no} = \tfrac{1}{3}(v_{ao} + v_{bo} + v_{co}) \tag{2.7}$$

Substituting again into Equation (2.5) this last expression, we finally obtain

$$v_{an} = \tfrac{2}{3}v_{ao} - \tfrac{1}{3}(v_{bo} + v_{co}) \tag{2.8}$$

$$v_{bn} = \tfrac{2}{3}v_{bo} - \tfrac{1}{3}(v_{ao} + v_{co}) \tag{2.9}$$

$$v_{cn} = \tfrac{2}{3}v_{co} - \tfrac{1}{3}(v_{bo} + v_{ao}) \tag{2.10}$$

Or more simply, directly from the order commands,

$$v_{an} = \frac{V_{bus}}{3}\left(2S_{a\_g} - S_{b\_g} - S_{c\_g}\right) \tag{2.11}$$

$$v_{bn} = \frac{V_{bus}}{3}\left(2S_{b\_g} - S_{a\_g} - S_{c\_g}\right) \tag{2.12}$$

$$v_{cn} = \frac{V_{bus}}{3}\left(2S_{c\_g} - S_{a\_g} - S_{b\_g}\right) \tag{2.13}$$

There are eight different combinations of output voltages, according to the eight permitted switching states of S_{a_g}, S_{b_g}, and S_{c_g}. Table 2.1 and Figure 2.5 show all

TABLE 2.1   Different Output Voltage Combinations of 2L-VSC

S_{a_g}	S_{b_g}	S_{c_g}	v_{ao}	v_{bo}	v_{co}	v_{an}	v_{bn}	v_{cn}
0	0	0	0	0	0	0	0	0
0	0	1	0	0	V_{bus}	$-\dfrac{V_{bus}}{3}$	$-\dfrac{V_{bus}}{3}$	$2\dfrac{V_{bus}}{3}$
0	1	0	0	V_{bus}	0	$-\dfrac{V_{bus}}{3}$	$2\dfrac{V_{bus}}{3}$	$-\dfrac{V_{bus}}{3}$
0	1	1	0	V_{bus}	V_{bus}	$-2\dfrac{V_{bus}}{3}$	$\dfrac{V_{bus}}{3}$	$\dfrac{V_{bus}}{3}$
1	0	0	V_{bus}	0	0	$2\dfrac{V_{bus}}{3}$	$-\dfrac{V_{bus}}{3}$	$-\dfrac{V_{bus}}{3}$
1	0	1	V_{bus}	0	V_{bus}	$\dfrac{V_{bus}}{3}$	$-2\dfrac{V_{bus}}{3}$	$\dfrac{V_{bus}}{3}$
1	1	0	V_{bus}	V_{bus}	0	$\dfrac{V_{bus}}{3}$	$\dfrac{V_{bus}}{3}$	$-2\dfrac{V_{bus}}{3}$
1	1	1	V_{bus}	V_{bus}	V_{bus}	0	0	0

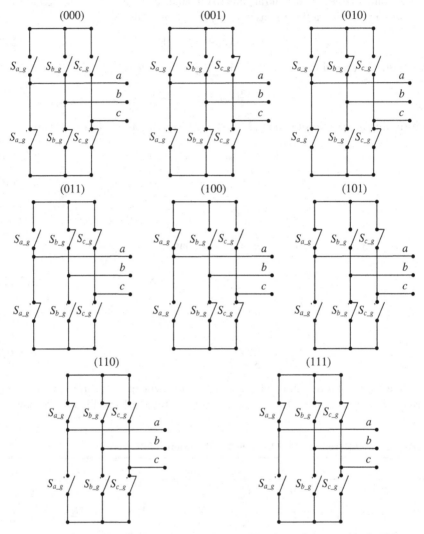

Figure 2.5 Eight different output voltage combinations of the two-level VSC.

these voltage combinations. As mentioned previously, the output voltages v_{ao}, v_{bo}, and v_{co} take only two different voltage levels: V_{bus} and 0; that is why this converter is identified as a "two-level converter." On the other hand, output voltages v_{an}, v_{bn}, and v_{cn} take five different voltage levels: $-2V_{bus}/3$, $-V_{bus}/3$, 0, $V_{bus}/3$, $2V_{bus}/3$. With a simple six pulse generation schema, the output voltage waveforms take the shape depicted in Figure 2.6.

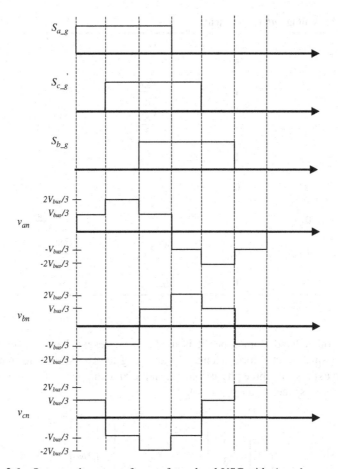

Figure 2.6 Output voltage waveforms of two-level VSC with six pulse generation.

By applying Fourier analysis, the amplitude of the harmonics of the output voltage are expressed as

$$\langle v_{an}\rangle_x = \langle v_{bn}\rangle_x = \langle v_{cn}\rangle_x = \frac{2V_{bus}}{3x\pi}\left[2 + \cos\left(x\frac{\pi}{3}\right) - \cos\left(x\frac{2\pi}{3}\right)\right]$$

$$x = 1, \ 5, \ 7, \ 11, \ 13, \ \ldots$$

(2.14)

Thanks to this harmonics decomposition, it is possible to know the amplitude of the fundamental component of the output voltage created, $\langle v_{an}\rangle_1 = \langle v_{bn}\rangle_1 = \langle v_{cn}\rangle_1$, together with the rest of the harmonics. With the constant DC bus voltage, this pulse generation strategy achieves fixed fundamental component amplitude $(2V_{bus}/\pi)$, a value that cannot be modified. On the contrary, by modifying the period of the pulses, output voltages of different periods or frequencies are also achieved.

TABLE 2.2   Voltage $\alpha\beta$ Components

S_{a_g}	S_{b_g}	S_{c_g}	v_α	v_β	Vector
0	0	0	0	0	V_0
0	0	1	$-\dfrac{V_{bus}}{3}$	$-\dfrac{\sqrt{3}V_{bus}}{3}$	V_5
0	1	0	$-\dfrac{V_{bus}}{3}$	$\dfrac{\sqrt{3}V_{bus}}{3}$	V_3
0	1	1	$-\frac{2}{3}V_{bus}$	0	V_4
1	0	0	$\frac{2}{3}V_{bus}$	0	V_1
1	0	1	$\dfrac{V_{bus}}{3}$	$-\dfrac{\sqrt{3}V_{bus}}{3}$	V_6
1	1	0	$\dfrac{V_{bus}}{3}$	$\dfrac{\sqrt{3}V_{bus}}{3}$	V_2
1	1	1	0	0	V_7

On the other hand, sometimes it is useful to represent these converter output voltages in space vector form. Hence, by applying the Clarke transformation, it is possible to express the three phase voltages in the $\alpha\beta$ components (see the Appendix for the basis of space vector notation):

$$\begin{bmatrix} v_\alpha \\ v_\beta \end{bmatrix} = \frac{2}{3} \cdot \begin{bmatrix} 1 & -\dfrac{1}{2} & -\dfrac{1}{2} \\ 0 & \dfrac{\sqrt{3}}{2} & -\dfrac{\sqrt{3}}{2} \end{bmatrix} \cdot \begin{bmatrix} v_{an} \\ v_{bn} \\ v_{cn} \end{bmatrix} \tag{2.15}$$

Table 2.2 shows all the voltage possibilities in $\alpha\beta$. Representing these $\alpha\beta$ voltages in the space vector diagram, the eight combinations configure a hexagon, as shown in Figure 2.7. Two of the voltage combinations, V_0 and V_7, are called zero vectors, since they produce zero voltage at each phase output. On the other hand, the rest of the voltages appear 60° phase shifted in the space vector diagram.

In subsequent sections, we show how, by using the eight different combinations of output voltage vectors, the two-level converter is able to create three-phase output voltages of different characteristics such as frequency and fundamental component amplitude.

2.2.1.2 Converter, Inductive Filter, and Grid Model   Once the two-level converter model has been presented, the rest of the grid side system components are analyzed next: the inductive filter and the grid voltage model. Coming back

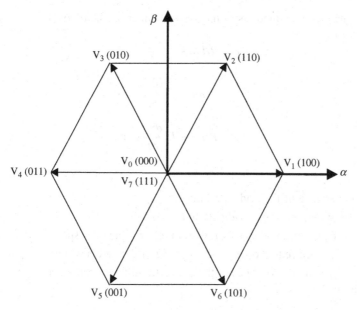

Figure 2.7   Space vector representation of the eight different output voltage combinations.

again to the grid side system model shown before in Figure 2.2, notice that a three-phase filter is located between the grid voltage and the converter's outputs.

A simple and reliable solution adopts an inductive filter, locating an inductance in each phase. The grid voltage is modeled as an ideal three-phase balanced voltage. The three-phase system can be modeled as three independent, but equivalent, single-phase systems as depicted in Figure 2.8. Note that the output AC voltages of the converter, referred to the neutral point are named with the sub-index 'f'.

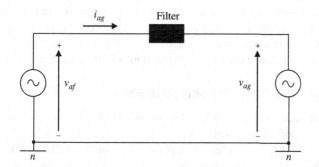

Figure 2.8   Simplified equivalent single-phase grid circuit (a phase).

Thus, the electric equations of the system can easily be derived as

$$v_{af} = R_f i_{ag} + L_f \frac{di_{ag}}{dt} + v_{ag} \tag{2.16}$$

$$v_{bf} = R_f i_{bg} + L_f \frac{di_{bg}}{dt} + v_{bg} \tag{2.17}$$

$$v_{cf} = R_f i_{cg} + L_f \frac{di_{cg}}{dt} + v_{cg} \tag{2.18}$$

where

L_f = inductance of the grid side filter (H)
R_f = resistive part of the grid side filter (Ω)
v_{ag}, v_{bg}, v_{cg} = grid voltages (V), with ω_s electric angular speed in (rad/s)
i_{ag}, i_{bg}, i_{cg} = currents flowing thorough the grid side converter's output (A)
v_{af}, v_{bf}, v_{cf} = output voltages of the converter referred to the neutral point of the load n (V)

Note again that there has been a modification of notation, since these are just the voltages defined in expressions (2.8), (2.9), and (2.10).

Consequently, for modeling purposes, it is necessary to isolate the first derivative of the currents:

$$\frac{di_{ag}}{dt} = \frac{1}{L_f}\left(v_{af} - R_f i_{ag} - v_{ag}\right) \tag{2.19}$$

$$\frac{di_{bg}}{dt} = \frac{1}{L_f}\left(v_{bf} - R_f i_{bg} - v_{bg}\right) \tag{2.20}$$

$$\frac{di_{cg}}{dt} = \frac{1}{L_f}\left(v_{cf} - R_f i_{cg} - v_{cg}\right) \tag{2.21}$$

Thus, the model of the grid side system is represented in Figure 2.9. The converter voltage outputs are created with the help of a modulator, to be studied in subsequent sections. The grid ideal sinusoidal voltages are generated at constant amplitude and frequency. Then the currents exchanged with the grid are calculated, taking into consideration the filter, according to expressions (2.19), (2.20), and (2.21).

2.2.2 Rotor Side Converter and *dv/dt* Filter

The rotor side converter that supplies the rotor of the DFIM, in general terms, is equal to the grid side converter shown in the previous section. Figure 2.10 illustrates the converter and the *dv/dt* filter used to supply the rotor of the DFIM. In this case also, a two-level VSC feeds the rotor. Between the rotor and the converter, in general, a *dv/dt* filter is located mainly with the objective to protect the machine from the harmful

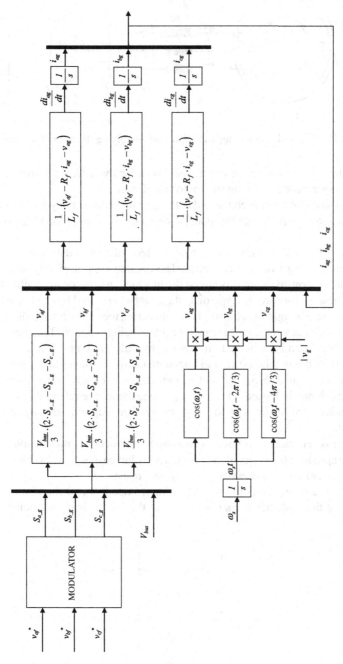

Figure 2.9 Simplified converter, filter, and grid model.

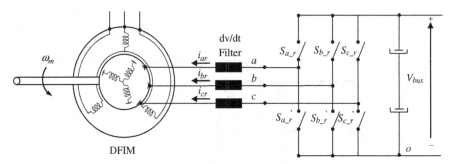

Figure 2.10 Rotor side converter and *dv/dt* filter supplying the rotor at the machine.

effects of the voltage source converter, such as capacitive leakage currents, bearing currents, and increased stress on the motor insulation.

The rotor side converter is connected to the grid side converter by the DC link. The model of the rotor together with the model of the machine is studied in the next chapter in detail.

The *dv/dt* filter mainly tries to attenuate the step voltages in the rotor terminals of the machine, coming from the converter. The combination of mainly three factors determine how harmful the effects are on the machine, which the *dv/dt* filter tends to mitigate. These factors are the type of voltage steps generated by the converter, the characteristics and length of cable used for connecting the converter and the machine, and finally the characteristics of the machine that is being supplied. The attenuation of the step voltages can be achieved, in general, by different types of filters.

Therefore, one possible solution to attenuate the overvoltages at the terminals of the motor is to locate a resistance and an inductance in parallel at the output of the converter, as shown in Figure 2.11. The resistance damps the reflection in the cable, while the inductance is necessary to reduce the voltage drop and the losses due to low frequencies.

This filter is generally composed of two passive elements. However, some authors propose to reproduce the effect of the resistance by increasing the power losses of the inductance, avoiding the necessity of a physical resistance.

Filters whose objective is to couple the input impedance of the motor with the impedance of the cable are normally located at the terminals of the motor. Among

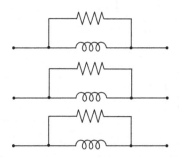

Figure 2.11 The *dv/dt* filter at the output of the converter.

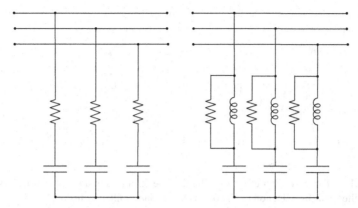

Figure 2.12 Possible filters for coupling the impedances at motor terminals.

different solutions, it is possible to locate an *RC* or an *RLC* filter, as shown in Figure 2.12.

Finally, at the output of the converter it is also possible to locate filter networks, as shown in Figure 2.13, to reduce the *dv/dt* at the converter itself.

Hence, the effect that produces the inclusion of a *dv/dt* filter is graphically depicted in Figure 2.14. It is possible to see how, without the filter, the overvoltage that occurs at the terminals of the motor is quite significant, while thanks to the inclusion of the proper filter this overvoltage is reduced.

2.2.3 DC Link

The DC part of the back-to-back converter is typically called the DC link. Thanks to the energy stored in a capacitor (or combination of several capacitors), it tries to maintain a constant voltage in its terminals. It is the linkage between the grid side and rotor side converters. Figure 2.15 shows a possible simplified model of a DC link. It is composed of a capacitor in parallel with a high resistance.

In order to derive the model of the DC link, the DC bus voltage must be calculated. This voltage is dependent on the current through the capacitor:

$$V_{bus} = \frac{1}{C_{bus}} \int i_c dt \qquad (2.22)$$

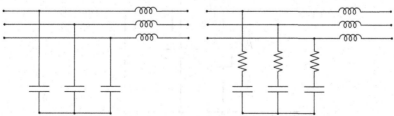

Figure 2.13 The *dv/dt* filter at the output of the converter.

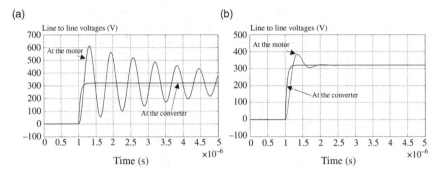

Figure 2.14 The *dv/dt* filter effect on voltage at the terminals of the motor: (a) overvoltages without filter and (b) reduction of the overvoltages due to filter inclusion.

The current through the capacitor can be found as

$$i_c = i_{r\_dc} - i_{g\_dc} - i_{res} \tag{2.23}$$

where

i_{res} = current through the resistance (A)
i_{g_dc} = DC current flowing from the DC link to the grid (A)
i_{r_dc} = DC current flowing from the rotor to the DC link (A)

On the other hand, the DC currents can be calculated as follows from the output AC currents of the converters:

$$i_{g\_dc} = S_{a\_g}i_{ag} + S_{b\_g}i_{bg} + S_{c\_g}i_{cg} \tag{2.24}$$

$$i_{r\_dc} = -S_{a\_r}i_{ar} - S_{b\_r}i_{br} - S_{c\_r}i_{cr} \tag{2.25}$$

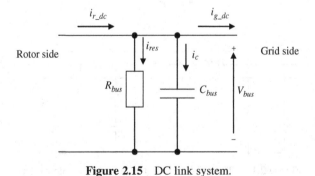

Figure 2.15 DC link system.

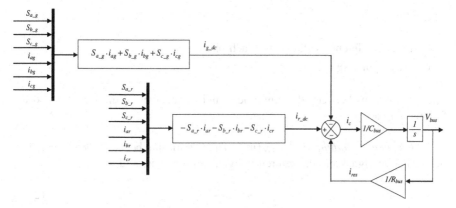

Figure 2.16   DC link model.

The current through the resistance is

$$i_{res} = \frac{V_{bus}}{R_{bus}} \tag{2.26}$$

Consequently, with all these equations, the model of the DC link system can be illustrated in Figure 2.16.

2.2.4 Pulse Generation of the Controlled Switches

The order commands for the controlled semiconductors of the two-level converter can be generated according to different laws called "Modulation." This section presents some of the most commonly used modulation techniques.

- Sinusoidal pulse width modulation (PWM) technique
- Sinusoidal pulse width modulation (PWM) technique with injection of third harmonic
- Space vector modulation (SVM) technique

For simplicity the modulations are studied for a general two-level converter, eliminating in the notation of the order commands the subindexes referring to the gird or rotor side converters: S_a, S_b, and S_c. Thus, everything in this section can be applied at both the rotor and grid side converters.

2.2.4.1 *Sinusoidal Pulse Width Modulation (PWM)*   This modulation is a widely extended modulation technique among voltage sourced converters. By using a triangular signal and comparing it with the reference signal (an image of the output voltage that is wanted at the output of the converter), it creates the output voltage according to the following law:

$$S_j = 1 \text{ if } v_j^* > v_{tri} \text{ with } j = a, b, c \tag{2.27}$$

where

v_a^*, v_b^*, v_c^* = reference voltages for each phase

v_{tri} = triangular signal

The relationships between the amplitudes and frequencies of the signals need the following indexes:

1. *Frequency Modulation Index* (m_f). The relationship between the frequency of the triangular signal and the reference signal can be expressed as

$$m_f = \frac{f_{tri}}{f_{ref}} \tag{2.28}$$

In general, in order to create a good quality output voltage, m_f should be a high number. However, since the imposed frequency of the triangular signal (f_{tri}) determines the switching frequency of the switches of the converter, this frequency should not be too high, in order not to produce high switching losses in the semiconductors. Therefore, when it comes to choosing a value for m_f, it is necessary to find a compromise between the quality of the voltage created and the power losses in the converter.

2. *Amplitude Modulation Index* (m_a). The relationship between the amplitude of the reference signal and the triangular signal can be expressed as

$$m_a = \frac{|v^*|}{|v_{tri}|} \tag{2.29}$$

Under ideal conditions, the amplitude relationship between the fundamental component of the achieved output voltage and the DC bus voltage is given by the amplitude modulation index as follows:

$$\langle v_{an} \rangle_1 = m_a \cdot \frac{V_{bus}}{2} \quad \text{if } m_a \leq 1 \tag{2.30}$$

Therefore, the obtained output voltages according to this PWM law are shown in Figure 2.17. The schematic block diagram for the implementation is shown in Figure 2.18. Note that the maximum achievable fundamental voltage without over-modulation $(m_a \leq 1)$ is $V_{bus}/2$. It is possible to find several alternative sinusoidal PWM laws from the presented version. In the literature, this sinusoidal PWM technique is typically known as unipolar PWM.

Note that, as seen before, the output voltages v_{ao}, v_{bo}, and v_{co} take only two different voltage levels—V_{bus} and 0—while output voltages v_{an}, v_{bn}, and v_{cn} take five different voltage levels: $-2V_{bus}/3$, $-V_{bus}/3$, 0, $V_{bus}/3$, $2V_{bus}/3$.

In addition, the normalized spectrum of the phase to neutral v_{an} voltage is shown in Figure 2.19. The fundamental voltage appears with amplitude 0.9 at f_{ref}, accompanied by dominant harmonics in multiples of f_{tri}. Note that as the frequency f_{tri} gets higher

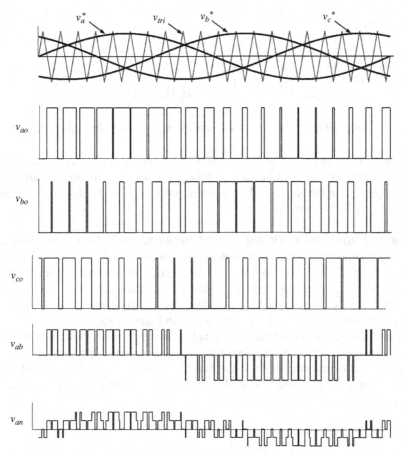

Figure 2.17   Output voltages of two-level converter with sinusoidal PWM: $m_f = 20, m_a = 0.9$.

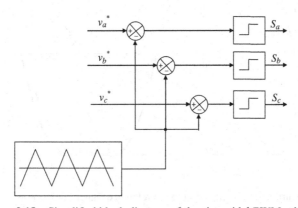

Figure 2.18   Simplified block diagram of the sinusoidal PWM schema.

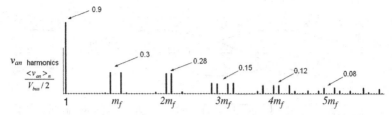

Figure 2.19   Spectrum of the output voltage v_{an}: $m_f = 20$, $m_a = 0.9$.

(switching frequency of switches), the group of harmonics can be displaced further away from the fundamental frequency. This fact is advantageous in obtaining good quality currents at the output of the converter; however, as said before, this increases power losses in the switches.

2.2.4.2 Sinusoidal PWM with Third Harmonic Injection   The sinusoidal PWM technique allows different modifications in order to achieve several specific improvements. In this case, an increase of the maximum achievable output voltage amplitude is studied. Hence, by simply adding a third harmonic signal to each of the reference signals, it is possible to obtain a significant amplitude increase at the output voltage without loss of quality, as represented in Figure 2.20.

It is remarkable that the reference signal resulting from the addition of the third (V_3) and first harmonic (V_1) is smaller in amplitude than the first harmonic. At the output, the obtained amplitude of the first harmonic is equal to the amplitude of the first harmonic reference. Note also that the third harmonic is not seen at the output voltage.

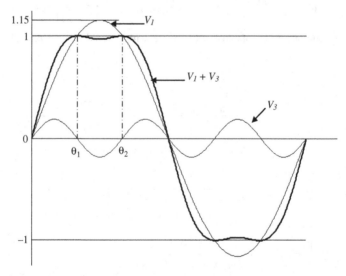

Figure 2.20   Third harmonic injection to the reference signal, for amplitude increase of the fundamental component.

On the other hand, the reference signal takes two maxima at $\theta_1 = \pi/3$ and $\theta_2 = 2\pi/3$, equal to 1. The first and third harmonic equations are given by

$$V_1 = V_{1\max}\sin\theta \tag{2.31}$$

$$V_3 = V_{3\max}\sin(3\theta) \tag{2.32}$$

Therefore, when $\theta_1 = \pi/3$, the first harmonic of the output voltage (line to neutral) takes the value $V_{bus}/2$. By substituting in Equation (2.31), we find

$$\frac{V_{bus}}{2} = V_{1\max}\sin\left(\frac{\pi}{3}\right) = \frac{\sqrt{3}}{2}V_{1\max} \tag{2.33}$$

Consequently, the amplitude of the first harmonic yields

$$V_{1\max} = \frac{V_{bus}}{\sqrt{3}} = \frac{V_{bus}}{1.73} \tag{2.34}$$

In this case, the amplitude modulation index is still defined as in Equation (2.29); however, since the reference is composed of the addition of two signals (fundamental component and injected third harmonic), m_a is referred only to the fundamental component:

$$m_a = \frac{|v_1^*|}{|v_{tri}|} \tag{2.35}$$

Thanks to the third harmonic injection, not considering the overmodulation region, the maximum achievable fundamental output voltage amplitude is $2/\sqrt{3} = 1.15$ times greater than the classic PWM. Consequently, the amplitude modulation index can be extended to

$$\text{Linear modulation region:}\quad 0 \le m_a \le 1.15 \tag{2.36}$$

The resulting output voltages with $m_a = 0.9$ and $m_f = 20$ are represented in Figure 2.21. The spectrum is also illustrated in Figure 2.22. In addition, if the rest of harmonics are compared to the spectrum in Figure 2.19, we see that the quality of the voltage waveform has not been significantly degraded.

For each phase reference, the third harmonic injected is equal:

$$V_{1\max}\sin(wt) + V_{3\max}\sin(3wt) \tag{2.37}$$

$$V_{1\max}\sin\left(wt - \frac{2\pi}{3}\right) + V_{3\max}\sin(3wt) \tag{2.38}$$

$$V_{1\max}\sin\left(wt + \frac{2\pi}{3}\right) + V_{3\max}\sin(3wt) \tag{2.39}$$

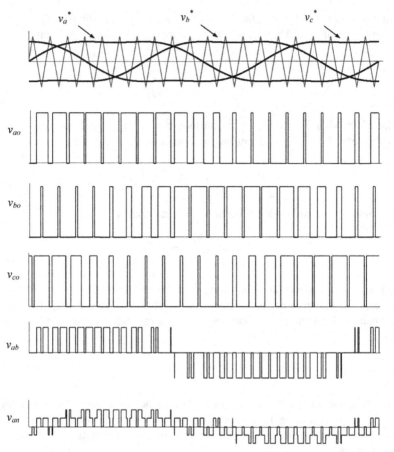

Figure 2.21 Output voltages of two-level converter with sinusoidal PWM: $m_f = 20$, $m_a = 0.9$, with third harmonic injection.

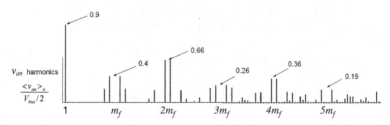

Figure 2.22 Spectrum of the output voltage v_{an}: $m_f = 20$, $m_a = 0.9$, with third harmonic injection.

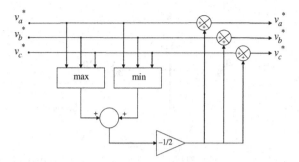

Figure 2.23 Block diagram for simplified third harmonic injection.

An easy procedure to inject the third harmonic is given by the expression

$$V_3 = -\frac{\max\{v_a^*, v_b^*, v_c^*\} + \min\{v_a^*, v_b^*, v_c^*\}}{2} \tag{2.40}$$

The block diagram for implementation is depicted in Figure 2.23.

Note that, with this procedure, not only the third harmonic is injected, since the resulting added signal takes a triangular shape, as illustrated in Figure 2.24.

2.2.4.3 Space Vector Modulation (SVM)
The space vector modulation (SVM) generates the AC output converter voltage, based on the space vector representation principle. As an alternative to the PWM method seen in the previous sections, with SVM the pulses for the controlled switches of the converter are generated by a slightly different philosophy.

Therefore, as seen in Section 2.2.1, the two-level converter provides eight different output voltage combinations on the AC side (V_0, V_1, V_2, V_3, V_4, V_5, V_6, V_7) that can be represented in a space vector diagram as illustrated in Figure 2.25. The region in the plane covered by the voltage vector disposition is a hexagon divided into six different

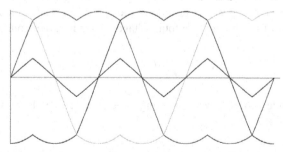

Figure 2.24 Resulting injected signal and voltage reference, with simplified third harmonic generation.

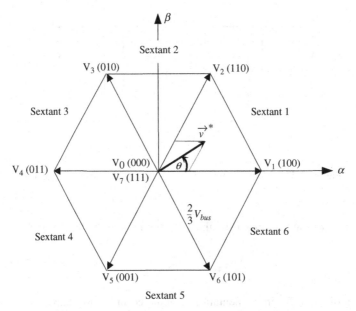

Figure 2.25 Space vector representation for SVM.

sextants. On the other hand, the three-phase output AC voltage reference can also be represented in the plane, as a rotating space vector of circular trajectory: $\vec{v}^*$. (See the Appendix for more information.)

The required voltage space vector can be obtained by switching the nearest three voltage vectors of the converter at the correct time during one constant switching period. Depending on where the voltage reference space vector is located, two of the used voltage vectors are different; however, the third injected voltage vector is always the zero vector (two possibilities—V_0 and V_7).

For instance, when the reference voltage vector is located in the first sextant, the active voltage vectors used are V_1 and V_2 (Figure 2.26). Each voltage vector is injected t_1 and t_2 time intervals in a constant switching period h. Combined with the injection of the zero vectors, it is possible to modify the generated space vector.

The synthesis of the required output voltage vector is carried out as follows:

$$v^* = V_1 d_1 + V_2 d_2 + V_0 (1 - d_1 - d_2) \tag{2.41}$$

The duty cycles (normalized injecting time) for each voltage vector can be calculated as follows:

$$d_1 = m_n \left(\cos\theta_n - \frac{\sin\theta_n}{\sqrt{3}} \right) \tag{2.42}$$

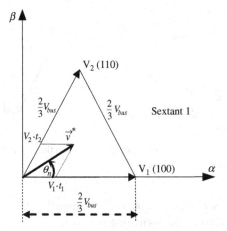

Figure 2.26 Space vector representation for SVM.

$$d_2 = m_n \cdot 2 \frac{\sin\theta_n}{\sqrt{3}} \tag{2.43}$$

Therefore, the injecting time for the zero vectors is (the remaining time)

$$d_{0,7} = 1 - d_1 - d_2 \tag{2.44}$$

where

m_n = normalized voltage amplitude of the reference space vector,

$$m_n = \frac{|\vec{v}|^*}{\frac{2}{3} V_{bus}} \tag{2.45}$$

θ_n = equivalent angle of the voltage space vector in the first sextant. It as the relationship given in Table 2.3.

Thus, by applying symmetry, the duty cycles are simply calculated by means of an equivalent space vector in the first sextant of the hexagon, as illustrated in the graphical examples of Figure 2.27.

TABLE 2.3 Equivalent Angle θ_n in the First Sextant ($\theta_n \to (0, \pi/3)$)

Sextant					
1	2	3	4	5	6
$\theta \to (0, \pi/3)$	$\theta \to (\pi/3, 2\pi/3)$	$\theta \to (2\pi/3, \pi)$	$\theta \to (\pi, 4\pi/3)$	$\theta \to (4\pi/3, 5\pi/3)$	$\theta \to (5\pi/3, 2\pi)$
$\theta_n = \theta$	$\theta_n = -\theta + \dfrac{2\pi}{3}$	$\theta_n = \theta - \dfrac{2\pi}{3}$	$\theta_n = -\theta + \dfrac{4\pi}{3}$	$\theta_n = \theta - \dfrac{4\pi}{3}$	$\theta_n = -\theta + \dfrac{6\pi}{3}$

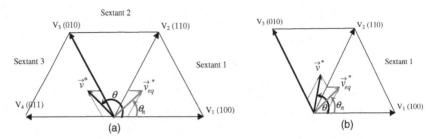

Figure 2.27 Equivalent space vector in sextant 1: (a) when the space vector is in sextant 3 and (b) when the space vector is in sextant 2.

Then, once the duty cycles are obtained, the injecting time can be calculated as follows, taking into account the switching period h:

$$t_1 = d_1 h \tag{2.46}$$

$$t_2 = d_2 h \tag{2.47}$$

$$t_{0,7} = d_{0,7} h \tag{2.48}$$

Consequently, when the injecting times are calculated, the next step is to create the corresponding pulses for the controlled switches of the two-level converter (S_a, S_b, S_c). This can be achieved with a comparison of the calculated injecting times, with a carrier (triangular) signal. The voltage and pulse generation in one switching period is shown in Figure 2.28.

Notice that the time dedicated to the zero vector injection is divided into two equal parts: $(h - t_1 - t_2)/2$. One part is dedicated to the V_0 zero vector and is injected symmetrically at the beginning and at the end of the switching period. The other part is dedicated to the V_7 zero vector and is injected just in the middle of the switching period.

On the other hand, the pulses can be generated with only one carrier signal (triangular) and, by comparing it to three constant levels, we can represent the applied duty cycles. The result of these three comparisons corresponds to the pulses for each controlled semiconductor of each phase, in the equivalent first sextant: $S_{a\_eq}, S_{b\_eq}, S_{c\_eq}$.

Thus, according to the voltage reference space vector rotation, the injecting times t_1 and t_2 are altered, modifying the injecting times of the vectors and the corresponding voltage vectors as well.

Finally, the resulting v_{ab} and v_{an} voltages are also represented in Figure 2.28. It must be highlighted that there are several alternative ways to inject the zero vector at each switching period. Depending on how they are placed and distributed in the voltage vector sequence injection, different performances can be obtained: output voltage properties, switching behavior of the switches, and so on.

Consequently, once the pulses are created in a fictitious equivalent space vector located in the first sextant, the real vectors must be created. Depending on the real

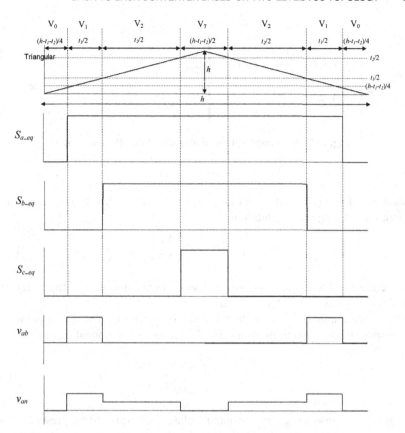

Figure 2.28 Pulses and output voltages of two-level converter with SVM in one switching period of sextant 1.

sextant where the space vector is located, it is only necessary to perform a change of phases as defined in Table 2.4.

To conclude, the simplified block diagram of Figure 2.29 illustrates the implementation schema of the SVM.

On the other hand, from the maximum achievable fundamental output voltage point of view, the SVM is equivalent to the PWM with the third harmonic injection

TABLE 2.4 Real Vector Generation by Changing the Phases, Depending on the Sextant

			Sextant			
1	2	3	4	5	6
$S_a \rightarrow S_{a\_eq}$	$S_a \rightarrow S_{b\_eq}$	$S_a \rightarrow S_{c\_eq}$	$S_a \rightarrow S_{c\_eq}$	$S_a \rightarrow S_{b\_eq}$	$S_a \rightarrow S_{a\_eq}$
$S_b \rightarrow S_{b\_eq}$	$S_b \rightarrow S_{a\_eq}$	$S_b \rightarrow S_{a\_eq}$	$S_b \rightarrow S_{b\_eq}$	$S_b \rightarrow S_{c\_eq}$	$S_b \rightarrow S_{c\_eq}$
$S_c \rightarrow S_{c\_eq}$	$S_c \rightarrow S_{c\_eq}$	$S_c \rightarrow S_{b\_eq}$	$S_c \rightarrow S_{a\_eq}$	$S_c \rightarrow S_{a\_eq}$	$S_c \rightarrow S_{b\_eq}$

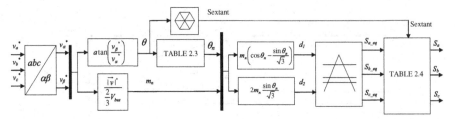

Figure 2.29 Simplified block diagram of the SVM schema.

modulation. The maximum achievable output voltage (phase to neutral point n), without considering overmodulation, is

$$V_{1\max} = \frac{V_{bus}}{\sqrt{3}} = \frac{V_{bus}}{1.73} \qquad (2.49)$$

which correspond to the amplitude of the space vector inscribed in the hexagon as graphically represented in Figure 2.30.

Hence, the amplitude modulation index, in this case, normalizes the output voltage with respect to the maximum achievable fundamental component, not considering overmodulation:

$$m_a = \frac{|v|^*}{V_{bus}/\sqrt{3}} \qquad (2.50)$$

Figure 2.31 shows the generated output voltages according to the presented SVM schema during one output AC voltage cycle. The used switching frequency, that is, the

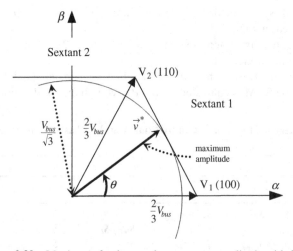

Figure 2.30 Maximum fundamental component amplitude with SVM.

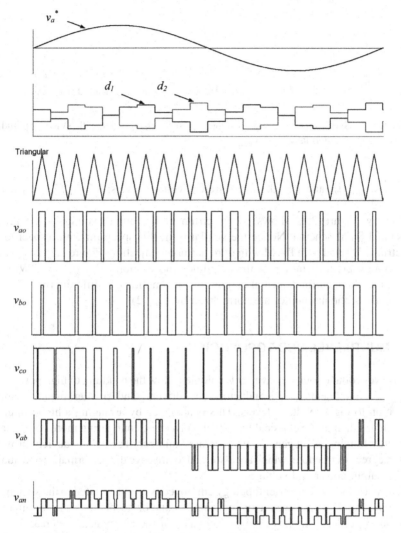

Figure 2.31 Output voltages of two-level converter with SVM: $m_f = 20$, $m_a = 0.9$, with symmetrical zero vectors ($t_{V0} = t_{V7}$).

frequency of the triangular shape employed for the pulse generation ($1/h$) is 20 times higher than the output frequency. Hence, the frequency modulation index defined in the PWM method is still valid, in the SVM yielding $m_f = 20$. On the other hand, the amplitude modulation index (m_a) is set to 0.9.

In addition, the calculated duty cycles are modified every switching period, resulting in an output voltage waveform similar to the one obtained with PWM.

By means of the SVM as well as by sinusoidal PWM schemas, the output voltages v_{ao}, v_{bo}, and v_{co}, take only two different voltage levels: V_{bus} and 0, while output

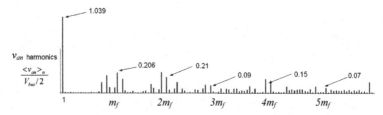

Figure 2.32   Spectrum of the output voltage v_{an}: $m_f = 20$, $m_a = 0.9$, with SVM (amplitude of the harmonics are normalized by $V_{bus}/2$).

voltages v_{an}, v_{bn}, and v_{cn} take five different voltage levels: $-2V_{bus}/3$, $-V_{bus}/3$, 0, $V_{bus}/3$, $2V_{bus}/3$.

Finally, Figure 2.32 shows the spectrum of the output voltage v_{an}, with the presented SVM schema. Notice the similarity of this spectrum with respect to the spectrum obtained with PWM with third harmonic injection (Figure 2.22). However, due to a different definition of the amplitude modulation index (m_a) in PWM and SVM, the obtained fundamental output voltage amplitude is different (note that the amplitude of the harmonics are normalized by $V_{bus}/2$).

2.3 MULTILEVEL VSC TOPOLOGIES

Multilevel voltage source converter topologies allow the handling of high power with standard components (semiconductors and drivers) and smaller filters, enabling operation to higher voltage levels. This is achieved by arranging a higher number of switches than in the classical two-level converters, configuring more complicated converter topologies. In general, the increased number of switching devices provides extra degrees of freedom, which are also used to improve the performance and quality of the output power and voltage.

The nature of the converter topology employed is closely related to the switch that is used. The silicon based existing semiconductor technology is continuously evolving in order to improve the performance of the components, increasing their voltage and current rating, reliability, modularity, and so on. A clear trend in the recent evolution of semiconductors imposed by market demands has been the increase of their voltage rating. For instance, IGBT modules have reached experimentally 8 kV of collector–emitter voltage (V_{ce}) while the IGBT press-packs have been discovered as a good solution for 6.5 kV of V_{ce}. Moreover, 10 kV IGCTs have also been developed experimentally. Today, IGBTs and IGCTs are commercially available with a maximum voltage rating of 6.5 kV. Added to this, new semiconductor technologies are being developing, such as SiC based semiconductors or diamond based ones. (See Figure 2.33.)

On the other hand, considering the trend of using voltage source converters (VSCs) with increasing converter output voltage and power, the restriction of the classical

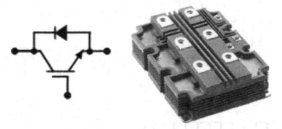

Figure 2.33   IGBT module (ABB – 5SNA1000G450300) of 4.5 kV and 1 kA. (*Source*: ABB.)

two-level converter imposed by the voltage rating of the semiconductors and the need to use series connected semiconductors, has led to the development of multilevel topologies. Although the most common topology is the 3L-NPC VSC, some other topologies are emerging, such as the FC (flying capacitor) and cascaded H bridge as well as the SMC (stacked multicell) and MMC (modular multilevel converter). Each of them focuses on strengthening different converters' characteristics, in order to be used under different conditions.

Hence, this section presents some of the most suitable multilevel VSC topologies for the DFIM based wind turbines. However, only the 3L-NPC VSC topology is studied in detail. Thus, Figures 2.34 and 2.35 show one leg of the most classical multilevel topologies (3L, 4L, and 5L) and their output voltages: the three-level neutral point clamped converter (3L-NPC), the four-level flying capacitor converter (4L-FC), and the five-level cascaded H-bridge converter (5L-CHB).

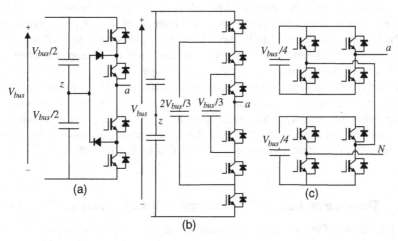

Figure 2.34   One leg of most classical multilevel topologies: (a) 3L-NPC, (b) 4L-FC, and (c) 5L-CHB.

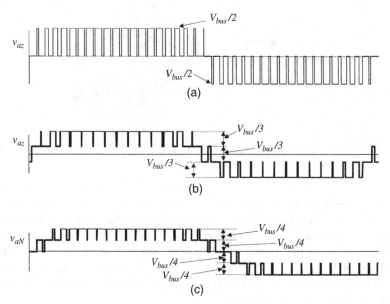

Figure 2.35 Output voltages of (a) 3L-NPC, (b) 4L-FC, and (c) 5L-CHB topologies.

It is possible to find newer multilevel converter topologies, suitable for the back-to-back converter of a doubly fed induction machine based wind turbine. These topologies effectively allow increasing the operating voltages and powers with given semiconductors, by only increasing the levels of the converter—thus avoiding the necessity of isolating transformers and keeping their capacity of power reversibility. There are reports of many innovative multilevel topologies in the literature, but this text only shows four of those converters: the three-level neutral point piloted (3L-NPP) topology, also reported as the three-level stacked multicell (3L-SMC) topology, the extension of this concept to five levels (i.e., the 5L-SMC topology), the five-level active neutral point clamped (5L-ANPC) topology, and the modular multilevel converter (MMC) concept. Figures 2.36 to 2.38 illustrate these topologies. There are many different advantages and disadvantages between them; however, it is possible that different wind turbine manufacturers could adopt these topologies in the future. Nevertheless, as mentioned before, this section only covers the detailed description of one of the multilevel topology, that is, the 3L-NPC VSC.

2.3.1 Three-Level Neutral Point Clamped VSC Topology (3L-NPC)

2.3.1.1 Basic Description The three-level neutral point clamped multilevel converter (Figure 2.39) is a VSC from the family of diode clamped multilevel converters. It generates output phase voltage waveforms (v_{ao}, v_{bo}, v_{co}) comprising three switching levels of amplitude $V_{bus}/2$ (0, $V_{bus}/2$, and V_{bus}). These voltages are

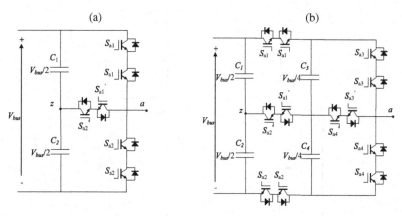

Figure 2.36 Stacked multicell topologies: (a) 3L-SMC or 3L-NPP and (b) 5L-SMC.

created on the AC side by switching the controlled switches (S_{a1}, S_{a1}', S_{a2}, S_{a2}', S_{b1}, S_{b1}', S_{b2}, S_{b2}', S_{c1}, S_{c1}', S_{c2}, S_{c2}'), connecting the DC bus to the AC output. The DC bus is created by two DC sources (C_1 and C_2 capacitors in practical cases), providing a neutral point z. This topology connects clamping diodes (D_a, D'_a, D_b, D'_b, D_c, D'_c) from the neutral point to each arm (a, b, c) to achieve the medium voltage level $V_{bus}/2$.

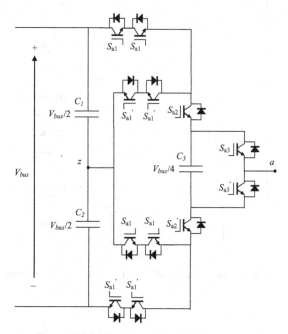

Figure 2.37 One leg of 5L-ANPC topology.

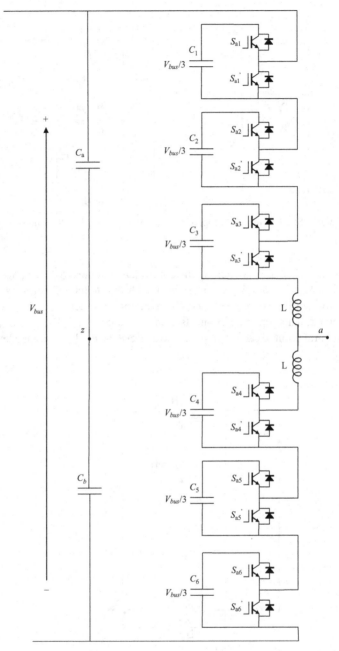

Figure 2.38 One leg of the modular multilevel converter topology with six modules.

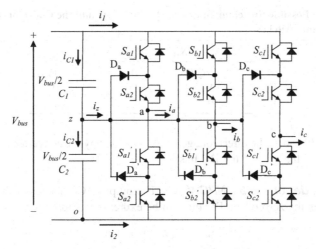

Figure 2.39   3L-NPC VSC topology.

To perform a bidirectional flow of the current through the converter, each controlled switch is accompanied by one diode in antiparallel, as done in two-level converters.

Ideally, the switching states of the converter at each phase are restricted in such a way that each switch presents its complementary switch. Thus, in order to avoid nonpermitted states that would result in improper operation of the converter, the following switching status is imposed:

$$S'_{a1} = \overline{S_{a1}} \text{ and } S'_{a2} = \overline{S_{a2}} \tag{2.51}$$

$$S'_{b1} = \overline{S_{b1}} \text{ and } S'_{b2} = \overline{S_{b2}} \tag{2.52}$$

$$S'_{c1} = \overline{S_{c1}} \text{ and } S'_{c2} = \overline{S_{c2}} \tag{2.53}$$

The commutation of the converter in normal operation produces a current flow through the neutral point z, of different characteristics depending on the operating conditions. This current flow makes the DC bus capacitors (C_1 and C_2) charge and discharge, yielding to an imbalance in the voltage share for each capacitor. In general, with a finite value of the capacitors, this converter requires a special modulation algorithm that achieves DC bus voltage balance.

Compared with the classic 2L-VSC studied in previous sections, this multilevel converter topology is suitable to achieve higher ranges of power and voltage. Thus, using semiconductors with equal voltage ratio, the AC output voltage provided by the 3L-NPC VSC can be significantly increased, since the switches see only half of the DC bus, that is, $V_{bus}/2$.

Table 2.5 shows the three permitted switching states of one phase of the 3L-NPC VSC (only phase a is represented). Notice that state $S_{a1} = 1$, $S_{a2} = 0$ is not allowed in this converter. The rest of the phases work equivalently. Only in state $S_{a1} = 0$, $S_{a2} = 1$ do the clamping diodes work. Finally, the voltage at each controlled semiconductor (V_{ce}: collector to emitter voltage) is 0 or $V_{bus}/2$ under any switching state.

TABLE 2.5   Possible Switching States of 3L-NPC VSC and the Collector–Emitter Voltages of the Switches

S_{a1}	S_{a2}	v_{ao}	V_{ce} of S_{a1}	V_{ce} of S'_{a1}	V_{ce} of S_{a2}	V_{ce} of S'_{a2}
0	0	0	$V_{bus}/2$	0	$V_{bus}/2$	0
0	1	$V_{bus}/2$	$V_{bus}/2$	0	0	$V_{bus}/2$
1	1	V_{bus}	0	$V_{bus}/2$	0	$V_{bus}/2$

Figure 2.40 shows the three voltage levels provided by the converter at each phase (referred to as the zero of the DC bus).

2.3.1.2 Model of the 3L-NPC VSC   The output AC voltages of the converter referred to as the zero of the DC bus can be expressed as follows:

$$v_{ao} = S_{a1}v_{C1} + S_{a2}v_{C2} \tag{2.54}$$

$$v_{bo} = S_{b1}v_{C1} + S_{b2}v_{C2} \tag{2.55}$$

$$v_{co} = S_{c1}v_{C1} + S_{c2}v_{C2} \tag{2.56}$$

From expressions (2.8)–(2.10) derived for the two-level converter, but valid also for this topology, combined with expressions (2.54)–(2.56), the output AC voltage expressions are calculated depending on the order commands:

$$v_{an} = \frac{v_{C1}}{3}(2S_{a1} - S_{b1} - S_{c1}) + \frac{v_{C2}}{3}(2S_{a2} - S_{b2} - S_{c2}) \tag{2.57}$$

$$v_{bn} = \frac{v_{C1}}{3}(2S_{b1} - S_{a1} - S_{c1}) + \frac{v_{C2}}{3}(2S_{b2} - S_{a2} - S_{c2}) \tag{2.58}$$

$$v_{cn} = \frac{v_{C1}}{3}(2S_{c1} - S_{b1} - S_{a1}) + \frac{v_{C2}}{3}(2S_{c2} - S_{b2} - S_{a2}) \tag{2.59}$$

Note that the voltages of the capacitors in general are not necessarily exactly equal. On the other hand, the currents flowing through the DC bus capacitors can be calculated first by deducing the currents:

$$i_1 = S_{a1}i_a + S_{b1}i_b + S_{c1}i_c \tag{2.60}$$

$$i_z = (S_{a2} - S_{a1})i_a + (S_{b2} - S_{b1})i_b + (S_{c2} - S_{c1})i_c \tag{2.61}$$

$$i_2 = -(i_1 + i_z) = -(S_{a2}i_a + S_{b2}i_b + S_{c2}i_c) \tag{2.62}$$

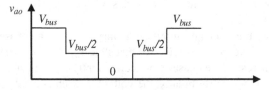

Figure 2.40   Three voltage levels.

Thus, depending on the state of the switches, the output phase currents (*abc*) determine the current though the capacitors as follows:

$$i_{C1} = -i_1 = -S_{a1}i_a - S_{b1}i_b - S_{c1}i_c \tag{2.63}$$

$$i_{C2} = i_2 = -S_{a2}i_a - S_{b2}i_b - S_{c2}i_c \tag{2.64}$$

Thanks to these expressions, it is possible to know how the DC bus capacitors C_1 and C_2 are charged and discharged, depending on the switching states and the output AC currents.

In a similar way, the reader can find the currents through the diodes or controlled switches (only phase *a*).

Controlled switches:

$$i_{IGBT1} = S_{a1} \cdot (i_a > 0) \cdot i_a \tag{2.65}$$

$$i_{IGBT2} = S_{a2} \cdot (i_a > 0) \cdot i_a \tag{2.66}$$

$$i_{IGBT1'} = \overline{S_{a1}} \cdot (i_a < 0) \cdot i_a \tag{2.67}$$

$$i_{IGBT2'} = \overline{S_{a2}} \cdot (i_a < 0) \cdot i_a \tag{2.68}$$

Diodes:

$$i_{Diode1} = i_{Dioide2} = (S_{a1} \cdot S_{a2}) \cdot (i_a < 0) \cdot i_a \tag{2.69}$$

$$i_{Diode1'} = i_{Diode2'} = (\overline{S_{a1}} \cdot \overline{S_{a2}}) \cdot (i_a > 0) \cdot i_a \tag{2.70}$$

Clamp diodes:

$$i_{Diode\_clamp} = (\overline{S_{a1}} \cdot S_{a2}) \cdot (i_a > 0) \cdot i_a \tag{2.71}$$

$$i_{Diode\_clamp'} = (\overline{S_{a1}} \cdot S_{a2}) \cdot (i_a < 0) \cdot i_a \tag{2.72}$$

Finally, Figure 2.41 shows the schematic block diagram of the converter model.

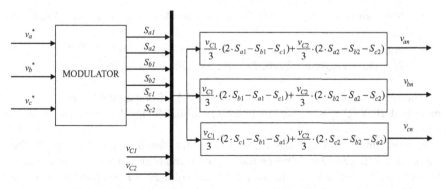

Figure 2.41   Ideal output voltage model of the 3L-NPC VSC.

2.3.1.3 Sinusoidal Pulse Width Modulation (PWM) with Third Harmonic
Injection The sinusoidal PWM philosophy of the 3L-NPC VSC is very similar to a 2L-VSC PWM. The scalar PWM pattern employed per phase uses a single voltage reference (v^*), which is compared to two-level shifted triangular carriers $(v_{tri\_1}$ and $v_{tri\_2})$.

The output voltages for each phase are created according to the following law (only a phase represented):

$$S_{a1} = 1 \text{ if } v_a^* \geq v_{tri\_1} \tag{2.73}$$

$$S_{a1} = 0 \text{ if } v_a^* < v_{tri\_1} \tag{2.74}$$

$$S_{a2} = 1 \text{ if } v_a^* \geq v_{tri\_2} \tag{2.75}$$

$$S_{a2} = 0 \text{ if } v_a^* < v_{tri\_2} \tag{2.76}$$

A third harmonic signal can also be injected into the reference signal, in order to increase by 1.15 the maximum available output for a given DC bus voltage, as was done in 2L-VSC PWM. Therefore, the maximum achievable voltage of the fundamental component becomes

$$V_{1\max} = \frac{V_{bus}}{\sqrt{3}} = \frac{V_{bus}}{1.73} \tag{2.77}$$

Figure 2.42 illustrates the PWM pulse generation philosophy together with the output voltages with $m_f = 20$, $m_a = 0.9$. The following can be observed:

- During half of the period, only one controlled switch (and its complement) commutates. In the positive semicycle of v_a^*, only S_{a1} and S_{a1}' commutate, while S_{a2} and S_{a2}' are maintained in the ON and OFF states, respectively. On the contrary, in the negative semicycle of v_a^*, only S_{a2} and S_{a2}' commutate, while S_{a1} and S_{a1}' are maintained in the OFF and ON states, respectively.
- As mentioned before, the output voltage v_{ao} presents three voltage levels (0, $V_{bus}/2$, and V_{bus}).
- The output voltage v_{ab} presents five voltage levels ($-V_{bus}$, $-V_{bus}/2$, 0, $V_{bus}/2$ and V_{bus}).
- Finally, the output voltage v_{an} presents nine voltage levels ($-4V_{bus}/6$, $-3V_{bus}/6$, $-2V_{bus}/6$, $-V_{bus}/6$, 0, $V_{bus}/6$, $2V_{bus}/6$, $3V_{bus}/6$, and $4V_{bus}/6$).

The spectrum of v_{an} is also shown in Figure 2.43. Compared with the previously presented 2L VSC spectra, we see that this multilevel signal shows less harmonic content.

Finally, this PWM technique does not ensure the balancing of the DC bus capacitor's voltages. Depending on the operating conditions of the converter (modulation index, power factor of the load, etc.), the voltages of the capacitors will drift, until the output AC voltages are too distorted and the converter does not operate

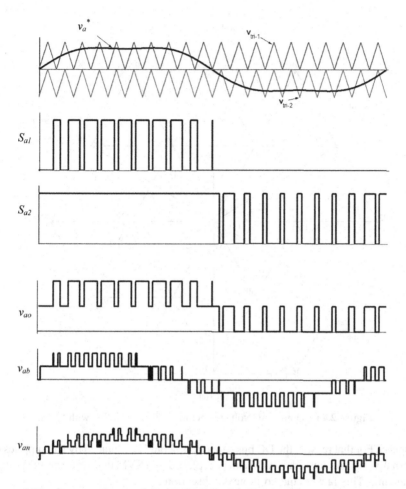

Figure 2.42 Output voltages of 3L-NPC VSC with PWM and third harmonic injection: $m_f = 20$, $m_a = 0.9$.

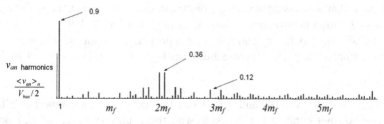

Figure 2.43 Spectrum of the output voltage v_{an} of 3L-NPC VSC with PWM and third harmonic injection: $m_f = 20$, $m_a = 0.9$, with SVM.

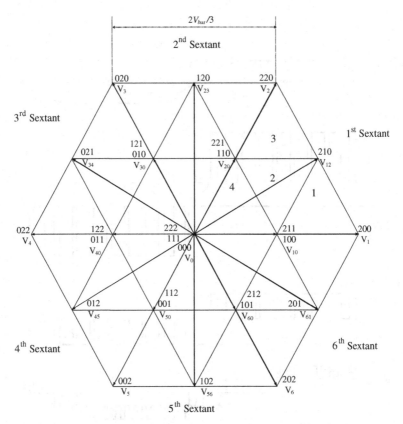

Figure 2.44   Output AC voltage vectors of 3L-NPC VSC, with SVM.

properly. For that reason, the DC bus capacitors should be replaced by DC supplies or, alternatively, the sinusoidal PWM can be replaced by a SVM that performs the voltage balancing. This fact is studied in next subsection.

2.3.1.4 *SVM of 3L-NPC VSC [17]*   The SVM for a 3L-NPC VSC is able to incorporate an algorithm to balance the DC bus capacitor's voltages. It is able to exploit all the voltage combinations provided by the converter, allowing multilevel output voltages on the AC side as well as balancing the capacitor's voltages. Compared with the 2L-VSC that presents eight voltage combinations at the AC side, the 3L-NPC VSC provides 27 allowable switching states, as exhibited in Figure 2.44.

Table 2.6 shows all the switching states that provide an output AC voltage vector in the first sextant of the plane. It is possible to distinguish voltage vectors with different natures:

- *Large vectors:* $V_1(200)$ and $V_2(220)$ are the vectors with the highest amplitude. They do not connect the AC side with the neutral point z, so they do not contribute to the charge or discharge of the capacitors of the DC bus.

TABLE 2.6 Different Output Voltage Combinations of 3L-NPC VSC in the First Sextant

	$V_1(200)$	$V_{10}(100)$	$V_{10}(211)$	$V_2(220)$	$V_{20}(110)$	$V_{20}(221)$	$V_{12}(210)$	$V_0(000)$	$V_0(111)$	$V_0(222)$
S_{a1}	1	0	1	1	0	1	1	0	0	1
S_{a2}	1	1	1	1	1	1	1	0	1	1
S_{b1}	0	0	0	1	0	1	0	0	0	1
S_{b2}	0	0	1	1	1	1	1	0	1	1
S_{c1}	0	0	0	0	0	0	0	0	0	1
S_{c2}	0	0	1	0	0	1	0	0	1	1
Vector	$V_1(200)$	$V_{10}(100)$	$V_{10}(211)$	$V_2(220)$	$V_{20}(110)$	$V_{20}(221)$	$V_{12}(210)$	$V_0(000)$	$V_0(111)$	$V_0(222)$
v_{ao}	V_{bus}	$\dfrac{V_{bus}}{2}$	V_{bus}	V_{bus}	$\dfrac{V_{bus}}{2}$	V_{bus}	V_{bus}	0	$\dfrac{V_{bus}}{2}$	V_{bus}
v_{bo}	0	0	$\dfrac{V_{bus}}{2}$	V_{bus}	$\dfrac{V_{bus}}{2}$	V_{bus}	$\dfrac{V_{bus}}{2}$	0	$\dfrac{V_{bus}}{2}$	V_{bus}
v_{co}	0	0	$\dfrac{V_{bus}}{2}$	0	0	$\dfrac{V_{bus}}{2}$	0	0	$\dfrac{V_{bus}}{2}$	V_{bus}
v_{an}	$2\dfrac{V_{bus}}{3}$	$\dfrac{V_{bus}}{3}$	$\dfrac{V_{bus}}{3}$	$\dfrac{V_{bus}}{3}$	$\dfrac{V_{bus}}{6}$	$\dfrac{V_{bus}}{6}$	$\dfrac{V_{bus}}{2}$	0	0	0
v_{bn}	$-\dfrac{V_{bus}}{3}$	$-\dfrac{V_{bus}}{6}$	$-\dfrac{V_{bus}}{6}$	$\dfrac{V_{bus}}{3}$	$\dfrac{V_{bus}}{6}$	$\dfrac{V_{bus}}{6}$	0	0	0	0
v_{cn}	$-\dfrac{V_{bus}}{3}$	$-\dfrac{V_{bus}}{6}$	$-\dfrac{V_{bus}}{6}$	$-2\dfrac{V_{bus}}{3}$	$-\dfrac{V_{bus}}{3}$	$-\dfrac{V_{bus}}{3}$	$-\dfrac{V_{bus}}{2}$	0	0	0
v_{α}	$2\dfrac{V_{bus}}{3}$	$\dfrac{V_{bus}}{3}$	$\dfrac{V_{bus}}{3}$	$\dfrac{V_{bus}}{3}$	$\dfrac{V_{bus}}{6}$	$\dfrac{V_{bus}}{6}$	$\dfrac{V_{bus}}{2}$	0	0	0
v_{β}	0	0	0	$\dfrac{\sqrt{3}V_{bus}}{3}$	$\dfrac{\sqrt{3}V_{bus}}{6}$	$\dfrac{\sqrt{3}V_{bus}}{6}$	$\dfrac{\sqrt{3}V_{bus}}{6}$	0	0	0

- *Medium vectors:* The amplitude of the $V_{12}(210)$ medium vector defines the maximum achievable output voltage vector in liner modulation. These vectors connect the AC side with the neutral point so they charge or discharge the capacitors of the DC bus.
- *Short vectors:* $V_{10}(100)$, $V_{10}(211)$, $V_{20}(110)$, and $V_{20}(221)$ have half the amplitude of the large vectors. They charge and discharge the capacitors. They always appear in pairs; this means that two different switching combinations of the converter produce the same AC output voltage vector. This is advantageous in balancing the voltages of the DC bus capacitors.
- *Zero vectors:* $V_0(000)$, $V_0(111)$, and $V_0(222)$ produce a zero AC output voltage and zero current through the neutral point z.
- In the rest of the sextants of the hexagon, due to symmetry, we can find voltage vectors with the same properties as the vectors that appear in the first sextant.
- Each sextant of the hexagon formed by the space vector can be divided into four different regions. Each region is delimited by the different voltage vector locations.

As we did for the 2L-VSC SVM and the 3L-NPC VSC SVM, for simplicity in implementation, we perform all the calculations of duty cycles and define the injecting vectors in the first sextant; later, we can extrapolate to the real sextant where the reference voltage vector is located.

First, the voltage reference vector is normalized as we did in the 2L-VSC SVM:

$$m_n = \frac{|\vec{v}|^*}{V_{bus}/3} \tag{2.78}$$

Note that the base voltage for the normalization is two times smaller than in the 2L-SVM. In this way, by applying the same normalization to the hexagon produced by the converter voltage vectors, the normalized first sextant yields the results in Figure 2.45.

The duty cycles or the projections of the normalized reference space vectors can be calculated as follows:

$$d_1 = m_n \left(\cos\theta_n - \frac{\sin\theta_n}{\sqrt{3}} \right) \tag{2.79}$$

$$d_2 = m_n \cdot 2 \frac{\sin(\theta_n)}{\sqrt{3}} \tag{2.80}$$

θ_n is derived from Table 2.3 as seen in the 2L-VSC SVM. These expressions are only useful when the reference voltage lies is region 4. In this case, the voltage vectors used to reproduce the reference vectors are the nearest three converter's voltage vectors: $V_{10}(211)$-(100), $V_0(000)$-(111)-(222), and $V_{20}(221)$-(110). Therefore, the

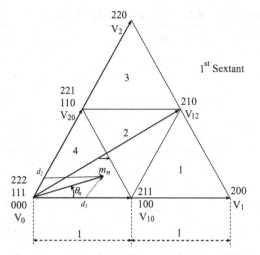

Figure 2.45 Normalized first sextant and voltage reference vector.

duty cycles of the voltage vectors used can be calculated directly as

$$d_{100-211} = d_1 \tag{2.81}$$

$$d_{110-221} = d_2 \tag{2.82}$$

$$d_{111} = 1 - d_1 - d_2 \tag{2.83}$$

Note that only vector (111) is used as the zero vector due to the fact that it reduces the commutations when the voltage vectors are applied (this fact is better seen later in this section).

When the voltage reference vector lies in the rest of the regions of the first sextant, the situation is graphically represented as in Figure 2.46.

Therefore, the duty cycle calculations, once Equations (2.79) and (2.80) are computed, for the choice of the nearest three vectors and in all the regions of the first sextant are summarized in Table 2.7.

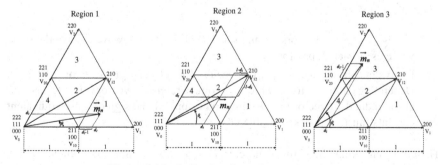

Figure 2.46 Projections in regions 1, 2, and 3.

TABLE 2.7 Duty Cycle Relations in Regions 1, 2, 3, and 4 of the First Sextant

Case	Region	Duty Cycles
$d_1 > 1$	1	$d_{200} = d_1 - 1$ $d_{210} = d_2$ $d_{100-211} = 2 - d_1 - d_2$
$d_1 \leq 1$ $d_2 \leq 1$ $d_1 + d_2 > 1$	2	$d_{100-211} = 1 - d_2$ $d_{110-221} = 1 - d_1$ $d_{210} = d_1 + d_2 - 1$
$d_1 > 1$	3	$d_{210} = d_1$ $d_{220} = d_2 - 1$ $d_{110-221} = 2 - d_1 - d_2$
$d_1 \leq 1$ $d_2 \leq 1$ $d_1 + d_2 \leq 1$	4	$d_{100-211} = d_1$ $d_{110-221} = d_2$ $d_{111} = 1 - d_1 - d_2$

From the calculation of the duty cycles themselves (Table 2.7), the region where the voltage vector is located is deduced at the same time. Therefore, once the duty cycles are computed, it is necessary to choose the actual redundant vector that is going to be injected. As seen before, from the AC output voltage point of view, there is no difference in the choice. However, the voltage of the DC bus capacitor is affected in a different manner, depending on which redundant vector is employed. Therefore, this fact is exploited in order to balance the voltages of the DC bus capacitors.

As graphically illustrated in Figure 2.47, the redundant vector provokes different current flows through the neutral point z. The current i_z is an image of the AC output currents (i_a, i_b, and i_c). Thus, for instance, if vector 100 is injected, the current i_z is equal to i_a output current (it goes entirely through the C_2 capacitor), assuming a balanced load supply of the converter. On the other hand, the injection of vector 211 makes $i_z = -i_a$—just the opposite effect in the charge and discharge of the capacitors (it goes entirely through the C_1 capacitor).

Exploiting this fact, the medium vectors can be selected according to Table 2.8. The decision for choosing the medium voltage vector is made after considering the following information:

- Voltages of the capacitors v_{c1} and v_{c2}. The objective is always to have the same voltages at each capacitor.
- Equivalent phase currents i_{a_eq} (for vectors 100-211), i_{c_eq} (for vectors 110-221). Depending on the sector in which the voltage reference vector is located as well as the redundant vector that is going to be applied, the sign of these equivalent currents must be checked (Table 2.9).

Therefore, checking the sign of the currents (i_{a_eq}, i_{a_eq}) and the voltages of the capacitors (v_{c1}, v_{c2}), the redundant vector choice in made in order to equalize the voltages of the capacitors v_{c1} and v_{c2}.

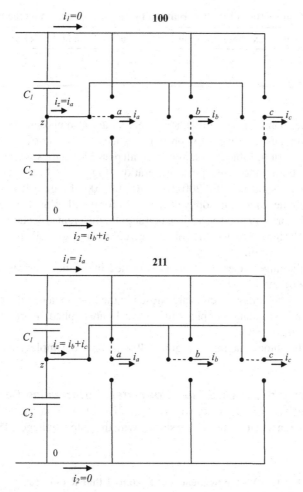

Figure 2.47 Example of control of the neutral point z current (i_z). Vector 100 produces $i_z = i_a$ while vector 211 produces $i_z = -i_a$.

TABLE 2.8 Medium Vector (V_{10} and V_{20}) Selection to Perform the Capacitor Voltage Balancing

$v_{c1} > v_{c2}$	$i_{a\_eq} > 0$	Vector V_{10}	$v_{c1} > v_{c2}$	$i_{c\_eq} > 0$	Vector V_{20}
0	0	100	0	0	221
0	1	211	0	1	110
1	0	211	1	0	110
1	1	100	1	1	221

TABLE 2.9 Equivalence of the Currents to Perform Capacitor Voltage Balancing

Equivalent Current	Sextant 1	Sextant 2	Sextant 3	Sextant 4	Sextant 5	Sextant 6
i_{a_eq}	i_a	i_b	i_b	i_c	i_c	i_a
i_{c_eq}	i_c	i_c	i_a	i_a	i_b	i_b

On the other hand, once the redundant vectors are also chosen, the sequence of vector injection must be checked in order to reduce the number of commutations at each switching period. Table 2.10 summarizes all possible vector sequences in the first sextant, for a triangular based pulse generation. Figure 2.48 illustrates the pulse generation for the sequence 100-200-210–210-200-100 of region 1. In this 3L-NPC SVM, although the same philosophy as the SVM for the 2L-VSC has been used, the sequence of vectors is created according to a slightly different procedure. In this case, depending on the regions where the voltage vector lies (1, 2, or 3), there is no necessity to inject the zero vector.

Note that the duty cycles must be transformed into time intervals according to Equations (2.46) and (2.47).

Finally, once the voltage vectors are chosen in the first sextant and the duty cycles are also selected, the changes of phase in Table 2.11 are applied, in order to obtain the real output voltage vector.

The simplified block diagram of Figure 2.49 illustrates the implementation schema of the SVM.

2.3.1.5 Example 2.1: Grid Side Converter Performance Based on 3L-NPC VSC with SVM
This example shows a 3L-NPC VSC connected to the grid, exchanging currents that are 60° phase shifted with the output voltages. The switching

TABLE 2.10 Vector Sequences of Sextant 1 that Reduce the Commutations of the Switches

Region	Double Vector	Sequence
1	100	100-200-210–210-200-100
	211	200–210–211–211–210-200
2	100-110	100-110-210–210-110-100
	100-221	100-210-221–221-210-100
	211-110	110-210-211–211-210-110
	211-221	210-211-221–221-211-210
3	110	110-210-220–220-210-110
	221	210-220-221–221-220-210
4	100-110	100-110-111–111-110-100
	100-221	100-111-221–221-111-100
	211-110	110-111-211–211-111-110
	211-221	111-211-221–221-211-111

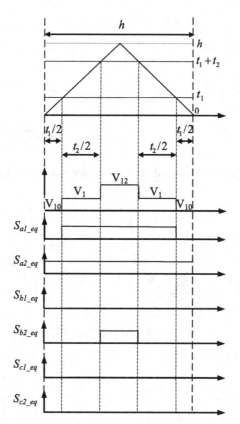

Figure 2.48 Triangular based pulse generation schema: 100-200-210–210-200-100 sequence of vector injections.

frequency is set to 2.5 kHz and the modulation is the SVM presented in this section. Figure 2.50 shows the most interesting moments of the simulation experiment.

In Figure 2.50a, two and a half cycles of the output phase to neutral voltage is illustrated. In this signal, it is possible to distinguish the nine levels of voltage achieved by the 3L-NPC VSC.

TABLE 2.11   Real Vector Generation by Changing the Phases, Depending on the Sextant in which the Reference Voltage Vector is Located

Sextant					
1	2	3	4	5	6
$S_a \rightarrow S_{a\_eq}$	$S_a \rightarrow S_{b\_eq}$	$S_a \rightarrow S_{c\_eq}$	$S_a \rightarrow S_{c\_eq}$	$S_a \rightarrow S_{b\_eq}$	$S_a \rightarrow S_{a\_eq}$
$S_b \rightarrow S_{b\_eq}$	$S_b \rightarrow S_{a\_eq}$	$S_b \rightarrow S_{a\_eq}$	$S_b \rightarrow S_{b\_eq}$	$S_b \rightarrow S_{c\_eq}$	$S_b \rightarrow S_{c\_eq}$
$S_c \rightarrow S_{c\_eq}$	$S_c \rightarrow S_{c\_eq}$	$S_c \rightarrow S_{b\_eq}$	$S_c \rightarrow S_{a\_eq}$	$S_c \rightarrow S_{a\_eq}$	$S_c \rightarrow S_{b\_eq}$

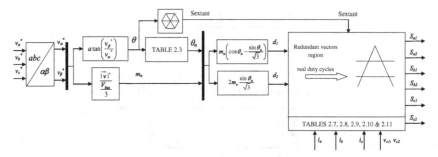

Figure 2.49 Simplified block diagram of the SVM schema for 3L-NPC VSC.

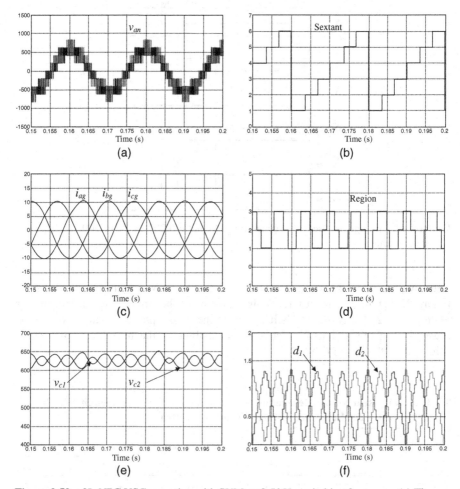

Figure 2.50 3L-NPC VSC operating with SVM, at 2.5 kHz switching frequency. (a) The v_{an} output voltage, (b) sextant, (c) *abc* converter output currents, (d) region, (e) v_{c1} and v_{c2} voltages of capacitors, and (f) duty cycles in the first sextant according to Equations (2.79) and (2.80).

On the other hand, Figure 2.50b shows the sextant in which the space vector voltage reference lies. In an equivalent way, the converter's output currents are shown in Figure 2.50c.

In order to generate this type of voltage and current performance, Figure 2.50d shows the region that the voltage reference space vector crosses in its rotation. In this case, the regions are 1, 2, and 3.

By paying attention to the DC side of the converter, Figure 2.50e illustrates the voltages of the DC bus capacitors. Under these operating conditions, the voltage balancing algorithm ensures a proper voltage balance. Notice that there exist voltage oscillations above the medium voltage that depends on the operating conditions of the converter as well as the capacitors' value of the DC bus.

Finally, Figure 2.50f shows the duty cycles in the first sextant calculated according to Equations (2.79) and (2.80).

2.4 CONTROL OF GRID SIDE SYSTEM

This section tackles the control of the grid side converter. For that purpose, it is necessary first to understand the model of the grid side system. As done later when we study the DFIM in subsequent chapters, the model of the grid side system is studied in two steps.

First, the steady state model of the grid side system is developed by using phasor theory, allowing the reader to understand the basic relationships between active and reactive powers, voltage, and currents at different operating modes.

Second, control of the grid side converter is studied, developing first the grid side dynamic model based on space vector theory. Thanks to these modeling developments, the reader is ready to understand the philosophy of the control strategy. Therefore, the control studied is the vector control technique, which employs a rotatory reference frame (dq), aligned with the grid voltage space vector. By means of this control strategy, it is possible to achieve the main two objectives of the grid side converter: control of the DC bus voltage and assured transmission of power through the converter, with controlled reactive power exchange.

2.4.1 Steady State Model of the Grid Side System

As studied in previous sections, the system configured by the grid side converter, filter, and grid voltage can ideally be represented as shown in Figure 2.51. The grid voltage (v_{ag}, v_{bg}, v_{cg}) is sinusoidal with constant amplitude and frequency. The voltage imposed by the grid side converter (v_{af}, v_{bf}, v_{cf}) can be modified in amplitude and phase as seen in previous sections. The filter considered in this section is the simplest solution, that is, a pure inductive filter (L_f). Note that a parasitic resistance (R_f) is also considered within the model of the filter.

For analysis purposes, if an ideal grid side system is considered, it is equivalent to a single-phase grid system as illustrated in Figure 2.52. In this way, it is only necessary to analyze one phase (e.g., a phase) to then extrapolate to the other two phases.

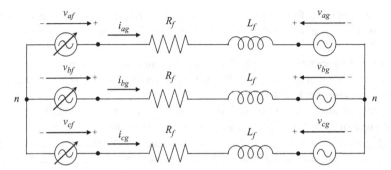

Figure 2.51 Simplified representation of the three-phase grid system.

The output voltage generated by the converter depends on the characteristics of the converter itself (two-level, multilevel topology, etc.) and the modulation technique employed. Under steady state operation conditions, all the magnitudes of the system (v_{af}, i_{ag}, v_{ag}) have constant amplitude phase shift between each other's and frequency. As seen in previous sections, the maximum achievable fundamental output voltage by the converter (in linear modulation, by SVM or by sinusoidal PWM with third harmonic injection) depends on the DC bus voltage:

$$Maximum \ \langle v_{af} \rangle_1 = \frac{V_{bus}}{\sqrt{3}} \qquad (2.84)$$

Thus, the voltage imposed by the converter can be modified depending on the requirements of the application. So under steady state operation, the electric equations of the system represented in Figure 2.52 are:

$$\underline{V}_{af} = \underline{V}_{ag} + (R_f + jL_f\omega_s)\underline{I}_{ag} \qquad (2.85)$$

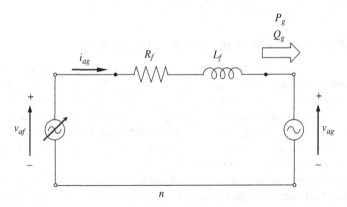

Figure 2.52 Simplified model of single-phase grid side system.

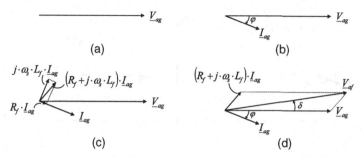

Figure 2.53   Phasor diagram construction of the grid side system.

Note that phasor representation is used ($\underline{V}_{af}$, $\underline{V}_{ag}$, and $\underline{I}_{ag}$) for the steady state. The general phasor diagram is constructed in Figure 2.53. It is built up in the following next steps:

- First, the grid voltage phasor is drawn (Figure 2.53a). The phase angle can be arbitrarily chosen.
- Second, the current is drawn (Figure 2.53b). Note that the phase shift between the grid voltage and current must be known.
- Third, once the current is drawn, the voltage drop in the filter is calculated and accordingly drawn (Figure 2.53c).
- Fourth, by means of Equation (2.85), the voltage of the converter is calculated and drawn (Figure 2.53d). Note that $\underline{V}_{af}$ is considered to be the phasor representation of the fundamental voltage of the converter output voltage ($\langle v_{af} \rangle_1$).

In general, it is possible to understand the process in a real system in the opposite direction. First, the grid voltage is established, then one specific voltage is imposed with the converter (specific voltage amplitude and phase shift), to finally establish the corresponding current (phase and amplitude).

The power flow study is normally carried out in the grid voltage ($\underline{V}_{ag}$) rather than in the converter ($\underline{V}_{af}$). Hence, the active and reactive powers calculated in the grid voltage point are (single phase)

$$P_g = |\underline{V}_{ag}| \cdot |\underline{I}_{ag}| \cdot \cos \varphi \qquad (2.86)$$

$$Q_g = |\underline{V}_{ag}| \cdot |\underline{I}_{ag}| \cdot \sin \varphi \qquad (2.87)$$

We adopt the power sign convention shown in Figure 2.54. When $P_g > 0$, the converter is delivering power to the grid; whereas when $P_g < 0$, the converter is receiving power from the grid.

Two particular operating conditions exist for the grid side converter: transmitting active power in both directions (positive and negative), but with unity power factor, that is, zero reactive power exchange at the grid point. Figure 2.55 shows these two

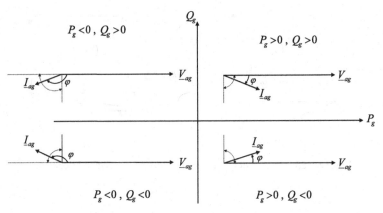

Figure 2.54   Power sign convention.

particular cases, where the current is shifted 0° (grid receives power from the converter) or 180° (grid delivers power to the converter) from the grid voltages.

On the other hand, if the voltage drop in R_f is neglected compared with the voltage drop in L_f, the active and reactive power expressions (2.86) and (2.87) can be reformulated, eliminating the dependence on the current (the derivation of these expressions is not addressed):

$$P_g = \frac{|\underline{V}_{ag}| \cdot |\underline{V}_{af}|}{\omega_s L_f} \sin \delta \ \ \text{with} \ \ R_f \to 0 \tag{2.88}$$

$$Q_g = |\underline{V}_{ag}| \cdot \frac{|\underline{V}_{af}|\cos \delta - |\underline{V}_{ag}|}{\omega_s L_f} \ \ \text{with} \ \ R_f \to 0 \tag{2.89}$$

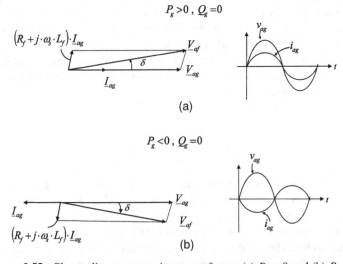

Figure 2.55   Phasor diagrams at unity power factor: (a) $P_g > 0$ and (b) $P_g < 0$.

Therefore, it is possible to rapidly know the active and reactive power flow through the grid, if only the converter voltage amplitude and phase shift with respect to the grid voltage is known.

2.4.1.1 Example 2.2: Operation at cos $\varphi = 1$ and cos $\varphi = -1$ The following practical case is studied:

Filter	$R_f = 0.01\ \Omega$; $L_f = 0.0005\ \text{H}$;
Grid voltage amplitude	690 V (50 Hz) (line to line rms voltage)
Grid current amplitude	400 A (rated rms current)
Power factor	$\cos \varphi = 1$

Therefore, the required converter voltage amplitude is (Figure 2.55a):

$$|\underline{V}_{af}| = \sqrt{\left(\underline{V}_{ag} + R_f\underline{I}_{ag}\right)^2 + \left(L_f\omega_s\underline{I}_{ag}\right)^2} =$$
$$= \sqrt{(563.4 + 0.01 \cdot 565.7)^2 + (0.0005 \cdot 314.16 \cdot 565.7)^2}$$
$$= 575.5\ \text{V}$$

Note that this is the fundamental component voltage amplitude. Consequently, the required DC bus voltage for proper operation of the grid side converter is (margin of 10% is added for control, not considering nonlinearities, losses, etc)

$$V_{bus} = (1.1)|\underline{V}_{af}| \cdot \sqrt{3}$$
$$= 1096.5\ \text{V}$$

In a similar way, we can consider that the grid is delivering active power to the converter:

$$\text{Power factor:}\quad \cos \varphi = -1$$

Therefore, the required converter voltage amplitude is (Figure 2.55b)

$$\underline{V}_{af}| = \sqrt{\left(\underline{V}_{ag} - R_f\underline{I}_{ag}\right)^2 + \left(L_f\omega_s\underline{I}_{ag}\right)^2}$$
$$= \sqrt{(563.4 - 0.01 \cdot 565.7)^2 + (0.0005 \cdot 314.16 \cdot 565.7)^2}$$
$$= 564.7\ \text{V}$$

Consequently, the required DC bus voltage for proper operation of the grid side converter is slightly less than in the previous example. However, if the converter can operate under both studied conditions, the DC bus voltage is imposed by the previous case, that is, $V_{bus} = 1096.5$ V.

2.4.1.2 *Example 2.3: Operation at cos* $\varphi = 0$ There is also a different case when the active power exchange with the grid is minimum. With an equal grid side converter as in the previous example,

$$\text{Power factor:} \quad \cos \varphi = 0$$

It is possible to distinguish two situations (neglecting the voltage drop in the resistance for a simpler exposition).

As seen in Figure 2.56, these two situations depend on the current phase (leading or lagging). In addition, the grid voltage and the converter voltage are in phase. The reader can note that these are the two extreme cases (maximum and minimum) in terms of the required converter voltage $|\underline{V}_{af}|$ for grid side system operation.

The required converter voltage amplitude is (Figure 2.56a):

$$|\underline{V}_{af}| = \sqrt{\left(\underline{V}_{ag} + L_f \omega_s \underline{I}_{ag}\right)^2 + (0)^2}$$

$$= \sqrt{(563.4 + 0.0005 \cdot 314.16 \cdot 565.7)^2}$$

$$= 652.3 \text{ V}$$

The required DC bus voltage for proper operation of the grid side converter, assuming the same criteria of the previous two examples, is

$$V_{bus} = (1.1)|\underline{V}_{af}| \cdot \sqrt{3}$$

$$= 1242.7 \text{ V}$$

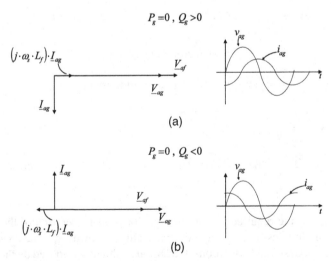

(a)

(b)

Figure 2.56   Phasor diagrams at zero power factor: (a) $Q_g > 0$ and (b) $Q_g < 0$.

On the contrary, for the case of Figure 2.56b,

$$|\underline{V}_{af}| = \sqrt{\left(\underline{V}_{ag} - L_f\omega_s\underline{I}_{ag}\right)^2 + (0)^2}$$
$$= \sqrt{(563.4 + 0.0005 \cdot 314.16 \cdot 565.7)^2}$$
$$= 474.5 \text{ V}$$

Therefore, the required DC bus voltage is

$$V_{bus} = (1.1)|\underline{V}_{af}| \cdot \sqrt{3}$$
$$= 904.1 \text{ V}$$

Note that for the final choice of the DC bus voltage, at the design stage, one must consider the operating conditions for the grid side converter and then, accordingly, find the case demanding the higher DC bus voltage.

2.4.2 Dynamic Modeling of the Grid Side System

Prior to the exposition of the vector control based schema, in this section, the grid side system was represented in a space vector form. This representation serves as the mathematical basis for understanding the dynamic behavior of the grid side system (dynamic model) and then deriving the vector control. The space vector representation (see the Appendix for more information) allows the derivation of a dynamic model of the grid side system. By means of the space vector tool, it is possible to use the differential equations defining the behavior of the grid side system's variables, such as current and voltages. Hence, once the dynamic model is studied, we will proceed to analyze the control in the next section, accompanied by a simulation based illustrative example.

2.4.2.1 $\alpha\beta$ Model   As studied in the previous section, the system configured by the grid side converter, filter, and grid voltage can ideally be represented as shown in Figure 2.51. By applying the space vector notation to the *abc* modeling equations (2.16)–(2.18), it is possible to represent the electric equations in $\alpha\beta$ components (for more information see the Appendix). Simply multiplying expression (2.16) by $\frac{2}{3}$, then Equation (2.17) by $\frac{2}{3}a$, and then Equation (2.18) by $\frac{2}{3}a^2$, the addition of the resulting equations yields

$$v_{\alpha f} = R_f i_{\alpha g} + L_f\frac{di_{\alpha g}}{dt} + v_{\alpha g} \tag{2.90}$$

$$v_{\beta f} = R_f i_{\beta g} + L_f\frac{di_{\beta g}}{dt} + v_{\beta g} \tag{2.91}$$

Or in a compact version, referred to a stationary reference frame,

$$\vec{v}_f^s = R_f\vec{i}_g^s + L_f\frac{d\vec{i}_g^s}{dt} + \vec{v}_g^s \tag{2.92}$$

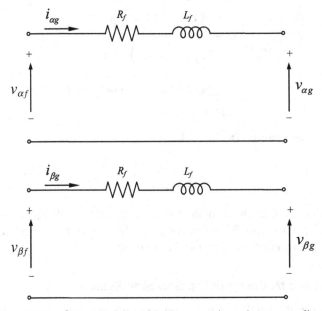

Figure 2.57 $\alpha\beta$ Model of the grid side system in stationary coordinates.

where

$$\vec{v}_f^s = v_{\alpha f} + jv_{\beta f} \tag{2.93}$$

$$\vec{v}_g^s = v_{\alpha g} + jv_{\beta g} \tag{2.94}$$

$$\vec{i}_g^s = i_{\alpha g} + ji_{\beta g} \tag{2.95}$$

The schematic representation of the resulting electric circuit is depicted in Figure 2.57.

2.4.2.2 *dq* Model   In a similar way, by multiplying Equation (2.92) by $e^{-j\theta}$, from the $\alpha\beta$ expressions the dq expressions are also derived (rotating frame):

$$\vec{v}_f^s \cdot e^{-j\theta} = R_f \vec{i}_g^s e^{-j\theta} + L_f \frac{d\vec{i}_g^s}{dt} e^{-j\theta} + \vec{v}_g^s e^{-j\theta} \tag{2.96}$$

yielding

$$\vec{v}_f^a = R_f \vec{i}_g^a + L_f \frac{d\vec{i}_g^a}{dt} + \vec{v}_g^a + j\omega_a L_f \vec{i}_g^a \tag{2.97}$$

Note that since $\theta = \omega_a t$, the angular position of the rotatory reference frame, we have

$$\frac{d\vec{i}_g^s}{dt}e^{-j\theta} = \frac{d\left(\vec{i}_g^s e^{-j\theta}\right)}{dt} + j\omega_a\vec{i}_g^s e^{-j\theta} \tag{2.98}$$

With dq components,

$$\vec{v}_f^a = v_{df} + jv_{qf} \tag{2.99}$$

$$\vec{v}_g^a = v_{dg} + jv_{qg} \tag{2.100}$$

$$\vec{i}_g^a = i_{dg} + ji_{qg} \tag{2.101}$$

Therefore, by decomposing into dq components, the basic equations for vector orientation are obtained:

$$v_{df} = R_f i_{dg} + L_f \frac{di_{dg}}{dt} + v_{dg} - \omega_a L_f i_{qg} \tag{2.102}$$

$$v_{qf} = R_f i_{qg} + L_f \frac{di_{qg}}{dt} + v_{qg} + \omega_a L_f i_{dg} \tag{2.103}$$

The schematic representation of the equivalent electric circuit is depicted in Figure 2.58.

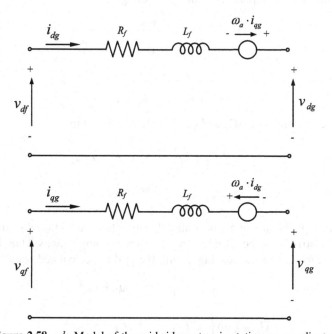

Figure 2.58   *dq* Model of the grid side system in stationary coordinates.

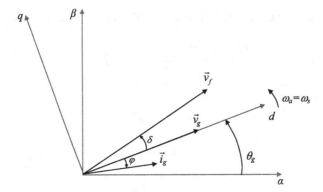

Figure 2.59 Alignment with d axis of the grid voltage space vector.

2.4.2.3 Alignment of the dq **Reference Frame** These last two expressions are the dq equations of the electric system, referred to a reference frame dq rotating at speed ω_a. In order to decouple and further simplify the system, typically the synchronous angular speed ω_a is chosen equal to the angular speed of the grid voltage ω_s, and the d axis of the rotating frame is aligned with the grid voltage space vector $\vec{v}_g^a$ as shown in Figure 2.59. This choice corresponds to one of the most important requirements, in order to perform the vector control technique discussed in the next section. Note that in the graphical representation of Figure 2.59, the superindex of the space vectors are omitted, but both reference frames ($\alpha\beta$ and dq) are represented.

The resulting dq components of the grid voltage yield

$$v_{dg} = |\vec{v}_g^a| \tag{2.104}$$

$$v_{qg} = 0 \tag{2.105}$$

$$\omega_a = \omega_s \tag{2.106}$$

$$\theta = \omega_a t \Rightarrow \theta = \theta_g = \omega_s t \tag{2.107}$$

Therefore, expressions (2.102) and (2.103) are simplified to

$$v_{df} = R_f i_{dg} + L_f \frac{di_{dg}}{dt} + v_{dg} - \omega_s L_f i_{qg} \tag{2.108}$$

$$v_{qf} = R_f i_{qg} + L_f \frac{di_{qg}}{dt} + \omega_s L_f i_{dg} \tag{2.109}$$

This grid voltage alignment not only slightly simplifies the voltage equations of the system, but also reduces the active and reactive power computations. Thus, if the total active and reactive powers exchanged with the grid are calculated, we find

$$P_g = \frac{3}{2}\text{Re}\left\{\vec{v}_g \cdot \vec{i}_g^*\right\} = \frac{3}{2}\left(v_{dg}i_{dg} + v_{qg}i_{qg}\right) \tag{2.110}$$

$$Q_g = \frac{3}{2}\text{Im}\left\{\vec{v}_g \cdot \vec{i}_g^*\right\} = \frac{3}{2}\left(v_{qg}i_{dg} - v_{dg}i_{qg}\right) \tag{2.111}$$

Considering relations (2.104) and (2.105), the power calculation can be simplified to

$$P_g = \tfrac{3}{2} v_{dg} i_{dg} = \tfrac{3}{2} |\vec{v}_g^a| i_{dg} \tag{2.112}$$

$$Q_g = -\tfrac{3}{2} v_{dg} i_{qg} = -\tfrac{3}{2} |\vec{v}_g^a| i_{qg} \tag{2.113}$$

Note that the voltage terms of these last two expressions are constant under ideal conditions; this means that a decoupled relationship between the current dq components and active and reactive powers has been obtained. Thus, the i_{dg} current is responsible for the P_g value, while the i_{qg} current is responsible for the Q_g value. This fact is exploited in the control of the grid side system, as studied later.

On the other hand, the active and reactive power calculated in the terminals of the converter is not the same as the power in the grid terminals. The power at the converter output is calculated as follows:

$$P_f = \tfrac{3}{2} \mathrm{Re}\left\{ \vec{v}_f \cdot \vec{i}_g^* \right\} = \tfrac{3}{2} \left(v_{df} i_{dg} + v_{qf} i_{qg} \right) \tag{2.114}$$

$$Q_f = \tfrac{3}{2} \mathrm{Im}\left\{ \vec{v}_f \cdot \vec{i}_g^* \right\} = \tfrac{3}{2} \left(v_{qf} i_{dg} - v_{df} i_{qg} \right) \tag{2.115}$$

Substituting the voltage expression (2.108) and (2.109) into these last two equations, we find

$$P_f = \tfrac{3}{2} \left(R_f |\vec{i}_g|^2 + L_f \frac{di_{dg}}{dt} i_{dg} + L_f \frac{di_{qg}}{dt} i_{qg} + v_{dg} i_{dg} \right) \tag{2.116}$$

$$Q_f = \tfrac{3}{2} \left(L_f \omega_s |\vec{i}_g|^2 - L_f \frac{di_{dg}}{dt} i_{qg} + L_f \frac{di_{qg}}{dt} i_{dg} - v_{dg} i_{qg} \right) \tag{2.117}$$

Assuming that, at steady state, the first derivatives of the dq current components are zero, we find

$$P_f = \tfrac{3}{2} \left(R_f |\vec{i}_g|^2 + v_{dg} i_{dg} \right) \tag{2.118}$$

$$Q_f = \tfrac{3}{2} \left(L_f \omega_s |\vec{i}_g|^2 - v_{dg} i_{qg} \right) \tag{2.119}$$

Notice that the converter also provides the active and reactive power of the filter (Figure 2.60). In the converters' power expressions, the power part assumed by the filter also appears.

Now that the steady state and dynamic models of the grid side system have been studied, we turn to the control.

2.4.3 Vector Control of the Grid Side System

Control is a necessary part of the grid side system. Without having control of some of the magnitudes of the grid side part, it is not possible to make it work properly. In this

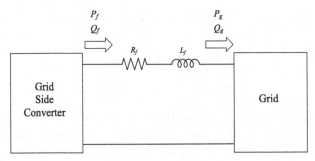

Figure 2.60   Power flow diagram.

section, a vector control based schema is studied. This control technique is widely extended among the control strategies for grid connected converters. It provides good performance characteristics with reasonably simple implementation requirements. The vector control technique follows the philosophy of representing the system that is going to be controlled—in our case the grid side system—in a space vector form. Thanks to this basis, reasonably good decoupled control of currents and power is achieved. In chapter 7, the same basic principles of vector control are applied to the machine's control.

2.4.3.1 Grid Voltage Oriented Vector Control

The grid side converter is in charge of controlling part of the power flow of the DFIM. The power generated by the wind turbine is partially delivered though the rotor of the DFIM as advanced in the previous chapter. This power flow that goes through the rotor flows also through the DC link and finally is transmitted by the grid side converter to the grid. The simplified block diagram of the grid side system, together with a schematic of its control block diagram, is given in Figure 2.61.

Subsequent paragraphs give the basic principles of control of the grid side converter. In this case, for a simpler exposition, the VSC topology chosen to present the control is a 2L-VSC. However, from a vector control point of view, nothing would change in the control if a multilevel VSC topology is used.

The pulses for the controlled switches (S_{a_g}, S_{b_g}, S_{c_g}) of the 2L-VSC, that is, the output voltage of the converter, are generated in order to control the DC bus voltage (V_{bus}) of the DC link and the reactive power exchanged with the grid (Q_g). This is done, in general, according to a closed loop control law. Some typical controls are vector control or direct power control. However, this section only studies the grid voltage oriented vector control (GVOVC).

On the other hand, control of V_{bus} is necessary since, as seen in previous sections, the DC link is mainly formed by a capacitor. Thus, the active power flow through the rotor must cross the DC link and then it must be transmitted to the grid. Therefore, by only controlling the V_{bus} variable to a constant value, this active power flow through the converters is ensured, together with a guarantee that both grid and rotor side converters have available the required DC voltage to work properly.

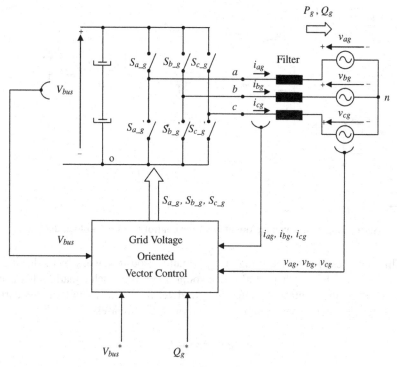

Figure 2.61 Grid side system control.

In a similar way, one variable that can also be controlled with this schema is the reactive power exchange with the grid (Q_g). In general, it can take different values depending on which current are to be minimized: the current through the back-to-back converter or the current through the stator of the machine.

Since, in general, a sensorless strategy is not typically adopted, for control the magnitudes that must be measured are the grid side current and voltage, together with the DC link voltage.

Therefore, the grid voltage oriented vector control (GVOVC) block diagram is shown in Figure 2.62. From the V_{bus} and Q_g references, it creates pulses for the controlled switches S_{a_g}, S_{b_g}, and S_{c_g}.

Thus, the modulator creates the pulses S_{a_g}, S_{b_g}, S_{c_g} from the *abc* voltage references for the grid side converter: v_{af}^*, v_{bf}^*, and v_{cf}^*. The modulator can be based on any of the schemas presented in this chapter for the 2L-VSC.

In this way, these *abc* voltage references are first created in *dq* coordinates (v_{df}^*, v_{af}^*), then transformed to $\alpha\beta$ coordinates (v_{af}^*, $v_{\beta f}^*$), and finally generate the *abc* voltage references.

Then, the *dq* voltage references (v_{df}^*, v_{af}^*) are independently created by the *dq* current (i_{dg}^*, i_{qg}^*) controllers. Note that this cause–effect law comes from expressions (2.108) and (2.109).

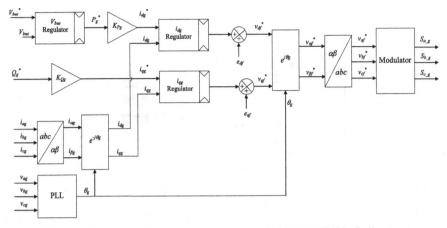

Figure 2.62   Grid voltage oriented vector control (GVOVC) block diagram.

This indicates that by modifying v_{df}, i_{dg} is mainly modified; while by modifying v_{qf}, i_{qg} is mainly modified. There is also one coupling term in each equation that is best considered in the control as a feed-forward term (at the output of the current controllers), for better performance in the dynamic responses:

$$e_{df} = -\omega_s L_f i_{qg} \qquad (2.120)$$

$$e_{qf} = \omega_s L_f i_{dg} \qquad (2.121)$$

Note that, under ideal conditions, the term v_{dg} is equal to the grid voltage amplitude and is constant as mentioned before.

It must be highlighted that the current references (i_{dg}^*, i_{qg}^*) are totally decoupled from the active and reactive powers, thanks to the alignment of the grid voltage space vector and the d axis of the rotating reference frame. Thus, i_{dg}^* control implies P_g control, while i_{qg}^* control implies Q_g control. The constant terms needed are easily deduced form Equations (2.112) and (2.113):

$$K_{Pg} = \frac{1}{\frac{3}{2} v_{dg}} \qquad (2.122)$$

$$K_{Qg} = \frac{1}{-\frac{3}{2} v_{dg}} \qquad (2.123)$$

As mentioned before, the power P_g reference is created by the V_{bus} regulator. Indirectly, by this loop, active power flow through the back-to-back converter is ensured.

Finally, for voltage and current coordinate transformations, the angle of the grid voltage is needed: θ_g. In general, this angle is estimated by a phase locked loop (PLL). Its closed loop nature provides stability and perturbation rejections to the angle estimation.

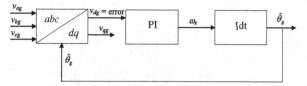

Figure 2.63   Classic PLL structure block diagram.

Consequently, the presented control strategy is able to control the variables (V_{bus}, Q_g) as specified, providing also good dynamic response performance due to its vector control structure.

2.4.3.2 Phase Locked Loop (PLL) for the Grid Angle Estimation
As introduced in the previous section, a PLL is used to estimate the angle of the grid voltage θ_g, in a closed loop way. Many different PLL structures have been proposed in the specialized literature; however, in this section, only a simple but efficient solution is presented.

The PLL seeks synchronization to a sinusoidally varying three-phase variable, in this case the grid voltage. The presented PLL is synchronized by using the dq coordinates of the voltage with which it is required to be synchronized. Thus, the d component of the grid voltage (v_{dg}) must be aligned with the d rotating reference frame; this means that the estimated θ_g must be modified until the q component of the voltage (v_{qg}) is zero. At that moment, it can be said that the rotating reference frame dq and the grid voltage space vector are synchronized and aligned to the d axis.

The closed loop PLL structure is illustrated in Figure 2.63. It takes the input abc voltages and transforms them into dq coordinates by using its own estimated θ_g angle. Then the calculated d component is passed through a PI controller, modifying the estimated angular speed until the d component is made zero; this means that the synchronization process has been stabilized (check Figure 2.64 for an illustrative graphical example).

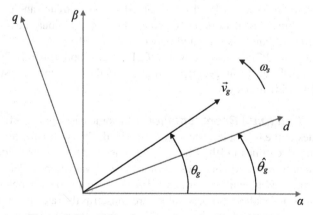

Figure 2.64   Grid voltage angle estimation with PLL.

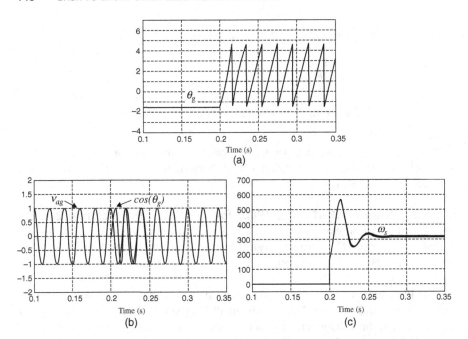

Figure 2.65   PLL performance: (a) estimated angle θ_g, (b) $\cos \theta_g$ and v_{ag}, and (c) ω_s.

This PLL must be continuously running with the control strategy, since the angle does not have a constant magnitude but is modified according to the angular speed of the grid voltage. The dynamic of the synchronization can be altered by tuning the PI controller. Unfortunately, the synchronization dynamics depend on the amplitude of the input voltages. Consequently, with this PLL structure, it is highly recommended that the input voltages be given in p.u. (see next section for details) so it is not necessary to use a new tuning process of the PI, if this PLL is used in a system with different input voltage amplitudes.

Figure 2.65 shows the proposed PLL performance when it is connected to a 380 V grid voltage. Until second 0.2, the PLL is disabled, so the estimated angle $\hat{\theta}_g$ is zero, as illustrated in Figure 2.65a. In order to reduce the computational cost of the PLL implementation, the angle is saturated between $3\pi/2$ and $-\pi/2$.

Figures 2.65b and 2.65c show how the PLL needs approximately 2.5 cycles to synchronize with the grid voltage. Depending on how the PI is tuned, a faster dynamic could be achieved if necessary.

2.4.3.3 Per Unit (p.u.) Representation
In some cases, it is useful to express the magnitudes of the grid side system as normalized. This normalization is done by selecting some base values and then dividing the magnitudes of the grid side system by these selected base values. In this book the normalizing values are called base values, while the normalized magnitudes are called per unit (p.u.) magnitudes.

In general, the normalizing or base values are chosen as the rated magnitudes of the system (voltage, currents, etc.); therefore, this normalization is often useful to give a quick view of how far one instantaneous magnitude is varying, for instance, the grid current from its rated value. On the other hand, the p.u. values are also useful to compare magnitudes of different systems (e.g., currents) rated to different power.

In the next chapter, when the model of the DFIM is presented, a more detailed analysis of the p.u. representation is made, applying p.u. to the model of the machine itself, to its parameters, and, of course, to the varying magnitudes of the machine (voltages, currents, fluxes, etc.). However, in this chapter, the p.u. is only applied to the magnitudes present in the grid side system.

Base Values   As mentioned before, any magnitude of the grid side system can be normalized by dividing the magnitude by a base value. For example,

$$Voltage \qquad \text{(V)}$$

$$Voltage_{pu} = Voltage/V_{base} \quad \text{(no units)}$$

As mentioned earlier in this book, this normalization is called the p.u. transformation. The p.u. values of magnitudes will be accompanied by the subscript "pu".

The base values for the grid side system are chosen as the **rated AC rms or DC values**:

$$\text{Base AC voltage:} \quad V_{base\_ac} = |\underline{V}_g|_{rated}(\text{V}) \qquad (2.124)$$

$$\text{Base current:} \quad I_{base} = |\underline{I}_g|_{rated}(\text{A}) \qquad (2.125)$$

$$\text{Base DC voltage:} \quad V_{base\_dc} = V_{bus\ rated}(\text{V}) \qquad (2.126)$$

where

$|\underline{V}_g|_{rated} = $ rated rms grid voltage (V_{rms})
$|\underline{I}_g|_{rated} = $ rated rms grid current (A_{rms})
$V_{bus\ rated} = $ rated DC bus voltage (V)

For the base values of AC magnitudes, the rms is chosen instead of the peak values, basically because the p.u. transformation is going to be applied to phasors expressed in rms values. If phasors were expressed in peak values, the base values accordingly could be chosen as peak values.

From these basic base values, it is possible to derive the base value for the AC power:

$$\text{Base power:} \quad S_{base} = 3V_{base\_ac}I_{base} \qquad (2.127)$$

It must be remarked that the choice for $V_{bus\ rated}$ can be made in different ways. The required DC bus voltage depends on the operating conditions of the grid side system (see Sections 2.4.1.1 and 2.4.1.2) as well as the type of modulation employed, the grid side filter, and so on. In this case, supposing that the grid side converter can operate at any operating condition, the DC bus voltage required would be (covers the maximum output AC voltage demand neglecting R_f)

$$V_{bus\ rated} = (1.1)\left(V_{base\_ac} + L_f\omega_s I_{base}\right) \cdot \sqrt{3}$$

Note that a 10% higher voltage is imposed (for transient control purposes) than the required theoretical DC bus voltage for the steady state. Several different choices could also be valid for $V_{bus\ rated}$.

Per Unit Transformation of Magnitudes   Consequently, from the base values, it is possible to derive the magnitudes in p.u., according to the following expressions:

$$\text{Grid voltage in p.u.:}\quad \underline{V}_{g\ pu} = \frac{\underline{V}_g}{V_{base\_ac}} \qquad (2.128)$$

$$\text{Grid current in p.u.:}\quad \underline{I}_{g\ pu} = \frac{\underline{I}_g}{I_{base}} \qquad (2.129)$$

$$\text{Active power in p.u.:}\quad P_{g\ pu} = 3\frac{\text{Re}\left\{\underline{V}_g \cdot \underline{I}_g^*\right\}}{S_{base}} \qquad (2.130)$$

$$\text{Reactive power in p.u.:}\quad Q_{g\ pu} = 3\frac{\text{Im}\left\{\underline{V}_g \cdot \underline{I}_g^*\right\}}{S_{base}} \qquad (2.131)$$

$$\text{DC bus voltage in p.u.:}\quad V_{bus\ pu} = \frac{V_{bus}}{V_{base\_dc}} \qquad (2.132)$$

Therefore, by following these guidelines for p.u. transformation, it is possible to normalize the magnitudes of the grid side system. Note that only the amplitude of the phasors is modified by this p.u. transformation; the angle remains unaltered.

2.4.3.4 Example 2.4: Back-to-Back Converter Supplying a 15 kW and 380 V DFIM   This example shows the most interesting magnitudes of a back-to-back converter supplying a 15 kW DFIM. The grid side converter is connected to a grid of 380 V and its nominal power is rated to 5 kW. The simulation experiment shows the steady state and the transient performance of the grid side system when it is controlled by the GVOVC presented in the previous section. The converter used is a 2L-VSC. Figure 2.66 shows the captured variables.

Thus, Figure 2.66a illustrates the grid voltage where the stator of the DFIM as well as the grid side converter is connected in p.u. In Figure 2.66b, the power exchange with the grid is shown. Notice that, during the experiment, the active power (P_g) delivered to the grid is 5 kW, due to the fact that the DFIM is generating through the rotor to its nominal power. On the other hand, in the middle of the experiment, the reactive power exchange with the grid is modified from 0 VAR (unity power factor) to −2 kVAR (0.4 p.u.).

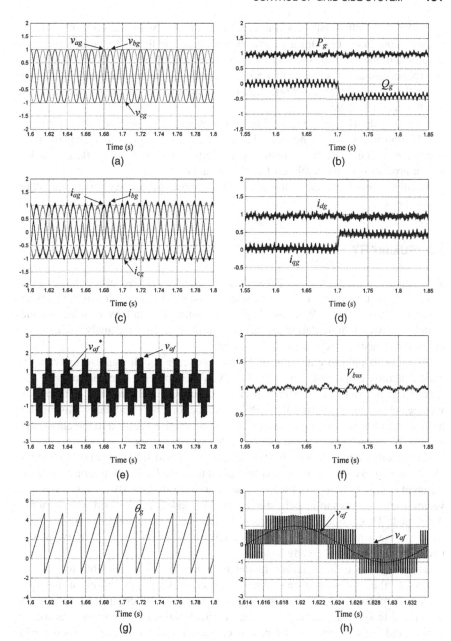

Figure 2.66 The 5 kW power exchange with the grid of a 2L-VSC, back-to-back converter supplying a 15 kW DFIM: (a) grid voltages in p.u., (b) power exchange with the grid in p.u., (c) *abc* grid side converter currents in p.u., (d) *dq* grid side converter currents in p.u., (e) grid side converter output voltage in p.u., (f) DC bus voltage in p.u, (g) estimated grid angle, and (h) zoom of grid side converter output voltage.

Under these circumstances, the current exchanges with the grid are shown in Figure 2.66c. These currents are controlled according to the vector control principles presented in the previous sections in dq coordinates. Figure 2.66d shows the current control performance. A reasonably good decoupling is achieved during the transient.

On the other hand, Figure 2.66e shows the converter's voltage reference of phase a, and the actual voltage.

Checking the DC side, Figure 2.66f illustrates the DC bus voltage performance. The Q_g transient is almost not seen due to a reasonably good and decoupled control performance. The PLL performance is exhibited in Figure 2.66g. Its closed loop nature produces a very stable grid angle estimation.

Finally, Figure 2.66h shows the detail of the a phase of the converter output voltage, in one cycle.

2.5 SUMMARY

This chapter provides the background and basic knowledge of the back-to-back converter for supplying DFIM based wind turbines. The principles of the converter based on the two-level VSC topology are studied in detail by considering ideal switches. Different modulation techniques for this converter are analyzed, exploring the improvement capacities of each of them.

Attention is paid to emerging multilevel converter topologies, by analyzing the most generally employed one—the 3L-NPC VSC topology. The particularities and special characteristics of this converter are described, including one example of a voltage balancing algorithm for the DC bus capacitors. In recent years, this topology has gained popularity as a good and effective alternative solution to the classic two-level converter when this must handle higher power ranges.

On the other hand, the steady state model of the grid side system is studied with phasor theory. By means of this mathematical tool, the basic behavior as well as the different operating modes of the grid side system are studied. By means of the space vector form, the dynamic behavior (dynamic model) of the grid side system is developed. This model, in contrast with the steady state model, permits one to represent the behavior of the grid side system variables (voltages, currents, powers) under the dynamic and steady states.

Therefore, once the dynamic model is understood, it is possible to study the control of the grid side system. The vector control strategy, based on the space vector representation of the system that is going to be controlled, in our case the grid side system, is revealed as an effective control that provides reasonably simple and easy to understand characteristics, accompanied by a good performance. Hence, by means of this control, the main objectives of the grid side converter are achieved: control of the DC bus voltage and control of the power flow through the grid side converter.

Finally, once the basic knowledge about the back-to-back converter is acquired by the reader, it is possible to proceed to study the DFIM with the help of subsequent chapters. It is necessary to keep in mind that the rotor of the DFIM is not supplied by an ideal sinusoidal voltage source, but is supplied by the back-to-back converter studied in this chapter, accompanied by all of its restrictions and limitations.

REFERENCES

1. M. P. Kazmierkowski, R. Krishnan, and F. Blaabjerg, *Control in Power Electronics Selected Problems*. Academic Press, 2002.
2. D. G. Holmes and T. A. Lipo, *Pulse Width Modulation for Power Converters*. John Wiley & Sons. Inc., 2006.
3. B. K. Bose, *Power Electronics and Drives*. Elsevier, 2006.
4. B. Wu, *High Power Converters and AC Drives*. John Wiley & Sons. Inc., 2006.
5. N. Mohan, T. Undeland, and R. Robbins, *Power Electronics*. John Wiley & Sons. Inc., 1989.
6. W. Hart, *Introduction to Power Electronics*, Prentice Hall, 1997.
7. R. H. Park,*Two Reactions Theory of Synchronous Machines*. AIEE transactions. Vol 48, pp 716–730, 192.
8. A. F. Moreira, T. A. Lipo, G. Venkataramanan, and S. Bernet,"Modeling and Evaluation of *dv/dt* Filters for AC Drives with High Switching Speed," EPE 2001, Graz.
9. Mahesh M. Swany, and John A. Houdek,"Low Cost Motor Protection Techniques for Use with PWM Driven Machines," MTE Corporation, Menomonee Falls, WI 53051. Intelligent Motion–June 1997 proceedings.
10. A. Fratta, G. M. Pellegrino, F. Scapino, and F. Villata, *Cost-Effective Line Termination Net for IGBT PWM VSI AC Motor Drives*. in Conf. ISIE, 2000.
11. G. Skibinski, R. Kerkman, D. Leggate, J. Pankau, and D. Schlegel, *Reflected Wave Modeling Techniques for PWM AC Motor Drives*. in Conf. APEC, 1998.
12. R. C. Portillo, M. M. Prats, J. I. Leon, J. A. Sanchez, J. M. Carrasco, E. Galvan, and L. G. Franquelo, "Modeling Strategy for Back-to-Back Three-Level Converters Applied to High-Power Wind Turbines," *IEEE Trans. Ind. Electron.*, Vol. 53, No. 5, pp. 1483–1491, October 2006.
13. J. Rodriguez, S. Bernet, W. Bin, J. O. Pontt, and S. Kouro, "Multilevel Voltage-Source-Converter Topologies for Industrial Medium-Voltage Drives," *IEEE Trans. Ind. Electron.* Vol. 54, No. 6, pp. 2930–2945, December 2007.
14. S. Aurtenechea, M. A. Rodríguez, E. Oyarbide, and J. R. Torrealday, "Predictive Control Strategy for DC/AC Converters Based on Direct Power Control," *IEEE Trans. Ind. Electron.*, Vol. 54, No. 3, June 2007.
15. S. Bernet, "State of the Art and Developments of Medium Voltage Converters—An Overview," *International Conference on PELINCEC*, 2005.
16. J. Pou, J. Zaragoza, P. Rodríguez, S. V. Ceballos, V. M. Sala, R. P. Burgos, and D. Boroyevich, "Fast-Processing Modulation Strategy for the Neutral-Point-Clamped Converter with Total Elimination of Low-Frequency Voltage Oscillations in the Neutral Point," *IEEE Trans. Ind. Electron.*, Vol. 54, No. 4, pp. 2288–2294, August 2007.

17. J. Pou, D. Boroyevich, and R. Pindado, "New Feedforward Space-Vector PWM Method to Obtain Balanced AC Output Voltages in a Three-Level Neutral-Point-Clamped Converter," *IEEE Trans. Ind. Electron.*, Vol. 49, No. 5, pp. 1026–1034, October 2002.

18. J. Rodriguez, S. Bernet, P. K. Steimer, and I. E. Lizama, "A Survey on Neutral-Point-Clamped Inverters," *IEEE Trans. Ind. Electron.*, Vol. 57, pp. 2219–2230, 2010.

19. S. Kouro, M. Malinowski, K. Gopakumar, G. Franquelo, J. Pou, J. Rodriguez, B. Wu, and A. Perez, "Recent Advances and Industrial Applications of Multilevel Converters," *IEEE Trans. Ind. Electron.*, Vol. 57, no. 8, pp. 2563–2580, 2010.

20. A. Nabae, I. Takahashi, and H. Akagi, "A New Neutral-Point-Clamped PWM Inverter," *IEEE Trans. Ind. Appl.*, Vol. IA-17, pp. 518–523, 1981.

21. J. Pou, R. Pindado, D. Boroyevich, and P. Rodriguez, "Limits of the Neutral-Point Balance in Back-to-Back-Connected Three-Level Converters," *IEEE Trans. Power Electron.*, Vol. 19, pp. 722–731, 2004.

22. M. Hiller, D. Krug, R. Sommer, and S. Rohner, "A New Highly Modular Medium Voltage Converter Topology for Industrial Drive Applications," in *13th European Conference on Power Electronics and Applications (EPE '09)* 2009, pp. 1–10.

23. T. A. Meynard, H. Foch, F. Forest, C. Turpin, F. Richardeau, L. Delmas, G. Gateau, and E. Lefeuvre, "Multicell Converters: Derived Topologies," *IEEE Trans. Ind. Electron*, Vol. 49, pp. 978–987, 2002.

24. R. Marquardt, "Current Rectification Circuit for Voltage Source Inverters with Separate Energy Stores Replaces Phase Blocks with Energy Storing Capacitors," German Patent (DE10103031A1) 25 July 2002.

25. S. Allebrod, R. Hamerski, and R. Marquardt, "New Transformerless, Scalable Modular Multilevel Converters for HVDC-Transmission," in *IEEE Power Electronics Specialists Conference*, pp. 174–179, June 2008.

CHAPTER 3

Steady State of the Doubly Fed Induction Machine

3.1 INTRODUCTION

Chapter 1 presented the basic background of wind turbines, introducing the most general aspects related to wind energy generation applications. It revealed that the generator and the converter in charge of feeding and controlling the machine are very important parts of the wind turbine and of the energy generation process itself.

Thus, once the converter was studied in Chapter 2 in detail, this chapter analyzes the generator: the doubly fed induction generator (DFIG) or, more generally, the doubly fed induction machine (DFIM). This crucial element of the wind turbine is treated as a machine, in general, that can operate in both modes—generating and motoring. However, in the normal operation of a wind turbine, it works as a generator, delivering energy from the wind to the electric grid.

In this chapter we study the DFIM, considering only the steady state. First, the steady state electric circuit of the machine is developed, deriving the steady state model electric equations along with its physics. Then, by means of these model equations, the modes of operation of the machine are presented and analyzed. Finally, based also on the model equations, a detailed performance evaluation of the machine is carried out, developing steady state performance curves that can reveal current, voltage, or different magnitude requirements, depending on the specific operating conditions of the machine.

Hence, this chapter serves as the first modeling approach of the machine, focusing analysis only on the steady state. Later, Chapters 4, 5, and 6 complete the material dedicated to the model of the machine, by means of dynamic modeling analysis, the testing procedure of the machine, and the dynamic behavior of the machine during voltage dips.

Doubly Fed Induction Machine: Modeling and Control for Wind Energy Generation,
First Edition. By G. Abad, J. López, M. A. Rodríguez, L. Marroyo, and G. Iwanski.
© 2011 the Institute of Electrical and Electronic Engineers, Inc. Published 2011 by John Wiley & Sons, Inc.

3.2 EQUIVALENT ELECTRIC CIRCUIT AT STEADY STATE

3.2.1 Basic Concepts on DFIM

3.2.1.1 Induced Force/Torque As advanced in Chapter 1, the DFIM consist of two three-phase winding sets: one placed in the stator and the other in the rotor. These two three-phase windings need to be supplied independently and also both windings can be bidirectionally energy supplied. The rotor three-phase windings can be connected in a star or delta configuration and they are supplied thanks to the brushes and the slip ring assembly. This machine is similar to a cage rotor induction machine; however, from the construction perspective, the rotor of the DFIM is bigger and requires maintenance due to the deterioration of the brushes and the slip rings.

Therefore, the stator is composed of three windings 120° spatially shifted and p pairs of poles. When these three stator windings are supplied by a balanced three-phase voltage of frequency f_s, the stator flux is induced. This stator flux rotates at constant speed. That is, the synchronous speed (n_s) is given by the expression

$$n_s = \frac{60 f_s}{p} \ (\text{rev}/\text{min}) \tag{3.1}$$

In principle, this rotational stator flux induces an emf in the rotor windings according to Faraday's law:

$$e_{ind} = (v \times B)L \tag{3.2}$$

where

e_{ind} = induced emf in one conductor of the rotor

v = speed of the conductor in relation to the stator flux rotation

B = stator flux density vector

L = length of the conductor

Due to this induced voltage in the rotor windings and the voltage that can be injected externally through the brushes, a current is induced in the rotor windings. This current, according to Laplace's law, creates an induced force in the rotor of the machine:

$$F = i \cdot (L \times B) \tag{3.3}$$

where

F = induced force (in relation to the induced torque of the machine)

i = current of the rotor conductor

B = stator flux density vector

L = length of the conductor

3.2.1.2 Slip Concept As described in the previous subsection, the induced voltage in the rotor depends on the relation between the stator flux rotational speed and the rotational speed of the rotor. In fact, the angular frequency of the induced rotor voltages and currents is given by the relation

$$\omega_r = \omega_s - \omega_m \qquad (3.4)$$

where

ω_r = angular frequency of the voltages and currents of the rotor windings (rad/s)
ω_s = angular frequency of the voltages and currents of the stator windings (rad/s)
ω_m = angular frequency of the rotor (rad/s)

and

$$\omega_m = p\Omega_m \qquad (3.5)$$

where

Ω_m = mechanical rotational speed at the rotor (rad/s)

Note that in normal operation at steady state, since the induced voltages and currents in the rotor windings have ω_r angular frequency, the externally supplied voltage in the rotor should also have ω_r angular frequency.

Hence, the commonly used term to define the relation between the speed of the stator and the rotor angular frequency is the slip, s:

$$s = \frac{\omega_s - \omega_m}{\omega_s} \qquad (3.6)$$

By combining expressions (3.4) and (3.6), the relation between the slip, the stator, and the rotor angular frequency is given by

$$\omega_r = s\omega_s \qquad (3.7)$$

From this last expression, equivalently, the relation between the frequencies can also be derived:

$$f_r = sf_s \qquad (3.8)$$

Depending on the sign of the slip, it is possible to distinguish three different operating modes for the machine:

$\omega_m < \omega_s \Rightarrow \omega_r > 0 \Rightarrow s > 0 \Rightarrow$ Subsynchronous operation
$\omega_m > \omega_s \Rightarrow \omega_r < 0 \Rightarrow s < 0 \Rightarrow$ Hypersynchronous operation
$\omega_m = \omega_s \Rightarrow \omega_r = 0 \Rightarrow s = 0 \Rightarrow$ Synchronous operation

3.2.2 Steady State Equivalent Circuit

3.2.2.1 Equivalent Circuit with Different Stator and Rotor Frequencies

Based on the principles presented in Section 3.2.0, the steady state equivalent electric circuit of the DFIM can be ideally simplified, as depicted in Figure 3.1, with the following assumptions:

- It is assumed that both the stator and the rotor are connected in the star configuration; however, only one phase of the stator and rotor three-phase windings is represented.
- The stator is supplied by the grid at constant and balanced three-phase AC voltage amplitude and frequency.
- The rotor is supplied also at constant and balanced AC voltage amplitude and frequency, independently from the stator, for instance, by a back-to-back voltage source converter.
- To represent steady state voltage and current magnitudes, the analysis is carried out using classical phasor theory:

  $\underline{V}_s$ = supplied stator voltage

  $\underline{V}'_r$ = supplied rotor voltage

  $\underline{I}_s$ = induced stator current

  $\underline{I}'_r$ = induced rotor current

  $\underline{E}_s$ = induced emf in the stator

  $\underline{E}'_{rs}$ = induced emf in the rotor

- The electric parameters of the stator and rotor are:

  R_s = stator resistance

  R'_r = rotor resistance

  $X_{\sigma s}$ = stator leakage impedance

  $X'_{\sigma r}$ = rotor leakage impedance

  N_s = stator winding's number of turns per phase

  N_r = rotor winding's number of turns per phase

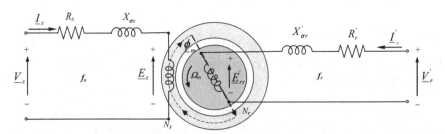

Figure 3.1   One phase steady state equivalent electric circuit of the DFIM with different stator and rotor frequencies.

It must be highlighted again that the voltage and currents of the stator and the rotor circuits have different frequencies. The stator frequency (f_s) is fixed if the stator is connected directly to the grid, while the frequency of the rotor voltages and currents is variable (f_r) and depends on the speed of the machine according to expression (3.8). Consequently, the impedance $X'_{\sigma r}$ of the rotor will also be dependent on the speed of the machine.

Analyzing the stator and rotor separately, the steady state model electric equations may be derived as follows.

Stator

$$\underline{V}_s - \underline{E}_s = (R_s + X_{\sigma s})\underline{I}_s \text{ at } f_s \tag{3.9}$$

where

$\underline{V}_s$ = supplied stator voltage with frequency f_s (V_{rms})
$\underline{E}_s$ = induced emf in the stator windings with frequency f_s (V_{rms})
$\underline{I}_s$ = induced stator current with frequency f_s (A_{rms})
$X_{\sigma s} = j \cdot \omega_s \cdot L_{\sigma s}$ = stator leakage impedance (Ω)

Similarly, for the rotor circuit, we can obtain the needed equations.

Rotor

$$\underline{V}'_r - \underline{E}'_{rs} = (R'_r + X'_{\sigma r})\underline{I}'_r \text{ at } f_r \tag{3.10}$$

where

$\underline{E}'_{rs}$ = induced emf in the rotor with frequency f_r, due to the slip between the stator and rotor fields, that is, the induced voltage in the rotor (V_{rms}); later we show that this voltage depends on the speed of the rotor.
$\underline{V}'_r$ = supplied rotor voltage; its frequency should be f_r at steady state, that is, the same frequency as the induced $\underline{E}'_{rs}$ voltage in the rotor (V_{rms})
$\underline{I}'_r$ = induced rotor current with frequency f_r (A_{rms})
$X'_{\sigma r} = j\omega_r L'_{\sigma r}$ = rotor leakage impedance (Ω); or, according to expression (3.7), $X'_{\sigma r} = js\omega_s L'_{\sigma r}$

3.2.2.2 Equivalent Circuit Referring the Rotor to the Stator

In general, the equivalent circuit of Figure 3.1 together with the model equations (3.9) and (3.10) are used as a first step to derive the final steady state circuit of the DFIM. The fact that the stator and rotor equivalent circuits and their model equations operate at different frequencies (f_s and f_r) make them not very practical beyond analysis of the machine.

Consequently, a much more practical and useful approach is to transform the equivalent circuit of Figure 3.1 to an equivalent circuit in which all the rotor and stator currents and voltages operate at the same frequency.

Referring Rotor emf to the Stator   For that purpose, it is first necessary to find the relation between the induced emfs of the stator and rotor. Hence, according to the basic form of Lenz's law:

$$E = N\frac{d\phi}{dt} \qquad (3.11)$$

The induced voltage E depends on the flux ϕ variation and the number of turns N. Consequently, the induced voltage $\underline{E}_s$ in the stator windings is

$$\underline{E}_s = \sqrt{2}\pi K_s N_s f_s \underline{\phi}_m \quad (V_{rms}) \qquad (3.12)$$

where

$\underline{\phi}_m$ = magnetizing flux
K_s = stator winding factor; in general, slightly smaller than 1 due to the geometry of the machine

Similarly, in the rotor windings, due to the slip speed, the induced emf $\underline{E}'_{rs}$ is

$$\underline{E}'_{rs} = \sqrt{2}\pi K_r N_r f_r \underline{\phi}_m \quad (V_{rms}) \qquad (3.13)$$

where

K_r = rotor winding factor; in general, as occurs in the stator, slightly smaller than 1 due to the geometry of the machine

Hence, by equating Equations (3.12) and (3.13) and taking into account expression (3.8), the relation between the induced emf in the sator and rotor yields

$$\frac{\underline{E}'_{rs}}{\underline{E}_s} = s\frac{K_r N_r}{K_s N_s} \qquad (3.14)$$

We often define a constant factor, u, that relates the stator and rotor induced voltages at zero speed ($s = 1$):

$$\frac{1}{u} = \frac{K_r N_r}{K_s N_s} \qquad (3.15)$$

In general and particularly also in wind energy generation applications, the machines are specially designed, in such a way that the u factor is mostly defined by the stator and rotor turns ratio:

$$\frac{K_r}{K_s} \simeq 1, \ \Rightarrow u \simeq \frac{N_s}{N_r} \qquad (3.16)$$

Referring the Rotor Impedances, Currents, and Voltages to the Stator According the u Factor   In order to refer the rotor circuit to the stator, it is necessary to refer the parameters, currents, and voltages of the rotor to the stator. Before we consider the difference in frequencies between the rotor and the stator, the influence of the u factor will be taken into account first.

As in ideal transformer theory, the stator and rotor of a DFIM can equivalently be treated as the primary and secondary of a transformer, respectively. Hence, the impedances of the rotor, referred to the stator side, can be calculated as

$$R_r = R'_r u^2 \tag{3.17}$$

$$L_{\sigma r} = L'_{\sigma r} u^2 \tag{3.18}$$

where

R_r = rotor resistance referred to the stator due to the u factor

$L_{\sigma r}$ = rotor leakage inductance referred to the stator due to the u factor

We do the same thing for the rotor voltages and currents:

$$\underline{I}_r = \frac{\underline{I}'_r}{u} \tag{3.19}$$

$$\underline{V}_r = \underline{V}'_r u \tag{3.20}$$

$$\underline{E}_{rs} = \underline{E}'_{rs} u \tag{3.21}$$

where

$\underline{I}_r$ = rotor current, referred to the stator side due to the u factor

$\underline{V}_r$ = supplied rotor voltage, referred to the stator side due to the u factor

$\underline{E}_{rs}$ = induced emf in the rotor, referred to the stator side due to the u factor

Consequently, the equivalent steady state circuit of Figure 3.1 is transformed into the equivalent circuit of Figure 3.2. Note that the frequencies of the stator and rotor magnitudes are still different.

According to the adopted notation, the prime superscript (′) for a rotor magnitude or parameter means real or measurable. On the contrary, when there is no prime superscript on the parameter or magnitude, it means that it has been referred to the stator. For stator magnitudes or parameters this distinction is unnecessary.

Final Equivalent Circuit   Finally, the equivalent steady state stator referred circuit is derived. The steady state model equation of the rotor winding, from expression (3.10), is now expressed as

$$\underline{V}_r - \underline{E}_{rs} = (R_r + js\omega_s L_{\sigma r})\underline{I}_r \quad \text{at} f_r \tag{3.22}$$

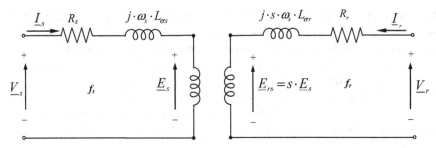

Figure 3.2 One-phase steady state equivalent electric circuit of the DFIM with different stator and rotor frequencies and rotor parameters, current, and voltages referred to the stator.

By substituting the relation between the stator and rotor emf, Equation (3.14), we find

$$\underline{V}_r - s\underline{E}_s = (R_r + js\omega_s L_{\sigma r})\underline{I}_r \text{ at } f_r \tag{3.23}$$

In order to transform this equation, to stator frequency, it must be divided by the slip, s:

$$\frac{\underline{V}_r}{s} - \underline{E}_s = \left(\frac{R_r}{s} + j\omega_s L_{\sigma r}\right)\underline{I}_r \text{ at } f_s \tag{3.24}$$

Consequently, this last expression represents one phase of the rotor winding entirely referred to the stator. Now, by combining it with Equation (3.9) of the stator winding:

$$\underline{V}_s - \frac{\underline{V}_r}{s} - (R_s + j\omega_s L_{\sigma s})\underline{I}_s + \left(\frac{R_r}{s} + j\omega_s L_{\sigma s}\right)\underline{I}_r = 0 \text{ at } f_s \tag{3.25}$$

It is now possible to derive the steady state equivalent circuit of the DFIM, completely referred to the stator, as illustrated in Figure 3.3.

The stator induced emf is equal to

$$\underline{E}_s = j\omega_s L_m(\underline{I}_s + \underline{I}_r) \text{ at } f_s \tag{3.26}$$

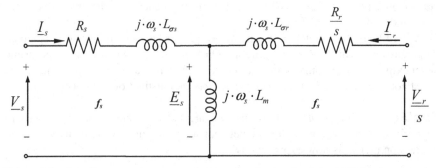

Figure 3.3 One-phase steady state equivalent electric circuit of the DFIM referred to the stator.

where L_m is the magnetizing inductance of the machine, normally measured at the stator side.

3.2.3 Phasor Diagram

From the steady state equivalent circuit of Figure 3.3, it is possible to derive the phasor diagram of the machine for specific operating conditions. In this subsection, we show how the phasor diagram is deduced, by using the model equations of the machine. For that purpose, from the stator and rotor currents, the stator and rotor fluxes are calculated according to the following expressions:

$$\underline{\Psi}_s = L_m(\underline{I}_s + \underline{I}_r) + L_{\sigma s}\underline{I}_s = L_s\underline{I}_s + L_m\underline{I}_r \tag{3.27}$$

$$\underline{\Psi}_r = L_m(\underline{I}_s + \underline{I}_r) + L_{\sigma r}\underline{I}_r = L_m\underline{I}_s + L_r\underline{I}_r \tag{3.28}$$

where L_s and L_r are the stator and rotor inductances:

$$L_s = L_m + L_{\sigma s} \tag{3.29}$$

$$L_r = L_m + L_{\sigma r} \tag{3.30}$$

Hence, substituting the fluxes into the voltage expressions (3.9) and (3.24), new voltage equations related to the fluxes can be derived:

$$\underline{V}_s - R_s\underline{I}_s = j\omega_s\underline{\Psi}_s \tag{3.31}$$

$$\underline{V}_r - R_r\underline{I}_r = js\omega_s\underline{\Psi}_r \tag{3.32}$$

Thus, by using expressions (3.27) to (3.32), the phasor diagram can be derived. Note that all these equations are referred to the stator and frequency f_s. Figure 3.4 shows the phasor diagram derivation procedure, for a DFIM operating in motor mode, with $Q_s = 0$ (Section 3.3.3 analyzes how to derive it) at subsynchronous speed.

Assuming that the parameters of the machine are known, we have the following steps:

Step a. Since the machine is operating as a motor with $Q_s = 0$, the stator voltage and current phasors are placed (supposed known amplitudes).

Step b. From expression (3.31), the stator flux phasor is placed.

Step c. From expression (3.27), the rotor current phasor is placed.

Step d. Once the rotor current is known, with expression (3.28), the rotor flux phasor is derived. Note that, in this example, the amplitude of the rotor phasor is greater than the stator flux's amplitude.

Step e. Once the rotor flux is placed, from expression (3.32), finally the rotor voltage phasor is derived. Note that at $s > 0$ or at $s < 0$, the location of the rotor voltage vector would be modified.

Step f. The final phasor diagram is shown.

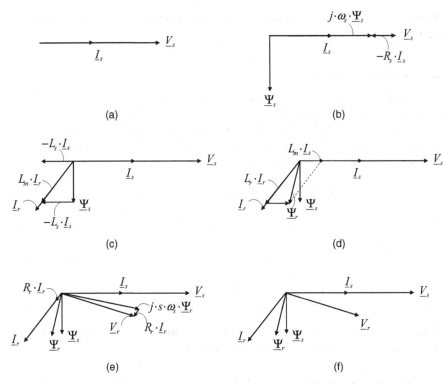

Figure 3.4   Phasor diagram of a DFIM operating as a motor, with $Q_s = 0$ at subsynchronous speed, following steps (a) through (f).

If the phase shift angles between the phasors are noted according to Figure 3.5, it is possible to distinguish five different phase angles:

δ: Phase shift between the stator and rotor flux phasors

γ_v: Phase shift between the stator voltage and stator flux phasors

γ_i: Phase shift between the stator current and stator flux phasors

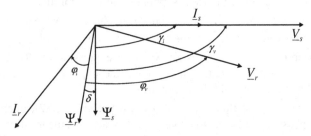

Figure 3.5   Phasor diagram of a DFIM operating as a motor, with $Q_s = 0$ at subsynchronous speed.

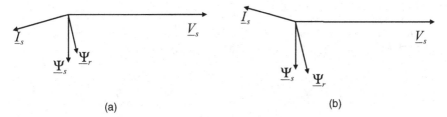

Figure 3.6 Phasor diagram in generator mode: (a) $Q_s > 0$ and (b) $Q_s < 0$.

φ_v: Phase shift between the rotor voltage and rotor flux phasors

φ_i: Phase shift between the rotor current and rotor flux phasors

For this particular case, the resulting angles have the following meanings:

δ: Stator flux leading the rotor flux implies motoring operation of the machine.

γ_v: Always very close to 90°.

γ_i: Since it is equal to γ_v the stator reactive power is zero, $Q_s = 0$, and positive stator active power, $P_s > 0$ (the grid is providing power to the stator).

φ_v: Always very close to 90° (leading $\underline{V}_r$ to $\underline{\Psi}_r$ at subsynchronous speed, lagging at hypersynchronous speed).

φ_i: Since $\varphi_v + \varphi_i > 90°$, leading $\underline{V}_r$ to $\underline{I}_r$, the rotor active power is negative, $P_r < 0$ (the rotor is absorbing power) and $Q_r > 0$ (inductive, the DFIM is magnetized through the rotor).

Finally, Figure 3.6 shows the phasor diagrams of two typical operating modes, when the DFIM operates in wind energy generation; generating ($P_s < 0$) with lagging ($Q_s > 0$) and leading ($Q_s < 0$) power factors. Notice that the relative position between the two stator and rotor fluxes is not altered, but the amplitude relations change.

3.3 OPERATION MODES ATTENDING TO SPEED AND POWER FLOWS

3.3.1 Basic Active Power Relations

From the steady state equivalent circuit of Figure 3.3, it is possible to study the phasor diagrams of the machine as shown in the previous subsection. It is also important to analyze the power relations of the DFIM, since it would allow us to study different operation modes of this machine. Hence, in this subsection, the basic power and torque expressions are derived from the steady state equivalent circuit. However, for that purpose, it is more suitable to modify the equivalent circuit of Figure 3.3, to the slightly modified equivalent circuit illustrated in Figure 3.7.

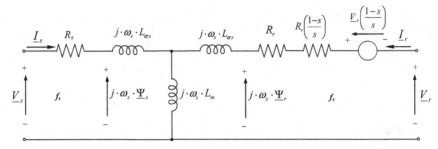

Figure 3.7 Modified one-phase steady state equivalent electric circuit of the DFIM referred to the stator.

The only modification consists of redefining the rotor resistance and rotor voltage terms, to two new terms—one nondependent on the slip and the other dependent:

$$\frac{R_r}{s} \Rightarrow R_r + R_r\left(\frac{1-s}{s}\right) \tag{3.33}$$

$$\frac{V_r}{s} \Rightarrow \underline{V}_r + \underline{V}_r\left(\frac{1-s}{s}\right) \tag{3.34}$$

In this way, it is easy to know which electric element is responsible for power losses and which element is responsible for generating mechanical power. Hence, the three phase active power losses of the machine are given in terms of the stator and rotor resistances:

$$P_{cu\_s} = 3R_s|\underline{I}_s|^2 \tag{3.35}$$

$$P_{cu\_r} = 3R_r|\underline{I}_r|^2 \tag{3.36}$$

Note that these copper losses always take a positive sign. Connected to this, the active power balance of the machine can be represented graphically as depicted in Figure 3.8.

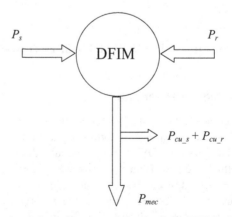

Figure 3.8 Schematic of the active power balance of the DFIM with motor convention.

Mathematically,

$$P_s + P_r = P_{cu\_s} + P_{cu\_r} + P_{mec} \qquad (3.37)$$

where

P_s = stator active power (W)

$$P_s = 3\mathrm{Re}\{\underline{V}_s \cdot \underline{I}_s^*\} = 3|\underline{V}_s| \cdot |\underline{I}_s|\cos(\gamma_v - \gamma_i) \qquad (3.38)$$

If $P_s > 0 \Rightarrow$ the machine is receiving power through the stator.
If $P_s < 0 \Rightarrow$ the machine is delivering power through the stator.

where

P_r = rotor active power (W)

$$P_r = 3\mathrm{Re}\{\underline{V}_r \cdot \underline{I}_r^*\} = 3|\underline{V}_r| \cdot |\underline{I}_r|\cos(\varphi_v - \varphi_i) \qquad (3.39)$$

If $P_r > 0 \Rightarrow$ the machine is receiving power through the rotor.
If $P_r < 0 \Rightarrow$ the machine is delivering power through the rotor.

where

P_{mec} = mechanical power (W), that is, the power transmitted though the electrical circuit to the mechanical shaft

The sign of this power defines the operation of the machine as a motor or a generator.

If $P_{mec} > 0 \Rightarrow$ the machine is delivering power through the shaft: MOTORING.
If $P_{mec} < 0 \Rightarrow$ the machine is receiving power through the shaft: GENERATING.

This convention is called the "motor convention." It gives a positive sign to the motor operation. When the mechanical power losses are neglected, the mechanical power gives us the next relation:

$$P_{mec} = T_{em}\Omega_m = T_{em}\frac{\omega_m}{p} \qquad (3.40)$$

where

T_{em} = electromagnetic torque in the shaft of the machine

Consequently, the mechanical power can be calculated from the electrical circuit in two different ways.

1. *Directly from Expression. (3.37).*

$$P_{mec} = P_s + P_r - P_{cu\_s} - P_{cu\_r} \tag{3.41}$$

yielding

$$P_{mec} = 3|\underline{V}_s| \cdot |\underline{I}_s|\cos(\gamma_v - \gamma_i) + 3|\underline{V}_r| \cdot |\underline{I}_r|\cos(\varphi_v - \varphi_i)$$
$$-3R_s \cdot |\underline{I}_s|^2 - 3R_r|\underline{I}_r|^2 \tag{3.42}$$

2. *From the Equivalent Circuit of Figure 3.7.* Since the inductances are only responsible for the reactive power, the elements that contain the slip term (*s*) are associated with the mechanical power:

$$P_{mec} = 3R_r\left(\frac{1-s}{s}\right)|\underline{I}_r|^2 - 3\left(\frac{1-s}{s}\right) \cdot \text{Re}\left(\underline{V}_r \cdot \underline{I}_r^*\right) \tag{3.43}$$

Finally, the efficiency of the machine can be defined as

$$\eta = \frac{P_{mec}}{P_s + P_r} \quad \text{if } P_{mec} > 0$$

$$\eta = \frac{P_s + P_r}{P_{mec}} \quad \text{if } P_{mec} < 0 \tag{3.44}$$

3.3.2 Torque Expressions

Accordingly, from expressions (3.43) and (3.40), the electromagnetic torque can be calculated by means of the electric magnitudes:

$$T_{em} = \frac{3pR_r}{\omega_m}\left(\frac{1-s}{s}\right)|\underline{I}_r|^2 - \frac{3p}{\omega_m}\left(\frac{1-s}{s}\right) \cdot \text{Re}\left(\underline{V}_r \cdot \underline{I}_r^*\right) \tag{3.45}$$

or more simply by taking into account that

$$\left(\frac{1-s}{s}\right) = \frac{\omega_m}{\omega_r}$$

yielding

$$T_{em} = \frac{3pR_r}{\omega_r}|\underline{I}_r|^2 - \frac{3p}{\omega_r}|\underline{V}_r| \cdot |\underline{I}_r|\cos(\varphi_v - \varphi_i) \tag{3.46}$$

In addition, there is a very useful alternative approach for calculating the torque. It consist of deriving the power expression by making the voltages disappear. Hence, from the steady state equivalent circuit of Figure 3.3, the stator voltage can be expressed as

$$\underline{V}_s = (R_s + j\omega_s L_{\sigma s})\underline{I}_s + j\omega_s L_m(\underline{I}_s + \underline{I}_r) \tag{3.47}$$

Substituting this into the stator active power expression (3.38), we find

$$
\begin{aligned}
P_s &= 3\mathrm{Re}\{\underline{V}_s\underline{I}_s^*\} \\
&= 3\mathrm{Re}\{R_s\underline{I}_s \cdot \underline{I}_s^* + j\omega_sL_{\sigma s}\underline{I}_s \cdot \underline{I}_s^* + j\omega_sL_m(\underline{I}_s + \underline{I}_r) \cdot \underline{I}_s^*\} \\
&= 3\mathrm{Re}\{R_s|\underline{I}_s|^2 + j\omega_sL_m\underline{I}_r \cdot \underline{I}_s^*\} \\
&= 3R_s|\underline{I}_s|^2 + 3\mathrm{Re}\{j\omega_sL_m\underline{I}_r \cdot \underline{I}_s^*\}
\end{aligned}
\tag{3.48}
$$

Equivalently, with the rotor voltage expression (3.32) combined with (3.28), we find

$$
\underline{V}_r = (R_r + js\omega_sL_{\sigma r})\underline{I}_r + js\omega_sL_m(\underline{I}_s + \underline{I}_r)
\tag{3.49}
$$

The rotor active power expression (3.39) yields

$$
\begin{aligned}
P_r &= 3\mathrm{Re}\{\underline{V}_r \cdot \underline{I}_r^*\} \\
&= 3\mathrm{Re}\{R_r\underline{I}_r \cdot \underline{I}_r^* + js\omega_sL_{\sigma r}\underline{I}_r \cdot \underline{I}_r^* + js\omega_sL_m(\underline{I}_s + \underline{I}_r) \cdot \underline{I}_r^*\} \\
&= 3\mathrm{Re}\{R_r|\underline{I}_r|^2 + js\omega_sL_m\underline{I}_s \cdot \underline{I}_r^*\} \\
&= 3R_r|\underline{I}_r|^2 + 3\mathrm{Re}\{js\omega_sL_m\underline{I}_s \cdot \underline{I}_r^*\}
\end{aligned}
\tag{3.50}
$$

Hence, by substituting these last two expressions, (3.48) and (3.50), into expression (3.41), the new mechanical power equation yields

$$
\begin{aligned}
P_{mec} &= P_s + P_r - P_{cu\_s} - P_{cu\_r} \\
&= 3\mathrm{Re}\{j\omega_sL_m\underline{I}_r \cdot \underline{I}_s^*\} + 3\mathrm{Re}\{js\omega_sL_m\underline{I}_s \cdot \underline{I}_r^*\} \\
&= -3\omega_sL_m(1 - s) \cdot \mathrm{Im}\{\underline{I}_r \cdot \underline{I}_s^*\} \\
&= -3\omega_mL_m \cdot \mathrm{Im}\{\underline{I}_r \cdot \underline{I}_s^*\} = 3\omega_mL_m \cdot \mathrm{Im}\{\underline{I}_r^* \cdot \underline{I}_s\}
\end{aligned}
\tag{3.51}
$$

Finally, based on expression (3.40), the electromagnetic torque yields

$$
T_{em} = 3pL_m \cdot \mathrm{Im}\{\underline{I}_r^* \cdot \underline{I}_s\}
\tag{3.52}
$$

As mentioned in Section 3.2.1, $\underline{I}_r$ and $\underline{I}_s$ phasors are defined as rms amperes.

Note that if instead of the currents we use the fluxes in the torque expression, we can achieve several equivalent torque equations:

$$
\begin{aligned}
T_{em} &= 3p\frac{L_m}{L_s} \cdot \mathrm{Im}\{\underline{\Psi}_s \cdot \underline{I}_r^*\} = 3p \cdot \mathrm{Im}\{\underline{\Psi}_s^* \cdot \underline{I}_s\} = 3p \cdot \mathrm{Im}\{\underline{\Psi}_r \cdot \underline{I}_r^*\} \\
&= 3\frac{L_m}{L_r}p \cdot \mathrm{Im}\{\underline{\Psi}_r^* \cdot \underline{I}_s\} = 3\frac{L_m}{\sigma L_rL_s}p \cdot \mathrm{Im}\{\underline{\Psi}_r^* \cdot \underline{\Psi}_s\}
\end{aligned}
\tag{3.53}
$$

where $\sigma = 1 - L_m^2/L_sL_r$ is the leakage coefficient.

3.3.3 Reactive Power Expressions

As we did for the active powers, in this subsection the stator and rotor reactive power expressions are derived. Similar to expression (3.48), the stator reactive power can be calculated as

$$
\begin{aligned}
Q_s &= 3\mathrm{Im}\{\underline{V}_s \cdot \underline{I}_s^*\} \\
&= 3\mathrm{Im}\{R_s \cdot \underline{I}_s \cdot \underline{I}_s^* + j\omega_s L_{\sigma s} \underline{I}_s \cdot \underline{I}_s^* + j\omega_s L_m (\underline{I}_s + \underline{I}_r) \cdot \underline{I}_s^*\} \\
&= 3\mathrm{Im}\{j\omega_s L_s |\underline{I}_s|^2 + j\omega_s L_m \underline{I}_r \cdot \underline{I}_s^*\} \\
&= 3\omega_s L_s |\underline{I}_s|^2 + 3\mathrm{Im}\{j\omega_s L_m \underline{I}_r \cdot \underline{I}_s^*\} \\
&= 3\omega_s L_s |\underline{I}_s|^2 + 3\omega_s L_m \cdot \mathrm{Re}\{\underline{I}_r \cdot \underline{I}_s^*\}
\end{aligned}
\tag{3.54}
$$

Equivalently, for the rotor side reactive power:

$$
\begin{aligned}
Q_r &= 3\mathrm{Im}\{\underline{V}_r \cdot \underline{I}_r^*\} \\
&= 3\mathrm{Im}\{R_r \underline{I}_r \cdot \underline{I}_r^* + js\omega_s L_{\sigma r} \underline{I}_r \cdot \underline{I}_r^* + js\omega_s L_m (\underline{I}_s + \underline{I}_r) \cdot \underline{I}_r^*\} \\
&= 3\mathrm{Im}\{js\omega_s L_r |\underline{I}_r|^2 + js\omega_s L_m \underline{I}_s \cdot \underline{I}_r^*) \\
&= 3s\omega_s L_r |\underline{I}_r|^2 + 3\mathrm{Im}\{js\omega_s L_m \underline{I}_s \cdot \underline{I}_r^*\} \\
&= 3s\omega_s L_r |\underline{I}_r|^2 + 3s\omega_s L_m \cdot \mathrm{Re}\{\underline{I}_s \cdot \underline{I}_r^*\}
\end{aligned}
\tag{3.55}
$$

3.3.4 Approximated Relations Between Active Powers, Torque, and Speeds

From expressions (3.48) and (3.50), by neglecting the copper power losses in the stator and rotor resistances, the relation between the stator and rotor power is immediate:

$$
P_r \cong -sP_s
\tag{3.56}
$$

Hence, from expression (3.41), the mechanical power can also be expressed as

$$
P_{mec} \cong P_s - sP_s = (1-s)P_s
\tag{3.57}
$$

On the other hand, by substituting Equation (3.40), we find

$$
P_{mec} \cong \frac{\omega_m}{\omega_s} P_s \cong T_{em} \frac{\omega_m}{p}
\tag{3.58}
$$

Consequently, the relation between the torque and the stator power yields

$$
P_s \cong T_{em} \frac{\omega_s}{p}
\tag{3.59}
$$

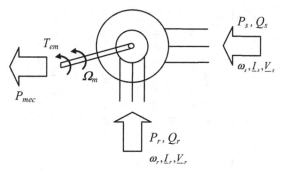

Figure 3.9 Schematic of the electric and mechanical inputs and outputs of the DFIM.

Similarly, it easy to deduce that

$$P_r \cong T_{em} \frac{\omega_r}{p} \tag{3.60}$$

3.3.5 Four Quadrant Modes of Operation

From the power relations (3.56)–(3.60) of the previous subsection, it is easy to see that the DFIM can operate under different conditions depending on the power and the speed. Figure 3.9 shows a simplified scheme of the power flow.

Table 3.1 shows the four possible combinations. Note that only in generator mode at hypersynchronism is it possible to deliver power through both the stator and rotor side to the grid. In an equivalent way, Figure 3.10 graphically illustrates the four quadrant operation modes.

For instance, Figure 3.11 shows the power balance of the DFIM at constant electromagnetic torque in operation modes 1 and 4, that is, motoring at subsynchronism and hypersynchronism.

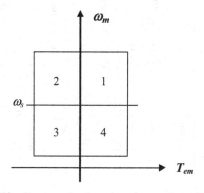

Figure 3.10 Four quadrant modes of operation of the DFIM.

TABLE 3.1 Four Modes of Operation attending to the speed and powers

	Mode	Speed	P_{mec}	P_s	P_r
1	Motor $(T_{em} > 0)$	$s < 0$ $(\omega_m > \omega_s)$ (hypersynchronism)	> 0 (the machine delivers mechanical power)	> 0 (the machine receives power through stator)	> 0 (the machine receives power through rotor)
2	Generator $(T_{em} < 0)$	$s < 0$ $(\omega_m > \omega_s)$ (hypersynchronism)	< 0 (the machine receives mechanical power)	< 0 (the machine delivers power through stator)	< 0 (the machine delivers power through rotor)
3	Generator $(T_{em} < 0)$	$s > 0$ $(\omega_m < \omega_s)$ (subsynchronism)	< 0 (the machine receives mechanical power)	< 0 (the machine delivers power through stator)	> 0 (the machine receives power through rotor)
4	Motor $(T_{em} > 0)$	$s > 0$ $(\omega_m < \omega_s)$ (subsynchronism)	> 0 (the machine delivers mechanical power)	> 0 (the machine receives power through stator)	< 0 (the machine delivers power through rotor)

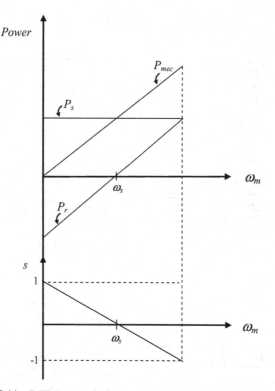

Figure 3.11   DFIM power balance at constant electromagnetic torque.

Finally, Figure 3.12 summarizes the active power flow schematic as well as the phasor diagrams of all the operating modes of the DFIM. Positive stator reactive power consumption ($Q_s > 0$) has been chosen for the phasor diagram representations. The reader must pay special attention to the relative positions between phasors at each operating mode.

3.4   PER UNIT TRANSFORMATION

In some cases, it is useful to express the model of the DFIM in per unit (p.u.) as we did in Chapter 2 with the grid side system. In general, the machine's parameters as well as the magnitudes (currents, voltages, fluxes, torque, powers, etc.) can be transformed into p.u. with the following advantages:

- Eases the comparison between different machines' parameters and magnitudes.
- The magnitudes expressed in p.u. automatically provide information on how far it is from the base or rated values, thus avoiding the necessity of knowing the nominal value at first glance.

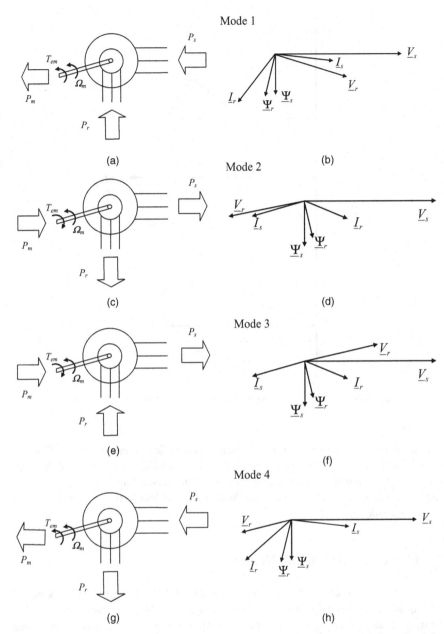

Figure 3.12   Active power flow representations and phasor diagrams (with $Q_s > 0$) of the DFIM at different modes of operation: (a, b) mode 1, (c, d) mode 2, (e, f) mode 3, (g, h) mode 4.

Hence, this section shows the p.u. transformation of the machine parameters and magnitudes, as well as the machinés alternative p.u. steady state model equations.

3.4.1 Base Values

As presented in the previous chapter when the grid side system was studied, any magnitude of the machine can be normalized by dividing that magnitude by a base value. For example,

$$Voltage\ (V)$$

$$Voltage_{pu} = Voltage/V_{base}\ (no\ units)$$

In this chapter also, this normalization is called the p.u. transformation. The p.u. values of magnitudes are accompanied by the subscript "pu," as was done in Chapter 2 with the grid side system. From the different base choice possibilities, this analysis selects *stator rated rms values of voltages and currents*:

$$\text{Base voltage: } V_{base} = |\underline{V}_s|_{rated}\ (V) \tag{3.61}$$

$$\text{Base current: } I_{base} = |\underline{I}_s|_{rated}\ (A) \tag{3.62}$$

$$\text{Base angular frequency: } \omega_{base} = \omega_{s\ rated}\ (rad/s) \tag{3.63}$$

where

$|\underline{V}_s|_{rated}$ = line to neutral rated rms stator voltage (V_{rms})
$|\underline{I}_s|_{rated}$ = rated rms stator current (A_{rms})
$\omega_{s\ rated}$ = rated stator angular frequency (rad/s)

The rms is chosen instead of the peak values, basically because the p.u. transformation is going to be applied to phasors expressed in rms values in this chapter. If phasors were expressed in peak values, the base values accordingly could be chosen as peak values. From these three basic base values, it is possible to derive the rest of the base values.

$$\text{Base flux: } \Psi_{base} = \frac{V_{base}}{\omega_{base}} \tag{3.64}$$

$$\text{Base impedance: } Z_{base} = \frac{V_{base}}{I_{base}} \tag{3.65}$$

$$\text{Base inductance: } L_{base} = \frac{\Psi_{base}}{I_{base}} \tag{3.66}$$

$$\text{Base power: } S_{base} = 3V_{base}I_{base} \tag{3.67}$$

$$\text{Base speed: } \Omega_{m\,base} = \frac{\omega_{base}}{p} \tag{3.68}$$

$$\text{Base torque: } T_{base} = \frac{S_{base}}{\Omega_{m\,base}} \tag{3.69}$$

3.4.2 Per Unit Transformation of Magnitudes and Parameters

Consequently, from the base values, it is possible to derive the parameters in p.u., according to the following expressions:

$$\text{Stator voltage in p.u.: } \underline{V}_{s\,pu} = \frac{\underline{V}_s}{V_{base}} \tag{3.70}$$

$$\text{Stator current in p.u.: } \underline{I}_{s\,pu} = \frac{\underline{I}_s}{I_{base}} \tag{3.71}$$

$$\text{Stator flux in p.u.: } \underline{\Psi}_{s\,pu} = \frac{\underline{\Psi}_s}{\Psi_{base}} \tag{3.72}$$

$$\text{Stator active power in p.u.: } P_{s\,pu} = 3\frac{\mathrm{Re}\{\underline{V}_s \cdot \underline{I}_s^*\}}{S_{base}} \tag{3.73}$$

$$\text{Stator reactive power in p.u.: } Q_{s\,pu} = 3\frac{\mathrm{Im}\{\underline{V}_s \cdot \underline{I}_s^*\}}{S_{base}} \tag{3.74}$$

Note that only the moduli of the phasors is transformed; that is, the angle is unaltered when transforming to p.u. Similarly, for the rotor, we have the following expressions:

$$\text{Rotor voltage in p.u.: } \underline{V}_{r\,pu} = \frac{\underline{V}_r}{V_{base}} = \frac{\underline{V}'_r \cdot u}{V_{base}} \tag{3.75}$$

$$\text{Rotor current in p.u.: } \underline{I}_{r\,pu} = \frac{\underline{I}_r}{I_{base}} = \frac{\underline{I}'_r}{I_{base} \cdot u} \tag{3.76}$$

$$\text{Rotor flux in p.u.: } \underline{\Psi}_{r\,pu} = \frac{\underline{\Psi}_r}{\Psi_{base}} \tag{3.77}$$

$$\text{Rotor active power in p.u.: } P_{r\,pu} = 3\frac{\mathrm{Re}\{\underline{V}_r \cdot \underline{I}_r^*\}}{S_{base}} \tag{3.78}$$

$$\text{Rotor reactive power in p.u.: } Q_{r\,pu} = 3\frac{\mathrm{Im}\{\underline{V}_r \cdot \underline{I}_r^*\}}{S_{base}} \tag{3.79}$$

Note that for the rotor voltage and currents, the p.u. values are derived from the stator referred voltages and currents.

Therefore, for the torque and speed, we find the following:

$$\text{Torque in p.u.:}\quad T_{em\,pu} = \frac{T_{em}}{T_{base}} \tag{3.80}$$

$$\text{Speed in p.u.:}\quad \Omega_{m\,pu} = \frac{\Omega_m}{\Omega_{base}} \tag{3.81}$$

Finally, the parameters of the machine are transformed to p.u. as follows:

$$\text{Resistances in p.u.:}\quad r = \frac{R}{Z_{base}} = \frac{RI_{base}}{V_{base}} \tag{3.82}$$

$$\text{Inductances in p.u.:}\quad l = \frac{L}{L_{base}} = \frac{LI_{base}}{V_{base}}\omega_{base} \tag{3.83}$$

Note that the parameters transformed into p.u. are written in lowercase. Both stator and rotor parameters are transformed to p.u. using the same base, as occurs with the magnitudes.

3.4.3 Steady State Equations of the DFIM in p.u

By considering the p.u. transformation studied in the previous two subsections, it is also possible to transform the steady state equations of the DFIM into p.u. First, we consider the stator voltage in Equation (3.84):

$$\underline{V}_s = R_s\underline{I}_s + j\omega_s\underline{\Psi}_s \tag{3.84}$$

We can transform the voltage, current, and flux to p.u. as follows:

$$\underline{V}_{s\,pu}V_{base} = R_s\underline{I}_{s\,pu}I_{base} + j\omega_s\underline{\Psi}_{s\,pu}\cdot\Psi_{base} \tag{3.85}$$

This expression is equivalent to

$$\underline{V}_{s\,pu} = R_s\frac{I_{base}}{V_{base}}\underline{I}_{s\,pu} + j\omega_s\underline{\Psi}_{s\,pu}\frac{V_{base}}{\omega_{base}}\cdot\frac{1}{V_{base}} \tag{3.86}$$

Finally, the p.u. expression yields

$$\underline{V}_{s\,pu} = r_s\underline{I}_{s\,pu} + j\underline{\Psi}_{s\,pu} \tag{3.87}$$

Equivalently, from the rotor voltage, Equation (3.88), the p.u. equation yields

$$\underline{V}_{r\,pu} = r_r\underline{I}_{r\,pu} + js\underline{\Psi}_{r\,pu} \tag{3.88}$$

In a similar manner, with the fluxes it is possible also to derive the p.u. expressions. From the stator flux expression (3.27):

$$\underline{\Psi}_s = L_s \underline{I}_s + L_m \underline{I}_r \tag{3.89}$$

Transforming the fluxes and currents to p.u.:

$$\underline{\Psi}_{s\,pu} \frac{V_{base}}{\omega_{base}} = L_s I_{base} \underline{I}_{s\,pu} + L_m I_{base} \underline{I}_{r\,pu} \tag{3.90}$$

This expression is equivalent to

$$\underline{\Psi}_{s\,pu} = L_s \frac{I_{base}}{V_{base}} \omega_{base} \underline{I}_{s\,pu} + L_m I_{base} \frac{I_{base}}{V_{base}} \omega_{base} \underline{I}_{r\,pu} \tag{3.91}$$

Therefore, the p.u. expression yields

$$\underline{\Psi}_{s\,pu} = l_s \underline{I}_{s\,pu} + l_m \underline{I}_{r\,pu} \tag{3.92}$$

Equivalently, for the rotor flux, from expression (3.28):

$$\underline{\Psi}_r = L_m \underline{I}_s + L_r \underline{I}_r \tag{3.93}$$

The rotor flux p.u. expression yields

$$\underline{\Psi}_{r\,pu} = l_m \underline{I}_{s\,pu} + l_r \underline{I}_{r\,pu} \tag{3.94}$$

On the other hand, for the electromagnetic torque in p.u., from one of the valid expressions (3.53):

$$T_{em\,pu} = \frac{3p \mathrm{Im}\{\underline{\Psi}_r \cdot \underline{I}_r^*\}}{T_{base}} \tag{3.95}$$

Is equivalent to

$$T_{em\,pu} = \frac{3p \mathrm{Im}\{\underline{\Psi}_r \cdot \underline{I}_r^*\}}{3 V_{base} I_{base} \frac{p}{\omega_{base}}} \tag{3.96}$$

So the torque in p.u. yields

$$T_{em\,pu} = \mathrm{Im}\left\{\underline{\Psi}_{r\,pu} \cdot \underline{I}_{r\,pu}^*\right\} \tag{3.97}$$

Finally, for the power expression in p.u., from Equations (3.73), (3.74), (378) and (3.79):

$$\begin{aligned}
P_{s\,pu} &= \frac{3|\underline{V}_s||\underline{I}_s|}{3 V_{base} I_{base}} \cos(\gamma_v - \gamma_i) \\
&= |\underline{V}_{s\,pu}||\underline{I}_{s\,pu}| \cos(\gamma_v - \gamma_i)
\end{aligned} \tag{3.98}$$

$$Q_{s\,pu} = |\underline{V}_{s\,pu}||\underline{L}_{s\,pu}|\sin(\gamma_v - \gamma_i) \qquad (3.99)$$

$$P_r = |\underline{V}_{r\,pu}||\underline{L}_{r\,pu}|\cos(\varphi_v - \varphi_i) \qquad (3.100)$$

$$Q_r = |\underline{V}_{r\,pu}||\underline{L}_{r\,pu}|\sin(\varphi_v - \varphi_i) \qquad (3.101)$$

Therefore, the following set of three examples applies some of the basics we have seen up to now in this chapter.

3.4.4 Example 3.1: Parameters of a 2 MW DFIM

Table 3.2 shows the characteristics and parameters of a 2 MW DFIM. All these characteristics and parameter values are subject to how the manufacturer has designed the machine; however, it is typical to find some basic common features among the brands.

The rated active power of the machine is often associated with the rated generated stator active power. Note that the machine can be generating or motoring more active power than the rated power, depending on the speed or slip ($P_s + P_r$). In essence, the rated power can be understood as the rated power of the machine at synchronous speed, that is, with $P_r = 0$ at $s = 0$.

TABLE 3.2 Multi-MW DFIM Characteristics

Characteristic	Value	Features
Synchronism	1500 rev/min	Synchronous speed at 50 Hz
Rated power	2 MW	Nominal stator three-phase active power
Rated stator voltage	690 V$_{rms}$	Line-to-line nominal stator voltage in rms
Rated stator current	1760 A$_{rms}$	Each phase nominal stator current in rms
Rated torque	12.7 k·Nm	Nominal torque at generator or motor modes
Stator connection	Star	
p	2	Pair of poles
Rated rotor voltage	2070 V$_{rms}$	Line-to-line nominal rotor voltage in rms (reached at speed near zero)
Rotor connection	Star	
u	0.34	
R_s	2.6 mΩ	Stator resistance
$L_{\sigma s}$	87 μH	Stator leakage inductance
L_m	2.5 mH	Magnetizing inductance
R'_r	26.1 mΩ	Rotor resistance
$L'_{\sigma r}$	783 μH	Rotor leakage inductance
R_r	2.9 mΩ	Rotor resistance referred to the stator
$L_{\sigma r}$	87 μH	Rotor leakage inductance referred to the stator
L_s	2.587 mH	Stator inductance: $L_s = L_m + L_{\sigma s}$
L_r	2.587 mH	Rotor inductance: $L_r = L_m + L_{\sigma r}$

However, when we are speaking about a wind turbine based on the DFIM, the rated active power is associated with the total rated power ($P_{mec} = P_s + P_r$) that can be generated by both the stator and rotor. This point is located at the maximum allowed slip at hypersynchronism (typically $s = \pm 0.3$ or less). Thus, a 2 MW DFIM, with a back-to-back converter prepared to operate at a maximum of $s = 0.3$, for instance, would lead to a wind turbine of

$$P_{mec} = P_s + P_r \cong P_s \cdot (1 - s) \xrightarrow{s = -0.3} P_{mec} \cong 2(1 + 0.3) \cong 2.6 \text{ MW}$$

On the other hand, the rated stator current is associated with the thermal limit of the machine, that is, the current that would result in the rated temperature rise. For this machine, the exchange of the rated stator power at the rated stator current corresponds to a stator power factor equal to 0.95:

$$P_s = \sqrt{3} V_{rated} I_{rated} \cos\phi \rightarrow \frac{2000000}{\sqrt{3} \cdot 690 \cdot 1760} \cong 0.95$$

Since this machine presents a factor $u = 0.34$, the nominal voltage of the real rotor voltage is nearly three times bigger than the stator voltage. This amplitude of voltage for the rotor voltage is reached at $s = \pm 1$, in this case, near zero speed ($s = 1$) or 3000 rev/min ($s = -1$).

Nevertheless, in a typical wind turbine, the speed of the machine never ranges too far away from the synchronous speed (typically $s = \pm 0.3$ or less), so the rotor voltage never reaches those high voltage amplitudes (expression (3.32)), thus reducing the scale of the power electronic converter in terms of the required voltage amplitude, and thus operating at similar voltage amplitudes as the grid/stator voltages.

In addition, because this typical slip range operation ($s = \pm 0.3$) implies that the rated rotor power exchange is $\pm 30\%$ of the rated stator power, this leads to a rated rotor current smaller than the rated stator current, thus also reducing the scale of the power electronic converter in terms of current amplitude (see subsequent Example 3.3, for detailed analysis).

3.4.5 Example 3.2: Parameters of Different Power DFIM

Table 3.3 shows the real characteristics and parameters of several doubly fed induction machines of different powers. The machines presented have different power ranges, voltages, and u factors.

In units, we can see how the resistances get smaller as the machine gets larger. The same occurs with the leakage inductances and magnetizing inductances. However, when the parameters are expressed in p.u. according to the base values choices presented in the previous section (rated stator voltages and currents), this tendency seems to be inverted for the inductances (magnetizing and leakages). However, as mentioned before, these tendencies depend on the manufacturers' designs. Thus, for instance, the reader can see that the 250 kW machine shown in this example does not follow the p.u. parameter tendency of leakage inductances and resistances.

TABLE 3.3   Different Power DFIM Characteristics

Characteristic	5 kW	15 kW	250 kW	2 MW
Synchronous speed (rev/min)	1500	1500	1500	1500
Rated power (kW)	5	15	250	2000
Rated line-to-line stator voltage (V_{rms})	380	380	400	690
Rated stator current (A_{rms})	8.36	32	370	1760
Rated torque (N·m)	31.8	95.5	1591	12732
Stator connection	Star	Star	Star	Star
p	2	2	2	2
Rated $\underline{V}_r$ (V_{rms})	205	380	400	2070
Rotor connection	Star	Star	Star	Star
u	0.54	1	1	0.34
R_s (mΩ)	720	161	20	2.6
$L_{\sigma s}$ (mH)	5.8	3	0.2	0.087
L_m (mH)	85.8	46.5	4.2	2.5
R'_r (mΩ)	2566	178	20	26.1
$L'_{\sigma r}$ (mH)	19.85	3	0.2	0.783
R_r (mΩ)	750	178	20	2.9
$L_{\sigma r}$ (mH)	6	3	0.2	0.087
L_s (mH)	91.6	49.5	4.4	2.587
L_r (mH)	91.6	49.5	4.4	2.587
V_{base}	220	220	231	398.4
I_{base}	8.36	32	370	1760
r_s	0.027	0.023	0.032	0.011
$l_{\sigma s}$	0.069	0.137	0.1	0.12
l_m	0.976	2.12	2.11	3.45
r_r	0.028	0.025	0.032	0.012
$l_{\sigma r}$	0.071	0.137	0.1	0.12

3.4.6   Example 3.3: Phasor Diagram of a 2 MW DFIM and p.u. Analysis

This example illustrates the steady state of a 2 MW DFIM. The operating point is chosen as the rated stator power generating, with zero stator reactive power, 50 Hz of grid frequency, and $s = -0.25$ slip (hypersynchronous speed).

First, the machine is supposed to be delivering 2 MW to the grid:

$$P_s = -2\,\text{MW}$$

with rated stator voltage (line to neutral)

$$\underline{V}_s = 398.4\,\underline{/0°}\,\text{V}$$

This indicates that when $Q_s = 0$ is chosen, the stator current phasor yields

$$P_s = 3\text{Re}\{\underline{V}_s \cdot \underline{I}_s^*\} \Rightarrow \underline{I}_s = 1673.8\,\underline{/180°}\,\text{A}$$

On the other hand, the base values are found as

$$V_{base} = 398.4\,\text{V}, \quad I_{base} = 1760\,\text{A}, \quad S_{base} = 3 \cdot 398.4 \cdot 1760 = 2.1\,\text{MW}$$

$$\Psi_{base} = \frac{V_{base}}{\omega_{base}} = \frac{398.4}{314.16} = 1.26\,\text{Wb}, \quad T_{base} = \frac{S_{base}}{\omega_{base}/p} = \frac{2.1e6}{314.16/2} = 13.3\,\text{k}\cdot\text{Nm}$$

Consequently, the stator flux can be calculated directly:

$$\underline{V}_s - R_s\underline{I}_s = j\omega_s\underline{\Psi}_s \Rightarrow$$

$$\underline{\Psi}_s = \frac{398.4\,\underline{/0^\circ} - 0.0026 \cdot 1673.8\,\underline{/180^\circ}}{j \cdot 314.16} = 1.28\,\underline{/-90^\circ}\,\text{Wb} \Rightarrow$$

$$\underline{\Psi}_{s\,pu} = \frac{1.28}{1.26}\,\underline{/-90^\circ} = 1.01\,\underline{/-90^\circ}$$

Continuing, the rotor current is found as

$$\underline{I}_r = \frac{\underline{\Psi}_s - L_s\underline{I}_s}{L_m} \Rightarrow$$

$$\underline{I}_r = \frac{1.28\,\underline{/-90^\circ} - 0.002587 \cdot 1673.8\,\underline{/180^\circ}}{0.0025} = 1807.4\,\underline{/-16.5^\circ}\,\text{A} \Rightarrow$$

$$\underline{I}_{r\,pu} = \frac{1807.4}{1760}\,\underline{/-16.5^\circ} = 1.02\,\underline{/-16.5^\circ}$$

Thus, the rotor flux can be derived:

$$\underline{\Psi}_r = L_m\underline{I}_s + L_r\underline{I}_r \Rightarrow$$

$$\underline{\Psi}_r = 0.0025 \cdot 1673.8\,\underline{/180^\circ} + 0.002587 \cdot 1807.4\,\underline{/-16.5^\circ}$$

$$\underline{\Psi}_r = 1.358\,\underline{/-77.4^\circ}\,\text{Wb} \Rightarrow$$

$$\underline{\Psi}_{r\,pu} = \frac{1.358}{1.26}\,\underline{/-77.4^\circ} = 1.07\,\underline{/-77.4^\circ}$$

Finally, all the phasors are obtained with the rotor voltage:

$$\underline{V}_r = R_r\underline{I}_r + js\omega_s\underline{\Psi}_r \Rightarrow$$

$$\underline{V}_r = 0.0029 \cdot 1807.4\,\underline{/-16.5^\circ} + j(-0.25) \cdot 314.16 \cdot 1.358\,\underline{/-77.4^\circ}$$

$$\underline{V}_r = 102.2\,\underline{/-165.9^\circ}\,\text{V} \Rightarrow$$

$$\underline{V}_{r\,pu} = \frac{102.2}{398.4}\,\underline{/-165.9^\circ} = 0.25\,\underline{/-165.9^\circ}$$

Finally, from the obtained phasors, the rotor active and reactive powers as well as the torque can be found:

$$P_r = 3\mathrm{Re}\{\underline{V}_r \cdot \underline{I}_r^*\} \Rightarrow P_r = -0.55\,\mathrm{MW} \Rightarrow P_{r\,pu} = \frac{P_r}{S_{base}} = -0.26$$

$$Q_r = 3\mathrm{Im}\{\underline{V}_r \cdot \underline{I}_r^*\} \Rightarrow Q_r = 23.4\,\mathrm{kVAR} \Rightarrow Q_{r\,pu} = \frac{Q_r}{S_{base}} = 0.011$$

This shows that the machine is being magnetized though the rotor.

$$T_{em} = 3p\frac{L_m}{L_s} \cdot \mathrm{Im}\{\underline{\Psi}_s \cdot \underline{I}_r^*\} \Rightarrow T_{em} = -12.9\,\mathrm{k \cdot Nm} \Rightarrow T_{em\,pu} = \frac{T_{em}}{T_{base}} = -0.97$$

It is seen that the total power generated by the doubly fed machine is

$$P_r + P_s = -2 - 0.55 = -2.55\,\mathrm{MW}$$

Therefore, the phasor diagram is depicted in Figure 3.13.

Note that all the rotor magnitudes are referred to the stator, so if real current and voltage are needed, the u factor must be taken into account as follows (the rotor powers remain unaltered):

$$\underline{V}'_r = \frac{\underline{V}_r}{u} = \frac{102.2}{0.34}\underline{/-165.9^\circ} = 300.6\underline{/-165.9^\circ}\,\mathrm{V}\ \text{with } f_r = 12.5\,\mathrm{Hz}$$

$$\underline{I}'_r = u\underline{I}_r = 0.34 \cdot 1807.4\underline{/-16.5^\circ} = 614.5\underline{/-16.5^\circ}\,\mathrm{A}\ \text{with } f_r = 12.5\,\mathrm{Hz}$$

while the stator magnitudes have $f_s = 50\,\mathrm{Hz}$. Finally, as seen in the previous chapter, from the required DC bus voltage perspective, taking into account that these phasors are rms values, ideally, the required minimum DC bus voltage to handle this rotor voltage amplitude would be (space vector modulation)

$$V_{bus} = \sqrt{3}|\underline{V}'_r| \cdot \sqrt{2} = \sqrt{3} \cdot 300.6 \cdot \sqrt{2} = 736\,\mathrm{V}$$

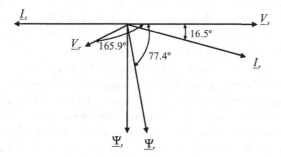

Figure 3.13 Phasor diagram in p.u. of a DFIM operating as a generator, with $Q_s = 0$, $P_s = -2\,\mathrm{MW}$, and $s = -0.25$.

On the other hand, normally a larger voltage is given to the DC bus, in order to take into account voltage drops in the switches of the converter, other operating points at higher slip, margin of voltage for transient control, and so on.

Note by checking Examples 2.1 and 2.2 of Chapter 2 that the grid side converter, due to the presence of the grid side inductive filter, in general will require more DC bus voltage than the rotor side converter at this same speed ($s = -0.25$).

Finally, this example shows that with this machine design, to generate 2 MW and 0.55 MW of power through the stator and rotor, the rotor side converter must be able to handle approximately $300\,V_{rms}$ and $600\,A_{rms}$ if the machine is magnetized through the rotor. Therefore, the converter operates with AC voltages of similar amplitudes as the stator/grid voltages, but with approximately three times smaller current amplitude than the stator currents. This equalized voltage amplitude operation at both sides of the back-to-back converter eliminates the necessity of a transformer dedicated only to the converter.

3.5 STEADY STATE CURVES: PERFORMANCE EVALUATION

This section studies the steady state performance of the DFIM at different operating points. It has been seen that depending on how is the machine supplied, the achieved steady state point can be different. Therefore, three different supplying philosophies for this machine are analyzed in detail. In all the cases, the DC bus voltage is controlled by the grid side converter to a predetermined value, for proper operation of both grid side and rotor side converters. However, the grid side variables are not examined in this section.

1. *Rotor Voltage Variation: Different Frequency, Amplitude, and Phase Shift Angle.* In this case, the supplied rotor voltage can be modified, independently varying its amplitude, frequency, and phase shift angle. Thus, all the magnitudes of the machine (torque, currents, efficiency, powers, etc.) are evaluated when the machine is driven at different rotor voltages in the open loop configuration as represented in Figure 3.14.

2. *Rotor Voltage Variation: Constant Voltage–Frequency (V-F) Ratio.* In this case, the rotor voltage presents a constant relation between its amplitude and frequency. The phase shift angle can be independently modified. Hence, the same analysis as in the previous example is carried out, but a constant rotor voltage–frequency ratio is maintained (Figure 3.14).

3. *Rotor Voltage Variation: Control of Stator Reactive Power and Torque.* In most of the real applications where this machine is used, the closed loop control strategy imposes the required rotor voltage (amplitude, frequency, and phase shift), in order to achieve a certain desired operating point. In general, this operating point is defined by the desired torque and stator reactive power values. Hence, in this case, the reachable steady state points will be evaluated as a function of the desired torque (T_{em}) and stator reactive power (Q_s) references (Figure 3.14).

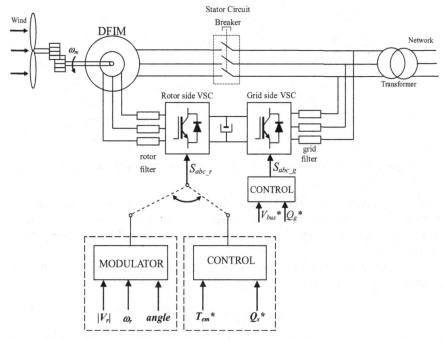

Figure 3.14 Different supplying philosophies for the rotor side converter (open loop and closed loop control).

In all three cases, the mathematical expressions providing the steady state magnitude value are derived and then graphically evaluated for a 2 MW DFIM. Thanks to this analysis, for instance, it is possible to immediately know all the steady state magnitudes of the machine (efficiency, currents, powers, etc.), resulting in the desired torque and reactive power.

3.5.1 Rotor Voltage Variation: Frequency, Amplitude, and Phase Shift

The DFIM is supplied by the stator and the rotor. By changing the stator and rotor voltages, the machine can operate at different operation points of torque and stator and rotor active and reactive powers. In general, the stator of the machine will be connected directly to the grid, what means that the stator voltage is fixed. On the contrary, the rotor voltage is supplied by a bidirectional power electronic converter, allowing modification of the rotor voltage amplitude, frequency and phase with respect to the stator voltage as required; Figure 3.15 graphically shows this.

If we want, to control the speed of the machine, as shown by Equation (3.4), the angular frequency of the rotor voltage must be modified accordingly. However, to achieve a certain power exchange with the grid or, for instance, a specific torque, the rotor voltage amplitude and phase shift must also be selected accordingly.

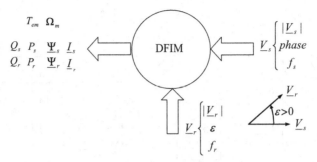

Figure 3.15 Steady state as a function of different stator and rotor voltages.

In the following, the relation between the stator and rotor voltages and the electromagnetic torque is derived, allowing us to understand how it is possible to modify the torque by changing the rotor voltage, when the machine is connected to a fixed grid voltage.

Choosing the expression that relates the torque and the stator and rotor fluxes—Equation (3.53)—we have

$$T_{em} = 3 \frac{L_m}{\sigma L_r L_s} p \operatorname{Im}\{\underline{\Psi}_r^* \cdot \underline{\Psi}_s\} \tag{3.102}$$

It is necessary to modify this expression and relate the torque to the stator and rotor voltages. Hence, by using expressions (3.31) and (3.32), we find

$$\underline{V}_s = R_s \underline{I}_s + j\omega_s \underline{\Psi}_s \tag{3.103}$$

$$\underline{V}_r = R_r \underline{I}_r + j\omega_r \underline{\Psi}_r \tag{3.104}$$

The stator and rotor currents must be eliminated from these equations. From the relation of the fluxes and currents—Equations (3.27) and (3.28)—we find

$$\underline{I}_s = \frac{1}{\sigma L_s} \underline{\Psi}_s - \frac{L_m}{\sigma L_s L_r} \underline{\Psi}_r \tag{3.105}$$

$$\underline{I}_r = -\frac{L_m}{\sigma L_s L_r} \underline{\Psi}_s + \frac{1}{\sigma L_r} \underline{\Psi}_r \tag{3.106}$$

Hence, by substituting these last expressions into the voltage equations (3.103) and (3.104), the direct relation between the fluxes and the voltages is derived:

$$\begin{bmatrix} \underline{V}_s \\ \underline{V}_r \end{bmatrix} = \begin{bmatrix} \dfrac{R_s}{\sigma L_s} + j\omega_s & -\dfrac{R_s L_m}{\sigma L_s L_r} \\ -\dfrac{R_r L_m}{\sigma L_s L_r} & \dfrac{R_r}{\sigma L_r} + j\omega_r \end{bmatrix} \cdot \begin{bmatrix} \underline{\Psi}_s \\ \underline{\Psi}_r \end{bmatrix} \tag{3.107}$$

The inverse relation is

$$
\begin{bmatrix} \underline{\Psi}_s \\ \underline{\Psi}_r \end{bmatrix} = (F_1 - jF_2) \cdot \begin{bmatrix} \dfrac{R_r}{\sigma L_r} + j\omega_r & \dfrac{R_s L_m}{\sigma L_s L_r} \\[2mm] \dfrac{R_r L_m}{\sigma L_s L_r} & \dfrac{R_s}{\sigma L_s} + j\omega_s \end{bmatrix} \cdot \begin{bmatrix} \underline{V}_s \\ \underline{V}_r \end{bmatrix}
\tag{3.108}
$$

with

$$
F_1 = \frac{K_1}{K_1^2 + K_2^2}, \quad F_2 = \frac{K_2}{K_1^2 + K_2^2}
\tag{3.109}
$$

$$
F = \sqrt{F_1^2 + F_2^2}, \quad f = a\,\tan\!\left(-\frac{F_2}{F_1}\right)
\tag{3.110}
$$

and

$$
K_1 = \frac{R_s R_r}{\sigma^2 L_s L_r}\left(1 - \frac{L_m^2}{L_s L_r}\right) - \omega_s \omega_r, \quad K_2 = \frac{\omega_s R_r}{\sigma L_r} + \frac{\omega_r R_s}{\sigma L_s}
\tag{3.111}
$$

In this way, by developing expression (3.108), the stator and rotor fluxes can be found:

$$
\begin{aligned}
\underline{\Psi}_s &= \big[FC_s|\underline{V}_r|\cos(f + \varepsilon) + FA_r|\underline{V}_s|\cos(f + B_r)\big] \\
&\quad + j\big[FC_s|\underline{V}_r|\sin(f + \varepsilon) + FA_r|\underline{V}_s|\sin(f + B_r)\big]
\end{aligned}
\tag{3.112}
$$

$$
\begin{aligned}
\underline{\Psi}_r &= \big[FC_r|\underline{V}_s|\cos(f) + FA_s|\underline{V}_r|\cos(f + B_s + \varepsilon)\big] \\
&\quad - j\big[FC_r|\underline{V}_s|\sin(f) + FA_s|\underline{V}_r|\sin(f + B_s + \varepsilon)\big]
\end{aligned}
\tag{3.113}
$$

with

$$
C_s = \frac{R_s L_m}{\sigma L_s L_r}, \quad C_r = \frac{R_r L_m}{\sigma L_s L_r}
\tag{3.114}
$$

$$
A_s = \sqrt{\left(\frac{R_s}{\sigma L_s}\right)^2 + \omega_s^2}, \quad B_s = a\,\tan\!\left(\frac{\sigma L_s \omega_s}{R_s}\right)
\tag{3.115}
$$

$$
A_r = \sqrt{\left(\frac{R_r}{\sigma L_r}\right)^2 + \omega_r^2}, \quad B_r = a\,\tan\!\left(\frac{\sigma L_r \omega_r}{R_r}\right)
\tag{3.116}
$$

Hence, by substituting the fluxes into Equation (3.102), the torque is derived:

$$
\begin{aligned}
T_{em} = 3\frac{L_m}{\sigma L_r L_s}pF^2\big\{ &C_s C_r|\underline{V}_s||\underline{V}_r|\sin\varepsilon \\
&+ A_r C_r|\underline{V}_s|^2\sin B_r \\
&- A_s C_s|\underline{V}_r|^2\sin B_s \\
&- A_s A_r|\underline{V}_s||\underline{V}_r|\sin(\varepsilon + B_s - B_r)\big\}
\end{aligned}
\tag{3.117}
$$

Consequently, assuming a constant grid voltage, the torque depends on three variables and the parameters of the machine:

$$T_{em} = f\left(|\underline{V}_r|, \omega_r, \varepsilon\right) \tag{3.118}$$

By following a similar procedure, it is possible to derive equivalent expressions for the rest of the variables. The results are:

$$
\begin{aligned}
P_s = 3|\underline{V}_s| \frac{F}{\sigma L_s} &\left\{ C_s|\underline{V}_r|\cos(f+\varepsilon) + A_r|\underline{V}_s|\cos(f+B_r) \right. \\
&\left. - \frac{L_m}{L_r}\left[C_r|\underline{V}_s|\cos(f) + A_s|\underline{V}_r|\cos(f+B_r+\varepsilon) \right] \right\}
\end{aligned} \tag{3.119}
$$

$$
\begin{aligned}
P_r = 3|\underline{V}_r| \frac{F}{\sigma L_r} &\left\{ C_r|\underline{V}_s|\cos(f-\varepsilon) + A_s|\underline{V}_r|\cos(f+B_s) \right. \\
&\left. - \frac{L_m}{L_s}\left[C_s|\underline{V}_r|\cos(f) + A_r|\underline{V}_s|\cos(f+B_r-\varepsilon) \right] \right\}
\end{aligned} \tag{3.120}
$$

$$
\begin{aligned}
Q_s = 3|\underline{V}_s| \frac{F}{\sigma L_s} &\left\{ -C_s|\underline{V}_r|\sin(f+\varepsilon) - A_r|\underline{V}_s|\cos(f+B_r) \right. \\
&\left. + \frac{L_m}{L_r}\left[C_r|\underline{V}_s|c\sin(f) + A_s|\underline{V}_r|\sin(f+B_r+\varepsilon) \right] \right\}
\end{aligned} \tag{3.121}
$$

$$
\begin{aligned}
Q_r = 3|\underline{V}_r| \frac{F}{\sigma L_r} &\left\{ C_r|\underline{V}_s|\sin(-f+\varepsilon) - A_s|\underline{V}_r|\sin(f+B_s) \right. \\
&\left. + \frac{L_m}{L_s}\left[C_s|\underline{V}_r|\sin(f) - A_r|\underline{V}_s|\sin(f+B_r-\varepsilon) \right] \right\}
\end{aligned} \tag{3.122}
$$

$$
|\underline{\Psi}_s|^2 = F^2\left\{ C_s^2|\underline{V}_r|^2 + A_r^2|\underline{V}_s|^2 + 2A_rC_s|\underline{V}_s||\underline{V}_r|\cos(\varepsilon - B_r) \right\} \tag{3.123}
$$

$$
|\underline{\Psi}_r|^2 = F^2\left\{ C_r^2|\underline{V}_s|^2 + A_s^2|\underline{V}_r|^2 + 2A_sC_r|\underline{V}_s||\underline{V}_r|\cos(\varepsilon + B_s) \right\} \tag{3.124}
$$

$$
\begin{aligned}
|\underline{I}_s|^2 = &\left(\frac{F}{\sigma L_s}\right)^2 \left\{ C_s^2|\underline{V}_r|^2 + A_r^2|\underline{V}_s|^2 + 2A_rC_s|\underline{V}_s||\underline{V}_r|\cos(B_r - \varepsilon) \right\} \\
&+ \left(\frac{L_m F}{\sigma L_s L_r}\right)^2 \left\{ C_r^2|\underline{V}_s|^2 + A_s^2|\underline{V}_r|^2 + 2A_sC_r|\underline{V}_s||\underline{V}_r|\cos(B_s + \varepsilon) \right\} \\
&- \left(\frac{2 \cdot L_m F^2}{\sigma L_s \sigma L_s L_r}\right) \left\{ C_sC_r|\underline{V}_s||\underline{V}_r|\cos(\varepsilon) + C_sA_s|\underline{V}_r|^2\cos(B_s) \right. \\
&\left. + C_rA_r|\underline{V}_s|^2\cos(B_r) + A_sA_r|\underline{V}_s||\underline{V}_r|\cos(B_s + \varepsilon - B_r) \right\}
\end{aligned} \tag{3.125}
$$

$$|\underline{I}_r|^2 = \left(\frac{L_mF}{\sigma L_s L_r}\right)^2 \left\{ C_s^2|\underline{V}_r|^2 + A_r^2|\underline{V}_s|^2 + 2A_rC_s|\underline{V}_s||\underline{V}_r|\cos(B_r - \varepsilon)\right\}$$

$$+ \left(\frac{F}{\sigma L_r}\right)^2 \left\{ C_r^2|\underline{V}_s|^2 + A_s^2|\underline{V}_r|^2 + 2A_sC_r|\underline{V}_s||\underline{V}_r|\cos(B_s + \varepsilon)\right\}$$

$$- \left(\frac{2 \cdot L_mF^2}{\sigma L_r \sigma L_s L_r}\right)\left\{ C_sC_r|\underline{V}_s||\underline{V}_r|\cos(\varepsilon) + C_sA_s|\underline{V}_r|^2\cos(B_s) \right.$$

$$\left. + C_rA_r|\underline{V}_s|^2\cos(B_r) + A_sA_r|\underline{V}_s||\underline{V}_r|\cos(B_s + \varepsilon - B_r)\right\} \qquad (3.126)$$

Note that with all these expressions, it is possible to immediately know the steady state magnitudes of the machine, for a given supplied rotor voltage.

3.5.1.1 Example 3.4: Performance at Fixed Rotor Voltage   From the previous expressions, when the stator voltage is fixed due to a direct connection to the grid, if the rotor voltage is fixed at a certain amplitude and phase, the machine can operate at different speeds, reaching different operating points. This example shows the steady state curves of a 2 MW DFIM, when the rotor voltage amplitude is fixed (in p.u.) to

$$\underline{V}_{r\,pu} \begin{cases} |\underline{V}_r| = 0.1|\underline{V}_s| \\ \varepsilon = 1.5^\circ \end{cases}$$

Figure 3.16 illustrates the torque and stator and rotor active and reactive powers under these rotor voltage conditions, at different speeds.

In all the cases, it can be noticed that, depending on the speed that is chosen, different operating points will be reached. Take in mind that the speed will be fixed by the selected rotor frequency (f_r) or, equivalently, by the rotor angular frequency (ω_r) imposed by the bidirectional converter.

- *Torque.* Depending on the speed, it can take positive or negative values, operating as a motor or a generator. For instance, if the load imposes $T_{em} = -1$ p.u., the necessary speed is $\omega_m = 0.93$ p.u. (generating at subsynchronous speed). Note that if the chosen speed is out of the range [0.77, 1.015], the machine is in a nonstable operating range, since the breakdown torque is reached.
- *Stator Active Power.* As shown by Equation (3.59), the behavior of the stator active power is almost proportional to the torque.
- *Rotor Active Power.* Its behavior at steady state has an approximately inverse relation to the stator active power as predicted by Equation (3.60).

Figure 3.17 shows the rest of the magnitudes of the machine under the same rotor voltage operation. It is remarkable that the achieved stator and rotor currents are too high for a realistic operating point.

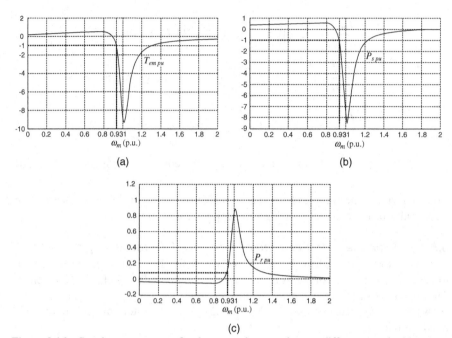

Figure 3.16   Steady state curves at fixed stator and rotor voltage at different speeds: (a) torque in p.u., (b) stator active power in p.u., and (c) rotor active power in p.u.

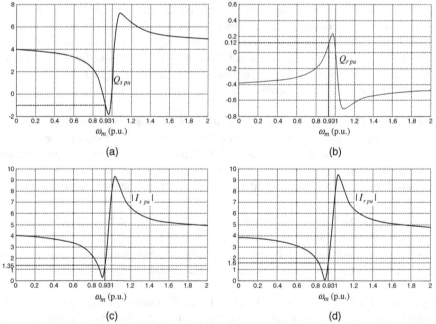

Figure 3.17   Steady state curves at fixed stator and rotor voltages at different speeds: (a) stator reactive power in p.u., (b) rotor reactive power in p.u., (c) stator current amplitude in p.u., and (d) rotor current amplitude in p.u.

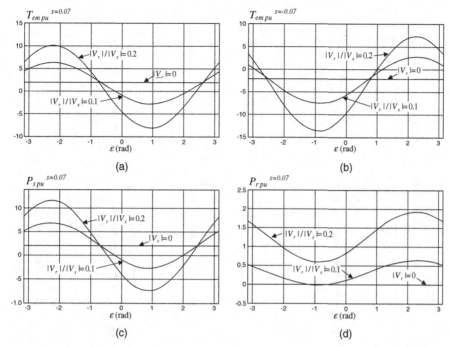

Figure 3.18 Steady state curves at constant speed and different rotor voltage amplitudes and phases: (a) torque in p.u. at $s = 0.07$, (b) torque in p.u. *at* $s = -0.07$, (c) stator active power in p.u. at $s = 0.07$, and (d) rotor active power in p.u. at $s = 0.07$.

3.5.1.2 *Example 3.5: Performance at Fixed Speed* This example shows the opposite case to the previous example. In this case, the speed of the machine is fixed ($s = 0.07$, that means $\omega_m = 0.93$), while the rotor voltage amplitude and phase are modified. The performance of the most representative magnitudes is illustrated in Figure 3.18.

At constant speed, the terms F, A_r, and B_r become constant. Hence, in all the cases, the magnitudes vary as a sinusoidal wave of constant amplitude. When the rotor voltage is zero, obviously all the magnitudes are constant. In all the cases, at higher rotor voltage amplitude, the torque and power capacities are increased (higher amplitude variation) even further above their rated values. Then as the phase shift changes, we can achieve any value within the maximum amplitude.

However, if we do not pay attention, these graphs can be evaluated at an unstable operating point, that is, outside the range of stability. That is why it is important to analyze the graphs of the previous example.

3.5.1.3 *Example 3.6: General Performance* This last example shows the general performance of the machine at steady state, when the three variables are modified simultaneously: rotor voltage amplitude, phase, and speed. Figures 3.19 and 3.20 illustrate the performance of the most relevant magnitudes. It is necessary to

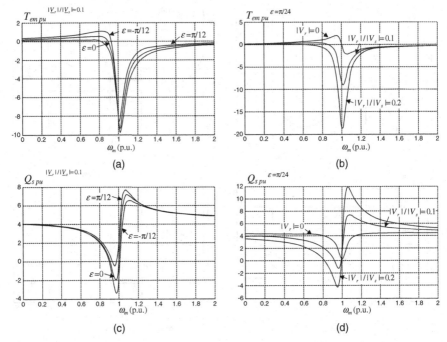

Figure 3.19 Steady state curves at different rotor voltage amplitude, phase, and speed: (a) torque in p.u. at constant rotor voltage amplitude, (b) torque in p.u. at constant phase, (c) stator reactive power in p.u. at constant rotor voltage amplitude, and (d) stator reactive power in p.u. at constant phase.

study these figures in detail to fully understand the behavior of the steady state of this machine. This task is left for the reader.

3.5.2 Rotor Voltage Variation: Constant Voltage–Frequency (V-F) Ratio

As noticed in the steady state evaluation of the previous section (Figures 3.19 and 3.20), if the machine is connected directly to the stator and the rotor is fed by voltages generated according to any rule, the established steady state can be inefficient, presenting too high stator and rotor currents. In addition, if the rotor voltage is not chosen properly, it is possible to have stability problems in the machine, due to operation close to the breakdown torque.

In summary, to avoid these two problems, the machine is normally supplied by a closed loop control that is in charge of imposing the required rotor voltage amplitude and phase shift, for given references (torque and stator reactive power in general).

However, in this section an intermediate supplying solution for the machine is studied. Following the same procedure utilized in the squirrel cage induction machines, that is, the constant voltage–frequency ratio, the equivalent philosophy is applied to the DFIM.

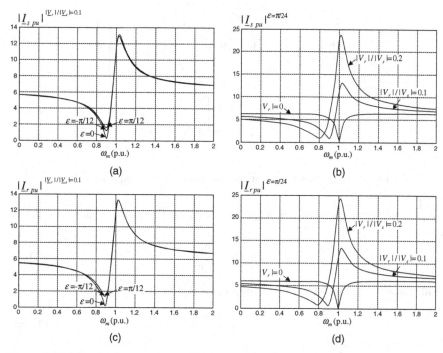

Figure 3.20 Steady state curves at different rotor voltage amplitude, phase, and speed: (a) stator current amplitude in p.u. at constant rotor voltage amplitude, (b) stator current amplitude in p.u. at constant phase, (c) rotor current amplitude in p.u. at constant rotor voltage amplitude, and (d) rotor current amplitude in p.u. at constant phase.

Hence, for simplicity in the analysis if we ignore the stator and rotor resistances, the next simple relations are derived using Equations (3.31) and (3.32):

$$|\underline{\Psi}_s| = \frac{|\underline{V}_s|}{\omega_s} \tag{3.127}$$

$$|\underline{\Psi}_r| = \frac{|\underline{V}_r|}{\omega_r} \tag{3.128}$$

It is possible to deduce the following:

- The stator flux amplitude is fixed by the grid.
- The rotor flux amplitude is directly fixed by the supplied rotor voltage amplitude and angular speed.

Hence, it is reasonable to maintain constant rotor flux amplitude at different operating conditions of the machine, as occurs with the stator flux, in order to establish equivalent rotor and stator flux behavior or magnetization levels. So for that purpose, the relation between the rotor voltage amplitude and the angular

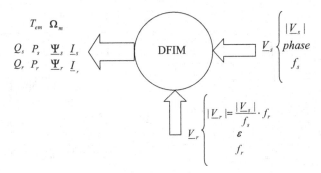

Figure 3.21 Steady-state at constant rotor voltage–frequency ratio.

frequency should constantly be kept as defined by Equation (3.132), in order to achieve

$$Factor = \frac{|\underline{\Psi}_r|}{|\underline{\Psi}_s|} \qquad (3.129)$$

With *Factor* = 1, the rotor voltage amplitude should be chosen as

$$|\underline{V}_r| = \frac{|\underline{V}_s|}{\omega_s}\omega_r \qquad (3.130)$$

Additionally, there exists one more degree of freedom, that is, the phase shift ε. This angle, in principle, can be independently chosen or even maintained constant, as will be shown in the subsequent analysis. The schematic block diagram showing this is illustrated in Figure 3.21.

Hence, by following the same philosophy as in the previous section, by paying attention to the supplied rotor voltage by means of the *Factor*, it is possible to achieve a more efficient steady state of the machine. Two subsequent examples will show how the steady states of the machine are achieved, when it is supplied according to the described procedure of this section.

3.5.2.1 Example 3.7: Performance at Constant Factor This example shows the steady state curves of a 2 MW DFIM, when the machine is supplied according to the rotor voltage law described in this subsection. Hence, if we apply a constant rotor voltage–frequency ratio, according to Equation (3.130), the torque, and stator and rotor currents can be calculated by means of expressions (3.125), (3.126), and (3.132). Figure 3.22 shows the steady state curves of these variables, at different phase shift angles.

It is noticeable that there is symmetry in torque behavior at different speeds as seen in Figures 3.22a and 3.22b. At the same speed, 0.92, for instance, 0.35 and −0.35 torque can be achieved at $-\pi/12$ and $\pi/12$ phase shift, respectively.

On the other hand, the smaller the phase angle, the smaller is the available torque. In addition, at synchronism, since the rotor voltage is created according to Equation (3.130), the rotor voltage amplitude is zero because $\omega_r = 0$, yielding $T_{em} = 0$ as

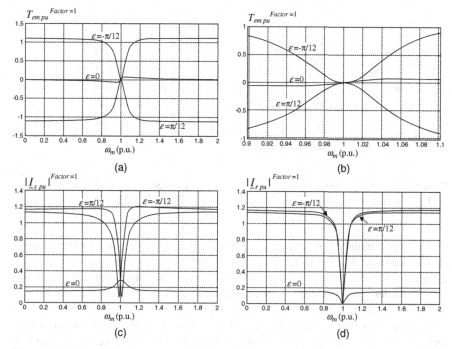

Figure 3.22 Steady state curves at fixed rotor voltage–frequency ratio: (a) T_{em} at different phase angles, (b) zoom of T_{em}, (c) stator current at different phase angles, and (d) rotor current at different phase angles.

well. The torque at synchronism can be increased, as is commonly done, by adding an offset rotor voltage to expression (3.130).

Finally, Figures 3.22c and 3.22d show that the required stator and rotor currents in this situation are reasonably reduced compared to the previous section (Figure 3.20). The proper choice of the rotor voltage amplitude yields a more efficient steady state of the machine.

When a *Factor* different from 1 is used, the behavior changes slightly. Figure 3.23a illustrates that at a given factor and angle shift, the achievable torque becomes greater than at *Factor* = 1. This increase in the torque provokes an increase in the stator current, as shown in Figure 3.23b.

3.5.3 Rotor Voltage Variation: Control of Stator Reactive Power and Torque

The open loop supply philosophy studied in the previous two subsections is not very representative of real applications. It is included in this chapter basically because it helps the reader to understand the machine and its behavior, especially if it is supplied according to the basic constant voltage and frequency (V-F) schemas often used in some other AC machines, for instance, the squirrel cage induction machine. However,

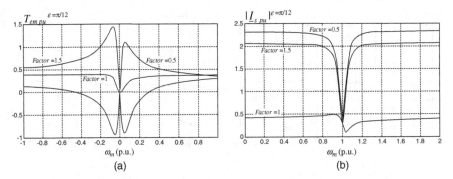

Figure 3.23 Steady state curves at fixed rotor voltage–frequency ratio with different factors: (a) T_{em} at phase angle $\pi/12$, and (b) stator current at phase angle $\pi/12$.

for the DFIM, mainly in order to avoid instability problems, it is much better to use a closed loop control strategy that guarantees the stability of the system. Not only due to this reason, but also in order to easily obtain efficient steady state operating points and good dynamic performance, in most practical applications, today a machine is normally driven by a closed loop control.

Consequently, this subsection presents the opposite study to the previous two sections. We will find the steady state point of the machine, when the control imposes the needed rotor voltage, in order to fulfill the reference values of the controlled variables. As shown in Chapter 1, in general, the controlled variables in a wind energy generation environment are torque (T_{em}) and stator reactive power (Q_s). Note that, depending on the control employed, several alternative variables as references can be used, P_s and $|\underline{\Psi}_r|$, for instance (this fact is clearly seen in Chapters 7 and 8).

Hence, here the most relevant magnitudes of the machine will be derived, when, by means of the correct control strategy, the torque (T_{em}) and the stator reactive power (Q_s) are controlled to the given reference values.

As schematically illustrated in Figure 3.24, for a fixed stator voltage and given T_{em} and Q_s references, the operating point defined by the unknown variables, such as rotor voltage (imposed by the control and the bidirectional converter), and stator and rotor fluxes and currents, will be derived.

For simplicity in the derivation of the expressions, it is necessary first to define the phasor diagram with stator flux at zero phase angle, as shown in Figure 3.25. Again for simplicity, if the voltage drop in the stator is neglected, the stator flux can be calculated from Equation (3.31):

$$|\underline{\Psi}_s| = \frac{|\underline{V}_s|}{\omega_s} \tag{3.131}$$

By substituting this last expression in the stator reactive power expression (3.54), we find

$$Q_{s\_ref} = 3\mathrm{Im}\{\underline{V}_s \cdot \underline{I}_s^*\} = 3\omega_s|\underline{\Psi}_s||\underline{I}_s|\cos\gamma_i \tag{3.132}$$

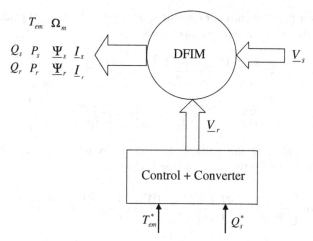

Figure 3.24 Steady state as function of different T_{em} and Q_s references.

Proceeding in the same way for the torque equation (3.53):

$$T_{em\_ref} = 3p \cdot \text{Im}\{\underline{\Psi}_s^* \cdot \underline{I}_s\} = 3p|\underline{\Psi}_s||\underline{I}_s|\sin \gamma_i \qquad (3.133)$$

Hence, by combining the last three expressions, the stator current can be calculated as

$$\underline{I}_s = \frac{Q_{s\_ref}}{3|\underline{V}_s|} + j\frac{\omega_s T_{em\_ref}}{3p|\underline{V}_s|} \qquad (3.134)$$

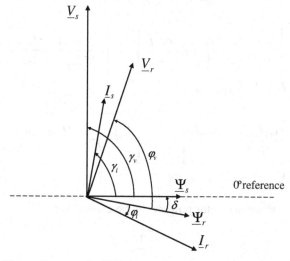

Figure 3.25 Phasor diagram of a DFIM operating as a motor, at subsynchronous speed.

Once the stator flux and current expressions are known, by using Equations (3.105) and (3.106), the rotor current and flux yields

$$\underline{\Psi}_r = \left[\frac{|\underline{V}_s|}{\omega_s} \cdot \frac{L_r}{L_m} - \frac{Q_{s\_ref}}{3|\underline{V}_s|} \cdot \frac{\sigma L_s L_r}{L_m}\right] - j\left[\frac{\omega_s T_{em\_ref}}{3p|\underline{V}_s|} \cdot \frac{\sigma L_s L_r}{L_m}\right] \qquad (3.135)$$

$$\underline{I}_r = \left[\frac{|\underline{V}_s|}{\omega_s} \cdot \frac{1}{L_m} - \frac{Q_{s\_ref}}{3|\underline{V}_s|} \cdot \frac{L_s}{L_m}\right] - j\left[\frac{\omega_s T_{em\_ref}}{3p|\underline{V}_s|} \cdot \frac{L_s}{L_m}\right] \qquad (3.136)$$

Once again, neglecting the voltage drop in the rotor resistance, from expressions (3.135) and (3.32) the rotor voltage is derived:

$$\underline{V}_r = \omega_r\left[\frac{\omega_s T_{em\_ref}}{3p|\underline{V}_s|} \cdot \frac{\sigma L_s L_r}{L_m}\right] + j\omega_r\left[\frac{|\underline{V}_s|}{\omega_s} \cdot \frac{L_r}{L_m} - \frac{Q_{s\_ref}}{3|\underline{V}_s|} \cdot \frac{\sigma L_s L_r}{L_m}\right] \qquad (3.137)$$

Consequently, the modules of the derived expressions are calculated according to the expressions:

$$|\underline{I}_s|^2 = \left[\frac{Q_{s\_ref}}{3|\underline{V}_s|}\right]^2 + \left[\frac{\omega_s T_{em\_ref}}{3p|\underline{V}_s|}\right]^2 = f(|\underline{V}_s|, Q_{s\_ref}, T_{em\_ref}) \qquad (3.138)$$

$$|\underline{I}_r|^2 = \left[\frac{|\underline{V}_s|}{\omega_s} \cdot \frac{1}{L_m} - \frac{Q_{s\_ref}}{3|\underline{V}_s|} \cdot \frac{L_s}{L_m}\right]^2 + \left[\frac{\omega_s T_{em\_ref}}{3p|\underline{V}_s|} \cdot \frac{L_s}{L_m}\right]^2$$
$$= f(|\underline{V}_s|, Q_{s\_ref}, T_{em\_ref}) \qquad (3.139)$$

$$|\underline{\Psi}_s| = \frac{|\underline{V}_s|}{\omega_s} = f(|\underline{V}_s|) \qquad (3.140)$$

$$|\underline{\Psi}_r|^2 = \left[\frac{|\underline{V}_s|}{\omega_s} \cdot \frac{L_r}{L_m} - \frac{Q_{s\_ref}}{3|\underline{V}_s|} \cdot \frac{\sigma L_s L_r}{L_m}\right]^2 + \left[\frac{\omega_s T_{em\_ref}}{3p|\underline{V}_s|} \cdot \frac{\sigma L_s L_r}{L_m}\right]^2$$
$$= f(|\underline{V}_s|, Q_{s\_ref}, T_{em\_ref}) \qquad (3.141)$$

$$|\underline{V}_r|^2 = \omega_r^2\left[\frac{\omega_s T_{em\_ref}}{3p|\underline{V}_s|} \cdot \frac{\sigma L_s L_r}{L_m}\right]^2 + \omega_r^2\left[\frac{|\underline{V}_s|}{\omega_s} \cdot \frac{L_r}{L_m} - \frac{Q_{s\_ref}}{3|\underline{V}_s|} \cdot \frac{\sigma L_s L_r}{L_m}\right]^2$$
$$= f(|\underline{V}_s|, Q_{s\_ref}, T_{em\_ref}, \omega_r) \qquad (3.142)$$

Accordingly, the stator and rotor powers can be expressed as

$$P_s = \frac{\omega_s T_{em\_ref}}{p} \qquad (3.143)$$

$$P_r = \omega_r T_{em\_ref} \frac{L_s L_r (\sigma - 1)}{pL_m^2}$$

$$= f(T_{em\_ref}\omega_r)$$

(3.144)

The following can be remarked:

- The stator current is decoupled in the sense that its real part is directly influenced by Q_{s_ref}, while its imaginary part is influenced by T_{em_ref} (or stator active power).
- The rotor current is influenced in a similar way by Q_{s_ref} and T_{em_ref}. However, there exists one difference: the real part consist of a fixed term and one additional term that depends negatively on Q_{s_ref}. If we add both real parts of the stator and rotor currents, the approximated result provides a constant term,

$$\left(\frac{|\underline{V}_s|}{\omega_s} \cdot \frac{1}{L_m} \right)$$

(3.145)

responsible for the magnetizing current (the mutual current). Hence, depending on how Q_{s_ref} is set, the magnetizing current is shared by both the stator and rotor currents, or alternatively, we can find the extreme cases:

  o Magnetization by the rotor: $Q_{s\_ref} = 0$, the real part of the stator current is zero and all the magnetizing current is provided by the rotor current.
  o Magnetization by the stator: $Q_{s\_ref} \cong 3|\underline{V}_s|^2/L_s\omega_s$, the real part of the rotor current is zero and all the magnetizing current is provided by the stator current.
  o Capacitive operation of the machine: $Q_{s\_ref} < 0$, for this purpose, both the stator and rotor currents must be increased.

- On the other hand, since the stator flux has constant behavior for a fixed grid, the rotor flux has very similar behavior to the rotor current. The real and imaginary parts of the rotor flux have very similar structure to the rotor current's real and imaginary parts.
- Finally, the rotor voltage amplitude has an equivalent structure as the rotor flux multiplied by ω_r. However, in this case, the real part of the flux corresponds to the imaginary part of the voltage and, similarly, the imaginary part of the flux corresponds to the real part of the voltage. In addition, depending on how the magnetization is chosen, that is, which is the Q_{s_ref} reference, the required rotor voltage can be larger or smaller, as occurs with the rotor currents.
- With regard to the powers, the stator active power depends only on T_{em_ref} for a fixed grid voltage. On the contrary, the rotor active and reactive powers' dependency on Q_{s_ref} and T_{em_ref} is more complex.

3.5.3.1 Example 3.8: Performance at Fixed T_{em} and Q_s

This example shows the steady state curves of a 2 MW DFIM, when torque and stator reactive power references are modified at different speeds. As shown by the equations derived previously, in general, the magnitudes will depend on the references and the speed of the machine for a fixed stator voltage. Consequently, the stator and rotor magnitudes with constant $Q_s = 0$ are shown first in Figure 3.26.

Figure 3.27 shows the same variables but at fixed torque ($T_{empu} = -1$). Notice that at fixed T_{em}, the higher current values are connected to negative values of Q_s, as described in this section.

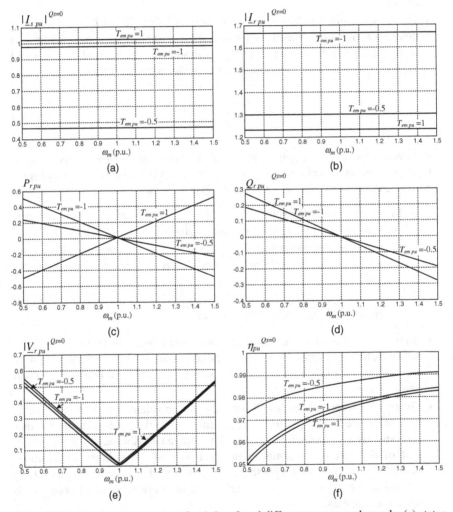

Figure 3.26 Steady state curves at fixed $Q_s = 0$ and different torques and speeds: (a) stator current, (b) rotor current, (c) rotor active power, (d) rotor reactive power, (e) rotor voltage, and (f) efficiency.

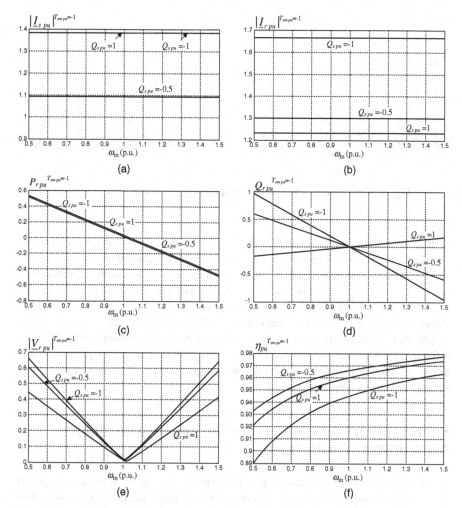

Figure 3.27 Steady state curves at fixed $T_{empu} = -1$ and different Q_s and speeds: (a) stator current, (b) rotor current, (c) rotor active power, (d) rotor reactive power, (e) rotor voltage, and (f) efficiency.

Also, the stator active power is connected to the T_{em}; therefore, the stator active power is not graphically represented in this evaluation. Similarly, the rotor active power depends mainly on the speed and the T_{em}, but is not affected by the Q_s (remember the simplified relation (3.56)).

On the contrary, the rotor reactive power strongly depends on all the variables: T_{em}, Q_s, and speed. At hypersynchronous speeds, the efficiency of the machine is increased, since both stator and rotor sides deliver electric energy to the grid.

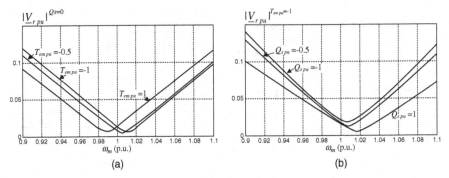

Figure 3.28 Zoom of the rotor voltages: (a) $Q_{spu} = 0$ and (b) $T_{empu} = -1$.

It is important to notice that the rotor voltage amplitude mainly depends on the speed of the machine. The torque does not strongly influence this variable.

To conclude, at synchronous speed, the required rotor voltage amplitude is different from zero, if some torque is needed. Note that expression (3.142) has neglected the voltage drop in the stator and rotor resistances. In consequence, as shown in Figure 3.28 (exact evaluation, not neglecting R_r), the required real voltage amplitude is slightly different from zero. This fact in real applications, when the rotor is fed by a back-to-back power electronic converter, yields to accuracy problems since at speeds near synchronism, a short deviation in the rotor voltage produces a strong variation in the desired operating point.

3.6 DESIGN REQUIREMENTS FOR THE DFIM IN WIND ENERGY GENERATION APPLICATIONS

A few years ago, control systems of wind turbines only allowed for a unity power factor of the generated power, as there were no requirements for reactive power generation given by the grid operators. However, recently introduced grid codes require reactive power generation or consumption (depending on the grid voltage rms value) by unconventional power systems like wind turbines. Typically, a requirement for the wind energy conversion systems is reactive power generation at nominal generated active power with a power factor equal to 0.95. However, other values of power factor may be required depending on the country or grid operator. Exemplary required power factors are 0.925 (EON—UK, Germany) or even 0.91 (Statnett—Norway) and 0.9 (AESO—Canada). It has to be noted, that for special conditions of wind turbine–grid connections, a lower power factor may be required by the grid operators. If the wind turbines are not able to meet operator requirements in the range of reactive power generation, additional reactive power compensators common to the entire wind farm are applied. Possibly, the required power factor in the future will be close to the power factor required for classic synchronous generators, that is, 0.8.

A doubly fed induction machine, already in use for decades, was designed for speed control by rotor connected resistors or for adjustable speed drive systems with rotor connected thyristor converters. For such applications, the machine was designed for reactive power consumption on the stator side. Almost all scientific publications on the DFIM, in which laboratory tests were presented, used the machine designed for the old type drives on the test bench. It is not important for the control system, but it is important for the DFIM concept itself, because in modern power systems, with fully controlled converters, the reactive current needed for machine excitation is provided on the rotor side, so the machine needs to be adequately designed.

In the generation mode, the rotor current is equal in value to the sum of the stator current flowing to the grid and the magnetization current. Active rotor current equals the value of the active stator current, while reactive rotor current is equal to the sum of the stator reactive current delivered to the grid and the magnetization current. The highest active power is generated by a wind turbine at maximum speed, which is typically 1.33 of synchronous speed. In this case, in an ideal machine, on the rotor side 33% of the stator active power is generated. In practice, a part of the rotor power is consumed by the power losses related to magnetization of the magnetic circuit. Sharing of reactive power supplied to the grid at maximum generation, between the stator and grid side converter, allows approximately equal loading of the rotor and grid side converter.

For induction machines in the megawatt range, designed for drive systems with reactive power consumption on the stator side, the magnetizing current is close to 20% of the nominal stator current. In Boldea [3], the electromagnetic design of a DFIM magnetized from the rotor side is described. The assumed ratio K_m of magnetizing current to stator nominal current is in the range of 0.1–0.3.

For the stator active power P_s and power factor $\cos(\gamma_v - \gamma_i)$ of the power delivered to the grid, the rms value of the stator current is

$$|\underline{I}_s| = \frac{S_s}{3|\underline{V}_s|} = \frac{P_s}{3|\underline{V}_s|\cos(\gamma_v - \gamma_i)} \tag{3.146}$$

Stator current can be divided into active and reactive parts:

$$\underline{I}_s = \frac{P_s}{3|\underline{V}_s|} + j\frac{P_s \sin(\gamma_v - \gamma_i)}{3|\underline{V}_s|\cos(\gamma_v - \gamma_i)} \tag{3.147}$$

Stator current can be described separately for the case of reactive power consumption,

$$\underline{I}_s = \frac{P_s}{3|\underline{V}_s|}\left(1 + j\frac{\sqrt{1 - \cos^2(\gamma_v - \gamma_i)}}{\cos(\gamma_v - \gamma_i)}\right) \tag{3.148}$$

or generation to the grid,

$$\underline{I}_s = \frac{P_s}{3|\underline{V}_s|}\left(1 - j\frac{\sqrt{1 - \cos^2(\gamma_v - \gamma_i)}}{\cos(\gamma_v - \gamma_i)}\right) \tag{3.149}$$

In both cases the rms of the stator current for a given active power and stator power factor is the same. The relation between the stator current rms values required for a given stator power factor and for unity stator power factor at the same active power is described by

$$\frac{|\underline{I}_s|}{|\underline{I}_s|^{PF=0}} = \frac{1}{\cos(\gamma_v - \gamma_i)} \tag{3.150}$$

For

$$K_m = \frac{|\underline{L}_m|}{|\underline{L}_s|_{rated}} \tag{3.151}$$

the rotor current equals

$$\underline{I}_r = \underline{I}_m - \underline{I}_s = jK_m |\underline{L}_s|_{rated} - |\underline{I}_s| \tag{3.152}$$

For the case of reactive power consumption and rated stator active power, the rotor current is described by

$$\underline{I}_r = -\frac{P_{s\,rated}}{3|\underline{V}_s|}\left(1 - j\frac{K_m - \sqrt{1 - \cos^2(\gamma_v - \gamma_i)}}{\cos(\gamma_v - \gamma_i)}\right) \tag{3.153}$$

Higher rotor current is required in the case of reactive power generation, as the rotor reactive current is needed for magnetization and for the reactive part of the stator current.

$$\underline{I}_r = -\frac{P_{s\,rated}}{3|\underline{V}_s|}\left(1 - j\frac{K_m + \sqrt{1 - \cos^2(\gamma_v - \gamma_i)}}{\cos(\gamma_v - \gamma_i)}\right) \tag{3.154}$$

Rotor current rms value for maximum (rated) active and maximum reactive stator power generation is given by

$$|\underline{I}_r| = \frac{P_{s\,rated}}{3|\underline{V}_s|}\sqrt{1 + \left(\frac{K_m + \sqrt{1 - \cos^2(\gamma_v - \gamma_i)}}{\cos(\gamma_v - \gamma_i)}\right)^2} \tag{3.155}$$

and its relation to the rated stator rms value is described by

$$K_{rs} = \frac{|\underline{I}_r|}{|\underline{L}_s|_{rated}} = \cos(\gamma_v - \gamma_i)\sqrt{1 + \left(\frac{K_m + \sqrt{1 - \cos^2(\gamma_v - \gamma_i)}}{\cos(\gamma_v - \gamma_i)}\right)^2} \tag{3.156}$$

The relation between rotor current rms values required for a given stator power factor and for unity stator power factor for rated stator active power and generated reactive power is given as

$$\frac{|\underline{I}_r|}{|\underline{I}_r|^{PF=0}} = \sqrt{\frac{1 + \left(\frac{K_m + \sqrt{1-\cos^2(\gamma_v - \gamma_i)}}{\cos(\gamma_v - \gamma_i)}\right)^2}{1 + K_m^2}} \qquad (3.157)$$

The ratio of K_{rs} factors for rated stator active power and for a given stator power factor and for unity stator power factor can also be found.

$$\frac{K_{rs}}{K_{rs}^{PF=0}} = \cos(\gamma_v - \gamma_i)\sqrt{\frac{1 + \left(\frac{K_m + \sqrt{1-\cos^2(\gamma_v - \gamma_i)}}{\cos(\gamma_v - \gamma_i)}\right)^2}{1 + K_m^2}} \qquad (3.158)$$

All equations were presented for all currents referred to the stator side.

Considering a stator to rotor turns ratio equal to 1/3, the rotor rms current referred to the rotor and K'_{rs} referred to the rotor are three times smaller. The rotor current rms to stator current rms ratio K'_{rs}, for $\cos(\gamma_v - \gamma_i)$ in the range of 0.8–1.0 at reactive power generation, is shown in Figure 3.29. This figure shows the requirement of the rotor current as well as the current of the rotor converter in relation to the stator current rms value. Simultaneously, based on this figure, the required current (or power) of the converter in the DFIM system can easily be compared to the required current (or power) of the full scale converter in series topologies with SCIG or SG. It can also be seen that the reactive power for magnetization requires a slightly higher rotor current than the current responsible for slip power at maximum speed. Simultaneous

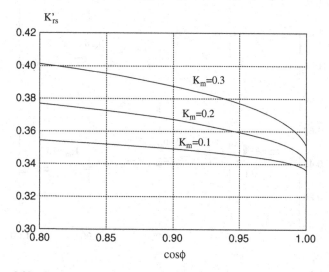

Figure 3.29 Dependence of rotor to stator current ratio on stator power factor.

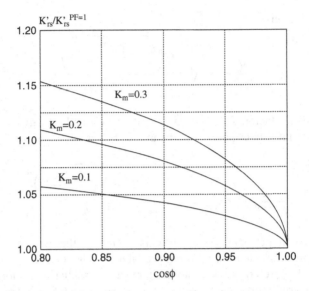

Figure 3.30 Ratio of the K'_{rs} factor to the K'_{rs} at unity stator power factor.

magnetization and generation of active and reactive power require significantly higher rotor current.

Figure 3.30 shows how the ratio of the rotor current to the stator current changes with the required power factor. For $K_m = 0.2$ and power factor equal to 0.9, the ratio of the rotor current to the stator current in relation to the case with unity stator power

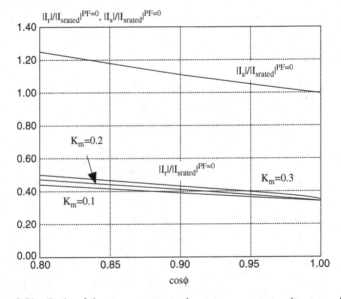

Figure 3.31 Ratio of the rotor current to the stator current at unity power factor.

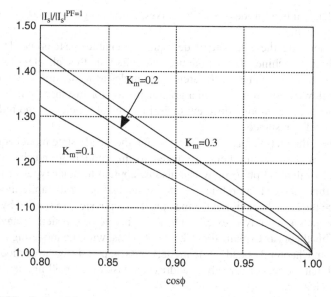

Figure 3.32   Ratio of the rotor current to the rotor current at unity stator power factor.

factor is 8% higher. It indicates that rotor and stator current requirements depend on the stator power factor.

Figure 3.31 shows the ratio of the stator current rms value and rotor current rms value for different $\cos(\gamma_v - \gamma_i)$ values to the active stator current at maximum generation. This figure also shows the requirements for a doubly fed induction machine stator and rotor, for a power factor in the range of 0.8–1. The reference is the stator current of the machine designed only for active power generation.

Figure 3.32 shows a comparison of the rotor current needed for generation of reactive power, to the rotor current needed at unity stator power factor. It can be seen, for example, that for $K_m = 0.2$ and the required $\cos(\gamma_v - \gamma_i) = 0.9$, the rotor side converter has to be designed for 20% higher current than in the case of unity stator power factor.

3.7 SUMMARY

The steady state model of the DFIM has been explained, showing in detail the derivation of the model electric equations. From this, we have demonstrated that the steady state model equations are useful for understanding the machine and as a tool for performance evaluation. The presented steady state modeling method is based on phasor theory, so it represents variables of the machine such as voltages, currents, and fluxes as enunciated in the theory of complexors, used for steady state analysis. Hence, these model equations provide a comprehensive approach, enabling one to understand the different operating modes and the behavior of the most representative electric or electromechanical magnitudes (currents, fluxes, torque, angular speeds, power flows, etc.) for the machine.

On the other hand, these derived model equations enable one to proceed to a deeper analysis of the machine. The evaluation presented in this chapter displays the performance of the machine for different magnitudes as a function of the operating condition, allowing one to decide in a graphical way, for instance, which operating point is more suitable among different ones, in terms of efficiency and reduction of rotor currents, for instance.

In addition, the method presented for deriving the steady state model equations is independent of the machine being analyzed so this modeling and the subsequent performance evaluation philosophy can also be applied to other types of machines.

Finally, this chapter is a preliminary step in analyzing the dynamic model of the DFIM. Once the basic behavior of the machine and it physics are studied by means of steady state analysis, study is extended to the dynamic or transient behavior of the machine. This theory is explained in Chapter 4. Thus, with the modeling knowledge obtained from Chapter 3 (steady state model) and Chapter 4 (dynamic model), it can be affirmed that the reader is ready to address the task of controlling the machine in subsequent chapters.

REFERENCES

1. S. J. Chapman, *Electrical Machines*. McGraw Hill, 1985.
2. J. F. Mora, *Electrical Machines*. McGraw Hill, 2003.
3. I. Boldea, *Variable Speed Generators*. CRC Press Taylor & Francis, 2006.
4. I. Boldea and A. Nasar, *The Induction Machine Handbook*. CRC Press, 2006.
5. B. K. Bose, *Modern Power Electronics and AC Drives*, Prenctice Hall, 2002.
6. W. Leonhard, *Control of Electrical Drives*. Springer-Verlag, 1985.
7. B. K. Bose, *Power Electronics and Motor Drives*. Elsevier, 2006.
8. M. P. Kazmierkowski, R. Krishnan, and F. Blaabjerg, *Control in Power Electronics: Selected Problems*. Academic Press, 2002.
9. P. Vas, *Sensorless Vector and Direct Torque Control*. Oxford University Press, 1998.
10. A. Veltman, D. W. J. Pulle, and R. W. De Doncker, *Fundamentals of Electric Drives*. Springer, 2007.
11. A. M. Trzynadlowski, *Control of Induction Motors*. Academic Press, 2001.
12. A. Hughes, *Electric Motors and Drives*. Elsevier, 1990.
13. M. Barnes, *Variable Speed Drives and Power Electronics*. Elsevier, 2003.
14. R. Datta,"Rotor Side Control of Grid-Connected Wound Rotor Induction Machine and Its Application to Wind Power Generation." Ph.D. thesis, Department of Electrical Engineering, Indian Institute of Science, Bangalore, India, February 2000.
15. A. Peterson."Analysis, Modeling and Control of Doubly-Fed Induction Generators for Wind Turbines." Ph.D. thesis, Chalmers University of Technology, Goteborg, Sweden, 2005.

CHAPTER 4

Dynamic Modeling of the Doubly Fed Induction Machine

4.1 INTRODUCTION

This chapter continues describing the model of the DFIM. In Chapter 3, the analysis focused on the steady state model; however, this knowledge is not sufficient to reach a reasonable level of understanding of the machine. The dynamic and transient behaviors of the DFIM must be examined for modeling purposes; and perhaps more importantly, for development of the subsequent machine control.

The validity of the mathematical model obtained in Chapter 3 is circumscribed to the steady state of the machine; however, it is also very important to know how a steady state can be achieved when the machine is in a different state. This dynamic behavior explains and defines the behavior of the machine's variables in transition periods as well as in the steady state. This dynamic behavior of machines is normally studied by a "dynamic model." By means of the dynamic model it is possible to know at all times the continuous performance (not only at steady state) of the variables of the machine, such as torque, currents, and fluxes, under certain voltage supplying conditions. In this way, by using the information provided by the dynamic model, it is possible to know how the transition from one state to another is going to be achieved, allowing one to detect unsafe behaviors, such as instabilities or high transient currents. On the other hand, the dynamic model provides additional information of the system during the steady state operation, such as dynamic oscillations, torque or current ripples, etc . . .

Consequently, the dynamic model, represented in general in differential equation form, is often structured as a compact set of model equations, allowing it to be simulated by computer based software and providing all the information related to the machine's variables. This is often called a "simulation model." It enables one to know the continuous behavior of all variables of the machine.

Thus, this chapter develops different dynamic models of the DFIM based on the space vector theory. By means of this powerful mathematical tool, the dynamic model equations (differential equations) of the DFIM are derived. Then, from the obtained

---

Doubly Fed Induction Machine: Modeling and Control for Wind Energy Generation,
First Edition. By G. Abad, J. López, M. A. Rodríguez, L. Marroyo, and G. Iwanski.

models, numerical examples as well as graphical representation of performances are obtained by dynamic simulation, providing the reader with practical information and a deeper understanding of the machine's behavior.

4.2 DYNAMIC MODELING OF THE DFIM

According to models of AC machines developed by several authors and as discussed at the beginning of the previous chapter, the simplified and idealized DFIM model can be described as three windings in the stator and three windings in the rotor, as illustrated in Figure 4.1.

These windings are an ideal representation of the real machine, which helps to derive an equivalent electric circuit, as shown in Figure 4.2.

Under this idealized model, the instantaneous stator voltages, current, and fluxes of the machine can be described by the following electric equations:

$$v_{as}(t) = R_s i_{as}(t) + \frac{d\psi_{as}(t)}{dt} \tag{4.1}$$

$$v_{bs}(t) = R_s i_{bs}(t) + \frac{d\psi_{bs}(t)}{dt} \tag{4.2}$$

$$v_{cs}(t) = R_s i_{cs}(t) + \frac{d\psi_{cs}(t)}{dt} \tag{4.3}$$

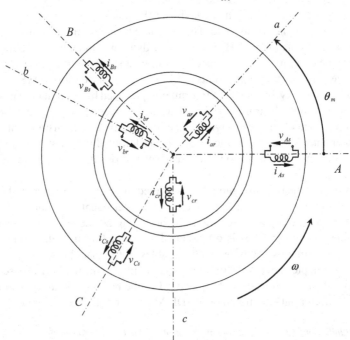

Figure 4.1 Ideal three-phase windings (stator and rotor) of the DFIM.

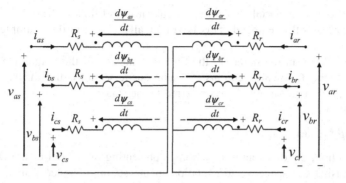

Figure 4.2   DFIM electric equivalent circuit.

where R_s is the stator resistance; $i_{as}(t)$, $i_{bs}(t)$, and $i_{cs}(t)$ are the stator currents of phases a, b, and c; $v_{as}(t)$, $v_{bs}(t)$, and $v_{cs}(t)$ are the applied stator voltages; and $\psi_{as}(t)$, $\psi_{bs}(t)$, and $\psi_{cs}(t)$ are the stator fluxes. The stator side electric magnitudes, at steady state, have a constant sinusoidal angular frequency, ω_s, the angular frequency imposed by the grid.

Similarly, the rotor magnitudes are described by

$$v_{ar}(t) = R_r i_{ar}(t) + \frac{d\psi_{ar}(t)}{dt} \tag{4.4}$$

$$v_{br}(t) = R_r i_{br}(t) + \frac{d\psi_{br}(t)}{dt} \tag{4.5}$$

$$v_{cr}(t) = R_r i_{cr}(t) + \frac{d\psi_{cr}(t)}{dt} \tag{4.6}$$

where R_r is the rotor resistance referred to the stator; $i_{ar}(t)$, $i_{br}(t)$, and $i_{cr}(t)$ are the stator referred rotor currents of phases a, b, and c; $v_{ar}(t)$, $v_{br}(t)$, and $v_{cr}(t)$ are the stator referred rotor voltages; and $\psi_{ar}(t)$, $\psi_{br}(t)$, and $\psi_{cr}(t)$ are the rotor fluxes. Under steady state operating conditions, the rotor magnitudes have constant angular frequency, ω_r.

In this chapter, assuming a general DFIM built with different turns in the stator and rotor, all parameters and magnitudes of the rotor are referred to the stator.

It was indicated in the previous chapter that the relation between the stator angular frequency and the rotor angular frequency is

$$\omega_r + \omega_m = \omega_s \tag{4.7}$$

where ω_m is the electrical angular frequency of the machine. Similarly, Ω_m is the mechanical angular speed, related to the electrical frequency by means of a pair of poles, p:

$$\omega_m = p\Omega_m \tag{4.8}$$

Hence, the rotor variables (voltages, currents, and fluxes) present a pulsation ω_r that varies with the speed. This is graphically shown in the examples of this chapter.

For simplicity in the notation, the time dependence of the magnitudes will be omitted in the following sections. In subsequent sections, the magnitudes and parameters of the rotor are always referred to the stator.

4.2.1 αβ Model

In this section, the differential equations representing the model of the DFIM are derived, using the space vector notation in the stator reference frame. (See the Appendix for basic details.) Hence, by multiplying Equations (4.1) and (4.4) by $\frac{2}{3}$, then multiplying Equations (4.2) and (4.5) by $\frac{2}{3}a$ and also multiplying Equations (4.3) and (4.6) by $\frac{2}{3}a^2$, the addition of the resulting equations yields the voltage equations of the DFIM, in space vector form:

$$\vec{v}_s^s = R_s \vec{i}_s^s + \frac{d\vec{\psi}_s^s}{dt} \tag{4.9}$$

$$\vec{v}_r^r = R_r \vec{i}_r^r + \frac{d\vec{\psi}_r^r}{dt} \tag{4.10}$$

where $\vec{v}_s^s$ is the stator voltage space vector, $\vec{i}_s^s$ is the stator current space vector, and $\vec{\psi}_s^s$ is the stator flux space vector. Equation (4.9) is represented in stator coordinates (αβ reference frame). $\vec{v}_r^r$ is the rotor voltage space vector, $\vec{i}_r^r$ is the rotor current space vector, and $\vec{\psi}_r^r$ is the rotor flux space vector. Equation (4.10) is represented in rotor coordinates (*DQ* reference frame).

Note that the superscripts "s" and "r" indicate that space vectors are referred to stator and rotor reference frames, respectively. On the other hand, the correlation between the fluxes and the currents, in space vector notation, is given by

$$\vec{\psi}_s^s = L_s \vec{i}_s^s + L_m \vec{i}_r^s \tag{4.11}$$

$$\vec{\psi}_r^r = L_m \vec{i}_s^r + L_r \vec{i}_r^r \tag{4.12}$$

where L_s and L_r are the stator and rotor inductances, L_m is the magnetizing inductance, and they are related to the stator leakage inductance $L_{\sigma s}$ and the rotor leakage inductance $L_{\sigma r}$, according to the following expressions:

$$L_s = L_{\sigma s} + L_m \tag{4.13}$$

$$L_r = L_{\sigma r} + L_m \tag{4.14}$$

Note again that Equation (4.11) is represented in the stator reference frame, while Equation (4.12) is in the rotor reference frame. Taking into account the coordinate transformation, the following relations hold (see Appendix, Section A.1 for more details):

$$\vec{\psi}_s^s = L_s \vec{i}_s^s + L_m \vec{i}_r^s = L_s \vec{i}_s^s + L_m e^{j\theta_m} \vec{i}_r^r \tag{4.15}$$

$$\vec{\psi}_r^r = L_m \vec{i}_s^r + L_r \vec{i}_r^r = L_m e^{-j\theta_m} \vec{i}_s^s + L_r \vec{i}_r^r \tag{4.16}$$

Consequently, by referring the corresponding space vectors to the stator reference frame (multiply Equation (4.10) by $e^{j\theta_m}$), the $\alpha\beta$ model of the DFIM is obtained by the next equations in stator coordinates:

$$\vec{v}_s^s = R_s \vec{i}_s^s + \frac{d\vec{\psi}_s^s}{dt} \tag{4.17}$$

$$\vec{v}_r^s = R_r \vec{i}_r^s + \frac{d\vec{\psi}_r^s}{dt} - j\omega_m \vec{\psi}_r^s \tag{4.18}$$

$$\vec{\psi}_s^s = L_s \vec{i}_s^s + L_m \vec{i}_r^s \tag{4.19}$$

$$\vec{\psi}_r^s = L_m \vec{i}_s^s + L_r \vec{i}_r^s \tag{4.20}$$

Note that to transform into Equation (4.18), it is necessary to consider

$$\frac{d\vec{\psi}_r^r}{dt} e^{j\theta_m} = \overbrace{\frac{d\left(\vec{\psi}_r^r e^{j\theta_m}\right)}{dt}}^{\vec{\psi}_r^s} - j\omega_m \overbrace{\vec{\psi}_r^r e^{j\theta_m}}^{\vec{\psi}_r^s} \tag{4.21}$$

Figure 4.3 shows the $\alpha\beta$ electrical model of the DFIM in stator coordinates.

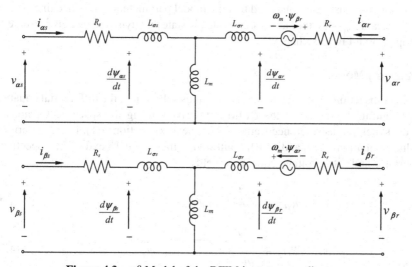

Figure 4.3 $\alpha\beta$ Model of the DFIM in stator coordinates.

Continuing with the model, the electric powers on the stator side and on the rotor side are calculated as follows:

$$P_s = \tfrac{3}{2}\,\mathrm{Re}\{\vec{v}_s\cdot\vec{i}_s^{\,*}\} = \tfrac{3}{2}(v_{\alpha s}i_{\alpha s} + v_{\beta s}i_{\beta s}) \qquad (4.22)$$

$$P_r = \tfrac{3}{2}\,\mathrm{Re}\{\vec{v}_r\cdot\vec{i}_r^{\,*}\} = \tfrac{3}{2}(v_{\alpha r}i_{\alpha r} + v_{\beta r}i_{\beta r}) \qquad (4.23)$$

$$Q_s = \tfrac{3}{2}\,\mathrm{Im}\{\vec{v}_s\cdot\vec{i}_s^{\,*}\} = \tfrac{3}{2}(v_{\beta s}i_{\alpha s} - v_{\alpha s}i_{\beta s}) \qquad (4.24)$$

$$Q_r = \tfrac{3}{2}\,\mathrm{Im}\{\vec{v}_r\cdot\vec{i}_r^{\,*}\} = \tfrac{3}{2}(v_{\beta r}i_{\alpha r} - v_{\alpha r}i_{\beta r}) \qquad (4.25)$$

where the superscript * represents the complex conjugate of a space vector as was used in phasors. Finally, the electromagnetic torque can be found from

$$T_{em} = \tfrac{3}{2}p\,\mathrm{Im}\{\vec{\psi}_r\cdot\vec{i}_r^{\,*}\} = \tfrac{3}{2}p(\psi_{\beta r}i_{\alpha r} - \psi_{\alpha r}i_{\beta r}) \qquad (4.26)$$

By substitution of Equations (4.19) and (4.20) into (4.26), the electromagnetic torque can also be calculated according to the following equivalent expressions:

$$T_{em} = \tfrac{3}{2}p\,\frac{L_m}{L_s}\,\mathrm{Im}\{\vec{\psi}_s\cdot\vec{i}_r^{\,*}\} = \tfrac{3}{2}p\,\mathrm{Im}\{\vec{\psi}_s^{\,*}\cdot\vec{i}_s\} = \tfrac{3}{2}\frac{L_m}{L_r}p\,\mathrm{Im}\{\vec{\psi}_r^{\,*}\cdot\vec{i}_s\}$$

$$= \tfrac{3}{2}\,\frac{L_m}{\sigma L_r L_s}p\,\mathrm{Im}\{\vec{\psi}_r^{\,*}\cdot\vec{\psi}_s\} = \tfrac{3}{2}L_m p\,\mathrm{Im}\{\vec{i}_s\cdot\vec{i}_r^{\,*}\} \qquad (4.27)$$

where $\sigma = 1 - L_m^2/L_s L_r$ is the leakage coefficient and p is the pair of poles of the machine. Note that for simplicity in the notation, the superscript "s" has been omitted from the space vector in the power and torque expressions.

Finally, it must be emphasized that the model parameters of the machine—R_s, R_r, $L_{\sigma s}$, $L_{\sigma r}$, and L_m—are the same for both steady state and dynamic models presented in Chapter 3 and in this chapter.

4.2.2 *dq* Model

In this subsection, in contrast with the previous subsection, the differential equations representing the model of the DFIM are derived, using the space vector notation in the synchronous reference frame (see Appendix, Section A.1). From the original voltage equations (4.9) and (4.10), multiplying them by $e^{-j\theta_s}$ and $e^{-j\theta_r}$, respectively, the stator and rotor voltage equations yields.

$$\vec{v}_s^{\,a} = R_s\vec{i}_s^{\,a} + \frac{d\vec{\psi}_s^{\,a}}{dt} + j\omega_s\vec{\psi}_s^{\,a} \qquad (4.28)$$

$$\vec{v}_r^{\,a} = R_r\vec{i}_r^{\,a} + \frac{d\vec{\psi}_r^{\,a}}{dt} + j\overbrace{(\omega_s - \omega_m)}^{\omega_r}\vec{\psi}_r^{\,a} \qquad (4.29)$$

In this case, the superscript "a" denotes space vectors referred to a synchronously rotating frame. From Equations (4.11) and (4.12), by using the same the flux expressions, we find

$$\vec{\psi}_s^a = L_s \vec{i}_s^a + L_m \vec{i}_r^a \tag{4.30}$$

$$\vec{\psi}_r^a = L_m \vec{i}_s^a + L_r \vec{i}_r^a \tag{4.31}$$

For a sinusoidal supply of voltages, at steady state, the dq components of the voltages, currents, and fluxes will be constant values, in contrast to the $\alpha\beta$ components that are sinusoidal magnitudes. Hence, the dq equivalent circuit model of the DFIM, in synchronous coordinates, is represented in Figure 4.4.

The torque and power expressions in the dq reference frame are equivalent to the $\alpha\beta$ equations:

$$P_s = \tfrac{3}{2}\text{Re}\{\vec{v}_s \cdot \vec{i}_s^*\} = \tfrac{3}{2}(v_{ds}i_{ds} + v_{qs}i_{qs}) \tag{4.32}$$

$$P_r = \tfrac{3}{2}\text{Re}\{\vec{v}_r \cdot \vec{i}_r^*\} = \tfrac{3}{2}(v_{dr}i_{dr} + v_{qr}i_{qr}) \tag{4.33}$$

$$Q_s = \tfrac{3}{2}\text{Im}\{\vec{v}_s \cdot \vec{i}_s^*\} = \tfrac{3}{2}(v_{qs}i_{ds} - v_{ds}i_{qs}) \tag{4.34}$$

$$Q_r = \tfrac{3}{2}\text{Im}\{\vec{v}_r \cdot \vec{i}_r^*\} = \tfrac{3}{2}(v_{qr}i_{dr} - v_{dr}i_{qr}) \tag{4.35}$$

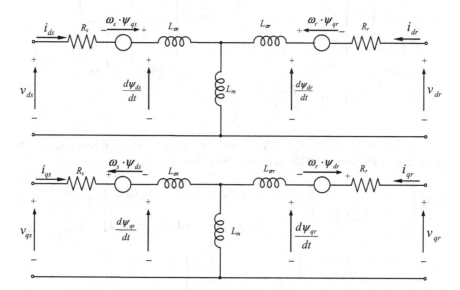

Figure 4.4 dq Model of the DFIM in synchronous coordinates.

Therefore, the torque expression also yields

$$T_{em} = \frac{3}{2} p \frac{L_m}{L_s} \operatorname{Im}\{\vec{\psi}_s \cdot \vec{i}_r^*\} = \frac{3}{2} p \frac{L_m}{L_s} \left(\psi_{qs} i_{dr} - \psi_{ds} i_{qr} \right) \tag{4.36}$$

In the same way, all the equivalent torque expressions of Equation (4.27) still hold. Again, for the sake of simplicity, the superscripts of the space vectors have been omitted in the power and torque expressions.

4.2.3 State-Space Representation of $\alpha\beta$ Model

A representation of the $\alpha\beta$ model in state-space equations is very useful for computer based simulation purposes. Rearranging Equations (4.17)–(4.20) and taking the fluxes as state-space magnitudes, the model of the DFIM is given by the next expression:

$$\frac{d}{dt} \begin{bmatrix} \vec{\psi}_s^s \\ \vec{\psi}_r^s \end{bmatrix} = \begin{bmatrix} \dfrac{-R_s}{\sigma L_s} & \dfrac{R_s L_m}{\sigma L_s L_r} \\[2mm] \dfrac{R_r L_m}{\sigma L_s L_r} & \dfrac{-R_r}{\sigma L_r} + j\omega_m \end{bmatrix} \cdot \begin{bmatrix} \vec{\psi}_s^s \\ \vec{\psi}_r^s \end{bmatrix} + \begin{bmatrix} \vec{v}_s^s \\ \vec{v}_r^s \end{bmatrix} \tag{4.37}$$

Expanding this last expression in the $\alpha\beta$ components, we obtain

$$\frac{d}{dt} \begin{bmatrix} \psi_{\alpha s} \\ \psi_{\beta s} \\ \psi_{\alpha r} \\ \psi_{\beta r} \end{bmatrix} = \begin{bmatrix} \dfrac{-R_s}{\sigma L_s} & 0 & \dfrac{R_s L_m}{\sigma L_s L_r} & 0 \\[2mm] 0 & \dfrac{-R_s}{\sigma L_s} & 0 & \dfrac{R_s L_m}{\sigma L_s L_r} \\[2mm] \dfrac{R_r L_m}{\sigma L_s L_r} & 0 & \dfrac{-R_r}{\sigma L_r} & -\omega_m \\[2mm] 0 & \dfrac{R_r L_m}{\sigma L_s L_r} & \omega_m & \dfrac{-R_r}{\sigma L_r} \end{bmatrix} \cdot \begin{bmatrix} \psi_{\alpha s} \\ \psi_{\beta s} \\ \psi_{\alpha r} \\ \psi_{\beta r} \end{bmatrix} + \begin{bmatrix} v_{\alpha s} \\ v_{\beta s} \\ v_{\alpha r} \\ v_{\beta r} \end{bmatrix} \tag{4.38}$$

If instead of the fluxes, the currents are chosen as state-space magnitudes, the equivalent model of the DFIM is derived as follows:

$$\frac{d}{dt} \begin{bmatrix} \vec{i}_s^s \\ \vec{i}_r^s \end{bmatrix} = \left(\frac{1}{\sigma L_s L_r} \right) \begin{bmatrix} -R_s L_r - j\omega_m L_m^2 & R_r L_m - j\omega_m L_m L_r \\ R_s L_m + j\omega_m L_m L_s & -R_r L_s + j\omega_m L_r L_s \end{bmatrix} \cdot \begin{bmatrix} \vec{i}_s^s \\ \vec{i}_r^s \end{bmatrix}$$

$$+ \left(\frac{1}{\sigma L_s L_r} \right) \begin{bmatrix} L_r & -L_m \\ -L_m & L_s \end{bmatrix} \cdot \begin{bmatrix} \vec{v}_s^s \\ \vec{v}_r^s \end{bmatrix} \tag{4.39}$$

Expanding in $\alpha\beta$ components, we have

$$
\frac{d}{dt}\begin{bmatrix} i_{\alpha s} \\ i_{\beta s} \\ i_{\alpha r} \\ i_{\beta r} \end{bmatrix} = \left(\frac{1}{\sigma L_s L_r}\right)\begin{bmatrix} -R_s L_r & \omega_m L_m^2 & R_r L_m & \omega_m L_m L_r \\ -\omega_m L_m^2 & -R_s L_r & -\omega_m L_m L_r & R_r L_m \\ R_s L_m & -\omega_m L_s L_m & -R_r L_s & -\omega_m L_r L_s \\ \omega_m L_s L_m & R_s L_m & \omega_m L_r L_s & -R_r L_s \end{bmatrix} \cdot \begin{bmatrix} i_{\alpha s} \\ i_{\beta s} \\ i_{\alpha r} \\ i_{\beta r} \end{bmatrix}
$$

$$
+\left(\frac{1}{\sigma L_s L_r}\right)\begin{bmatrix} L_r & 0 & -L_m & 0 \\ 0 & L_r & 0 & -L_m \\ -L_m & 0 & L_s & 0 \\ 0 & -L_m & 0 & L_s \end{bmatrix} \cdot \begin{bmatrix} v_{\alpha s} \\ v_{\beta s} \\ v_{\alpha r} \\ v_{\beta r} \end{bmatrix}
$$

$$\text{(4.40)}$$

Depending on the choice of the state-space magnitudes, different state-space models can be obtained.

4.2.3.1 Example 4.1: Simulation Block Diagram from the $\alpha\beta$ Model of the DFIM

For on-line simulation purposes, it is possible to use the $\alpha\beta$ model of the DFIM, as illustrated in Figure 4.5. This simulation structure is suitable for implementation in computer based software tools such as Matlab-Simulink. Hence, the simulation block diagram is composed of the following inputs, parameters, and outputs:

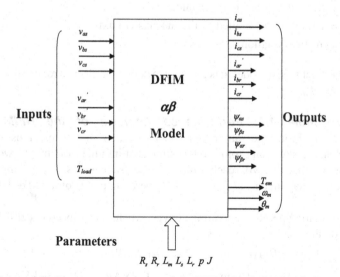

Figure 4.5   Input–output structure of the DFIM simulation block diagram.

- *Inputs.* Stator and rotor phase (*abc*) voltages and the load torque (T_{load}).
- *Parameters.* Constant parameters of the DFIM model, including the electric and mechanical parts of the machine.
- *Outputs.* Stator and rotor phase (*abc*) currents, stator and rotor $\alpha\beta$ fluxes, the electromagnetic torque, the speed, and the rotor angle.

This simulation model calculates the above-mentioned outputs, according to the imposed voltage inputs and the machine's parameters. A more detailed block diagram is presented in Figure 4.6, based on the state-space representation of Equations (4.39) and (4.40).

The real phase (*abc*) input voltages are transformed into voltages referred to the stationary frame ($\alpha\beta$). Note that rotational transformation and Clarke matrix transformations are required as well as the stator/rotor turns ratio provided by the factor *u*.

Once the stator and rotor voltages are in $\alpha\beta$ coordinates, the state-space solution of expression (4.40) is calculated. Hence, the stator and rotor currents $\alpha\beta$ components are derived. Once that is done the phase *abc* stator and rotor currents are calculated.

In an equivalent way, from the stator and rotor $\alpha\beta$ currents, the stator and rotor $\alpha\beta$ fluxes are calculated by means of expressions (4.19) and (4.20). Accordingly, the electromagnetic torque is also calculated from expression (4.27). After that, the mechanical model of the machine is implemented. In this case, a very simple mechanical model has been considered as depicted in Figure 4.7, mathematically represented by the following equation:

$$ T_{em} - T_{load} = J \frac{d\Omega_m}{dt} \tag{4.41} $$

where

J = equivalent inertia of the mechanical axis

T_{load} = external torque applied to the mechanical axis

Ω_m = mechanical rotational speed

From the mechanical model, it is possible to derive the electric rotational speed ω_m and the angle θ_m.

4.2.3.2 Example 4.2: Performance Analysis of the DFIM in $\alpha\beta$ Coordinates
From the simulation model studied in the previous example, depending on the stator and rotor voltages injected into the machine, as well as the load torque, the machine will reach different steady state points. This example shows the *abc* and $\alpha\beta$ magnitudes, at two different operating points of a 2 MW DFIM.

- First operating point—generating at hypersynchronous speed under the following conditions:

$$ \Omega_m = 2100 \text{ rev/min} \quad P_s = -1.2 \text{ MW} \quad P_r = -0.5 \text{ MW} $$
$$ T_{em} = -7400 \text{ N·m} \quad Q_s = 1.1 \text{ MVAR} \quad Q_r = 80 \text{ kVAR} $$

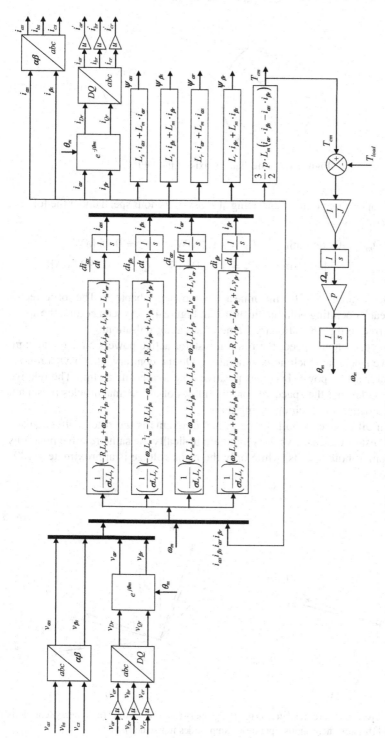

Figure 4.6 Simulation block diagram of the DFIM.

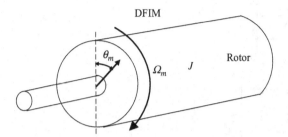

Figure 4.7   Mechanical axis of the DFIM.

- Second operating point—generating at subsynchronous speed under the following conditions:

$$\Omega_m = 900\,\text{rev/min} \qquad P_s = -1.2\,\text{MW} \qquad P_r = 0.45\,\text{MW}$$

$$T_{em} = -7400\,\text{N·m} \qquad Q_s = 1.1\,\text{MVAR} \qquad Q_r = -0.19\,\text{MVAR}$$

The model parameters of the machine are as shown in Chapter 3. The space vector diagram of each operating point is illustrated in Figure 4.8. As seen before, the space vectors referred to the $\alpha\beta$ stationary frame rotate at ω_s pulsation.

At hypersynchronous speed, the rotor and stator active powers have equal sign, both negative since the machine is generating. On the contrary, at subsynchronous speed the rotor active power becomes positive (due to a positive slip). The relative phase shift angles and the space vector amplitudes of all the magnitudes presented in Figure 4.8 determine each operating point.

The main difference in both space vector diagrams is the rotor voltage space vector. The rest of the space vectors maintain basically the same relative positions between them at both speeds, while only the rotor voltage is approximately 180° phase shifted.

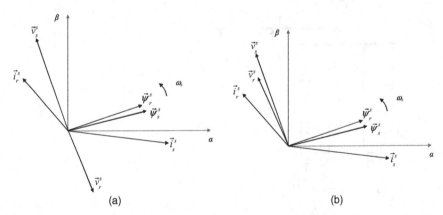

Figure 4.8   Space vector representation in $\alpha\beta$ stationary reference frame: (a) hypersynchronous operation and (b) subsynchronous operation (amplitudes not scaled).

Compared with the phasor diagrams examined in previous chapters for steady state analysis, in this case the space vectors are rotating vectors with constant rotating speed at steady state. Note also that the amplitude of the space vectors is adopted as the peak value, while for phasors we adopted rms values.

Similarly, the machine's steady state behavior is illustrated in Figures 4.9 and 4.11. The time evolution of the most representative magnitudes is presented in abc and $\alpha\beta$ coordinates, as given by the block diagram of Figure 4.6.

Therefore, the stator and rotor input voltages are imposed as shown in Figures 4.9a–4.9d. Approximately in the middle of the simulation experiment, the rotor voltage is 180° phase shifted, producing the speed change from 2100 rev/min to 900 rev/min. On the contrary, the stator voltage is not modified.

This quick speed change of the machine is achieved thanks to a closed loop control, which is not analyzed in this chapter. In fact, so severe a change is not realistic, however, it is useful to show the performance at hypersynchronous and subsynchronous speeds simultaneously in a graph.

Checking the voltages, note that the stator a phase and α component are the same waveforms. Both abc and $\alpha\beta$ stator voltages have $\omega_s = 314.16$ rad/s pulsation and equal amplitude. However, the phase rotor voltages have $\omega_r = -125.6$ rad/s pulsation at hypersynchronism (acb sequence) and $\omega_r = 125.6$ rad/s at subsynchronism (abc sequence). On the other hand, the $\alpha\beta$ rotor voltages have $\omega_s = 314.16$ rad/s pulsation (abc sequence) in all simulation experiments. In addition, the abc rotor voltages shown in Figure 4.9c are rotor referred, that is, the real rotor voltages. That is the reason for the different amplitude $\alpha\beta$ rotor voltages, in this case referred to the stator.

On the other hand, as noticed in Figures 4.9e and 4.9f, the torque is mainly constant, while the speed changes approximately in the middle of the experiment. As highlighted before, to achieve this performance, it is necessary to impose an appropriate control strategy. In this way, the stator is sinusoidally supplied, while the rotor is fed by a two-level back-to-back converter (see Chapter 2 for more details). The control strategy in charge of generating the required rotor voltages is a direct control technique (DTC) that will be studied in Chapter 8. For simplicity in the exposition, the rotor voltages are shown filtered, meaning that only the fundamental component of the voltage applied to the rotor is shown. In any case, the ripples due to this converter based supply, for the currents, powers, and torque, are still present in the simulation based graphics. This fact is graphically represented in Figure 4.10.

Finally, the stator and rotor active and reactive power behaviors are shown in Figures 4.9g and 4.9h.

Continuing with the analysis, in Figure 4.11, the rest of the machine's most interesting magnitudes are presented. Only $\alpha\beta$ fluxes are calculated. Stator and rotor fluxes are slightly phase shifted in relation to each other, remaining unaltered during during the experiment, as can be seen in Figures 4.11a and 4.11b.

The stator currents have very similar behavior to the stator voltages, Figures 4.11c and 4.11d. The same amplitude and pulsation ($\omega_s = 314.16$ rad/s) exist for both phase and $\alpha\beta$ currents. They remain unaltered during the experiment.

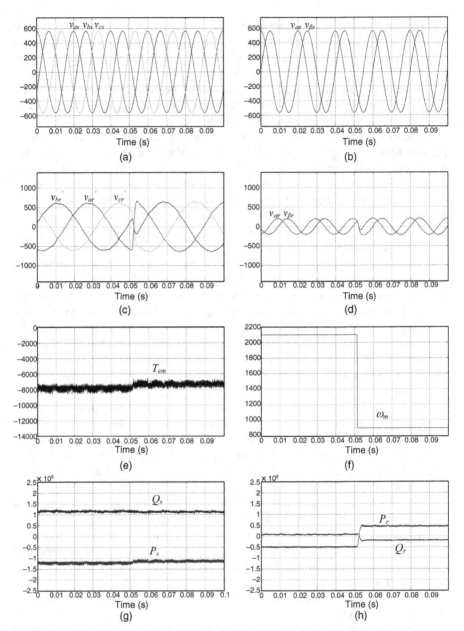

Figure 4.9 Generator operation of a 2 MW DFIM, at hypersynchronous and subsyncronous speeds: (a) *abc* stator voltages, (b) αβ stator voltages (V), (c) *abc* real rotor voltages (V), (d) αβ rotor voltages (stator referred) (V), (e) electromagnetic torque (N·m), (f) speed (rpm), (g) active and reactive powers (W and VAR), and (h) rotor active and reactive powers (W and VAR).

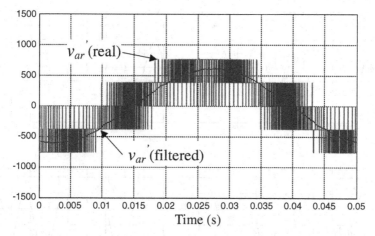

Figure 4.10   Converter's output a phase voltage and filtered voltage.

Finally, the rotor phase currents shown in Figure 4.11e have $\omega_r = -125.6\,\text{rad/s}$ pulsation at hypersynchronism (*acb* sequence) and $\omega_r = 125.6\,\text{rad/s}$ at subsynchronism (*abc* sequence) as also occurred with the rotor phase voltages. However, the rotor $\alpha\beta$ currents remain unaltered during the experiment, keeping a constant amplitude as well as pulsation ($\omega_s = 314.16\,\text{rad/s}$). Note again that, as the real phase currents are shown in Figure 4.11e, they have different amplitudes compared to the $\alpha\beta$ rotor currents of Figure 4.11f.

4.2.3.3 Example 4.3: Performance Analysis of the DFIM in DQ and dq Coordinates
From the simulation block diagram of Figure 4.6, it is possible to derive *DQ* and *dq* components, from the calculated *abc* or $\alpha\beta$ components, without needing a different model of the machine. So, by simply using reference frame transformations, the behavior of the machine is inferred in *DQ* and *dq* reference frames.

Hence, in Figure 4.12, the transformation into the *DQ* reference frame is illustrated. The "*x*" notation represents voltages or currents. For the rotor voltages and currents, only a rotational transformation is required. However, for stator voltages and currents, rotational and Clarke transformations become necessary.

Finally, for both stator and rotor fluxes, the *DQ* components are calculated by means of the rotational transformation.

Consequently, by applying these transformations, the space vector diagrams of Figure 4.8 are converted to the space vector diagrams of Figure 4.13. In this case, the *DQ* reference frame rotates at ω_m angular frequency.

The space vectors referred to this *DQ* rotating frame also rotate but at different speed, that is, ω_r angular frequency. This ω_r (Equation (4.7)) is negative at hypersynchronous speed and positive at subsynchronous speed.

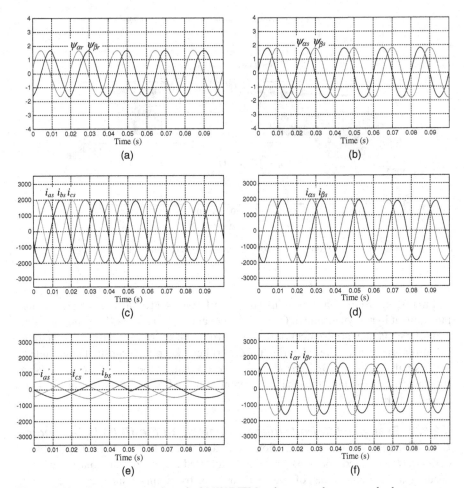

Figure 4.11 Generator operation of a 2 MW DFIM, at hypersynchronous and subsyncronous speeds: (a) $\alpha\beta$ rotor fluxes (Wb), (b) $\alpha\beta$ stator fluxes (Wb), (c) abc stator currents, (d) $\alpha\beta$ stator currents, (e) abc real rotor currents, and (f) $\alpha\beta$ rotor currents (stator referred).

At the same time, it can be noticed that the absolute rotating speed of the space vectors, referred to a stationary reference frame, is equal at both hypersynchronous and subsynchronous speeds (i.e., ω_s).

Finally, for the same reason of the previous remark, the disposition and phase shift angles of each space vector in both diagrams remain unaltered, compared to the diagram of Figure 4.8. The only difference is the rotating speeds.

In a similar way, the magnitudes in the dq reference frame are derived according to the block diagram of Figure 4.14. As before, the "x" notation represents voltages or currents. For the dq rotating reference frame, an alignment with the stator flux space vector is chosen by means of angle θ_ψ. Note that for this general modeling analysis, any other choice would also be valid.

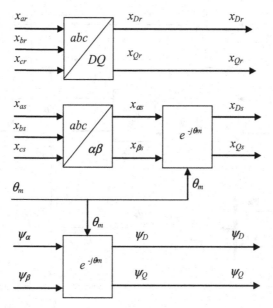

Figure 4.12 Transformation to DQ reference frame.

The stator and rotor voltages and currents are first transformed to $\alpha\beta$ components and then to dq components. The fluxes are directly transformed from $\alpha\beta$ components to dq components.

Consequently, by applying these last transformations, the space vector diagrams of Figure 4.8 are converted to the space vector diagrams of Figure 4.15. In this case, the dq reference frames rotate at ω_s electrical pulsation, so the space vectors referred to

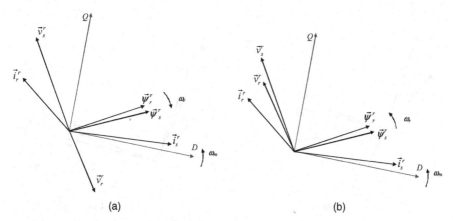

Figure 4.13 Space vector representation in DQ rotatory reference frame: (a) hypersynchronous operation and (b) subsynchronous operation (amplitudes not scaled).

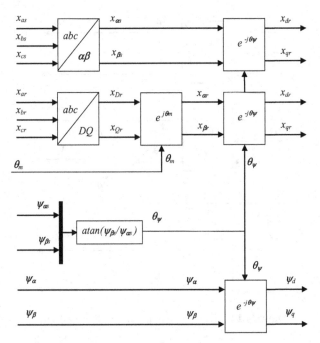

Figure 4.14 Transformation to dq reference frame.

this dq rotating frame are stationary. This leads to constant dq components of the space vector projections, due to this rotating dq axis.

The absolute rotating speed of the space vectors, referred to a stationary reference frame, is equal at both hypersynchronous and subsynchronous speeds (i.e., ω_s).

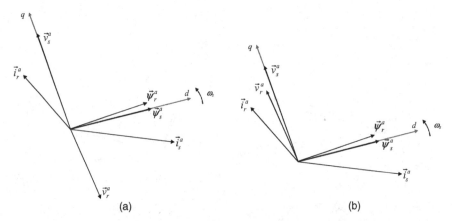

Figure 4.15 Space vector representation in dq rotatory reference frame: (a) hypersynchronous operation and (b) subsynchronous operation (amplitudes not scaled).

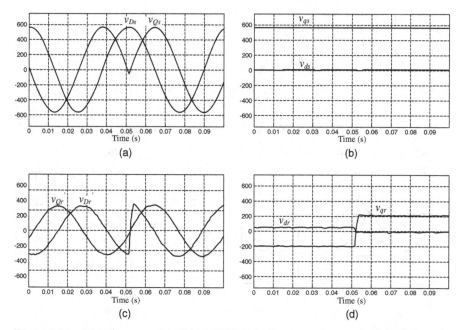

Figure 4.16 Generator operation of a 2 MW DFIM, at hypersynchronous and subsyncronous speeds: (a) DQ stator voltages (V), (b) dq stator voltages (V), (c) DQ real rotor voltages (V), and (d) dq rotor voltages (stator referred) (V).

Finally, in this case again, the disposition and phase shift angles of each space vector in both diagrams remain unaltered, compared to the diagram of Figure 4.8. The only difference is the rotating speeds.

Hence, the machine's magnitudes are illustrated in Figures 4.16 and 4.17. The time evolution of the most representative magnitudes is presented in DQ and dq coordinates. All these magnitudes are calculated from the behavior of the machine shown in Figures 4.9 and 4.11.

The stator and rotor voltages are illustrated in Figure 4.16. Due to the change of speed, the stator DQ voltages have a change in the pulsation, from $\omega_r = -125.6\,\text{rad/s}$ at hypersynchronism to $\omega_r = 125.6\,\text{rad/s}$ at subsynchronism, as noticed in Figure 4.16a. However, the dq stator voltage components remain constant during the entire simulation experiment, due to the fact that they do not depend on the speed.

In a similar way, the rotor DQ voltages shown in Figure 4.16c have an equal pulsation change as the stator voltages, due to the change of speed from hypersynchronism to subsynchronism. However, in this case, the dq rotor voltages illustrated in Figure 4.16d are also modified, not due to the change of the speed itself, but due to the relative position change from the rotor flux space vector, that is, from 90° delayed to 90° in advance.

Continuing with the analysis, the rest of the machine's magnitudes are shown in Figures 4.17a to 4.17d. As occurs with the stator voltages in Figure 4.16a, despite the fact that they remain unaltered during the speed change, the DQ components of the stator fluxes reflect this speed change as can be noticed in Figure 4.17a. On the

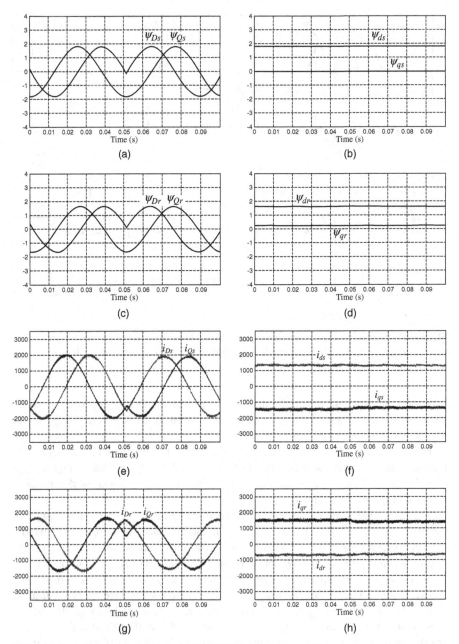

Figure 4.17 Generator operation of a 2 MW DFIM, at hypersynchronous and subsyncronous speeds: (a) *DQ* stator fluxes (Wb), (b) *dq* stator fluxes (Wb), (c) *DQ* rotor fluxes (Wb), (d) *dq* rotor fluxes (Wb), (e) *DQ* stator currents (A), (f) *dq* stator currents (A), (g) *DQ* rotor currents (stator referred) (A), and (h) *dq* rotor currents (stator referred) (A).

contrary, the stator dq fluxes are constant during the entire experiment: Figure 4.17b. In addition, since the dq reference frame has been aligned to the stator flux space vector, the q component of the flux is zero.

With the rotor fluxes, something equivalent as before occurs. The rotor and stator fluxe space vectors are smoothly phase shifted; consequently, they show a similar behavior to the stator flux, as illustrated in Figures 4.17c and 4.17d.

Finally, the stator and rotor DQ currents again have an equal pulsation variation due to the change of speed. However, the dq components remain unaltered, as shown in Figures 4.17e to 4.17h. All these rotor currents are referred to the stator.

4.2.3.4 Example 4.4: Numerical Expressions of the Variables of the DFIM at Steady State

To conclude, the mathematical expressions of this particular simulation experiment of each space vector at steady state can be represented. For simplicity, only the stator voltage and the rotor currents are mathematically expressed. Taking as reference the stator voltage, a summary is given in Tables 4.1 and 4.2. Note that these mathematical expressions, only consider the fundamental components, not taking into account the rotor current ripples for instance.

4.2.4 State-Space Representation of dq Model

Continuing with the dynamic model study, a representation of the dq model in state-space equations is also possible to obtain. Rearranging Equations (4.28)–(4.31), and

TABLE 4.1 Mathematical Expressions in $\alpha\beta$ and DQ Coordinates

Coordinates	Hypersynchronism	Subsynchronism
$\alpha\beta$	$\vec{v}_s = \|\vec{v}_s\|e^{j\omega_s t}$ $v_{\alpha s} = 563.4\cos(314.16t)$ $v_{\beta s} = 563.4\sin(314.16t)$	$\vec{v}_s = \|\vec{v}_s\|e^{j\omega_s t}$ $v_{\alpha s} = 563.4\cos(314.16t)$ $v_{\beta s} = 563.4\sin(314.16t)$
	$\vec{i}_r = \|\vec{i}_r\|e^{j(\omega_s t + \gamma)}$ $i_{\alpha r} = 1650\cos(314.16t + 0.45)$ $i_{\beta r} = 1650\sin(314.16t + 0.45)$	$\vec{i}_r = \|\vec{i}_r\|e^{j(\omega_s t + \gamma)}$ $i_{\alpha r} = 1650\cos(314.16t + 0.45)$ $i_{\beta r} = 1650\sin(314.16t + 0.45)$
DQ	$\vec{v}_s = \|\vec{v}_s\|e^{-j\omega_r t}$ $v_{Ds} = 563.4\cos(-125.6t)$ $\quad = 563.4\cos(125.6t)$ $v_{Qs} = 563.4\sin(-125.6t)$ $\quad = -563.4\sin(125.6t)$	$\vec{v}_s = \|\vec{v}_s\|e^{j\omega_r t}$ $v_{Ds} = 563.4\cos(125.6t)$ $v_{Qs} = 563.4\sin(125.6t)$
	$\vec{i}_r = \|\vec{i}_r\|e^{j(-\omega_r t + \gamma)}$ $i_{Dr} = 1650\cos(-125.6t + 0.45)$ $\quad = 1650\cos(125.6t - 0.45)$ $i_{Qr} = 1650\sin(-125.6t + 0.45)$ $\quad = -1650\sin(125.6t - 0.45)$	$\vec{i}_r = \|\vec{i}_r\|e^{j(\omega_r t + \gamma)}$ $i_{Dr} = 1650\cos(125.6t + 0.45)$ $i_{Qr} = 1650\sin(125.6t + 0.45)$

TABLE 4.2   Mathematical Expressions in *dq* Coordinates and *abc*

Coordinates	Hypersynchronism	Subsynchronism
dq	$\vec{v}_s = \|\vec{v}_s\| e^{j\pi/2}$	$\vec{v}_s = \|\vec{v}_s\| e^{j\pi/2}$
	$v_{ds} \cong 0$	$v_{ds} \cong 0$
	$v_{qs} \cong 563.4$	$v_{qs} \cong 563.4$
	$\vec{i}_r = \|\vec{i}_r\| e^{j(\pi/2+0.45)}$	$\vec{i}_r = \|\vec{i}_r\| e^{j(\pi/2+0.45)}$
	$i_{dr} \cong -736$	$i_{dr} \cong -736$
	$i_{qr} \cong 1482$	$i_{qr} \cong 1482$
abc	$v_{as} = 563.4\cos(314.16t)$	$v_{as} = 563.4\cos(314.16t)$
	$v_{bs} = 563.4\cos(314.16t - 2\pi/3)$	$v_{bs} = 563.4\cos(314.16t - 2\pi/3)$
	$v_{cs} = 563.4\cos(314.16t - 4\pi/3)$	$v_{cs} = 563.4\cos(314.16t - 4\pi/3)$
	$i'_{ar} = 550\cos(-125.6t + 0.45)$	$i'_{ar} = 550\cos(125.6t + 0.45)$
	$i'_{br} = 550\cos(-125.6t + 0.45 - 2\pi/3)$	$i'_{br} = 550\cos(125.6t + 0.45 - 2\pi/3)$
	$i'_{cr} = 550\cos(-125.6t + 0.45 - 4\pi/3)$	$i'_{cr} = 550\cos(125.6t + 0.45 - 4\pi/3)$

taking the fluxes as state-space magnitudes, the model of the DFIM is given by the next expression:

$$\frac{d}{dt}\begin{bmatrix}\vec{\psi}_s^a \\ \vec{\psi}_r^a\end{bmatrix} = \begin{bmatrix}\dfrac{-R_s}{\sigma L_s} - j\omega_s & \dfrac{R_s L_m}{\sigma L_s L_r} \\[2ex] \dfrac{R_r L_m}{\sigma L_s L_r} & \dfrac{-R_r}{\sigma L_r} - j\omega_r\end{bmatrix} \cdot \begin{bmatrix}\vec{\psi}_s^a \\ \vec{\psi}_r^a\end{bmatrix} + \begin{bmatrix}\vec{v}_s^a \\ \vec{v}_r^a\end{bmatrix} \qquad (4.42)$$

Expanding this last expression in the *dq* components, we obtain

$$\frac{d}{dt}\begin{bmatrix}\psi_{ds} \\ \psi_{qs} \\ \psi_{dr} \\ \psi_{qr}\end{bmatrix} = \begin{bmatrix}\dfrac{-R_s}{\sigma L_s} & \omega_s & \dfrac{R_s L_m}{\sigma L_s L_r} & 0 \\[2ex] -\omega_s & \dfrac{-R_s}{\sigma L_s} & 0 & \dfrac{R_s L_m}{\sigma L_s L_r} \\[2ex] \dfrac{R_r L_m}{\sigma L_s L_r} & 0 & \dfrac{-R_r}{\sigma L_r} & \omega_r \\[2ex] 0 & \dfrac{R_r L_m}{\sigma L_s L_r} & -\omega_r & \dfrac{-R_r}{\sigma L_r}\end{bmatrix} \cdot \begin{bmatrix}\psi_{ds} \\ \psi_{qs} \\ \psi_{dr} \\ \psi_{qr}\end{bmatrix} + \begin{bmatrix}v_{ds} \\ v_{qs} \\ v_{dr} \\ v_{qr}\end{bmatrix} \qquad (4.43)$$

Once again, if instead of the fluxes the currents are chosen as state-space magnitudes, the equivalent model of the DFIM is expressed as follows, in the synchronous reference frame:

$$
\frac{d}{dt}\begin{bmatrix} \vec{i}_s\,a \\ \vec{i}_r\,a \end{bmatrix} = \left(\frac{1}{\sigma\cdot L_s\cdot L_r}\right) \begin{bmatrix} -R_s\cdot L_r - j\cdot\omega_m\cdot L_m^2 - j\cdot\omega_s\cdot\sigma\cdot L_s\cdot L_r & R_r\cdot L_m - j\cdot\omega_m\cdot L_m\cdot L_r \\ R_s\cdot L_m + j\cdot\omega_m\cdot L_m\cdot L_s & -R_r\cdot L_s + j\cdot\omega_m\cdot L_r\cdot L_s - j\cdot\omega_s\cdot\sigma\cdot L_s\cdot L_r \end{bmatrix}
$$

$$
\cdot\begin{bmatrix} \vec{i}_s\,a \\ \vec{i}_r\,a \end{bmatrix} + \left(\frac{1}{\sigma\cdot L_s\cdot L_r}\right) \begin{bmatrix} L_r & -L_m \\ -L_m & L_s \end{bmatrix}\cdot\begin{bmatrix} \vec{v}_s\,a \\ \vec{v}_r\,a \end{bmatrix}
$$

$$(4.44)$$

Expanding in dq components, we find

$$
\frac{d}{dt}\begin{bmatrix} i_{ds} \\ i_{qs} \\ i_{dr} \\ i_{qr} \end{bmatrix} = \left(\frac{1}{\sigma.L_s.L_r}\right) \begin{bmatrix} -R_s.L_r & \omega_m.L_m^2 + \omega_s.\sigma.L_s.L_r & R_r.L_m & \omega_m.L_m.L_r \\ -\omega_m.L_m^2 - \omega_s.\sigma.L_s.L_r & -R_s.L_r & -\omega_m.L_m.L_r & R_r.L_m \\ R_s.L_m & -\omega_m.L_s.L_m & -R_r.L_s & -\omega_m.L_r.L_s + \omega_s.\sigma.L_s.L_r \\ \omega_m.L_s.L_m & R_s.L_m & \omega_m.L_r.L_s - \omega_s.\sigma.L_s.L_r & -R_r.L_s \end{bmatrix} \begin{bmatrix} i_{ds} \\ i_{qs} \\ i_{dr} \\ i_{qr} \end{bmatrix}
$$

$$
\cdot\begin{bmatrix} i_{ds} \\ i_{qs} \\ i_{dr} \\ i_{qr} \end{bmatrix} + \left(\frac{1}{\sigma.L_s.L_r}\right) \begin{bmatrix} L_r & 0 & -L_m & 0 \\ 0 & L_r & 0 & -L_m \\ -L_m & 0 & L_s & 0 \\ 0 & -L_m & 0 & L_s \end{bmatrix}\cdot\begin{bmatrix} v_{ds} \\ v_{qs} \\ v_{dr} \\ v_{qr} \end{bmatrix}
$$

$$(4.45)$$

State-space representations of the DFIM in the dq reference frame are especially useful in obtaining the steady state for given stator and rotor input voltages.

4.2.4.1 Example 4.5: Steady State Evaluation in dq Coordinates

By using the dq state-space representation of the DFIM, it is possible to simply derive the steady state magnitudes of the machine. For instance, considering the state-space representation of expression (4.43), at steady state, the derivatives of the fluxes are equal to zero:

$$
\frac{d}{dt}\begin{bmatrix} \psi_{ds} \\ \psi_{qs} \\ \psi_{dr} \\ \psi_{qr} \end{bmatrix} = 0
$$

$$(4.46)$$

Hence, the state-space expression yields

$$
\begin{bmatrix} 0 \\ 0 \\ 0 \\ 0 \end{bmatrix} = \begin{bmatrix} \dfrac{-R_s}{\sigma L_s} & \omega_s & \dfrac{R_s L_m}{\sigma L_s L_r} & 0 \\[3mm] -\omega_s & \dfrac{-R_s}{\sigma L_s} & 0 & \dfrac{R_s L_m}{\sigma L_s L_r} \\[3mm] \dfrac{R_r L_m}{\sigma L_s L_r} & 0 & \dfrac{-R_r}{\sigma L_r} & \omega_r \\[3mm] 0 & \dfrac{R_r L_m}{\sigma L_s L_r} & -\omega_r & \dfrac{-R_r}{\sigma L_r} \end{bmatrix} \cdot \begin{bmatrix} \psi_{ds} \\ \psi_{qs} \\ \psi_{dr} \\ \psi_{qr} \end{bmatrix} + \begin{bmatrix} v_{ds} \\ v_{qs} \\ v_{dr} \\ v_{qr} \end{bmatrix} \qquad (4.47)
$$

Rearranging the terms, the rotor and stator dq fluxes can be calculated as follows:

$$
\begin{bmatrix} \psi_{ds} \\ \psi_{qs} \\ \psi_{dr} \\ \psi_{qr} \end{bmatrix} = - \begin{bmatrix} \dfrac{-R_s}{\sigma L_s} & \omega_s & \dfrac{R_s L_m}{\sigma L_s L_r} & 0 \\[3mm] -\omega_s & \dfrac{-R_s}{\sigma L_s} & 0 & \dfrac{R_s L_m}{\sigma L_s L_r} \\[3mm] \dfrac{R_r L_m}{\sigma L_s L_r} & 0 & \dfrac{-R_r}{\sigma L_r} & \omega_r \\[3mm] 0 & \dfrac{R_r L_m}{\sigma L_s L_r} & -\omega_r & \dfrac{-R_r}{\sigma L_r} \end{bmatrix}^{-1} \cdot \begin{bmatrix} v_{ds} \\ v_{qs} \\ v_{dr} \\ v_{qr} \end{bmatrix} \qquad (4.48)
$$

This last expression (4.48) can be useful in deriving the rotor and stator fluxes of the machine, by knowing the imposed input stator and rotor voltages as well as the speed of the machine.

After that, once the fluxes are known, the remaining stator and rotor currents can be calculated by the following expression (derived from Equations (4.30) and (4.31)):

$$
\begin{bmatrix} i_{ds} \\ i_{qs} \\ i_{dr} \\ i_{qr} \end{bmatrix} = \begin{bmatrix} \dfrac{1}{\sigma L_s} & 0 & \dfrac{-L_m}{\sigma L_s L_r} & 0 \\[3mm] 0 & \dfrac{1}{\sigma L_s} & 0 & \dfrac{-L_m}{\sigma L_s L_r} \\[3mm] \dfrac{-L_m}{\sigma L_s L_r} & 0 & \dfrac{1}{\sigma L_r} & 0 \\[3mm] 0 & \dfrac{-L_m}{\sigma L_s L_r} & 0 & \dfrac{1}{\sigma L_r} \end{bmatrix} \cdot \begin{bmatrix} \psi_{ds} \\ \psi_{qs} \\ \psi_{dr} \\ \psi_{qr} \end{bmatrix} \qquad (4.49)
$$

4.2.4.2 Example 4.6: Steady State Evaluation in dq Coordinates; Numerical Solution

By using the procedure of Example 4.5, the synchronous operating mode can also be evaluated. It is known that the 2 MW DFIM is operating under the following conditions:

$v_{ds} = 4$ V

$v_{ds} = 563.4$ V

$v_{dr} = -2.5$ V

$v_{qr} = 7.5$ V

$\Omega_m = 1500$ rev/min (synchronous speed, $\omega_r = 0$)

Substituting these operating conditions and the parameters of the machine into expression (4.48), we obtain

$$\begin{bmatrix} \psi_{ds} \\ \psi_{qs} \\ \psi_{dr} \\ \psi_{qr} \end{bmatrix} = - \begin{bmatrix} -15.36 & 314.16 & 14.86 & 0 \\ -314.16 & -15.36 & 0 & 14.86 \\ 16.57 & 0 & -17.15 & 0 \\ 0 & 16.57 & 0 & -17.15 \end{bmatrix}^{-1} \cdot \begin{bmatrix} 4 \\ 563.4 \\ -2.5 \\ 7.5 \end{bmatrix} \quad (4.50)$$

Hence, the fluxes yield8?A3B2 tptxt=7pt?>

$$\begin{bmatrix} \psi_{ds} \\ \psi_{qs} \\ \psi_{dr} \\ \psi_{qr} \end{bmatrix} = \begin{bmatrix} 1.81 \\ 0 \\ 1.6 \\ 0.43 \end{bmatrix} \text{ (Wb)} \quad (4.51)$$

Once the fluxes are known, it is possible, for instance, to derive the electromagnetic torque of the machine with expression (4.27):

$$T_{em} = \frac{3}{2} \frac{L_m}{\sigma L_r L_s} p \, \text{Im}\{ \vec{\psi}_r^* \cdot \vec{\psi}_s \} = \frac{3}{2} \frac{L_m}{\sigma L_r L_s} p (\psi_{dr} \psi_{qs} - \psi_{qr} \psi_{ds}) \quad (4.52)$$

Numerically,

$$T_{em} = 17147(1.6 \cdot 0 - 0.43 \cdot 1.81) = -13.6 \, \text{k} \cdot \text{N} \cdot \text{m} \quad (4.53)$$

Finally, the stator and rotor currents can be calculated with expression (4.49):

$$\begin{bmatrix} i_{ds} \\ i_{qs} \\ i_{dr} \\ i_{qr} \end{bmatrix} = \begin{bmatrix} 5910 & 0 & -5715.7 & 0 \\ 0 & 5910 & 0 & -5715.7 \\ -5715.7 & 0 & 5914.6 & 0 \\ 0 & -5715.7 & 0 & 5914.6 \end{bmatrix} \cdot \begin{bmatrix} 1.81 \\ 0 \\ 1.6 \\ 0.43 \end{bmatrix} \quad (4.54)$$

Note that the rotor currents are referred to the stator side. Finally, the currents are

$$
\begin{bmatrix} i_{ds} \\ i_{qs} \\ i_{dr} \\ i_{qr} \end{bmatrix} = \begin{bmatrix} 1534.3 \\ -2499.2 \\ -862.1 \\ 2586.2 \end{bmatrix} (\text{A}) \tag{4.55}
$$

Thus, the stator and rotor powers can also be calculated from expressions (4.32)–(4.35), yielding

$$P_s = 2.1 \, (\text{MW})$$
$$Q_s = 1.3 \, (\text{MVAR})$$
$$P_r = 32.3 \, (\text{kW})$$
$$Q_s \cong 0$$

Consequently, the simulated voltages and currents under this situation are shown in Figure 4.18. Note that the *abc* stator voltages and currents have 50 Hz sinusoidal behavior (Figures 4.18a and 4.18c), while the *abc* rotor voltages and currents have DC constant values (Figures 4.18b and 4.18d). To conclude, the synchronous speed and the torque are illustrated in Figures 4.18e and 4.18f.

4.2.5 Relation Between the Steady State Model and the Dynamic Model

As stated in the introduction to this chapter, the reader can see that the steady state model developed in Chapter 3 with phasors is just a particular case of a more general model of the machine, that is, the dynamic model developed in this chapter with the help of space vector theory. Thus, while the steady state model can only deal with sinusoidal magnitudes once the steady state of the machine has been reached, the dynamic model can also represent more general behaviors, such as transient or dynamic phenomena. In addition, the dynamic model is not restricted to sinusoidal steady states; it can also consider nonsinusoidal supply approaches.

The close relation between these models can clearly be seen, for instance, by comparing the *dq* dynamic model Equations (4.28)–(4.36), with the steady state model equations derived in the previous chapter: Equations (3.31), (3.32), (3.27), (3.28), (3.48), (3.53) and (3.54):

$$\underline{V}_s = R_s \underline{I}_s + j\omega_s \underline{\Psi}_s \qquad \vec{v}_s^a = R_s \vec{i}_s^a + \frac{d\vec{\psi}_s^a}{dt} + j\omega_s \vec{\psi}_s^a$$

$$\underline{V}_r = R_r \underline{I}_r + j\omega_r \underline{\Psi}_r \qquad \vec{v}_r^a = R_r \vec{i}_r^a + \frac{d\vec{\psi}_r^a}{dt} + j\omega_r \vec{\psi}_r^a$$

$$\underline{\Psi}_s = L_s \underline{I}_s + L_m \underline{I}_r \qquad \vec{\psi}_s^a = L_s \vec{i}_s^a + L_m \vec{i}_r^a$$

$$\underline{\Psi}_r = L_m \underline{I}_s + L_r \underline{I}_r \qquad \vec{\psi}_r^a = L_m \vec{i}_s^a + L_r \vec{i}_r^a$$

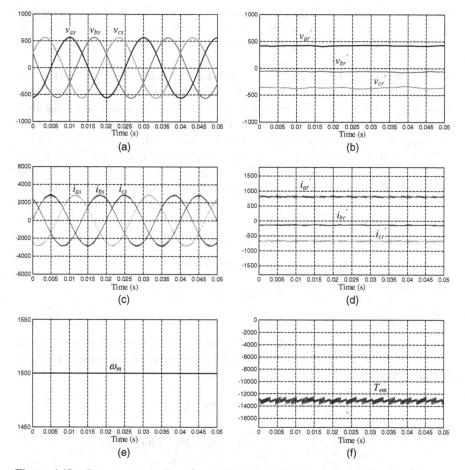

Figure 4.18 Generator operation of a 2 MW DFIM at synchronous speed: (a) *abc* stator voltages (V), (b) *abc* rotor voltages (V), (c) *abc* stator currents (A), (d) *abc* rotor currents (A), (e) speed (rpm), and (f) electromagnetic torque (N·m).

$$T_{em} = 3p\,\mathrm{Im}\{\underline{\Psi}_r \cdot \underline{I}_r^*\} \qquad T_{em} = \tfrac{3}{2}p\,\mathrm{Im}\{\vec{\psi}_r \cdot \vec{i}_r^*\}$$

$$P_s = 3\mathrm{Re}\{\underline{V}_s \cdot \underline{I}_s^*\} \qquad P_s = \tfrac{3}{2}\mathrm{Re}\{\vec{v}_s \cdot \vec{i}_s^*\}$$

$$Q_s = 3\mathrm{Im}\{\underline{V}_s \cdot \underline{I}_s^*\} \qquad Q_s = \tfrac{3}{2}\mathrm{Im}\{\vec{v}_s \cdot \vec{i}_s^*\}$$

Note that the rotor power relations are omitted from this comparison, due to their equivalence to the stator power expressions. It is clear that the appearance of these equations is very similar. The steady state model only can represent sinusoidal magnitudes by means of an amplitude and a phase. However, the dynamic model, if a non-sinusoidal supply is applied with a converter, it can also represent the harmonic performance in continuous or discrete time domin. The amplitudes of the sinusoidal magnitudes represented by the phasors are given in rms; however, the fundamental components amplitudes of the space vectors refer to peak values. The following example gives more details of the correlation between these two models.

4.2.5.1 Example 4.7: Steady State with Dynamic Model Equations   This

example is similar to Example 3.3 of Chapter 3 and derives the steady state of a 2 MW machine by means of the *dq* representation of space vectors supposing sinusoidal ideal supply. The operating point is chosen as a 2 MW stator power generating, with zero stator reactive power, 50 Hz of grid frequency and $s = -0.25$ slip (hypersynchronous speed).

An alternative resolution approach to the one presented in Example 4.6 is carried out. Hence, assuming a rated stator voltage supply (690 V_{rms} line to line) and taking the *d* axis of the reference frame aligned with the stator voltage space vector, we have

$$\vec{v}_s^a = v_{ds} + jv_{qs} = 563.4 + 0j \text{ (V)}$$

Since $Q_s = 0$ is chosen, the stator current space vector yields

$$
\left.
\begin{aligned}
P_s &= \tfrac{3}{2}\left(v_{ds}i_{ds} + v_{qs}i_{qs}\right) = \tfrac{3}{2}(563.4 i_{ds}) = 2000000 \\
Q_s &= \tfrac{3}{2}\left(v_{qs}i_{ds} - v_{ds}i_{qs}\right) = \tfrac{3}{2}(-563.4 i_{qs}) = 0
\end{aligned}
\right\}
\Rightarrow
$$

$$\vec{i}_s^a = i_{ds} + ji_{qs} = -2366.6 + 0j \text{ (A)}$$

Since it is assumed that the steady state has been reached, from Equation (4.28) the derivative term is zero (particular case of steady state) and it is being possible to derive the stator flux:

$$\vec{v}_s^a = R_s\vec{i}_s^a + \frac{d\vec{\psi}_s^a}{dt} + j\omega_s\vec{\psi}_s^a = R_s\vec{i}_s^a + j\omega_s\vec{\psi}_s^a \Rightarrow$$

$$\vec{\psi}_s^a = \frac{\vec{v}_s^a - R_s\vec{i}_s^a}{j\omega_s} = \frac{563.4 + 0j - 0.0026(-2366.6 + 0j)}{j314.16} \Rightarrow$$

$$\vec{\psi}_s^a = \psi_{ds} + j\psi_{qs} = 0 - j1.81 \text{ (Wb)}$$

Once the stator flux and currents are obtained, it is possible to derive the rotor flux:

$$\vec{\psi}_s^a = L_s\vec{i}_s^a + L_m\vec{i}_r^a \Rightarrow \vec{i}_r^a = \frac{\vec{\psi}_s^a - L_s\vec{i}_s^a}{L_m} \Rightarrow$$

$$\vec{i}_r^a = \frac{(0 - j1.81) - 0.002587(-2366.6 + 0j)}{0.0025} \Rightarrow$$

$$\vec{i}_r^a = i_{dr} + ji_{qr} = 2449 - 725.1j \text{ (A)}$$

Proceeding with the rotor flux,

$$\vec{\psi}_r^a = L_m\vec{i}_s^a + L_r\vec{i}_r^a = 0.0025(-2366.6 + 0j) + 0.002587(2449 - 725.1j) \Rightarrow$$

$$\vec{\psi}_r^a = \psi_{dr} + j\psi_{qr} = 0.4 - 1.8j \text{ (Wb)}$$

Finally, again assuming steady state, the rotor voltage is derived:

$$\vec{v}_r^a = R_r\vec{i}_r^a + \frac{d\vec{\psi}_r^a}{dt} + j\omega_r\vec{\psi}_r^a = R_r\vec{i}_r^a + j\omega_r\vec{\psi}_r^a \Rightarrow$$

$$\vec{v}_r^a = 0.0029(2449 - 725.1j) + j(-0.25 \cdot 314.16)\cdot(0.4 - 1.8j) \Rightarrow$$

$$\vec{v}_r^a = v_{dr} + jv_{qr} = -140.2 - 35j \, (V)$$

To conclude, the torque is found:

$$T_{em} = \tfrac{3}{2}p \, \text{Im}\{\vec{\psi}_r \cdot \vec{i}_r^*\} = \tfrac{3}{2}\cdot 2 \cdot \text{Im}\{(0.4 - 1.8j)\cdot(2449 + 725.1j)\} = -12900 \, (N\cdot m)$$

Note that by solving this example, we have followed exactly the same procedure as in Example 3.3 of Chapter 3; but in this case we use space vector notation, referred to a synchronous rotating frame.

The graphical representation of the derived space vectors is shown in Figure 4.19a. In a similar way, the phasor diagram representation corresponding to these operating conditions was derived in the previous chapter and now again is depicted in Figure 4.19b.

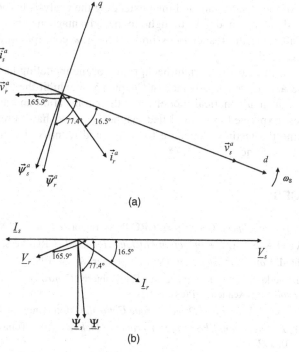

(a)

(b)

Figure 4.19 (a) Space vector representation at a given operating point. (b) Phasor diagram representation at a given operating point (not scaled amplitudes).

4.3 SUMMARY

This chapter has described the $\alpha\beta$ and dq dynamic models of the DFIM. Not only have the differential equations of these models been derived and presented, but also the state-space representation has also been developed.

The structure of these models has been adapted to create and implement simulation dynamic models, to perform computer based simulations. Practical numerical and simulated examples based on these models have emphasized the different peculiarities and characteristic aspects of the DFIM behavior, providing useful ideas and procedures that contribute to the reader is wider understanding of the machine. All these examples have dealt with steady states, obtained by means of dynamic simulation. The dynamic behavior of the machine is studied in subsequent chapters of this book. Thus, Chapter 6 studies the transient behavior of the machine, when it is not closed loop controlled and is subjected to voltage dips. Then, Chapters 7 and 8 show the achievable dynamic behavior of the machine when it is closed loop controlled by two control strategies—field oriented control and direct control.

In addition, the dynamic model has been revealed as a more generic model than the steady state model studied in Chapter 3, not only for the DFIM, but for AC electric machines in general. In the intrinsic nature of the dynamic model, information related to the dynamic and steady state behavior is found. However, first examining the specific case of the steady state and then extending the analysis to the general case showing the dynamic model is thought to be advantageous from a pedagogic perspective, allowing the reader to assimilate the new concepts in a more scaled and rationalized way.

Finally, once the most important background of the modeling has been studied in Chapters 3 and 4, we can advance in Chapter 5 to exhibit the linkage between the theoretical or mathematical models, and their translation and applicability to real machines. Explained in a simplified manner, once we have a real machine, it seeks to define the testing procedure to derive and identify the modeling parameters of the machine.

REFERENCES

1. I. Boldea, *Variable Speed Generators*. CRC Press Taylor & Francis, 2006.
2. I. Boldea and A. Nasar, *The Induction Machine Handbook*. CRC Press, 2006.
3. W. Leonhard, *Control of Electrical Drives*. Springer-Verlag, 1985.
4. M. P. Kazmierkowski, R. Krishnan, and F. Blaabjerg, *Control in Power Electronics: Selected Problems*. Academic Press, 2002.
5. P. Vas, *Sensorless Vector and Direct Torque Control*. Oxford University Press, 1998.
6. R. H. Park, *Two Reactions Theory of Synchronous Machines*. AIEE Transactions. Vol. 48, pp. 716–730, 1928.
7. A. Veltman, D. W. J. Pulle, and R.W. De Doncker, *Fundamentals of Electric Drives*. Springer, 2007.

8. B. K. Bose, *Power Electronics and Drives*. Elsevier, 2006.

9. A. M. Trzynadlowski, *Control of Induction Motors*. Academic Press, 2001.

10. A. Hughes, *Electric Motors and Drives*. Elsevier, 1990.

11. M. Barnes, *Variable Speed Drives and Power Electronics*. Elsevier, 2003.

12. S. J. Chapman, *Electrical Machines*. McGraw Hill, 1985.

13. J. F. Mora, *Electrical Machines*. McGraw Hill, 2003.

14. A. Peterson, "Analysis, Modeling and Control of Doubly-Fed Induction Generators for Wind Turbines." Ph.D. thesis, Chalmers University of Technology, Goteborg, Sweden, 2005.

CHAPTER 5

Testing the DFIM

5.1 INTRODUCTION

In the previous two chapters, modeling of the DFIM has been examined, by reviewing the fundamental steady state and dynamic models of the machine. These models are defined by means of equations derived from equivalent electric circuits of the machine. The equations for the model, among other objectives, mainly define the behavior of the machine a give a the basis for understanding the machine and its operation.

However, in a practical context, before we continue with the analysis of the machine and address its control, it is important for the reader to have a firm knowledge about how the model parameters of the machine can be obtained. Even though the studied models of the machine represent various versions of a unique model of the DFIM, the parameters of this model (values of inductances, resistances, etc.) mainly differentiate one machine from another by altering its behavior. In fact, for an effective use of the DFIM model, it is very important to have a reasonably accurate knowledge of its parameters, so that the deductions and conclusions obtained from their use can be as near as possible to reality.

For that purpose, in this chapter, the focus is on how the model parameters of the DFIM can be identified when we have available the real machine. The identification of these parameters are normally carried out by means of a set of experimental tests, considering the machine as a "black box," assuming that the model defining the machine's behavior is any of the versions utilized in Chapters 3 and 4. Therefore, making the machine operate under several specific conditions, defined by a concatenation of experimental tests, it is possible to obtain consecutively all the model parameters.

Therefore, this chapter presents three concatenated experimental tests that do not require any additional power supply than the back-to-back converter utilized in real applications, and effectively derives the model parameters of the machine. In this way, by performing these three off-line tests of the machine, the required information is available for a further mathematical model analysis or for subsequent control and operation in a real application.

Doubly Fed Induction Machine: Modeling and Control for Wind Energy Generation,
First Edition. By G. Abad, J. López, M. A. Rodríguez, L. Marroyo, and G. Iwanski.
© 2011 the Institute of Electrical and Electronic Engineers, Inc. Published 2011 by John Wiley & Sons, Inc.

5.2 OFF-LINE ESTIMATION OF DFIM MODEL PARAMETERS

This section describes some off-line testing methods, to estimate the model parameters of the DFIM. There exist many different methods and variants that can provide the model parameters; however, this section examines only the most general and probably some of the most used ones. Basically, it follows some of the guides defined by IEEE standards [3]; they are examined in this book as an example of existing alternative methods of estimation. On the basis of the presented estimation tests, it is possible to find a wide range of similar solutions or variants of these tests. On the other hand, it is also possible to find radically different tests that this chapter does not cover. The described tests in this chapter are:

- Stator and rotor resistances estimation test
- Leakage inductances estimation tests
- Magnetizing inductance and iron loss resistance estimation with no-load test

These estimation tests are necessary when, for instance, the manufacturer of the machine does not provide all the information needed, or when it is necessary to contrast some assumed parameters' values.

Hence, as studied in Chapter 3, the single-phase steady state model of the DFIM can be represented as illustrated in Figure 5.1 (stator referred). There are six parameters that must be estimated $(R_s, R_r, R_{fe}, L_m, L_{\sigma s}, L_{\sigma r})$. As will be studied in subsequent chapters, not all the parameters are always needed for control. Depending on the control strategy used or the estimator-observer structures employed, perhaps detailed knowledge of some parameters is unnecessary. However, for modeling purposes, all the model parameters are needed if the machine behavior is to be analyzed.

On the other hand, the tests presented in this chapter are carried out under the following assumptions:

- A special experimental test bench on which to perform the tests is unavailable. Hence, the tests are made in a real wind turbine in a stage prior to normal operation. This implies that the supply system is the back-to-back converter of

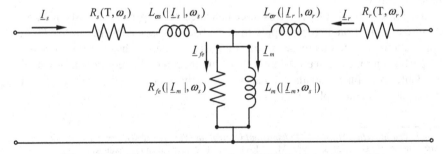

Figure 5.1 Single-phase stator referred DFIM steady state model.

the real application and the measurements done to perform the estimations are the measurements required to achieve control of the wind turbine in the real world.

- The machine is not coupled to the mechanical system of the wind turbine so it can rotate freely. This means that the tests can be performed without any restriction imposed by the mechanical system of the wind turbine. In addition, this situation allows one to perform no-load tests (the no-load test must be carried out before coupling the DFIM to the mechanical system).

- There is a possibility of connecting the back-to-back converter to the stator and to the rotor. Thus, some tests are performed with the DFIM operating as a squirrel cage induction machine: supplied through the stator with the rotor short-circuited.

- As studied in depth in subsequent chapters, the control strategy of the DFIM needs some of the parameters of the machine for proper operation. For that purpose, the parameter estimation tests presented in this chapter are mostly oriented to provide the information needed by the control strategy, although, of course, they can also be utilized for more extensive modeling analysis or other purposes.

5.2.1 Considerations About the Model Parameters of the DFIM

It is important to highlight that in the DFIM model shown in Figure 5.1, there are some significant differences from the steady state model presented in Chapter 3. For a detailed and accurate model of the DFIM, sometimes it is necessary to consider the following:

1. R_s. The stator resistance value, due to constructive and design aspects, in general is assumed capable of altering its value due to the temperature (T) and angular frequency of the current that flows (ω_s):

$$R_s \Rightarrow f(\text{T}, \omega_s)$$

In general, a resistance value determination due to a temperature variation is calculated as

$$R_y = R_x \frac{(\text{T}_y + k)}{(\text{T}_x + k)} \tag{5.1}$$

where

$R_x =$ known value of resistance at temperature $T_x (\Omega)$
$T_x =$ temperature at which the R_x was estimated (°C)
$T_y =$ temperature at which the resistance is to be corrected (°C)
$R_y =$ corrected value of resistance at temperature $T_y (\Omega)$
$k =$ conductivity (typical value for copper may be 235.5)

Thus, depending on the operating conditions and how the machine itself has been designed, the R_s value can be modified. In general, for wind generation applications, since the stator of the DFIM is connected directly to the stator, the frequency seen by this resistance is fixed.

However, due to the temperature variation (due to stator current variation), its value can change. Nevertheless, the variation of this resistance in general is not considered at the control stage, for instance, since the temperature measurement is unavailable and its variation is not so significant that it can distort the control behavior.

2. R_r. The rotor resistance value has equivalent behavior to the stator resistance:

$$R_r \Rightarrow f(\mathrm{T}, \omega_r)$$

Note that in this case, both ω_r and T can modify the resistance value. However, in many practical cases, when R_r takes part in the control, a constant value is considered in the control strategy.

3. $L_{\sigma s}$. The stator leakage inductance value, in general, is assumed capable of altering its value due to the amplitude and frequency of the current that flows:

$$L_{\sigma s} \Rightarrow f(|\underline{I}_s|, \omega_s)$$

Also, in this case, depending on the real variation (provided by the manufacturer or the estimation test explained in next section) that this parameter has, one must decide whether or not it is necessary to consider the leakage inductance as a variable in the control strategy.

4. $L_{\sigma r}$. The rotor leakage inductance value, in general, has equivalent behavior to the stator leakage inductance:

$$L_{\sigma r} \Rightarrow f(|\underline{I}_r|, \omega_r)$$

5. L_m. The magnetizing inductance value, in general, is assumed capable of altering its value mainly due to the amplitude and frequency of the current that flows:

$$L_m \Rightarrow f(|\underline{I}_m|, \omega_s)$$

Thus, depending on the voltage applied in the stator of the machine, a different level of magnetization is achieved, that is, a different L_m. In wind energy generation applications, since the machine is directly connected to a constant amplitude voltage (grid voltage), it means that, in normal operation, the L_m is maintained nearly constant. However, as shown later in the no-load test, all magnetization curves can be estimated.

6. R_{fe}. Resistance modeling of the iron losses has not been considered until now in this book. In many cases, this parameter is not considered at the control stage, although it is known that it is always present. In general, iron losses do not affect control performance very significantly if they are neglected. However, they can easily be estimated at the same time as the magnetizing L_m inductance estimation,

in the no-load test. Depending on the operating conditions, its value depends mainly on

$$R_{fe} \Rightarrow f(|\underline{L}_m|, \omega_s, f_{sw})$$

where $f_{sw} = 1/h$, the switching frequency of the two-level converter supplying the machine. Iron power losses P_{fe} are difficult to calculate and they could even require the use of finite element techniques for an accurate estimation. Although complex expressions have been found in the literature, Equation (5.2) provides an approximation that is valid for a qualitative approach:

$$P_{fe} = af\,B^2 + b(\Delta f\,B)^2 \tag{5.2}$$

where a and b depend on the material properties and Δ is the lamination thickness. Iron losses are due to hysteresis and eddy current losses. Both are proportional to the square of the flux density (B^2). Hysteresis losses are proportional to the frequency, whereas eddy losses are proportional to the square of the frequency. Hysteresis losses are usually predominant.

Finally, iron losses also depend on the nature of the supplying voltage employed. When the machine is fed by a VSC at a specific switching frequency f_{sw}, the iron losses are higher than if the supply is pure sinusoidal.

As occurs with the magnetizing inductance L_m, due to the direct connection of the stator to the grid voltage, the magnetizing current and stator frequency are constant. In addition, the switching frequency of the converter in general is also maintained constant; consequently, this resistance value modeling of the iron losses can be assumed to be also constant in wind energy generation applications.

More information about modeling aspects related to the iron losses is given in the Appendix of this book.

5.2.2 Stator and Rotor Resistances Estimation by VSC

As introduced before, the first test studied in this chapter is the stator and rotor resistances estimation tests. These tests are very similar in procedure, since they try to find a supply mode of the machine where the predominant parameter present is the resistance, which we want to measure. In this way, this test should be done first in the parameter estimation procedure. Once both the rotor and stator resistances are estimated, it is possible to proceed to the estimation of the rest of the parameters.

This test is performed under the assumption that the only supply system available is the back-to-back converter of the real wind energy generation application. First, the resistance estimation test is presented. Figure 5.2a illustrates a schematic of the system configuration for this test assuming that the machine is in a star connection. The rotor of the DFIM can be opened or short-circuited and only two phases of the stator side are supplied by a DC voltage.

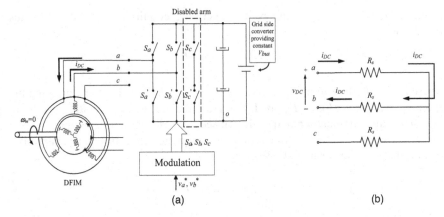

Figure 5.2 (a) System configuration for stator resistance estimation, (b) Equivalent circuit on stator resistance estimation.

In this situation, the induced DC current flows by only two windings of the machine—phases a and b. Due to the DC supply, the magnetizing circuit is seen as a short-circuit, so the only impedance present is the stator resistances of phases a and b, as illustrated in Figure 5.2b. Note that this situation is the same whether the rotor side is short-circuited or not.

Therefore, under these circumstances, there is no torque generated by the machine, so the speed is zero. The voltage references for this test must be very small compared to the nominal voltage of the machine, due to the small impedance present in the experiment. The experimenter must make sure that the current does not exceed its limit. Note that reference voltages v_a^* and v_b^* must be independently created for the modulator. If the three phases of the stator are supplied, the voltage references of each phase should be properly coordinated.

Thus, by measuring the DC current, the stator resistance can be calculated as follows:

$$R_s = \frac{v_{DC}}{i_{DC}} \cdot \frac{1}{2} \qquad (5.3)$$

It has been assumed that the three stator resistances are equal. If an imbalance between windings is expected, this test should be repeated three times across phases a-b, b-c, and c-a, so a set of three equations with three unknowns (R_s of phases a, b, and c) is derived.

Table 5.1 illustrates an example of this test applied to a real 15 kW DFIM. It is highly recommended to record several voltage and current levels to ensure the validity of the performed test. The accuracy of the estimation can be significantly affected depending on the accuracy of the measurements, temperature variation, and so on.

Finally, Figure 5.3 is an example of the oscilloscope recorded waves during the experiment. Since the VSC is supplying the stator, the averaged values of these

TABLE 5.1 Off-line Stator Resistance Estimation of a 15 kW DFIM

v_{dc} (V)	i_{dc} (A)	R_s (mΩ)
2.68	8.25	163
3.58	11.13	161
7.26	22.7	160

voltages and currents (v_{DC} and i_{DC}) must be substituted in Equation (5.3), especially the voltage. In general, this is not a special requirement for this test, since the wind turbine system incorporates these stator voltage and current measurements (prepared for filtering the switching frequency of the converter, as illustrated in Figure 5.4) in normal control and start-up of the wind turbine (see Chapters 8–10). In addition, the average values of voltages of this experiment (Table 5.1) are very low compared to the rated voltage of the converter, producing a difficulty in measuring accurately the exact value (note also the short thickness of the pulses of the voltage v_{ab} in Figure 5.3).

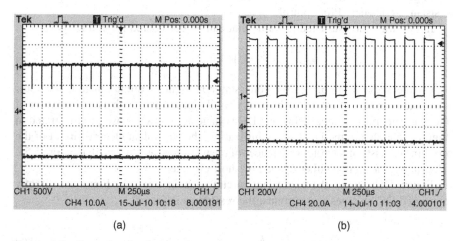

(a)   (b)

Figure 5.3 Example of voltage and current measurements of the experiment, at different current levels for a 15 kW DFIM: (a) v_{dc} (v_{ab}) and i_{dc} and (b) v_{ao} and i_{dc}.

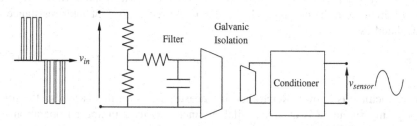

Figure 5.4 Simplified schematic example of a converter's voltage sensor.

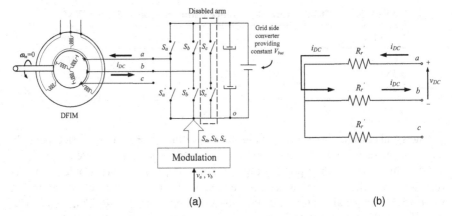

Figure 5.5 (a) System configuration for rotor resistance estimation. (b) Equivalent circuit for rotor resistance estimation.

In general, a weighted average R_s (Table 5.1) can be adopted as the unique stator resistance considered for control.

In this way, the estimated R_s value that can be considered for control is

$$R_s \cong 161 \text{ m}\Omega$$

On the same basis of the previous stator resistance estimation test, the rotor resistance can also be estimated by a very similar procedure. In this case, the rotor of the machine must be supplied with a DC voltage across two phases, as illustrated in Figure 5.5a. Due to the DC voltage utilization, the equivalent electric circuit of the machine consists of the rotor resistances of the windings, as illustrated in Figure 5.5a. The stator side can be either opened or short-circuited.

Figure 5.5b shows the equivalent circuit seen in this test. From the directly measured DC voltages and currents, the resistance is found:

$$R'_r = \frac{v_{DC}}{i_{DC}} \cdot \frac{1}{2} \tag{5.4}$$

Note that when real measurements of the rotor voltage and currents are used, the real rotor resistance (R'_r) is directly estimated. If a DFIM with a stator–rotor turn ratio (u) different from the one is used, the stator referred equivalent rotor resistance is calculated as

$$R_r = R'_r u^2 \tag{5.5}$$

where factor u is known. Regarding the measurements, in this case also the rotor current measurement is clearly available, since it is used in normal operation for control of the DFIM. However, the rotor voltage measurement is normally not used at

TABLE 5.2   Off-Line Rotor Resistance Estimation of a 15 kW DFIM

v_{dc} (V)	i_{dc} (A)	R'_r (mΩ)
2.06	5.7	181
5,47	15.3	179
7.03	19.8	177

the control stage so it is likely not available. Thus, in order to know the applied DC voltage, there are several alternatives:

- Reallocate the stator voltage sensors, placing them in the rotor only for this test. This solution is not always possible due to design reasons.
- Use the voltage references instead of voltage measurements. This solution requires a good compensation for the nonlinear properties of the VSC, such as dead times and voltage drops in the semiconductors. If they are not considered, the correspondence between the voltage reference and the real output voltage is especially affected at low voltages, affecting the accuracy of the estimation.

Nevertheless, if neither of these last two solutions can be adopted, the next test (leakage inductances estimation) can also provide an estimation of the rotor resistance.

Table 5.2 shows a specific example of a 15 kW DFIM rotor resistance estimation. In some machine designs, it could be that the rotor impedance varies significantly with the rotor position. This fact can be detected if the estimated resistance, according to the procedure described, is different at different rotor positions. Under such a situation, by slowly rotating the rotor, the measured current can be recorded until it covers one cycle, thus detecting its maximum and minimum values. Then, once the positions corresponding to the maximum and minimum currents are recorded, the maximum and minimum resistances can also be calculated, obtaining finally its averaged value as the adopted rotor resistance.

Therefore, the estimated R_r value that can be considered for control in this example is

$$R_r \cong 178 \, \text{m}\Omega$$

The following considerations are necessary in both resistance estimation methods:

- The temperature of the windings tested affects the estimation, so the tests should be performed at a known temperature that is reasonably close to the real operation temperature of the DFIM.
- The dv/dt filter normally is included in the experiment, for proper operation of the converter + machine system. Hence, the equivalent impedance of this element should be taken into consideration, if it can significantly affect the accuracy of the estimation.
- Commercially available equipment that can help:
  - Ohm-meter can accurately measure directly the resistances.

- o Special supply systems can feed the DFIM in an alternative way to the converter of the wind turbine.
- o Accurate sensors can provide more exact records of the generated voltage and currents.

Note that these last remarks are applicable also to subsequent estimation tests.

5.2.3 Leakage Inductances Estimation by VSC

Once the stator and rotor resistances are estimated, as seen in the previous section, it is possible to proceed to estimation of the leakage inductances. To perform this experiment, it is not necessary to know the resistance values; however, it is possible to corroborate the obtained resistance values of the previous experiment.

Basically, this test seeks to find an equivalent electrical circuit of the machine, where the magnetizing path (magnetizing inductances and iron losses) can be neglected, so the most dominant inductances are the inductances that we want to estimate—the stator and rotor leakage inductances.

As stated before, this test is performed under the assumption that the only supply system available is the back-to-back converter of the real wind energy generator. Figure 5.6a illustrates a schematic of the system configuration for this test assuming that the machine is in a star connection. The rotor must be short-circuited and the speed of the machine must be kept to zero ($s = 1$). In this situation, it does not matter how the voltage is applied to the stator of the machine, since the rotor ($R_r/s = R_r$ and $L_{\sigma r}$) is a much easier path for the current than the magnetizing circuit (L_m and R_{fe}). In general machine designs, the next impedance relation is typical:

$$|R_{fe}//jX_m| \gg |Rr + jX_{\sigma r}|$$

Consequently, if an AC voltage is applied to the stator, the majority of the current flow goes entirely through the stator resistance and inductance (R_s and $L_{\sigma s}$) and rotor resistance and inductance (R_r and $L_{\sigma r}$). This fact is shown graphically in Figure 5.6b. In order to achieve no rotation of the mechanical axis during this experiment ($s = 1$), the supply mode of the machine should not create torque. As depicted in Figures 5.6a and 5.6b, this is possible by feeding only two phases of the machine.

As in the previous experiment, the impedance present in this experiment is considerably low, so the AC voltages (v_{ab}) used must be small. To explore the behavior at different operating conditions, it is recommendable that one repeat the experiment at several AC voltage frequencies and amplitudes.

Therefore, by measuring the AC voltages and currents, the estimation can be performed as follows. First, the equivalent phase circuit is described by the steady state phasor equation:

$$\underline{V}_{ab} = 2(R_s + R_r)\underline{I}_{AC} + 2j\omega_s(L_{\sigma s} + L_{\sigma r})\underline{I}_{AC} \qquad (5.6)$$

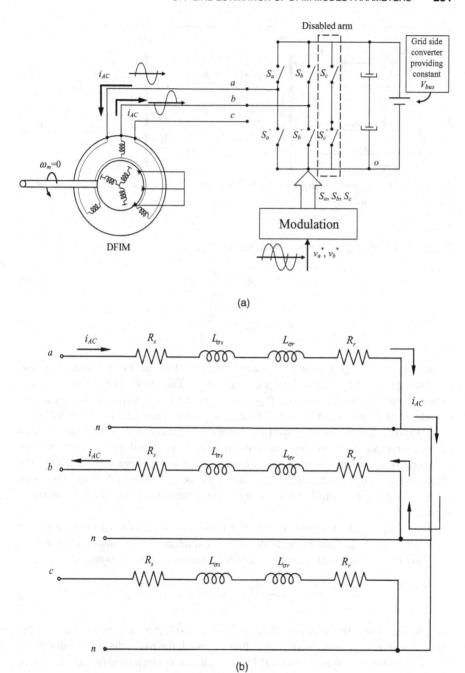

Figure 5.6 (a) System configuration for leakage inductances estimation. (b) Equivalent circuit for leakage inductances estimation.

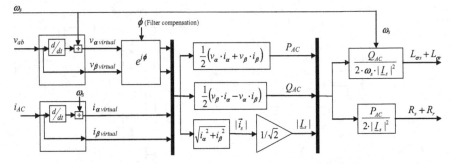

Figure 5.7 Leakage inductances and rotor and stator resistances estimation by single-phase P_{AC} and Q_{AC} calculation.

Thus, it is clear that, by a power balance analysis,

$$R_s + R_r = \frac{P_{AC}}{2|\vec{I}_{AC}|^2} \tag{5.7}$$

$$L_{\sigma s} + L_{\sigma r} = \frac{Q_{AC}}{2\omega_s|\vec{I}_{AC}|^2} \tag{5.8}$$

Hence, it is only possible to derive the addition of both resistances and inductances, but they cannot be found separately. The active and reactive powers to be measured in the experiment (P_{AC} and Q_{AC}) must be single phase, so they can be calculated as depicted in Figure 5.7. Note that a three-phase philosophy calculation of P, Q and the module is employed, by simply calculating a 90° phase lead signal of each voltage and current measurement (in this case virtual $\alpha\beta$ components). In addition, since a lowpass filter is used to measure the voltage created by the VSC, the phase variation (ϕ) produced by the filter must be compensated. This can easily be done by using the rotational transformation as depicted in the block diagram of Figure 5.7.

Depending on the design of the machine, different criteria exist to establish the values of $L_{\sigma s}$ and $L_{\sigma r}$ once the addition of both inductances is estimated according to Equation (5.8). In general, equal leakage inductances can be assumed [3]:

$$\frac{L_{\sigma s} + L_{\sigma r}}{2} \cong L_{\sigma s} \cong L_{\sigma r} \tag{5.9}$$

Note that since, in Equations (5.6), (5.7), and (5.8), the current measured in the stator is assumed to be the current that flows through the rotor, the stator referred R_r and $L_{\sigma r}$ are estimated. Again, if a DFIM with a stator–rotor turn ratio (u) different from the one being analyzed is used, the real rotor parameters are calculated as

$$R'_r = \frac{R_r}{u^2}, \qquad L'_{\sigma r} = \frac{L_{\sigma r}}{u^2} \tag{5.10}$$

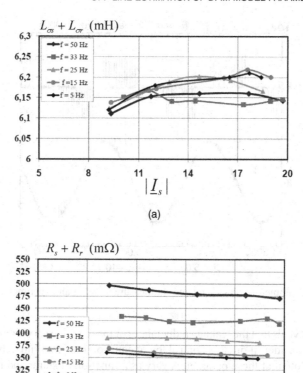

Figure 5.8   Leakage inductances and rotor and stator resistances estimation of a 15 kW DFIM.

Again, it has been assumed that all impedances of different phases are equal. Figure 5.8 illustrates an example of this test applied to a real 15 kW DFIM ($u = 1$). As in the previous experiment, several records of tests are kept for different frequencies and current amplitudes. A constant leakage inductance estimation is found, accompanied by a reasonably good correlation between the previously estimated (previous experiment) stator and rotor resistances. The estimated resistances have higher values at higher frequencies, since the rotor resistance takes a higher value due to skin effects [5]. Note that at normal operation in wind power generation applications, the rotor frequency should not exceed 15 Hz.

On the other hand, Figure 5.9 shows an example of voltage and current waveforms during a leakage estimation test. Notice that a very low AC voltage amplitude (low rms) is needed to have a high amplitude of current.

So the constant values of the estimated parameters considered for the control are

$$L_{\sigma s} \cong 3\,\text{mH} \quad \text{and} \quad L_{\sigma r} \cong 3\,\text{mH}$$

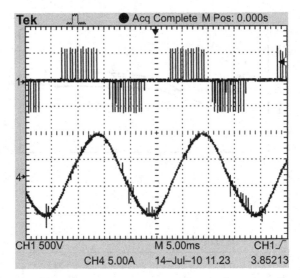

Figure 5.9 Example of voltage and current measurements in an experiment: v_{ab} and i_{AC}.

If the test reveals that the leakage inductances of the machine vary significantly, it is possible to consider it in the control by means of one look-up table for each inductance. Note that these parameters vary as a function of stator and rotor current amplitudes and stator and rotor frequencies as discussed in Section 5.2.1. This fact is advantageous since all magnitudes are measured by sensors during normal operation. One illustrative example is shown in Figure 5.10.

5.2.3.1 Rotor Locked Test There is an alternative solution for leakage inductances estimation that is very commonly used. This test is based on the same principles as the previously presented test, but the stator of the machine is supplied completely by the three phases.

Under these circumstances, the AC voltage applied at the three phases creates a torque on the axis of the machine that tends to move the rotor. As studied before, this movement would produce a slip, increasing the rotor resistance impedance (R_r/s) and making the current flow not only through the rotor but also through the magnetic circuit. As this situation complicates parameter estimations, in order to avoid movement of the rotor, normally the rotor is mechanically locked, as depicted in Figure 5.11.

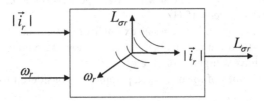

Figure 5.10 Consideration of $L_{\sigma r}$ parameter variation at control, with a look-up table.

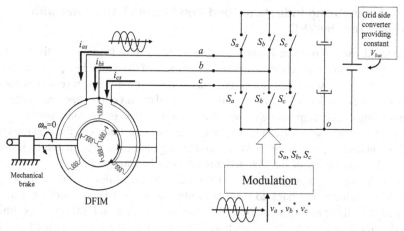

Figure 5.11   System configuration for leakage inductance estimations with rotor locked test.

Therefore, as was done in the previous experiment, feeding the stator with a three-phase AC voltage, it is possible to perform an equivalent test to derive the leakage inductance parameters.

It must be pointed out that, in this case, as three AC voltages and currents are available, the active and reactive powers can be measured as in a three-phase system, according to the equations

$$R_s + R_r = \frac{P_s}{3|\underline{I}_s|^2} \tag{5.11}$$

$$L_{\sigma s} + L_{\sigma r} = \frac{Q_s}{3\omega_s|\underline{I}_s|^2} \tag{5.12}$$

Hence, as opposed to a single-phase calculation of active and reactive powers, it is possible to perform the calculations with a three-phase basis using the Clarke transformation, as graphically illustrated in Figure 5.12.

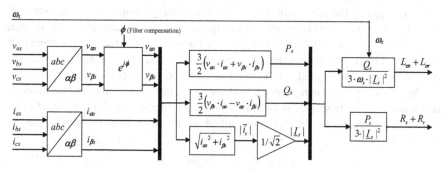

Figure 5.12   Parameter estimation by the rotor locked test.

5.2.4 Magnetizing Inductance and Iron Losses Estimation with No-Load Test by VSC

Once the stator and rotor resistances are estimated, and we assume that the obtained values are accurate enough for our purposes, the rest of the machine's parameters can be estimated. Following the same philosophy as in the previously presented two tests, this experiment seeks to find an equivalent electrical circuit of the machine, where the magnetizing path (magnetizing inductance and iron losses) is dominant so the current that goes through the rotor can be neglected. In this way, thanks to the previous two tests, knowledge of the stator resistance and leakage inductance is guaranteed, so the rest of the parameters of the magnetizing part can be estimated.

As stated before, this test is performed under the assumption that the only supply system available is the back-to-back converter of the real wind energy generator. Figure 5.13a illustrates a schematic of the system configuration for this test, assuming that the machine is in a star connection. The rotor must be short-circuited and the mechanical axis must be under no load, so it can rotate freely.

Therefore, if the machine is AC voltage supplied by the stator, it can operate as a squirrel cage induction machine (SCIM). By feeding the stator with a three-phase AC voltage, under no load torque, it can work at slip $= 0$ or nearly zero as depicted in the torque–slip curve of a SCIM (Figure 5.14).

In this situation, no matter how the voltage is applied to the stator of the machine, since the rotor impedance ($R_r/s \rightarrow \infty$) is much greater than the magnetizing circuit impedances, the majority of the current goes through the L_m and R_{fe}.

Consequently, if an AC voltage is applied to the stator, the current flow goes mostly through the stator resistance and inductance (R_s and $L_{\sigma s}$) and the iron loss resistance and magnetizing inductance (L_m and R_{fe}). This fact is shown graphically in Figure 5.13b.

Under this test, an acceleration torque is created when supplying through the stator, so the machine reaches a steady state of constant speed. The voltage amplitude of the experiment should be increased slowly at constant amplitude/frequency ratio until the final voltage value is attained. In that way, the machine accelerates progressively, avoiding high current demands until the steady state is reached.

Opposite to the previous experiments, the impedance in this experiment is considerably higher, so the AC voltages can get closer to the rated voltage of the converter. To explore the behavior at different operating conditions, we recommend repeating the experiment at several AC voltage frequencies and amplitudes. However, the normal operation of the DFIM in the wind turbine is at a rated stator voltage grid frequency. Hence, although different operating points can be tested and recorded, the only useful one corresponds to the rated voltage and frequency.

Therefore, by measuring voltage and currents, the estimation of the parameters is possible. First, based on the equivalent phase circuit described by the steady state phasor equation, the voltage applied to the iron loss resistance can be calculated as

$$\underline{V}_{fe} = \underline{V}_s - (R_s + j\omega_s L_{\sigma s})\underline{I}_s \tag{5.13}$$

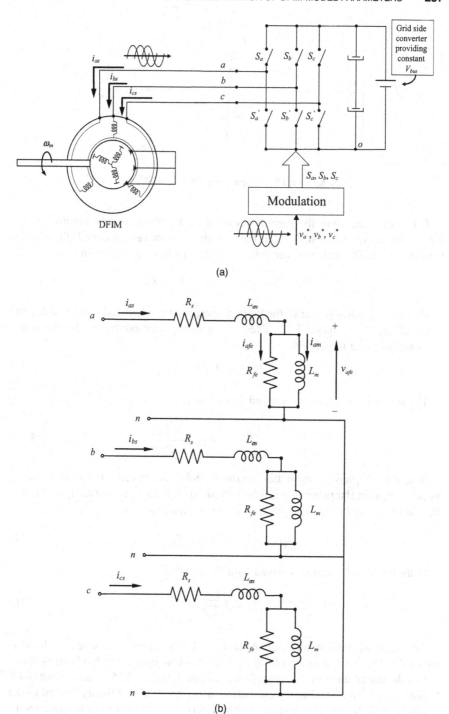

(a)

(b)

Figure 5.13 (a) System configuration for magnetizing inductance and iron losses estimation with no-load test. (b) Equivalent circuit for no-load test.

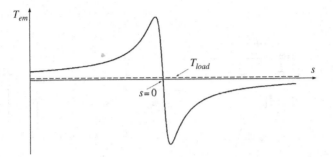

Figure 5.14   Torque–slip curve of a SCIM.

It is clear that, from the parameters obtained in previous experiments and the measurements of this test $(\underline{V}_s, \underline{I}_s)$, the voltage $\underline{V}_{fe}$ can be calculated directly. On the other hand, the active power balance of the machine during the test is

$$P_s = P_{fe} + P_{R_s} \rightarrow P_{fe} = P_s - P_{R_s} \tag{5.14}$$

P_s, the total stator power of the three phases, can be calculated as depicted in the block diagram of Figure 5.12. Thus, the active power consumed by the stator resistances can be found with

$$P_{R_s} = 3|\underline{I}_s|^2 R_s \tag{5.15}$$

The iron loss resistance is derived according to

$$P_{fe} = \frac{3|\underline{V}_{fe}|^2}{R_{fe}} \rightarrow R_{fe} = \frac{3|\underline{V}_{fe}|^2}{P_{fe}} \tag{5.16}$$

Note that all phasor amplitudes are rms. In addition, in order to estimate the L_m inductance, a similar procedure is followed based on reactive power analysis. Hence, the reactive power in the L_m inductance can be calculated as

$$Q_{L_m} = Q_s - 3\omega_s L_{\sigma s}|\underline{I}_s|^2 \tag{5.17}$$

Consequently, L_m can be derived with the equation

$$L_m = 3\frac{|\underline{V}_{fe}|^2}{\omega_s Q_{L_m}} \tag{5.18}$$

Again, Q_s, the total stator reactive power of the three phases, can be calculated as depicted in the block diagram of Figure 5.12. Also, in this case, it has been assumed that all the impedances of different phases are equal. Figure 5.15 illustrates the block diagram used in performing the estimation, using space vector theory instead of the phasor representation. The reader can simply derive the equivalences in space vector form of Equations (5.13)–(5.18).

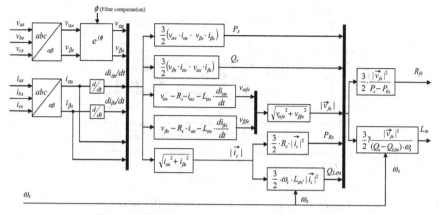

Figure 5.15 Parameter estimation by no-load test.

This section illustrates an example of this test applied to a real 15 kW DFIM. As in the previous experiment, several results of different tests are recorded at different frequencies and current amplitudes, reaching different steady state speeds and recording operating points at different stator fluxes. From the stator flux phasor expression, by neglecting the stator resistance, the stator flux can be approximated as

$$\frac{|V_s|}{\omega_s} = |\Psi_s| \tag{5.19}$$

No matter what the frequency of the test, the estimated L_m curve is repetitive, as graphically represented in Figure 5.16. On the other hand, depending on the magnetization level, L_m is different.

On the contrary, the estimated R_{fe} curves reveal that the iron losses vary with the voltage amplitude (stator flux) as well with the frequency, as illustrated in Figure 5.17.

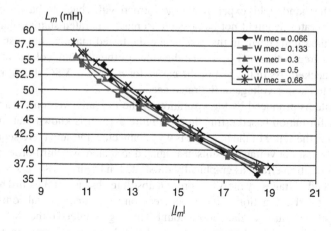

Figure 5.16 Estimated L_m curves at different speeds.

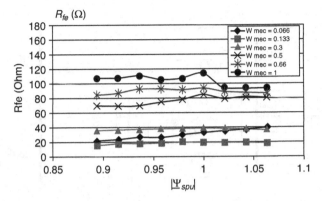

Figure 5.17   Estimated R_{fe} curves at different speeds.

Since the reactive powers measured are significantly higher than the active power measured, this fact often leads to inaccurate determinations of iron losses.

Therefore, for this machine operating as a DFIM, connected to a grid voltage of 380 V and 50 Hz, the estimated values would be

$$L_m \cong 46.5\,\text{mH} \quad \text{and} \quad R_{fe} \cong 110\,\Omega$$

Note that as the experiment is carried out by a VSC, the iron loss resistance when the machine is connected to the grid (sinusoidal voltage) would be significantly less. On the other hand, some machine designs also have mechanical losses (P_{mec}). In many cases, these losses are considered as electrical losses and they are directly considered in the iron losses. So the estimated R_{fe} would integrate both iron losses and mechanical losses.

However, as mentioned before, in general, high power machine designs are more efficient than low power machines, so the losses that occur are not as significant and they are not considered at the control stage.

On the other hand, if this experiment is performed with a back-to-back converter of the wind generator, it might not be possible to reach the rated stator voltage of the machine at the rated speed. In the case where the tested machine has a stator–rotor voltage ratio greater than 1 (i.e., $u > 1$), the value of the nominal voltage of the VSC should be checked. As in the rest of the cases, for $u \leq 1$, the VSC should be able to reach the rated stator voltage at the rated speed.

The problems produced by the saturation of the flux are overcome thanks to the Clarke transformation based estimation of Figure 5.15. It is known that, in general designs, when the stator of the machine is wye connected, the sum (zero sequence) of the three-phase stator voltages results in a signal dominated by a third-order harmonic due to the saturation effect, as graphically illustrated in Figure 5.18. This occurs even if the generated voltages by the VSC, present absolute absence of the third harmonic. Thus, for a proper estimation of the parameters, only the fundamental component of the phase voltages must be taken into account. This is guaranteed if the phase voltages are measured. Then the $\alpha\beta$ components are derived as was done in the estimation

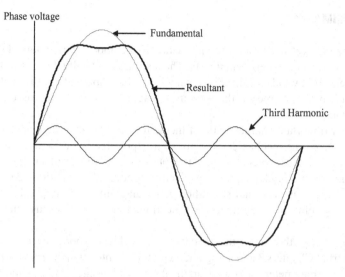

Phase voltage

Figure 5.18 Saturation effect on the stator phase voltages.

block diagram of Figure 5.15. For the $\alpha\beta$ components, the third harmonic is eliminated due to its zero sequence nature and the parameters are estimated based on first harmonic measurements. On the other hand, if the voltages used are the reference voltages or phase-to-phase measured voltages, this problem does not arise.

Finally, an example of the measured voltage and currents of the experiment is given in Figure 5.19. In this case, the accuracy of the measured voltage is not so critical since its amplitude is considerably higher than in the previously presented two tests.

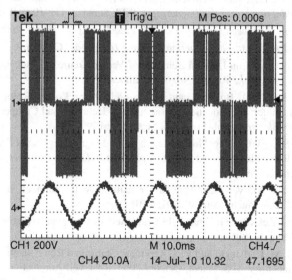

Figure 5.19 Example of voltage and current measurements of the experiment: v_{ab} and i_{as}.

5.3 SUMMARY

In this chapter, we studied a procedure to identify the model parameters of the DFIM by utilizing off-line experimental tests. The tests are especially designed to use only the elements that will be utilized in a real wind turbine application: the back-to-back converter as power source and the currents, voltage, and speed sensors used to perform the control.

Another important characteristic of the defined tests is that the machine must be decoupled from the mechanical system, so it is possible to perform a no-load test.

Therefore, this chapter explains the physics and requirements of each test, describing for the reader equivalent electric circuits, the equations derived from them for the parameters, and simplified block diagrams to be implemented in real hardware systems for automatic identification. Finally, we discuss the obtained results.

Consequently, thanks to the theoretical knowledge about machine modeling obtained from Chapters 3 and 4, together with this chapter, which gives the necessary information (parameters) for real applicability and usage of these models, it is possible to address a deeper analysis of the machine's behavior and its control. Prior to tackling control, the next chapter explores the performance of the DFIM, subject to a grid with disturbances such as voltage dips.

REFERENCES

1. S. J. Chapman, *Electrical Machines*. McGraw Hill, 1985.
2. J. F. Mora, *Electrical Machines*. McGraw Hill, 2003.
3. Institute of Electrical and Electronics Engineers (IEEE), *IEEE Standard Procedure for Polyphase Induction Motors and Generators*, IEEE Standard 112-1996. IEEE Press, 1996.
4. I. Boldea, *Variable Speed Generators*. CRC Press/Taylor & Francis, 2006.
5. I. Boldea and A. Nasar, *The Induction Machine Handbook*. CRC Press, 2006.
6. E. Levi, "Impact of Iron Loss on Behavior of Vector Controlled Induction Machines," *IEEE Trans. Ind. Appl.*, Vol. 31, No. 6, December 1995.
7. E. Levi, M. Sokola, A. Boglietti, and M. Pastorelli, "Iron Loss in Rotor-Flux-Oriented Induction Machines: Idenfitication, Assessment of Detuning and Compensation," *IEEE Trans. Power Electron.*, Vol. 11, No. 5, September 1996.
8. E. Levi and S. N. Vukosavic, "Identification of the Magnetising Curve During Commissioning of a Rotor Flux Oriented Induction Machine," *IEEE Proc. Power Appl.*, Vol. 146, No. 6, November 1999.
9. S. Heikkila and M. Vertanen,"Method for Estimating the Sum of Stator and Rotor Resistances," U.S. Patent, August 2007, ABB Oy.
10. G. Pace,"Rotor Resistance Estimation by Calibrated Measurement of Stator Temperature," U.S. Patent, March 2005.
11. S. Royak, M. M. Harbaugh, and R. J. Kerkman,"Stator and Rotor Identifier Using High Frequency Injection," U.S. Patent, October 2003, Rockwell Automation Technologies.

12. A. C. Smith, R. C. Healey, and S. Williamson, "A Transient Induction Motor Model Including Saturation and Deep Bar Effect," *IEEE Trans. Energy Conversion*, Vol. 11, No. 1, March 1996.

13. R. C. Healey, S. Williamson, and A. C. Smith, "Improved Cage Rotor Models for Vector Controlled Induction Motors," *IEEE Trans. Ind. Appl.*, Vol. 31, No. 4, July/August 1995.

14. F. Profumo, G. Griva, M. Pastorelli, and J. C. Moreira, "Universal Field Oriented Controller Based on Air Gap Flux Sensing via Third Harmonic Stator Voltage," *IEEE Trans. Ind. Appl.*, Vol. 30, No. 2, pp. 448–455, 1994.

15. J. S. Hsu, S. P. Liou, B. T. Lin, and W. F. Weldon, "Losses Influenced by Third-Harmonic Flux in Induction Motors," *IEEE Trans. Energy Conversion*, Vol. 6, No. 3, pp. 461–468, September 1991.

16. L. Kreindler, J. C. Moreira, A. Testa, and T. A. Lipo, "Direct Field Orientation Controller Using the Stator Phase Voltage Third Harmonic," *IEEE Trans. Ind. Appl.*, Vol. 30, No. 2, pp. 441–447, 1994.

CHAPTER 6

Analysis of the DFIM Under Voltage Dips

6.1 INTRODUCTION

One of the main advantages of the DFIM is that it provides variable speed using a small and economic power converter. These machines are controlled by a converter connected at the rotor, where the power is only a small fraction, approximately equal to the slip, of the stator power. This characteristic makes the DFIM especially suitable for applications where the slip is narrowly limited, such as wind turbines.

However, as wind energy penetration increased and the number of wind turbines using the DFIM expanded, the main DFIM disadvantage became more and more relevant: its excessive sensitivity to electric grid disturbances.

A drop of one or more phase voltages can be especially damaging for the electronic converter. A voltage dip causes overcurrents and overvoltages in the rotor windings, which would damage the converter connected to their terminals if no countermeasures were taken.

Today, a big effort is being made, by both industry and academia, to accurately predict these currents and voltages [1–4] and to design protection systems against these perturbations [5–9].

This chapter carries out a theoretical analysis of the dynamic behavior of the induction machine during grid disturbances. As will be shown, voltage dips and generally any perturbation of the grid voltage cause the induced rotor electromagnetic force (emf) to increase notably. Such overvoltage negatively affects the machine regulation loops and in the case of severe voltage dips, it may cause the rotor converter to saturate and hence lose control of the machine.

As a first step, normal operation of the machine will be analyzed: that is, operation with the stator voltages constant in amplitude, frequency, and phase. After that, the emf induced in a three-phase voltage dip will be deduced. Next, the analysis will be extended to asymmetrical voltage dips, during which the voltage drop is not the same in the three phases.

Doubly Fed Induction Machine: Modeling and Control for Wind Energy Generation,
First Edition. By G. Abad, J. López, M. A. Rodríguez, L. Marroyo, and G. Iwanski.
© 2011 the Institute of Electrical and Electronic Engineers, Inc. Published 2011 by John Wiley & Sons, Inc.

Finally, an equivalent electrical circuit to accurately describe the behavior of the DFIM from the rotor point of view will be proposed. This model will be used in the following chapters to analyze the machine–converter interactions in the case of grid perturbations.

6.2 ELECTROMAGNETIC FORCE INDUCED IN THE ROTOR

From the point of view of the rotor converter, the electromagnetic force is one of the most important variables, since it acts as a perturbation in the control loops and, given the case, it could saturate the converter (the interaction between the machine and the converter will be further analyzed in Chapters 7 and 8). The electromagnetic force induced in the rotor windings varies considerably under grid disturbances. This is the cause of the problematic behavior of DFIMs under perturbations. In this section, a generic expression for the emf induced in the rotor is obtained. In later sections, that expression will be tuned for the case of three-phase and asymmetrical voltage dips.

We rewrite the rotor voltage expression (4.10):

$$\vec{v}_r^r = R_r \vec{i}_r^r + \frac{d}{dt} \vec{\psi}_r^r \tag{6.1a}$$

The rotor flux is very similar to the stator flux imposed by the grid when the stator is directly connected to the grid. A slight difference between rotor flux and stator flux might appear due to leakage inductances. From Equations (4.11) and (4.12) the relation between both fluxes can be calculated as

$$\vec{\psi}_r^r = \frac{L_m}{L_s} \vec{\psi}_s^r + \sigma L_r \vec{i}_r^r \tag{6.1b}$$

where $\sigma = 1 - L_m^2 / L_s L_r$ is the leakage coefficient.

Combining Equations (6.1a) and (6.1b), the following expression is obtained:

$$\vec{v}_r^r = \frac{L_m}{L_s} \frac{d}{dt} \vec{\psi}_s^r + \left(R_r + \sigma L_r \frac{d}{dt} \right) \vec{i}_r^r \tag{6.2}$$

Thus, the rotor voltage might be divided into two terms. The first term corresponds to the emf induced by the stator flux in the rotor. It is the voltage in the rotor open-circuit terminals (where $i_r = 0$). Thus, its expression is

$$\vec{e}_r^r = \frac{L_m}{L_s} \frac{d}{dt} \vec{\psi}_s^r \tag{6.3}$$

Or, if the variables are expressed in a stator reference frame, we have

$$\vec{e}_r^s = \frac{L_m}{L_s} \left(\frac{d}{dt} \vec{\psi}_s^s - j\omega_m \vec{\psi}_s^s \right) \tag{6.4}$$

The second term only appears if there is a current in the rotor. It is due to the voltage drop in both the rotor resistance R_r and the rotor transient inductance, σL_r.

6.3 NORMAL OPERATION

During normal operation, the electrical grid is a source of three balanced voltages of constant amplitude and frequency. That is, the three phases have the same voltage and are shifted 120°:

$$v_a = \hat{V}_g \cos(\omega_s t + \varphi)$$
$$v_b = \hat{V}_g \cos(\omega_s t + \varphi - 2\pi/3)$$
$$v_c = \hat{V}_g \cos(\omega_s t + \varphi - 4\pi/3)$$

Thus, the space vector of the stator voltage is a rotating vector of constant amplitude $\hat{V}_g$ that rotates at synchronous speed ω_s:

$$\vec{v}_s^s = \hat{V}_g e^{j\varphi} e^{j\omega_s t} = \sqrt{2}\underline{V}_g e^{j\omega_s t} \tag{6.5}$$

where $\underline{V}_g$ is the phasor of the grid voltage in rms.

Since the voltage drop in the stator resistance, R_s, is significantly smaller than the stator voltage drop (about 1% in a multimegawatt machine), that voltage can be neglected. By doing so, the phasor of the stator flux is found to be

$$\underline{\psi}_s = \frac{V_g}{j\omega_s} \tag{6.6}$$

So, at steady state, the space vector of the stator flux is a rotating vector of constant amplitude, proportional to the grid voltage, and synchronous speed:

$$\vec{\psi}_s^s = \sqrt{2}\underline{\psi}_s e^{j\omega_s \cdot t} = \frac{\hat{V}_g e^{j\varphi}}{j\omega_s} e^{j\omega_s t} \tag{6.7}$$

From the point of view of the rotor windings, the rotational speed of the stator flux is the difference between the synchronous speed and the electrical rotor speed; that is, the slip frequency, ω_r. If, for example, the machine is operating at synchronous frequency, the rotor windings will see a constant flux and no emf will be induced in these windings. In any other case, the flux across the rotor is variable, as seen if it is expressed in a rotor reference frame:

$$\vec{\psi}_s^r = \vec{\psi}_s^s e^{-j\omega_m t} = \sqrt{2}\underline{\psi}_s e^{j(\omega_s - \omega_m)t} = \sqrt{2}\underline{\psi}_s e^{j\omega_r t} \tag{6.8}$$

This variable flux will thus induce an emf proportional to the slip frequency:

$$\vec{e}_r^r = \frac{L_m}{L_s}\frac{d}{dt}\vec{\psi}_s^r = j\omega_r \frac{L_m}{L_s}\vec{\psi}_s^r \tag{6.9}$$

The amplitude of this emf can be expressed as a function of the stator voltage:

$$|\hat{E}_r| = \omega_r \frac{L_m}{L_s} \frac{\hat{V}_g}{\omega_s} = \hat{V}_g \frac{L_m}{L_s} s \tag{6.10}$$

As the machine typically operates with slips up to 25%, this voltage is relatively low. It should be noted that this voltage is referred to the stator side. In real windings, the voltage is multiplied by the stator/rotor turns ratio, u. For low voltage machines (up to 2–3 MW) this ratio is usually chosen to obtain a voltage in the rotor similar to the stator voltage to avoid placing a transformer in the back-to-back converter connected to the rotor.

6.4 THREE-PHASE VOLTAGE DIPS

A voltage dip is a sudden drop of one or more voltage phases. Such a dip is said to be a three-phase, symmetrical, or balanced dip if the drop is the same in the three phases. This fault could be caused, for example, by inrushing currents at motor start-up or by a near short-circuit between the three phases and the ground.

Although the analysis can be applied to any type of voltage dip, we will now examine an abrupt voltage dip; that is, the voltage drops at time $t = 0$ from its initial value, $\hat{V}_{pre}$, to its final value, $\hat{V}_{fault}$.

$$\vec{v}_s^s(t < 0) = \hat{V}_{pre} e^{j\omega_s t}$$
$$\vec{v}_s^s(t \geq 0) = \hat{V}_{fault} e^{j\omega_s t} \tag{6.11}$$

Obviously, such an abrupt voltage is just a mathematical exercise as it is not possible in real systems. In actual systems, the voltage drops with a particular derivative, which depends on the grid and the characteristics of the fault that has caused the voltage dip. Then why have we considered the possibility of an instantaneous voltage dip?

- It is the worst case scenario. In any other situation, the voltages induced in the rotor would be lower.
- Current regulation codes usually establish a time lapse of 1 ms for the voltage to drop. Such an abrupt voltage dip might be considered instantaneous since differences in the induced rotor voltage between both kinds of dips are lower than 1%.

To analyze this case, first the situation in which the rotor is open-circuited will be examined. After that, the analysis will include the influence of the rotor current as well.

6.4.1 Total Voltage Dip, Rotor Open-Circuited

We talk about a total voltage dip when the voltage during the dip is zero, which might be caused by a short-circuit right across the terminals of the machine. It is often believed that, when the stator has no voltage, the machine is demagnetized; that is, there is no flux, and thus no emf is induced in the rotor windings. In fact, it is true that, in the steady state, the flux is proportional to the stator voltage, and therefore if the dip is long enough, the machine will demagnetize completely. However, the flux cannot be discontinuous as it is a state variable. On the contrary, the flux evolves from its initial value prefault to zero, resulting in a transient emf induced in the rotor terminals.

The evolution of the stator flux can be calculated mathematically from the dynamic expression of the stator, Equation (4.9). With the rotor being open-circuited that expression can be written as

$$\frac{d}{dt}\vec{\psi}_s^s = \vec{v}_s^s - \frac{R_s}{L_s}\vec{\psi}_s^s \tag{6.12}$$

In the case of a total voltage dip, $v_s = 0$, the solution of expression (6.12) is

$$\frac{d}{dt}\vec{\psi}_s^s = -\frac{R_s}{L_s}\vec{\psi}_s^s$$

$$\vec{\psi}_s^s = \vec{\Psi}_0 e^{-t/\tau_s} \tag{6.13}$$

where $\vec{\Psi}_0$ is the initial value of the flux (for $t = 0$) and $\tau_s = L_s/R_s$ is the time constant of the stator.

The value of $\vec{\Psi}_0$ can be calculated by considering the flux just before and after the dip appearance:

$$\vec{\psi}_s^s(t < 0) = \frac{\hat{V}_{pre}}{j\omega_s}e^{j\omega_s t}$$

$$\vec{\psi}_s^s(t \geq 0) = \vec{\Psi}_0 e^{-t/\tau_s} \tag{6.14}$$

As the flux must be continuous, both expressions must yield the same value at $t = 0$. Consequently, we have

$$\vec{\Psi}_0 = \frac{\hat{V}_{pre}}{j\omega_s} \tag{6.15}$$

One important characteristic of the flux during the dip is that it doesn't rotate: it is fixed with the stator. In other words, the flux, which was rotating at the grid frequency before the dip, freezes during the dip. Its amplitude decays exponentially from its initial value to zero with the time constant of the stator. In multimegawatt machines, this time constant ranges between 0.8 and 1.5 s, much longer than the average duration of a voltage dip.

Figure 6.1 shows the trajectory of the space vector before and during the dip. As can be observed, before the dip the space vector is a rotating vector and it follows a circular trajectory. At the beginning of the dip, the flux freezes and its amplitude slowly decreases to zero. During the dip, its trajectory is a straight line from the point on the circle where it was just before the dip, to the origin, where it ends at steady state.

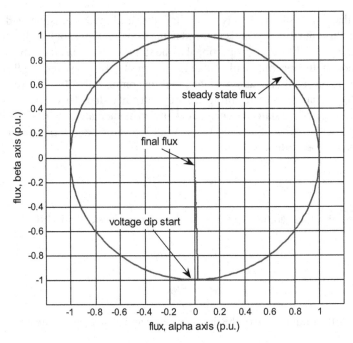

Figure 6.1   Stator flux trajectory during a total voltage dip.

From the expression of the stator flux, Equation (6.14), and taking into account that there is no current in the rotor, the stator current can be obtained before and during the dip:

$$\vec{i}_s^s(t < 0) = \frac{\vec{\psi}_s^s(t < 0)}{L_s} = \frac{\hat{V}_{pre}}{j\omega_s L_s}e^{j\omega_s t}$$

$$\vec{i}_s^s(t \geq 0) = \frac{\vec{\psi}_s^s(t \geq 0)}{L_s} = \frac{\vec{\Psi}_0 e^{-t/\tau_s}}{L_s} \tag{6.16}$$

Similar to the stator flux, the space vector of the currents stops rotating during the dip, which implies the circulation of DC currents in the stator of the machine. Even if they are DC, these currents are not constant as they decay exponentially with the same dynamics as the flux. As can be observed in Figure 6.2, the initial amplitude of the current of each phase corresponds to its instantaneous value at the time the voltage dip appears.

The previous figure was obtained from a small-size laboratory machine whose time constant was approximately 100 ms. The currents disappear in about 400 ms. A bigger machine, like the ones installed in wind turbines, usually has much less resistance, and its time constant is thus quite larger. The decaying of the flux and the currents for these machines might last for some seconds.

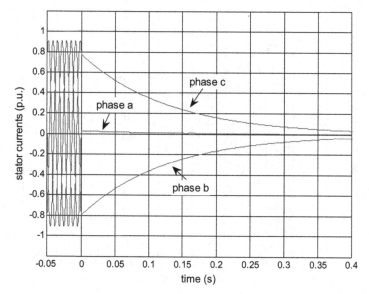

Figure 6.2   Stator currents during a total voltage dip.

As has been stated previously, during a total voltage dip there is flux inside the machine even if there is no voltage in the grid. This transitory flux induces an emf in the rotor windings in the same way as a steady state flux does during normal operation. However, both situations are quite different: the steady state flux rotates synchronously—at a speed very similar to the rotor speed—whereas the transitory flux during a dip is static and does not rotate. Thus, regarding the rotor windings, the steady state flux rotates very slowly, at the slip frequency. On the contrary, the transitory flux is seen by the rotor as rotating much faster, at rotor speed but in an opposite sense. Consequently, the emf induced by the transitory flux will be much higher than the emf induced by the steady state during normal operation.

Figure 6.3 shows the voltages in the rotor terminals of a machine operating with a slip, $s = -20\%$. Before the dip, the rotor voltages correspond to the normal operation with $s = -0.2$. Their peak value is around 80 V. When the full dip appears, the natural flux induces a voltage whose amplitude decreases exponentially, similarly to the stator flux (or current). The maximum amplitude is reached at the beginning of the dip, when the flux is maximal, and it is equal to 380 V, a value even higher than the peak grid voltage.

Besides the amplitude, the frequency of the voltage varies notably before and during the dip. During normal operation, the frequency is equal to the slip frequency, that is, $\omega_r = s\omega_s$. As the grid frequency was 50 Hz, the result is -10 Hz. During the dip, on the contrary, the flux is seen by the rotor windings as rotating at -60 Hz, and so this is the frequency of the voltage induced in the rotor.

The electromagnetic force induced by the transitory flux can be calculated using expression (6.3) if the stator flux is referred to a rotor reference frame. From

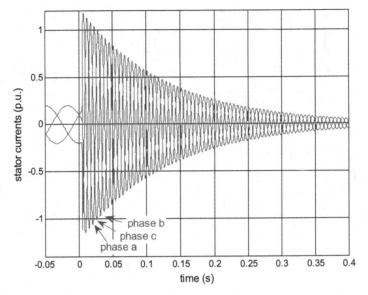

Figure 6.3 Electromagnetic force induced during a total voltage dip.

Equation (6.13) we obtain

$$\vec{\psi}_s^r = \vec{\Psi}_0 e^{-t/\tau_s} e^{-j\omega_m t} \tag{6.17}$$

Substituting this flux in Equation (6.3), we have

$$\vec{e}_r^r = -\frac{L_m}{L_s}\left(\frac{1}{\tau_s} + j\omega_m\right)\vec{\psi}_s^r = -\frac{L_m}{L_s}\left(\frac{1}{\tau_s} + j\omega_m\right)\vec{\Psi}_0 e^{-t/\tau_s} e^{-j\omega_m t} \tag{6.18}$$

It is interesting to note that the induced emf is proportional to the transitory flux, and thus the voltage induced decays likewise: exponentially with the stator time constant. The last expression also shows that the space vector is rotating at speed $-\omega_m$, and hence, as seen in Figure 6.3, the voltages at the stator terminals are in negative sequence, and its frequency is the same as the electrical rotor speed.

As previously stated, the maximum voltage is induced at the beginning of the dip. Its value can be calculated from Equations (6.18) and (6.15), neglecting the term $1/\tau_s$:

$$\left|\hat{E}_{r,\max}\right| \approx \frac{L_m}{L_s} \cdot \frac{\omega_m}{\omega_s} \hat{V}_{pre} = \hat{V}_{pre} \frac{L_m}{L_s}(1 - s) \tag{6.19}$$

This expression clearly shows that the emf induced during the dip is proportional to the factor $(1 - s)$, unlike the emf induced during normal operation, which is proportional to s, as stated in Equation (6.9). Taking into account that the slip in these machines usually ranges between -0.25 and 0.25, it can be deduced that, in the case of a voltage dip, the amplitude of the emf induced at the beginning of a total dip can be 3 to 5 times higher than during normal operation. This abnormal overvoltage in the rotor notably affects the operation of the rotor converter, usually saturating the converter and resulting therefore in the loss of current control.

6.4.2 Partial Voltage Dip, Rotor Open-Circuited

Generally, grid faults are partial voltage dips; that is, the stator voltage drops below its rated voltage but remains above zero. These dips are characterized by a relative drop or depth. A total voltage dip is then a voltage dip of depth equal to 1, or 100%. This section analyzes the influence of the depth of the voltage dip in the emf induced at the rotor windings. The generator is assumed to be working under normal operating conditions when, at a given moment $t = 0$, a voltage dip of depth (or profundity) p occurs:

$$\vec{v}_s^s = \begin{cases} \hat{V}_{pre} e^{j\omega_s t} & \text{for } t < 0 \\ (1-p)\hat{V}_{pre} e^{j\omega_s t} & \text{for } t \geq 0 \end{cases} \tag{6.20}$$

Figure 6.4 shows, for example, the evolution of the grid or stator voltages during a 50% partial voltage dip.

In steady state, the flux is proportional to the stator voltage, so the flux will experience the same drop as the voltage:

$$\vec{\psi}_s^s(steady\_state) = \begin{cases} \dfrac{\hat{V}_{pre}}{j\omega_s} e^{j\omega_s t} & \text{for } t < 0 \\ \dfrac{(1-p)\hat{V}_{pre}}{j\omega_s} e^{j\omega_s t} & \text{for } t \geq 0 \end{cases} \tag{6.21}$$

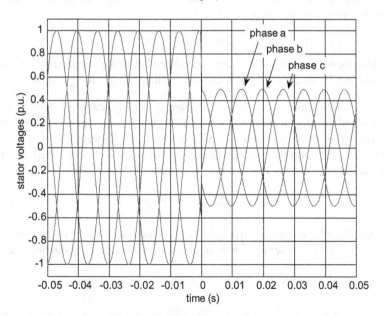

Figure 6.4 Grid phase voltages in case of a 50% voltage dip.

However, and just as it happened during total voltage dips, the flux does not change instantaneously because it is a state variable, and thus it is continuous. Instead, the flux changes progressively from one steady value to another. Its evolution can be described using expression (6.12).

$$\frac{d}{dt}\vec{\psi}_s^s = \vec{v}_s^s - \frac{R_s}{L_s}\vec{\psi}_s^s \tag{6.22}$$

As with any ordinary differential equation, this expression can be divided into two terms: the homogeneous and the particular solution.

- The homogeneous solution is the transient or the natural response of the equation. It depends on the initial conditions of the total solution.
- The particular solution is the steady state response. This solution is proportional to the input of the equation; that is, to the stator voltage.

From now on these terms will be referred to as "natural flux," ψ_{sn}, and "forced flux," ψ_{sf}. Their characteristics are very different:

- The forced flux amplitude is constant and proportional to the grid voltage, as previously stated in Equation (6.6):

$$\vec{\psi}_{sf}^s = \frac{\hat{V}_{fault}}{j\omega_s}e^{j\omega_s t} = \frac{\hat{V}_{pie}(1-\mathrm{p})}{j\omega_s}e^{j\omega_s t} \tag{6.23}$$

This flux is a space vector rotating at the grid frequency.

- The natural flux is a transient flux, caused by the voltage change. Its initial value is $\vec{\Psi}_{n0}$ and it decays progressively:

$$\vec{\psi}_{sn}^s = \vec{\Psi}_{n0}e^{-t/\tau_s} \tag{6.24}$$

This flux doesn't depend on the stator voltage but on the initial conditions of the machine. It can exist even if there is no voltage at the stator terminals.

During normal operation, the grid voltage is usually stable in amplitude and frequency and therefore there is only a forced flux in the stator. On the contrary, during a total voltage dip the stator voltage is null and hence the only flux present in the stator is the natural flux. During a partial voltage dip both natural flux and forced flux are present in the machine, so the total flux is the sum of both of them:

$$\vec{\psi}_s^s = \frac{\hat{V}_{fault}}{j\omega_s}e^{j\omega_s t} + \vec{\Psi}_{n0}e^{-t/\tau_s} \tag{6.25}$$

The initial flux $\vec{\Psi}_{n0}$ is obtained from the initial condition (at $t = 0$) by imposing that the stator flux must be identical immediately before and after the dip appearance.

$$\vec{\psi}_s^s(t < 0) = \frac{\hat{V}_{pre}}{j\omega_s}e^{j\omega_s t}$$

$$\vec{\psi}_s^s(t \geq 0) = \frac{\hat{V}_{fault}}{j\omega_s}e^{j\omega_s t} + \vec{\Psi}_{n0}e^{-t/\tau_s} \tag{6.26}$$

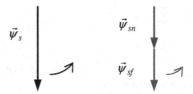

Figure 6.5 Decomposition of the flux at the beginning of the dip.

In order to avoid discontinuities, both expressions must be identical at $t = 0$. Hence,

$$\vec{\Psi}_{n0} = \frac{p\hat{V}_{pre}}{j\omega_s} \tag{6.27}$$

Figure 6.5 shows the space vector of the stator flux just before and after the occurrence of the fault. Before the fault the flux was rotating at the grid frequency. After the fault the flux is split into two components: one rotating at the grid frequency (forced flux) and another static with the stator (natural flux). The total flux, that results from the addition of the two fluxes, is exactly the same as the one existing before the fault.

During the dip, the forced flux continues rotating; meanwhile, the natural flux keeps constant as represented in Figure 6.6.

The trajectory of the flux in the alpha–beta plane can be explained using this decomposition: the forced flux term makes the space vector trace a circle and because of the natural flux term, the circle is not centered.

Figure 6.7 shows the trajectory of the flux as a consequence of a 50% dip. Before the dip, the flux traces a circle with a per unit radius equal to 1. When the dip takes places, the size of the circle decreases as the voltage diminishes, and the center of the circle is moved as a consequence of the natural flux. During the course of the dip, the circle moves to the center as the natural flux goes to zero. After the transitory finishes, the steady state natural flux disappears and the trajectory of the flux gets centered again.

This flux is created by a stator current composed of two components: an AC component for the forced flux and a decaying DC component for the natural flux:

$$\vec{i}_s^s(t \geq 0) = \frac{\hat{V}_{fault}}{jL_s\omega_s}e^{j\omega_s t} + \frac{\vec{\Psi}_{n0}e^{-t/\tau_s}}{L_s} \tag{6.28}$$

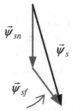

Figure 6.6 Evolution of the flux during the voltage dip.

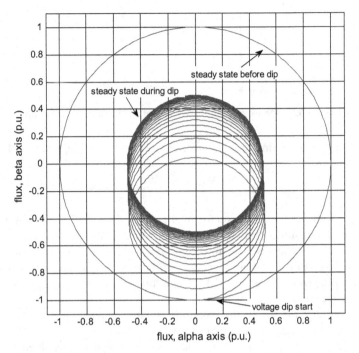

Figure 6.7 Stator flux trajectory consequence of a 50% voltage dip.

Each of the two terms of the flux induces a subsequent component in the emf induced in the rotor, $\vec{e}_r$:

$$\vec{e}_r^r = \vec{e}_{rf}^r + \vec{e}_{rn}^r \tag{6.29}$$

The first component of the emf is induced by the forced flux. It can be calculated by introducing the forced flux, Equation (6.23), into Equation (6.4). As this flux is synchronous, this voltage is similar to the voltage induced during normal operations, that is, an AC voltage proportional to the stator voltage and to the slip:

$$\vec{e}_{rf}^r = \frac{L_m}{L_s} \hat{V}_{fault} e^{\omega_r t} s \tag{6.30}$$

This emf is relatively small, and in any case smaller than the normal operating voltage since the stator voltage has dropped.

The second component of the emf is induced by the natural flux. It can be calculated from Equation (6.18) and neglecting the term $1/\tau_s$:

$$\vec{e}_{rn}^r = -\frac{L_m}{L_s} j\omega_m \vec{\Psi}_{n0} e^{-t/\tau_s} e^{-j\omega_m t} \tag{6.31}$$

Finally, the total emf induced in the rotor results from adding the two components. Expressing this voltage as a function of the dip depth yields.

$$\vec{e}_r^r = \frac{L_m}{L_s} \hat{V}_{pre} \left(s(1-p)e^{j\omega_r t} - (1-s)pe^{-j\omega_m t} e^{-t/\tau_s} \right) \tag{6.32}$$

The two terms of the previous equation are different in nature. The first one is generated by the new grid voltage and its amplitude is small as it is proportional to the slip. Its frequency is the difference between the synchronous and the rotor frequencies; therefore, it usually achieves a few hertz. The second term is a transient voltage caused by the natural flux. Its amplitude might be important as it is proportional to the depth of the dip $(\hat{V}_{pre} - \hat{V}_{fault})$, and its frequency is the rotor electrical speed, ω_m.

Figure 6.8 shows the voltage in the open-circuited rotor for a machine operating at a 20% slip during an 80% dip. It can be observed that the emf induced before the fault is sinusoidal with a frequency equal to 10%, corresponding to the slip frequency of the machine. The amplitude, referred to the stator side, is 0.8 p.u., this is the product of the stator voltage and the slip.

During the dip, high voltages are induced in the rotor mainly due to the natural flux. The maximum is reached during the first instants of the dip, when the natural flux is the highest. The shapes of the voltages might be quite odd as they result from the addition of two sinusoidals of different frequency. At the beginning of the dip only a sinusoidal of 40 Hz can be distinguished. It is related to the emf induced by the natural flux, which is seen as rotating at that frequency (the rotational speed) by the rotor windings. As in the case of total voltage dip, the sequence of the phase is inverted.

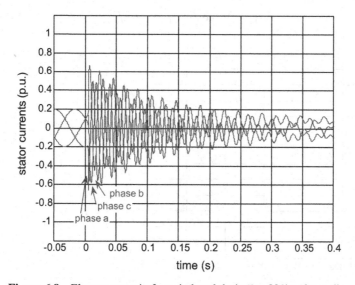

Figure 6.8 Electromagnetic force induced during an 80% voltage dip.

During the dip, as the natural flux decays, the 40 Hz sinusoidals disappear. Then, the other term is perceived: a sinusoidal of 10 Hz whose sequence of phases is positive. This corresponds to the emf induced by the forced flux. Its shape is similar to the prefault voltage, but its amplitude is now five times lower since this voltage is proportional to the new grid voltage.

6.5 ASYMMETRICAL VOLTAGE DIPS

Frequently, the faults that cause voltage dips don't affect the three phases in the same manner. The most common fault in electrical networks, for example, is a single line-to-ground short-circuit. Short-circuits between two lines might also be frequent. In such cases, the remaining voltage is not the same in the three lines and/or the phase shifts between the three voltages is no longer 120°. The system is thus said to be unbalanced or asymmetrical.

Traditionally, the effect of asymmetrical faults in power systems has been studied by means of the 1918 symmetrical component method, developed by Charles Legeyt Fortescue.

6.5.1 Fundamentals of the Symmetrical Component Method

In a balanced system (Figure 6.9), the voltages of the three phases have the same amplitude and their waveforms are 120° (or $2\pi/3$ radians) offset in time:

$$v_a = \hat{V}_g \cos(\omega t + \varphi)$$
$$v_b = \hat{V}_g \cos(\omega t + \varphi - 2\pi/3)$$
$$v_c = \hat{V}_g \cos(\omega t + \varphi - 4\pi/3)$$

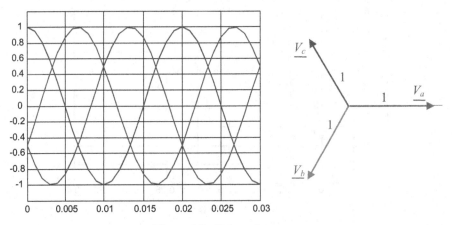

Figure 6.9 Balanced system.

Using the phasor notation, the phasors are three complex numbers with identical modulus but with their arguments shifted 120°:

$$\begin{bmatrix} \underline{V_a} \\ \underline{V_b} \\ \underline{V_c} \end{bmatrix} = \begin{bmatrix} \hat{V}_g e^{j\varphi} \\ \hat{V}_g e^{j\varphi - 2\pi/3} \\ \hat{V}_g e^{j\varphi - 4\pi/3} \end{bmatrix} = \hat{V} e^{j\varphi} \begin{bmatrix} 1 \\ a \\ a^2 \end{bmatrix}$$

were a is a unitary vector at an angle of 120°. It is easy to see that $a^2 = a* = 1\angle 240°$, $a^3 = 1$, and $1 + a + a^2 = 0$.

An unbalanced system is characterized by its phasors having different modulus and/or the differences between its arguments being other than 120°. According to the symmetrical component theory, any unbalanced three-phase system can be broken up into three sets of balanced three-phase systems. Thus, any phasor (voltage, current, flux, etc.) can be decomposed as the addition of three phasors of a balanced system. These three balanced systems are called:

- *The positive sequence.* often denoted with subscript 1, where the phases follow the regular sequence order; that is, phase $a \rightarrow$ phase $b \rightarrow$ phase c. It is the habitual three-phase balanced system present in an undistorted grid (Figure 6.10).

- *The negative sequence.* often denoted by subscript 2, where the sequence of the phasors has the opposite direction of the positive sequence: phase $a \rightarrow$ phase $c \rightarrow$ phase b. This sequence thus causes the motor to turn in the opposite direction (Figure 6.11).

- *The zero sequence.* often denoted by subscript 0, where phasors are all in phase. The phasors being identical in modulus and argument, the voltages in the three phases are identical at all times (Figure 6.12).

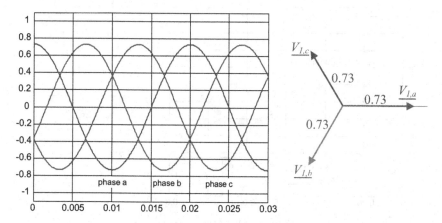

Figure 6.10 Positive sequence.

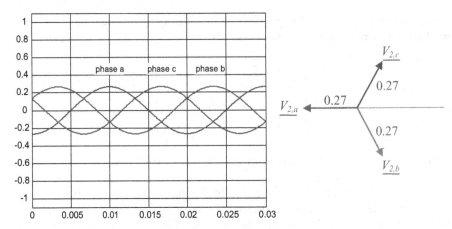

Figure 6.11 Negative sequence.

In an ordinary balanced system only the positive sequence is present. If an inverse system is built by switching two of the phases, only the negative sequence will be present. In any other situation, the three sequences can be present in the system. Then, the voltage in one phase is the sum of the same phase voltage in each of the three sequences.

For example, a line-to-ground short-circuit in phase a will originate a single-phase voltage dip where phases b and c remain more or less unchanged whereas phase a drops. According to the symmetrical component method this unbalanced system can be decomposed into the sequences represented in Figures 6.10, 6.11 and 6.12, so that the sum of the three components is equal to the original voltagess it is represented in Figure 6.13.

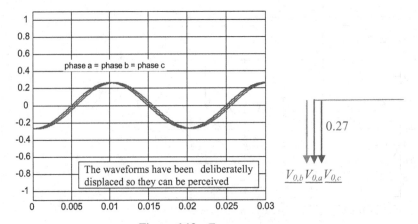

Figure 6.12 Zero sequence.

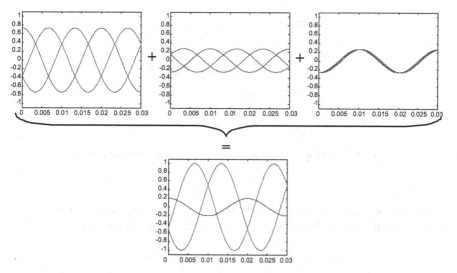

Figure 6.13 Addition of the three sequences for a single-phase dip.

The three sequences can be calculated from the phasors of the three phases of the original system:

$$
\begin{bmatrix} \underline{V}_{1,a} \\ \underline{V}_{2,a} \\ \underline{V}_{0,a} \end{bmatrix} = \frac{1}{3} \begin{bmatrix} 1 & a & a^2 \\ 1 & a^2 & a \\ 1 & 1 & 1 \end{bmatrix} \cdot \begin{bmatrix} \underline{V}_a \\ \underline{V}_b \\ \underline{V}_c \end{bmatrix}
\tag{6.33}
$$

where

$\underline{V}_a$, $\underline{V}_b$, $\underline{V}_c$ are the phasors of phases a, b, and c.

$\underline{V}_{1,a}$, $\underline{V}_{2,a}$, $\underline{V}_{0,a}$ are the phasors of phase a of the positive, negative, and zero sequences. Often the subscript a is omitted; thus, instead of $\underline{V}_{1,a}$ we will use $\underline{V}_1$, instead of $\underline{V}_{2,a}$, we will use $\underline{V}_2$, and so on.

6.5.2 Symmetrical Components Applied to the DFIM

In an electric machine, with symmetrical windings shifted 120°, a positive sequence of voltages implies a stator flux of constant amplitude and rotating at the grid frequency. In the machine represented in Figure 6.14 the flux would turn counterclockwise.

A negative sequence also implies a flux with constant amplitude, but in such a case, the flux would turn in the opposite direction, that is, clockwise in the machine of the figure.

Finally, the zero sequence can only create a leakage flux if the machine has its neutral grounded. In the vast majority of cases, however, the neutral is isolated and the zero sequence voltage has no influence at all on the behavior of the machine.

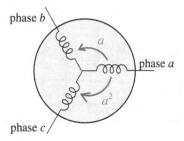

Figure 6.14   Typical configuration of windings in a three-phase machine.

Ignoring the zero sequence, the grid voltage can be decomposed as the sum of the positive and negative voltages. Using a space vector notation we have

$$\vec{v}_s^s = \sqrt{2}\underline{V}_1 e^{j\omega_s t} + \sqrt{2}\underline{V}_2 e^{-j\omega_s t} \qquad (6.34)$$

If the machine is not saturated, it can be considered a linear system. The steady state flux will then be the addition of two terms, one arising from the positive component of the stator voltage, ψ_{s1}, and another arising from the negative component, ψ_{s2}. Neglecting the stator resistance, they may be written as

$$\vec{\psi}_{s1}^s = \frac{\sqrt{2}\underline{V}_1}{j\omega_s} e^{j\omega_s t}$$

$$\vec{\psi}_{s2}^s = \frac{\sqrt{2}\underline{V}_2}{-j\omega_s} e^{-j\omega_s t} \qquad (6.35)$$

Additionally, and just like in a three-phase dip, a natural flux might surge at the beginning of the dip so that the total stator flux is continuous. The total flux is then the sum of the three fluxes:

$$\vec{\psi}_s^s = \vec{\psi}_{s1}^s + \vec{\psi}_{s2}^s + \vec{\psi}_{sn}^s \qquad (6.36)$$

The positive and negative fluxes are the steady state fluxes, a function of the grid voltage. The natural flux, in contrast, is a transitory flux, independent on the grid voltage. Its initial value Ψ_{n0} guarantees no discontinuity in the total flux. In three-phase voltage dips, this initial natural flux is a function of depth whereas in asymmetrical dips, it depends as well on the type of fault (single-phase, phase-to-phase, etc.) and on the time of appearance.

Each flux induces an emf in the rotor according to its amplitude and its relative speed with respect to the rotor windings. The total open-circuit rotor voltage is then the sum of the three terms:

$$\vec{e}_r = \vec{e}_{r1} + \vec{e}_{r2} + \vec{e}_{rm} \qquad (6.37)$$

The first two terms can be calculated by substituting into Equation (6.4) the fluxes obtained in Equation (6.35). So, expressing them in the rotor reference frame, we have

$$\vec{e}^r_{r1} = \sqrt{2}\underline{V}_1 \frac{L_m}{L_s} s e^{js\omega_s \cdot t}$$

$$\vec{e}^r_{r2} = \sqrt{2}\underline{V}_2 \frac{L_m}{L_s}(2-s)e^{-j(2-s)\omega_s t} \qquad (6.38)$$

The first voltage is small as it is proportional to the slip. Its frequency is equal to the slip frequency, that is a few hertz. The second voltage, however, is multiplied by a factor close to 2, and therefore its amplitude could be important if the asymmetrical ratio of the dip is large. Since the slip is usually small, its frequency is approximately twice the grid frequency.

The emf induced by the natural flux is exactly the same as in the three-phase dips, Equation (6.31), the only difference being that the initial value of this flux depends on other factors besides the depth of the dip:

$$\vec{e}^r_{rn} = -\frac{L_m}{L_s}j\omega_m\vec{\psi}^r_{sn} \qquad (6.39)$$

Since the natural flux is fixed with the stator, the frequency of this emf is equal to the rotor speed, that is, approximately the grid frequency.

In the following sections, the method of symmetrical components will be used to analyze the specific cases of single-phase and phase-to-phase dips. The method also resolves any other types of voltage dip.

6.5.3 Single-Phase Dip

The most frequent grid fault is a short-circuit between one line and the ground. Let's consider, for example, that a short-circuit in phase a causes the voltage in this phase to drop. Assuming that the positive and the negative sequence networks have equal impedances, the voltages in phases b and c would remain the same. The phasors of the phase voltages can be written as follows:

$$\sqrt{2}\underline{V}_a = \hat{V}_{pre}(1-p)$$
$$\sqrt{2}\underline{V}_b = \hat{V}_{pre}a^2 \qquad (6.40)$$
$$\sqrt{2}\underline{V}_c = \hat{V}_{pre}a$$

where the $\sqrt{2}$ term appears because the phasors are expressed in rms values.

The positive, negative, and zero components become

$$\begin{bmatrix} \sqrt{2}\underline{V}_1 \\ \sqrt{2}\underline{V}_2 \\ \sqrt{2}\underline{V}_0 \end{bmatrix} = \frac{1}{3}\begin{bmatrix} 1 & a & a^2 \\ 1 & a^2 & a \\ 1 & 1 & 1 \end{bmatrix} \cdot \begin{bmatrix} \hat{V}_{pre}(1-p) \\ \hat{V}_{pre}a^2 \\ \hat{V}_{pre}a \end{bmatrix} = \hat{V}_{pre} \cdot \begin{bmatrix} 1-p/3 \\ -p/3 \\ -p/3 \end{bmatrix} \qquad (6.41)$$

As previously reasoned, the zero voltage does not affect the behavior of the machine. The other two voltages cause a positive and negative flux in the stator. These two fluxes can be calculated by introducing the voltages of the last expression in Equation (6.35).

The natural flux can be obtained from the initial conditions by considering that the total flux must be continuous:

$$\vec{\psi}_s(t_0^-) = \vec{\psi}_s(t_0^+)$$
$$\vec{\psi}_s(t_0^-) = \vec{\psi}_{s1}(t_0^+) + \vec{\psi}_{s2}(t_0^+) + \vec{\psi}_{sn}(t_0^+) \tag{6.42}$$

By using the above equation we can obtain the initial natural flux:

$$\vec{\Psi}_{n0} = \vec{\psi}_s^s(t_0^-) - \vec{\psi}_{s1}^s(t_0^+) - \vec{\psi}_{s2}^s(t_0^+) \tag{6.43}$$

Unlike the case of a symmetrical fault, the amplitude of the natural flux depends on the timing of the fault. This is because Equation (6.43) is a subtraction of complex numbers and therefore the results depend on the phase of the positive and the negative fluxes at instant t_0^+. Since the rotating direction is different for the two fluxes, the phase shift between them changes over time. Hence, the result of Equation (6.43) is different depending on when the dip begins.

If the dip begins at $t_0 = 0$, the natural flux is zero because at this moment the positive and negative fluxes are aligned and their sum is equal to the flux before the fault (Figure 6.15):

$$\vec{\psi}_s^s(0^-) = \frac{V_{pre}}{j\omega_s}$$
$$\vec{\psi}_s^s(0^+) = \frac{\hat{V}_{pre}(1 - p/3)}{j\omega_s} + \frac{-\hat{V}_{pre}\, p/3}{-j\omega_s} = \frac{\hat{V}_{pre}}{j\omega_s} \tag{6.44}$$

Since there is no natural flux, there is no transitory and the stator flux remains in the steady state. This is shown in Figure 6.16 for an 80% dip depth.

As can be observed, the trajectory traced by the flux during the dip is elliptical. This is a common characteristic in asymmetric voltage dips and is due to the presence of two fluxes rotating in opposite directions. Hence, the two flux vectors add

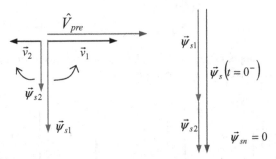

Figure 6.15 Space vectors of the stator flux for an 80% single-phase dip starting at $t_0 = 0$.

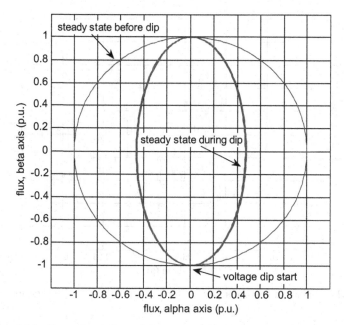

Figure 6.16   Stator flux trajectory for an 80% single-phase dip starting at $t_0 = 0$.

constructively and destructively twice per period, which gives rise to the major and minor axes of the ellipse, respectively.

Figure 6.17 shows the voltages in one of the rotor phases when the generator is operating at supersynchronous speed ($s = -20\%$) for the same dip as that in Figure 6.16.

If the dip begins at $\mathrm{t}_0 = T/4$ (with T being the grid period), the worst situation happens. In this case, the natural flux initial value is the largest since the positive and negative fluxes are at the beginning of the dip opposed and they sum destructively, as can be seen in Figure 6.18.

As the total flux must be the same just before and after the beginning of the fault, the initial value of the natural flux has to be equal to

$$\vec{\psi}_s^s(T/4^-) = \frac{j\hat{V}_{pre}}{j\omega_s}$$

$$\vec{\psi}_s^s(T/4^+) = \frac{j\hat{V}_{pre}(1 - p/3)}{j\omega_s} + \frac{j\hat{V}_{pre}p/3}{-j\omega_s} + \vec{\Psi}_{n0}$$

$$\vec{\Psi}_{n0} = \frac{\hat{V}_{pre}\frac{2}{3}p}{\omega_s} \tag{6.45}$$

Again, the positive and negative fluxes cause the stator flux trajectory to be elliptical. However, the natural flux now causes the ellipse to be off-center, so that the

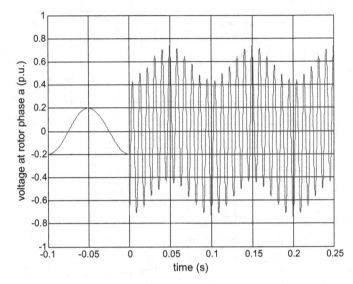

Figure 6.17 Voltage across rotor phase a for an 80% single-phase dip starting at $t_0 = 0$.

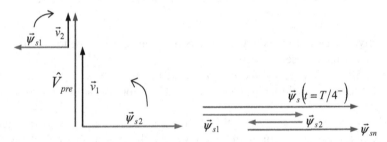

Figure 6.18 Space vectors of the stator flux for an 80% single-phase dip starting at $t_0 = T/4$.

trajectories before and after the fault are tangential. As the natural flux decays, the ellipse moves to the center. At steady state the ellipse is completely centered, as can be seen in Figure 6.19.

The voltage of one rotor phase is shown in Figure 6.20. As can be seen, the waveform is very similar to that corresponding to a dip at $t_0 = 0$ (see Figure 6.21), except for the presence of the transitory due to the term induced by the natural flux.

6.5.4 Phase-to-Phase Dip

In order to analyze the phase-to-phase faults, a short-circuit between phases b and c is now considered. The short-circuit current between these two phases makes their

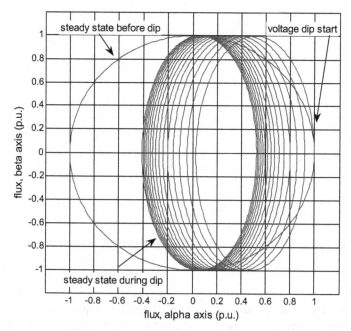

Figure 6.19   Stator flux trajectory for a 50% single-phase dip beginning at $t_0 = T/4$.

voltages get closer in value. In the extreme case, for example, of a total short-circuit, the two voltages are equal and their phasors are situated as represented in Figure 6.27 for the case $p = 1$. The composite voltage in this case is zero and the depth will be considered to be 1 as commonly done in grid codes.

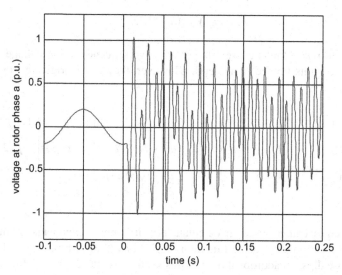

Figure 6.20   Voltage at rotor phase a, for an 80% single-phase dip starting at $t_0 = T/4$.

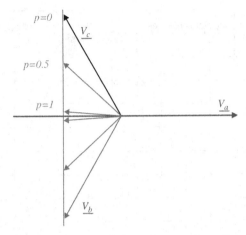

Figure 6.21   Phasors of the rotor voltages for a phase-to-phase dip with different depths.

In a general case the phasor voltage of phase b moves upwards while the phasor voltage of phase c moves downwards. The voltage of phase a remains the same if it is assumed that the positive and the negative sequence networks have equal impedance.

From Figure 6.21, the phasors of the three voltages can be obtained:

$$\sqrt{2}\underline{V}_a = \hat{V}_{pre}$$
$$\sqrt{2}\underline{V}_b = \hat{V}_{pre}\left(a^2 + j\sqrt{3}/2 \cdot p\right) \tag{6.46}$$
$$\sqrt{2}\underline{V}_c = \hat{V}_{pre}\left(a - j\sqrt{3}/2 \cdot p\right)$$

The phase-to-phase faults cause a larger negative sequence in the voltage than the single-phase faults for the same depth. However, they don't originate from a zero sequence:

$$\begin{bmatrix} \sqrt{2}\underline{V}_1 \\ \sqrt{2}\underline{V}_2 \\ \sqrt{2}\underline{V}_0 \end{bmatrix} = \frac{1}{3} \begin{bmatrix} 1 & a & a^2 \\ 1 & a^2 & a \\ 1 & 1 & 1 \end{bmatrix} \cdot \hat{V}_{pre} \begin{bmatrix} 1 \\ a^2 + j\sqrt{3}/2 \cdot p \\ a - j\sqrt{3}/2 \cdot p \end{bmatrix} = \hat{V}_{pre} \cdot \begin{bmatrix} 1 - p/2 \\ p/2 \\ 0 \end{bmatrix} \tag{6.47}$$

These voltages can be used to calculate the different components of the stator fluxes and rotor voltages using Equations (6.35) and (6.38), respectively. Just as in single-phase dips, an additional natural flux could arise depending on the timing of the fault.

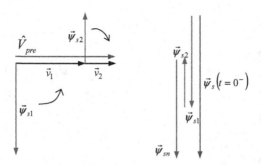

Figure 6.22 Space vectors of the stator flux for an 80% phase-to-phase dip starting at $t_0 = 0$.

If the dip begins at $t_0 = 0$, the largest natural flux arises in the stator of the machine since the positive and negative fluxes are opposed (Figure 6.22). Its initial value would be

$$\vec{\psi}_{sn}(0^+) = \Psi_{n0} = \frac{\hat{V}_{pre}}{j\omega_s} p \qquad (6.48)$$

Figure 6.23 shows the trajectory for the stator flux. As in the case of a single-phase fault beginning at $t_0 = T/4$, the flux is elliptical and its trajectory is not centered due to the natural flux. In this case, the elliptical waveform is more noticeable, since the imbalance in a phase-to-phase fault is greater than in a single-phase fault for the same depth. The larger asymmetry causes a higher voltage than in the case of a single-phase fault.

If the dip begins at $t_0 = T/4$, the addition of the positive and negative fluxes is equal to the prefault flux and consequently the natural flux is zero (Figure 6.24). In this case, the stator flux is in steady state from the beginning. For the same dip depth, the rotor voltage is smaller since no voltage is induced by the natural flux (Figure 6.25).

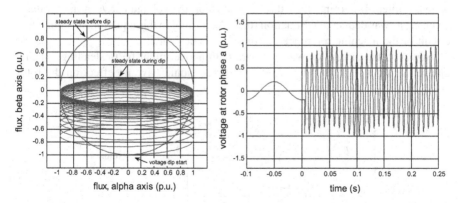

Figure 6.23 Stator flux and rotor voltage for an 80% phase-to-phase dip starting at $t_0 = 0$.

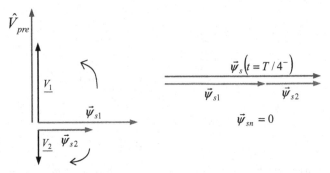

Figure 6.24 Space vectors of the stator flux for an 80% phase-to-phase dip starting at $t_0 = 0$.

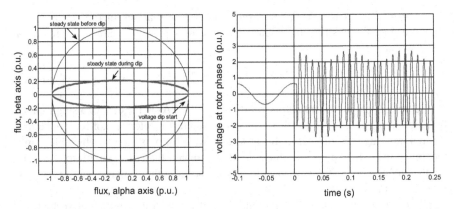

Figure 6.25 Stator flux and rotor voltage for an 80% phase-to-phase dip starting at $t_0 = T/4$.

6.6 INFLUENCE OF THE ROTOR CURRENTS

The previous sections analyzed the behavior of the machine considering the rotor to be open-circuited. If currents flow through the rotor, the voltage between its terminals changes, as was deduced in Equation (6.2). According to this equation, an equivalent circuit of the machine referenced to the rotor is designed.

However, the circuit shown in Figure 6.26 above might be misleading. At first glance, it might look as if the stator resistance would have no effect on the rotor

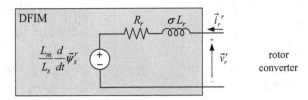

Figure 6.26 Equivalent circuit of the rotor.

voltage. This is correct when the rotor current is null. In such a case, the stator resistance solely damps the natural flux, as seen in Sections 6.4.1 and 6.4.2.

On the contrary, if current flows through the rotor, especially if it is a high current such as during a voltage fault, the stator current varies noticeably. As a consequence, the stator resistance voltage drop is not negligible and causes the rotor voltage to vary. As a matter of fact, as detailed in Chapter 9, there are various protection techniques based on increasing the stator resistance to constrain the voltage between the rotor terminals, hence protecting the rotor converter.

As a first step, the behavior of a machine under a total voltage dip is analyzed. After that, the analysis will be extended to other types of voltage dips.

6.6.1 Influence of the Rotor Current in a Total Three-Phase Voltage Dip

During a total voltage fault, the forced flux disappears as there is no current between the terminals of the machine. However, as shown in Section 6.4.1, the machine does not get demagnetized. On the contrary, a transitory flux appears in the stator, the so-called natural flux, which decays exponentially with the stator time constant. The natural flux is fixed to the stator, thus it induces high currents in the rotor. The expression of the natural flux induced emf was deduced in Section 6.4.1 for the case of the rotor current being zero. When currents flow through the rotor the voltage is modified in two different ways:

- Changing the decrease of the natural flux; generally, accelerating the decline and hence shortening the transitory voltage.
- Altering the rotor voltage as a consequence of the change of the stator resistance.

6.6.1.1 Natural Flux Evolution
The evolution of the natural flux can be obtained from the dynamic expression of the stator, Equation (4.9):

$$\vec{v}_s^s = R_s \vec{i}_s^s + d\frac{\vec{\psi}_{sn}^s}{dt} = 0$$

Expressing the stator current as a function of the stator flux and rotor current yields

$$\vec{i}_s^s = \frac{\vec{\psi}_{sn}^s}{L_s} - \frac{L_m}{L_s}\vec{i}_r^s$$

$$\frac{d}{dt}\vec{\psi}_{sn}^s = -\frac{R_s}{L_s}\vec{\psi}_{sn}^s + \frac{L_m}{L_s}R_s\vec{i}_r^s \tag{6.49}$$

The first term appeared in past sections while studying the behavior of the machine with no current in the rotor and is responsible for damping the natural flux when the rotor is in an open-circuit condition. The last term is new and expresses the influence of the rotor current in the evolution of the stator flux.

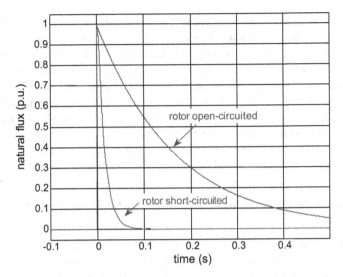

Figure 6.27 Evolution of the natural flux during a total voltage dip.

Depending on the phase shift between the rotor current and the natural flux, the damping could be accelerated or slowed. This is crucial for the behavior of the machine. If, for example, the control of the rotor converter injects a current in phase with the natural flux, the damping of the flux would be slowed down, and might even be canceled out or become negative. In the latest case, the natural flux would not disappear, but would rather increase until the converter goes out of control. This situation will be analyzed in Chapter 7 while studying the interactions between the control loops and the machine.

The opposite situation happens when a crowbar is connected to the rotor. The crowbar is a commonly used protection that operates by putting a short-circuit of low resistance across the rotor windings. The connection of the crowbar generates the circulation of large currents through the rotor and stator windings. Since the stator currents are almost in opposite phase to the natural flux, the flux damping would be accelerated, as shown in Figure 6.27.

6.6.1.2 Voltage at Rotor Terminals
The expression of the rotor voltage was already obtained in Section 6.2. In the case of a total voltage dip, when the only flux in the machine is the natural flux, we have

$$\vec{v}_r^r = R_r \vec{i}_r^r + \sigma L_r \frac{d\vec{i}_r^r}{dt} + \frac{L_m}{L_s} \frac{d}{dt} \vec{\psi}_{sn}^r \qquad (6.50)$$

The first term of the expression is the voltage drop in the rotor resistance. The second term is the voltage drop in the transitory rotor inductance. The last term is the emf induced by the stator flux in the rotor windings. As in the last sections, considering no current in the rotor, only this third term appeared. In a general case, not only the three terms must be considered but the induced emf must be recalculated since the current circulating through the rotor modifies this voltage.

The emf can be obtained if Equation (6.49) is expressed in a rotor reference frame:

$$\frac{d}{dt}\vec{\psi}_{sn}^r = -j\omega_m \vec{\psi}_{sn}^r - 1/\tau_s\,\vec{\psi}_{sn}^r + \frac{L_m}{L_s}R_s\vec{i}_r^r \tag{6.51}$$

Merging this expression with Equation (6.50) and neglecting the term $1/\tau_s$ yields

$$\vec{v}_r^r = R_r\vec{i}_r^r + \sigma L_r\frac{d}{dt}\vec{i}_r^r - j\omega_m\frac{L_m}{L_s}\vec{\psi}_{sn}^r + \left(\frac{L_m}{L_s}\right)^2 R_s\vec{i}_r^r \tag{6.52}$$

It can be observed that the voltage has two kinds of terms: terms that depend on the rotor current (and do not appear when the rotor is open-circuited) and terms that are independent of the rotor current. Adding terms of the same nature, we obtain

$$\vec{v}_r^r = \frac{L_m}{L_s}(-j\omega_m\vec{\psi}_{sn}^r) + \left[R_r + \left(\frac{L_m}{L_s}\right)^2 R_s\right]\vec{i}_r^r + \sigma L_r\frac{d}{dt}\vec{i}_r^r \tag{6.53}$$

That is, the rotor voltage is the addition of:

- The emf induced by the stator flux. Since the inductance L_m is almost equal to L_s,

$$\vec{e}_r^r = \frac{L_m}{L_s}(-j\omega_m\vec{\psi}_{sn}^r) \approx -j\omega_m\vec{\psi}_{sn}^r \tag{6.54}$$

- The voltage drop in the resistances of the machine. Note that, unlike in steady state, both stator and resistance have an influence on the rotor voltage:

$$R_r + (L_m/L_s)^2 R_s \approx R_r + R_s$$

- The voltage drop in the transitory inductance of the machine seen from the rotor:

$$\sigma L_r = L_r - \frac{L_m^2}{L_s} \approx L_{\sigma s} + L_{\sigma r}$$

From Equation (6.53) the equivalent circuit of the machine during a total voltage dip is depicted in Figure 6.28.

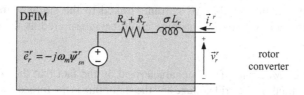

Figure 6.28   Equivalent circuit during a total voltage dip.

6.6.2 Rotor Voltage in a General Case

In the last section the rotor voltage was derived for cases of total three-phase dips, where only a natural flux would be present. In a general case, the flux is made up of three components.

These three terms involve their corresponding emf in the rotor. Correspondingly, these voltages will cause the circulation of three currents with similar characteristics. It can be seen that both the flux, and the currents and the voltages can be split into three terms:

- A synchronous rotating term, present during regular functioning conditions of the machine
- An inversely rotating term, that appears during asymmetrical voltage dips
- An additional term, originated by the voltage transient and fixed to the stator

Based on this distinction, applicable to all the machine variables, the super-position principle can be applied to interpret the machine as the sum of three machines: one where all the variables rotate synchronously, another where all the variables rotate inversely, and a third one where the reference frame of the variables is fixed to the stator.

6.6.2.1 Superposition Principle   The superposition principle states that the response caused by a stimulus u in a system in an initial state xi is the sum of the response that would have been caused by the stimulus u_1 and initial state xi_1 plus the response to stimulus u_2 and initial state xi_2 if $u = u_1 + u_2$ and $xi = xi_1 + xi_2$.

Application of the superposition principle requires the system to be linear. The DFIM fulfills that condition under the following assumptions:

- The machine is linear; that is, it does not saturate. Exceptionally, in cases of very deep dips, it is reasonable to apply the superposition principle despite the fact that the machine might indeed saturate.
- The converter behaves as a linear system; that is, it does not saturate. This assumption is generally true for most control laws, including field oriented control and direct torque control if the rotor converter does not saturate.

An example of how the superposition principle simplifies the study of a certain machine during a voltage fault could be the case of a three-phase dip. The natural flux generated involves the circulation of DC currents in the stator. These currents are superposed on the regular AC currents linked to the stator voltage and the forced flux. In such a case, the machine could be interpreted as the sum of two machines:

- *The Forced Machine.* This is characterized by the fact that the voltage at the stator terminals is the grid voltage and the stator flux is the forced flux, that is, the steady state flux for that voltage. From the beginning, this machine is at steady

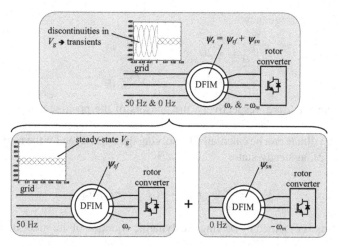

Figure 6.29   Superposition principle during three-phase dips.

state so there is no natural flux and all its variables are synchronous with the grid voltage. The emf induced in its rotor is the forced voltage, calculated in Equation (6.30):

$$\vec{e}_{rf}^{\,r} = \frac{L_m}{L_s} \hat{V}_{fault} e^{\omega_r t} s$$

- *The Natural Machine.* This has zero voltage at its stator terminals. Such a machine has no forced flux at the stator but only natural flux. All the variables of the machine are transitory and are fixed with the stator. The emf induced in its rotor is equal to the natural voltage, stated in Equation (6.31):

$$\vec{e}_{rn}^{\,r} = -\frac{L_m}{L_s} j\omega_m \vec{\Psi}_{n0} e^{-t/\tau_s} e^{-j\omega_m t}$$

The response of the machine during a three-phase voltage dip can be calculated as the behavior of the machine in steady state at the new grid voltage V_{fault} plus the response of the machine with an initial flux ψ_n but with the stator short-circuited. The separation of the machine into these two machines is graphically represented in Figure 6.29.

The superposition principle can also be applied to asymmetrical voltage dips. In those cases, the machine can then be analyzed as the addition of three machines: the positive, the negative, and the natural machine.

6.6.2.2 The Positive Machine
This is a machine operating at steady state. Its stator voltage is the positive sequence of the grid voltage $\underline{V}_1$. In this machine the stator flux is a space vector rotating at grid frequency. In a rotor reference frame, this vector

rotates at the slip frequency:

$$\vec{\psi}_{s1}^{s} = \sqrt{2}\underline{\psi}_{s1} e^{j\omega_s t}$$

$$\vec{\psi}_{s1}^{r} = \sqrt{2}\underline{\psi}_{s1} e^{j\omega_r t} = \sqrt{2}\underline{\psi}_{s1} e^{j\omega_s s t} \tag{6.55}$$

where $\underline{\Psi}_{s1}$ and $\underline{\Psi}_{s1}$ are, respectively, the phasors of the positive and negative flux (in rms).

The rotor voltage can be calculated by solving the expression for the rotor voltage, Equation (6.2), at steady state:

$$\underline{V}_{r1} = \frac{L_m}{L_s}\underline{\psi}_{s1} j\omega_s s + R_r \underline{I}_{r1} + \sigma L_r j\omega_r \underline{I}_{r1} \tag{6.56}$$

The stator flux is obtained by solving the dynamic equation of the stator in the steady state:

$$\underline{V}_1 = \underline{\psi}_{s1} j\omega_s + R_s \underline{I}_{s1} \tag{6.57}$$

As we are interested in ascertaining the influence of the rotor current, we might express the stator current as a function of the rotor current and the stator flux:

$$\underline{V}_1 = \underline{\psi}_{s1} j\omega_s + \frac{R_s}{L_s}\underline{\psi}_{s1} - \frac{L_m}{L_s}\underline{I}_{r1} \tag{6.58}$$

So the positive flux is

$$\underline{\psi}_{s1} = \frac{1}{j\omega_s + 1/\tau_s}\left(\underline{V}_1 + \frac{L_m}{j\omega_s L_s}\underline{I}_{r1}\right) \tag{6.59}$$

The term $1/\tau_s$ is significantly smaller than $j\omega_s$ so it is negligible. Introducing this value into Equation (6.56), we obtain the rotor voltage at the rotor terminals:

$$\underline{V}_{r1} = \frac{L_m}{L_s}\underline{V}_1 s + \left[R_r + \left(\frac{L_m}{L_s}\right)^2 R_s s\right]\underline{I}_{r1} + \sigma L_r j\omega_r \underline{I}_{r1}$$

Then, the positive sequence of the rotor voltage has three components: the emf originated by the grid voltage, a voltage drop in the stator and rotor resistance, and the voltage drop in the transitory rotor inductance. Note that the voltage drop of the stator resistance is multiplied by the slip. As the machine operates usually with a small slip, this term can be neglected, so the space vector of the rotor voltage remains then

$$\vec{v}_{r1}^{r} = \frac{L_m}{L_s}\underline{V}_1 s e^{j\omega_r t} + R_r \vec{i}_{r1}^{r} + \sigma L_r \frac{d}{dt}\vec{i}_{r1}^{r} \tag{6.60}$$

6.6.2.3 The Negative Machine

The negative machine also operates at steady state. Its stator voltage is the negative sequence of the grid voltage, $\underline{V}_2$. Its rotor voltage can be obtained in a similar way as in the case of the positive machine:

$$\underline{V}_{r2} = \frac{L_m}{L_s}\underline{V}_2(2-s) + \left[R_r + \left(\frac{L_m}{L_s}\right)^2 R_s(2-s)\right]\underline{I}_{r2} + \sigma L_r j\omega_s(2-s)\underline{I}_{r2}$$

Neglecting again the slip multiplication in the voltage drop of the stator resistance, we have.

$$\vec{v}_{r2}^{\,r} = \frac{L_m}{L_s}\underline{V}_2(2-s)e^{-j(2-s)\omega_s t} + \left[R_r + \left(\frac{L_m}{L_s}\right)^2 R_s 2\right]\vec{i}_{r2}^{\,r} + \sigma L_r \frac{d}{dt}\vec{i}_{r2}^{\,r} \qquad (6.61)$$

6.6.2.4 *The Natural Machine* The stator voltage of the natural machine is zero and thus there is no steady state flux, but only natural flux. All the variables in that machine are transitory. The rotor voltage in the case of a natural machine has already been obtained in Section 6.6.1:

$$\vec{v}_{rn}^{\,r} = \frac{L_m}{L_s}(-j\omega_m\vec{\psi}_{sn}^{\,r}) + \left[R_r + \left(\frac{L_m}{L_s}\right)^2 R_s\right]\vec{i}_{rn}^{\,r} + \sigma L_r \frac{d}{dt}\vec{i}_{rn}^{\,r} \qquad (6.62)$$

6.7 DFIM EQUIVALENT MODEL DURING VOLTAGE DIPS

Actually, the behavior of the machine under voltage dips can be predicted by means of the full-order model of the machine presented in Chapter 4. This model predicts the overcurrents and overvoltages that are generated by grid disturbances. A disadvantage of the model is its complexity to the point where it is impractical to see and predict the influence of each parameter of the machine during voltage dips. More to the point, the full-order model is not manageable or useful for designing possible solutions to avoid the consequences that voltage dips have on doubly fed induction machines.

Some authors have proposed using simplified models that neglect the stator flux dynamics [10]. However, the behavior of the machine is not accurately predicted since it is strongly influenced by a set of poorly damped poles that arise from the flux dynamics.

In this section, a reduced model of the machine is developed using the analysis stated along previous sections.

6.7.1 Equivalent Model in Case of Linearity

If the system comprised of the machine and the converter behaves as a linear system, it is usually very convenient to use the method based on the superposition principle presented in Section 6.6.2. According to this method, the machine is split into three: the positive, the negative, and the natural machines as represented in Figure 6.30. The behavior of each machine is independent, so it can be analyzed separately using

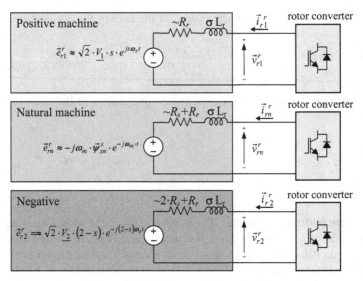

Figure 6.30 Rotor equivalent circuits.

its equivalent circuit and taking into account the control laws of the converter. After that, the currents, voltages, and fluxes of the real machine can be determined by adding the variables of the three machines. This procedure will be used intensively in Chapter 7 to analyze the interaction between the machine and the converter under grid disturbances.

As Figure 6.30 shows, the emf induced in the machine depends on the positive and negative sequences of the voltage grid and also on the natural flux. Table 6.1 compiles the per unit values for these voltages for the most common faults in a three-phase power system: three-phase (3ϕ), single line to ground (ϕn), phase-to-phase ($\phi\phi$), and two-phase-to-ground ($2\phi n$). The table also includes the initial value of the natural flux. From this value, the natural flux evolves following the dynamics found in Equation (6.49):

$$\frac{d}{dt}\vec{\psi}_{sn}^{s} = -\frac{R_s}{L_s}\vec{\psi}_{sn}^{s} + \frac{L_m}{L_s}R_s\vec{i}_r^{s}$$

TABLE 6.1 Per Unit Values of Voltages and Initial Natural Flux

Fault	V_1	V_2	ψ_{n0}
ϕn	$1 - p/3$	$p/3$	0 to $2p/3$
$\phi\phi$	$1 - p/2$	$p/2$	0 to p
$2\phi n$	$1 - 2p/3$	$p/3$	$p/3$ to p
3ϕ	$1 - p$	0	p

6.7.2 Equivalent Model in Case of Nonlinearity

Sometimes it is not possible to apply the superposition principle because the system cannot be considered to be linear. The most common causes are:

- The rotor converter does not behave linearly, as happens when using a reference frame oriented with the stator flux.
- The rotor converter saturates because the emf induced in the rotor is very high. This is the case during severe voltage dips.

In the first case the analysis of the machine becomes rather complicated and one often needs to use linearization methods as in Petersson et al. [11].

In the case of severe dips, the emf induced in the rotor makes the converter rise to its maximum voltage and the control is no longer linear. In these situations, the machines cannot be split into the positive, negative, and natural flux; instead, we must consider the machine as a whole. Since the equivalent circuit of the machine is slightly different depending on the flux analyzed, we can take the characteristics of the most prominent flux.

6.7.2.1 Equivalent Model for Normal Operation   During regular operation regimes, there is only a positive flux in the stator of the machine. The emf induced is proportional to the slip. Also, the resistance seen from the rotor is almost equal to the rotor resistance. The behavior of the machine can be studied by means of the equivalent circuit shown in Figure 6.31. Note that the term L_m/L_s has been omitted as it is near one.

6.7.2.2 Equivalent Model During Three-Phase Voltage Dips   The stator flux in a three-phase voltage dip is composed of a positive and a natural flux. The emf induced by the positive flux can be neglected as it is relatively small compared to the emf induced by the natural flux. From the rotor point of view, the machine behaves in a very similar way as during a total voltage dip. The equivalent circuit is thus obtained in Section 6.6.1.2 (Figure 6.32).

6.7.2.3 Equivalent Model During Asymmetrical Voltage Dips   An asymmetrical voltage dip generates a natural and a negative flux. Although a positive flux will remain in the stator, this flux can be ignored for the same aforementioned reasons. The behavior of the machine is therefore determined by

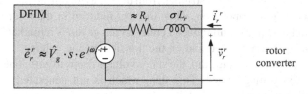

Figure 6.31 Equivalent circuit for normal operation.

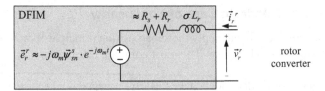

Figure 6.32 Equivalent circuit for three-phase voltage dips.

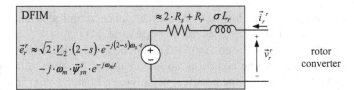

Figure 6.33 Equivalent circuit for asymmetrical voltage dips.

the negative flux. From Equation (6.67) the simplified circuit shown in Figure 6.33 is obtained. Note that the stator resistance affects the rotor converter and it appears in the circuit multiplied by 2.

6.7.3 Model of the Grid

Throughout this chapter, the grid voltage has been assumed to be known and to be independent of the behavior of the machine. This assumption can be made when the network is strong, so that the fault voltage is not affected by the generator current. However, usually the generator is connected to a step-up transformer whose resistances and reactances notably modify the behavior of the DFIM. In this case, the network and the transformer reactance are not negligible and they need to be added in series with the generator leakage inductance to all previous equations. The same can be said about the resistances that must be added to the stator resistance.

6.8 SUMMARY

This chapter has examined the behavior of the DFIM under voltage dips and has reasoned why its behavior differs so much from the behavior during normal operation.

Unlike under regular operation, when the emf induced in the rotor is small and proportional to the slip, the emf induced under a voltage dip is remarkably larger and depends on the rotational speed and on the transient of the voltage.

In a total three-phase voltage dip, the stator voltage drops to zero, but the machine remains transitorily magnetized. This flux, which is not generated by any stator voltage, is called *natural flux*.

The natural flux is fixed with the stator (completely fixed if the rotor has no current; almost fixed in any other case). From the point of view of the rotor windings, this flux is seen as a flux that rotates inversely at the rotor speed. As a consequence, the natural flux induces an emf much greater than those appearing during regular operation. This voltage is not proportional to the slip, but to the rotor speed.

In the case of a general voltage dip, the stator flux may be separated into three terms: the *positive*, the *negative*, and the *natural flux*. The positive flux rotates at synchronous speed and is always present, including during regular operation of the machine. It induces an emf proportional to the slip. The negative flux only appears in cases of asymmetrical faults since it is generated by the negative sequence of the grid voltage. It also rotates at synchronous speed but in the opposite direction. The natural flux is a transient flux that appears due to voltage variations. Its initial value depends on the fault type, and it decays exponentially. In the case of asymmetrical faults, it also depends on the timing of the fault.

The dynamics of the natural flux depend on the rotor current. In the absence of current the flux decays exponentially according to the stator time constant. For large generators this time constant ranges from 0.8 to 1.5 s, hence the transitory can lasts some seconds. In contrast, some protection techniques short-circuit the rotor, generating large currents in the rotor. In these cases, the transitory is accelerated by a factor of approximately 10.

As the natural and the negative fluxes rotate at relatively high speed with respect to the rotor (the natural flux close to the synchronous speed, the negative flux twice this speed), they induce voltages in the rotor that are significantly greater than those appearing under normal operation.

If these overvoltages exceed the limits of the rotor converter, control of the current is lost momentarily or even permanently. In this situation, overcurrents appear that can damage the converter.

REFERENCES

1. J. B. Ekanayake, L. Holdsworth, X. G. Wu, and N. Jenkins, "Dynamic Modeling of Doubly Fed Induction Generator Wind Turbines," *IEEE Trans. Power Systems*, Vol. 18, No. 2, pp. 803–809, May 2003.
2. A. Perdana, O. Carlson, and J. Persson,"Dynamic Response of Grid-Connected Wind Turbine with Doubly Fed Induction Generator During Disturbances," Nordic Workshop on Power and Industrial Electronics, Trondheim, Norway, 2004.
3. S. Seman, J. Niiranen, S. Kanerva, A. Arkkio, and J. Saitz, "Performance Study of a Double Fed Wind-Power Induction Generator Under Network Disturbances," *IEEE Trans. Energy Conversion*, Vol. 21, No. 4, pp. 883–890, December 2006.
4. J. Morren and S. W. H. de Haan, "Short-Circuit Current of Wind Turbines with Doubly Fed Induction Generator," *IEEE Trans. Energy Conversion*, Vol. 22, No. 1, pp. 174–180, March 2007.

5. John Godsk Nielsen(Vestas), "Method for Controlling a Power-Grid Connected Wind Turbine Generator During Grid Faults and Apparatus for Implementing Said Method," Worldwide Patent No. WO2004070936 August 2004.

6. Lorenz Feddersen(Vestas Wind Systems), "Circuit to Be Used in a Wind Power Plant," U.S. Patent No. US7102247 September 2006.

7. Joho Reinhard(Alstom), "Method for Operating a Power Plant," U.S. Patent No: US6239511 May 2001 and European Patent No: EP0984552 July 1999.

8. A. Stoev, and A. Dittrich(Diltec), "Generator System Having a Generator that Is Directly Coupled to the Mains, and Method for Controlling Mains Interruptions," European Patent No. EP1561275 July 2008.

9. J. I. Llorente and M. Visiers(Gamesa Innovation & Technology), "Method and Device for Preventing the Disconnection of an Electric Power Generating Plant from the Electric Grid," European Patent Application No: EP1803932 July 2007.

10. J. B. Ekanayake, L. Holdsworth, and N. Jenkins, "Comparison of 5th Order and 3rd Order Machine Models for Doubly Fed Induction Generator (DFIG) Wind Turbines," *Elect. Power Systems Res.*, Vol. 67, pp. 207–215, December 2003.

11. A. Petersson, T. Thiringer, L. Harnefors, and T. Petru, "Modeling and Experimental Verification of Grid Interaction of a DFIG Wind Turbine", *IEEE Trans. Energy Conversion*, Vol. 4, No. 4, pp. 878–886, December 2005.

CHAPTER 7

Vector Control Strategies for Grid-Connected DFIM Wind Turbines

7.1 INTRODUCTION

Once the most interesting elements of the DFIM based wind turbines have been studied separately, such as the back-to-back converter in Chapter 2, the modeling of doubly fed induction machines in Chapters 3 and 4, a practical procedure for the identification of the model parameters of a machine in Chapter 5, or, finally, the basis for understanding the machine behavior during voltage dips in Chapter 6, this chapter immerses the reader in control techniques of the doubly fed induction machine (DFIM).

Control itself plays a very important role in drives and consequently also in wind turbine technology. Control of the doubly fed induction generator when generating energy in a wind turbine is necessary and unavoidable. Control maintains magnitudes of the generator, such as torque, and active and reactive power, and also magnitudes related to the grid side converter, such as the reactive power and the DC bus voltage close to their optimum values, for proper and effective energy generation. In this way, control, together with the modulator if implemented, is in charge of generating the converter switch pulses according to the desired reference values.

This book deals with two widely used control philosophies for doubly fed induction machines: the vector control technique (also known as field oriented) and the direct control technique. This chapter covers vector control (field oriented control), while Chapter 8 focuses on direct control. Vector control has already been studied in this book, in Chapter 2 when applied to the control of the grid side converter. This chapter extends that basic vector control philosophy to the DFIM.

In addition, control of DFIM based wind turbines needs to pay special attention to the distorted voltage operation scenario. Apart from the normal operation of the wind turbine, the designed control must be prepared to tackle problematic situations derived from grid voltage disturbances. These grid voltage disturbances can be of different natures (voltage dips, imbalances, harmonics, etc.) depending on the

---

Doubly Fed Induction Machine: Modeling and Control for Wind Energy Generation,
First Edition. By G. Abad, J. López, M. A. Rodríguez, L. Marroyo, and G. Iwanski.
© 2011 the Institute of Electrical and Electronic Engineers, Inc. Published 2011 by John Wiley & Sons, Inc.

characteristics of the electric grid itself. However, the most frequently occurring disturbance is a voltage dip. In fact, it can be said that this is also the most problematic one, producing often disconnections of the wind turbines if appropriate countermeasures are not applied. This is a real problem that grid code requirements of many different countries with strong wind energy penetration address in order to avoid, among other problems, destabilization of the grid system.

In view of that, this chapter includes the control improvements required to handle this disturbance. This chapter also analyzes why the most severe voltage dips cannot be handled by only control techniques and the necessity of some additional hardware solutions, like crowbars or braking choppers. These solutions are discussed further in Chapter 9.

Toward the end of the chapter, voltage imbalances are also addressed by control solutions.

In summary, by the end of this chapter the reader will have gained basic but also advanced vector control knowledge, applied to the specific scenario of DFIM based wind turbines.

7.2 VECTOR CONTROL

Vector control of a grid connected DFIM is very similar to the widespread classical vector control of a squirrel cage machine. This machine is controlled in a synchronously rotating dq reference frame, with the d axis oriented along the rotor flux space vector position. The direct current is thus proportional to the rotor flux while the quadrature current is proportional to the electromagnetic torque. By controlling independently the two components of the current, a decoupled control between the torque and the rotor excitation current is obtained.

In a similar way, in vector control of a DFIM, the components of the d and the q axis of the rotor current are regulated. As will be shown, if a reference frame orientated with the stator flux is used, the active and reactive power flows of the stator can be controlled independently by means of the quadrature and the direct current, respectively (Figure 7.1).

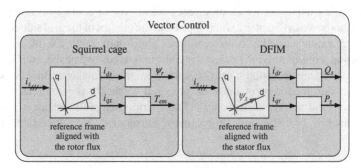

Figure 7.1 Comparison between the Vector Control of the squirrel cage machine and the DFIM.

In a first step, the relationship between the rotor current and the stator powers is obtained. Then, the closed control loops for the current are presented. Lastly, different alternatives for orienting the reference frame are discussed.

7.2.1 Calculation of the Current References

The generic expressions for the DFIM machine, Equations (4.28) to (4.36), can be simplified by using a reference frame aligned with the stator flux. Under stator-flux orientation, the relationship between the current and the fluxes may be written as

$$i_{ds}L_s + i_{dr}L_m = \psi_s$$
$$i_{qs}L_s + i_{qr}L_m = 0$$

(7.1)

Using the above equations, the relationship between the rotor and the stator currents is obtained:

$$i_{ds} = \frac{\psi_s}{L_s} - \frac{L_m}{L_s}i_{dr}$$
$$i_{qs} = -\frac{L_m}{L_s}i_{qr}$$

(7.2)

In steady state, the stator flux is proportional to the grid voltage, $\hat{V}_g$. Neglecting the small drop in the stator resistance yields

$$v_{ds} = 0$$
$$v_{qs} = \hat{V}_g \approx \omega_s \psi_s$$

(7.3)

Thus, when orienting the direct axis with the stator flux, the voltage aligns with the quadrature axis. The stator active and reactive power flow can then be written as

$$P_s = \tfrac{3}{2}v_{qs}i_{qs}$$
$$Q_s = \tfrac{3}{2}v_{qs}i_{ds}$$

(7.4)

Combining these equations with Equation (7.2), we obtain

$$P_s = -\tfrac{3}{2}\hat{V}_g\tfrac{L_m}{L_s}i_{qr}$$
$$Q_s = \tfrac{3}{2}\hat{V}_g\tfrac{\psi_s}{L_s} - \tfrac{3}{2}\hat{V}_g\tfrac{L_m}{L_s}i_{dr} = \tfrac{3}{2}\tfrac{\hat{V}_g^2}{\omega_s L_s} - \tfrac{3}{2}\hat{V}_g\tfrac{L_m}{L_s}i_{dr}$$

(7.5)

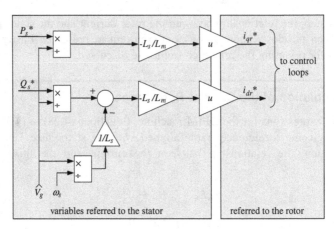

Figure 7.2   Calculation of the current references.

The above equations clearly show that under the stator flux orientation, the active and reactive powers are decoupled and can be controlled via the rotor currents. By means of the direct current, we can control the active power while the reactive power can be controlled via the quadrature current. Using the above equation, the reference currents can be calculated from the desired powers, as shown in Figure 7.2. Then, control loops are used to ensure that the actual currents follow these references.

The previous relationships are not completely exact, since the effect of the stator resistances has been neglected. The actual stator powers won't be exactly the desired values. To correct this error, even if it is not very relevant, and others due to inaccuracies in the values of the machine's parameters (L_s, L_m, turns ratio, etc.), usually two outer power loops are added: the first loop would regulate the active power by means of the direct current i_{dr}, and the second loop would regulate the reactive power by means of the direct current. In Figure 7.3, a schematic including this option is shown.

With these outer power loops an accurate control of the power flow in the stator is achieved, regardless of the inexactness in the machine parameters or small misalignments of the reference frame.

Typically, we don't desire to control the stator power but the electromagnetic torque generated by the machine, as is the case of the wind turbine control strategies discussed in Section 1.3.2. If this is the case, we can use the relationship between the torque, the stator flux, and the rotor current, Equation (4.36):

$$T_{em} = \frac{3}{2}p\frac{L_m}{L_s}\left(\psi_{qs}i_{dr} - \psi_{ds}i_{qr}\right) \tag{7.6}$$

Under stator flux orientation this expression may be written as

$$T_{em} = \frac{3}{2}p\frac{L_m}{L_s}\left(-\psi_{ds}i_{qr}\right) \tag{7.7}$$

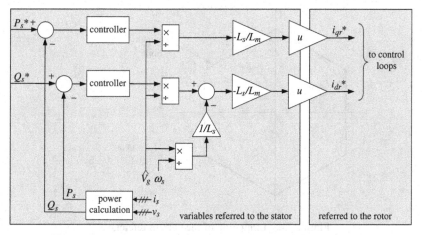

Figure 7.3 Reference calculation including external power loops.

And substituting the stator flux by the value obtained in Equation (7.3), we have

$$T_{em} = -\frac{3}{2}p\frac{L_m}{L_s}\frac{\hat{V}_g}{\omega_s}i_{qr} \qquad (7.8)$$

Hence, the electromagnetic torque is directly controlled by the quadrature current. Alternatively, we may use this expression to calculate the mechanical power:

$$P_{mec} = -\frac{3}{2}p\frac{L_m}{L_s}\frac{\hat{V}_g}{\omega_s}i_{qr}\omega_m = -\frac{3}{2}p\frac{L_m}{L_s}\hat{V}_g(1-s)i_{qr} \qquad (7.9)$$

Merging this expression with Equation (7.4), we find the relationship between the stator and the mechanical power:

$$P_{mec} = (1-s)P_s \qquad (7.10)$$

Using this last expression it is possible to translate any reference of electromagnetic torque into a reference of stator power to be introduced in the schemas of Figure 7.2 or 7.3.

$$P_s = \frac{1}{1-s}P_{mec} = \frac{\omega_s}{\omega_m}P_{mec} = \omega_s T_{em}$$

7.2.2 Limitation of the Current References

Once the current references have been calculated, they must be limited so that they do not surpass the maximum current allowed by the machine and the rotor converter.

We will denote the maximum peak value of the rotor converter as I_{phase_max}. If the current of the three phases is kept below this limit, the space vector of the total

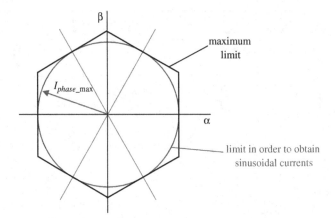

Figure 7.4 Limitation of the current references.

current must be within the area circumscribed by the hexagon represented in Figure 7.4, whose inner ratio is equal to I_{phase_max}, and outer ratio $I_{phase\_max} \sqrt{3/2}$. (Note that these values vary by a factor of $\sqrt{3/2}$ if a power invariant $dq0$ transformation is used instead of the Park transformation used in this book.)

The maximum modulus of the current space vector ranges between these two values depending on the vector angle. However, in order to assure a current sinusoidal waveform, the modulus of the space vector must be kept constant. Consequently, the maximum phase current should be lower than I_{phase_lim}.

The space vector is made up of two components: the values in the d axis and the q axis. If the current has to be limited, we can act on either of the two components. A commonly used option is to proportionally limit both components, thus keeping the angle of the space vector unchanged.

Another usual option is to prioritize one component over the other. If, for example, tracking of the reactive power is considered more important than tracking of the active power, we will give priority to the d-axis current: the limitation will be applied first to the q component, and only if this is not enough, to the d component.

7.2.3 Current Control Loops

Once the reference rotor currents are calculated, the rotor converter must ensure that the actual currents track these references. Generally, the rotor converter is a three-phase inverter that fixes the voltage of the rotor terminals but not the currents. Control loops must then be incorporated to ensure that the currents effectively follow their references.

Although it is possible to implement such control loops in any reference frame, the most practical solution is to use the same reference frame as the commands; that is, aligned with the stator flow d-axis. Furthermore, along this axis, the variables are constant during steady state, which helps the regulation of the currents.

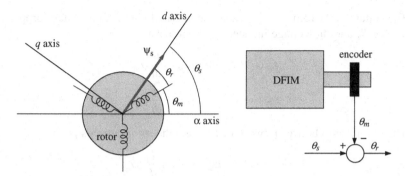

Figure 7.5 Calculation of the angles for the Park transformations.

Voltages calculated in this way are referred to as a synchronous reference system. The rotor inverter requires that the voltage of each of its three phases be calculated from v_{dr} and v_{qr} by means of Park's inverse transformation. Also, currents i_{dr} and i_{qr} are calculated according to the three phase currents using Park's transformation. Both functions require the angle between axis d and the rotor phases, θ_r, to be known. Hence, the position of the rotor phases, that is, the angular position of the rotor, θ_m, must be known. There are several methods to estimate this value gathered in the literature [1–3], although the most common solution put into practice by the industry is by means of an encoder coupled to the rotor (Figure 7.5).

7.2.3.1 *Plant of the Control Loop*
There are an infinity of different controllers that might be used for the control loop. The most widely used is the proportional-integral (PI) controller due to its simplicity.

To calculate the PI regulator parameters, the relationship between the rotor voltage and the rotor currents must be known. Such a relationship can be obtained from the dq axis model, shown in Figure 7.6.

Neglecting the stator resistances, the model simplifies significantly. By doing so, the voltage is aligned with the q axis, so its d component becomes zero. On the other hand, the model considers the d axis aligned with the flux, so the flux along the q axis is null.

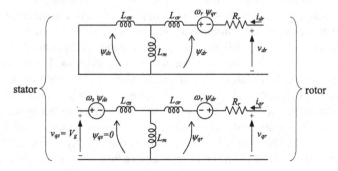

Figure 7.6 Equivalent circuit of the DFIM.

The voltage at the rotor terminals corresponds to the sum of the voltage drop in the resistance R_r and the voltage induced by the flux rotor:

$$v_{dr} = R_r i_{dr} - \omega_r \psi_{qr} + \frac{d}{dt} \psi_{dr}$$

$$v_{qr} = R_r i_{qr} + \omega_r \psi_{dr} + \frac{d}{dt} \psi_{qr}$$

(7.11)

The relationship between this rotor flux and the rotor currents is

$$\psi_{dr} = \left(L_r - L_m^2/L_s\right) i_{dr} + \frac{L_m}{L_s} \psi_{ds}$$

$$\psi_{qr} = \left(L_r - L_m^2/L_s\right) i_{qr}$$

(7.12)

The first terms of the above equations yield the flux directly related to the rotor current. The proportionality ratio between these fluxes and the currents is referred to as the transitory inductance of the rotor, $\sigma \cdot L_r$, and is the result of adding in series the rotor leakage inductance with the parallel magnetizing and the stator leakage inductance:

$$\left(L_r - L_m^2/L_s\right) = L_{\sigma r} + L_m // L_{\sigma s} = \sigma L_r$$

(7.13)

where σ is the leakage coefficient of the machine.

From expressions (7.11) and (7.12), the relationship between the rotor voltages and the currents is obtained:

$$v_{dr} = R_r i_{dr} - \omega_r \sigma L_r i_{qr} + \sigma L_r \frac{d}{dt} i_{dr} + \frac{L_m}{L_s} \frac{d}{dt} \psi_{ds}$$

$$v_{qr} = R_r i_{qr} + \omega_r \sigma L_r i_{dr} + \sigma L_r \frac{d}{dt} i_{qr} + \omega_r \frac{L_m}{L_s} \psi_{ds}$$

(7.14)

During regular operation, when the grid voltage is constant in amplitude, the derivative of the flux ψ_{ds} is zero and therefore the last term of the first equation disappears (this is not the case during a voltage dip, as will be analyzed in Section 7.4). The resulting equations are graphically represented in the block diagram of Figure 7.7.

From the control point of view, the term $(\omega_r L_m/L_s) \psi_{ds}$ is a perturbation, since it depends on the stator flux (i.e., on the grid voltage), an external variable independent of the loop. As it is constant, it will easily be compensated for by the controller. The terms depicted in the middle of the figure are known as the cross terms. These terms appear because the reference frame turns at a different speed than the rotor terminals. Although these terms are constant during a permanent regime and do not affect the functioning of the control loops, frequently they are estimated and compensated for by the control to notably reduce its negative effects during transitory stages, as shown in Figure 7.8.

By compensating the cross terms, the control loop is notably simplified. Both axes are now identical and their plant is now reduced to a first-order transfer function plus, eventually, a current sensor filter transfer function (Figure 7.9).

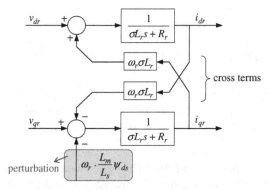

Figure 7.7 Plant of the current control loop.

7.2.4 Reference Frame Orientations

The basis of vector control is to refer the rotor currents in a synchronous reference frame, oriented so that its *d* axis is aligned with the stator flux. As explained above, in the steady state, this is almost equivalent to aligning the *q* axis with the stator voltage, because the flux is 90° shifted with respect to the stator. So, there are two options to align the axis: to estimate the flux and align the *d* axis to it, or to align the *q* axis with the voltage and delay by 90° the *d* axis. The first option is called flux orientation and the second option is voltage orientation (also known as grid flux orientation).

7.2.4.1 Flux Orientation   Flux orientation was the first and remains the classic option for DFIM control. It consists of estimating the machine flux and aligning the *d* axis of the reference frame to it. Initially, the air-gap-flux orientation was used [1,4]. Lately, the stator-flux orientation has been adopted. Current literature focuses mostly on this last option [2,5], and also on the first proposals of using the DFIM for wind power generation [6].

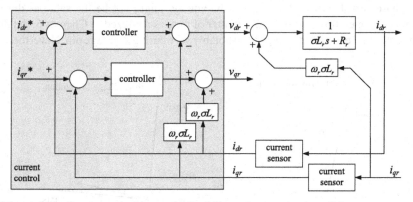

Figure 7.8 Current control loop with feed-forward compensation of the cross terms.

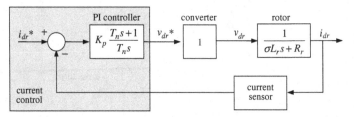

Figure 7.9 Block diagram of the current control system.

7.2.4.2 Grid Voltage Orientation

This option measures the stator voltages and aligns the q axis with the voltage space vector. Axis d would be shifted 90° from it. A phase locked loop (PLL) is commonly added to obtain the angle of the fundamental [7] and reduce the effects of voltage harmonics.

This option, also known as grid-flux orientation, was cited in Datta and Ranganathan [8] and has been lately studied by others [9,10].

7.2.4.3 Comparison

Both options are quite similar if the stator voltage and its corresponding flux are stable; meaning that they turn synchronously at constant amplitude. Under these circumstances, the flux is shifted 90° from the grid voltage. If the stator resistance is null, the misalignment would be precisely 90°, making both reference frames coincident. In practice, the voltage drop due to the stator resistance is very small in comparison with the total voltage of the machine, especially for high power ones because current machines tend to be more inductive and less resistive with the rise in power and voltage (Figure 7.10).

Indeed, it is hard to decide which option is best. Regarding the basic expression of vector control (Equations (7.1) and (7.3)), Equation (7.1) is mathematically exact for the flux orientation and, in turn, Equation (7.3) is exact for the voltage orientation. In any case, there is no difference between both referencing techniques during permanent regime operations and literature references might be found for either case.

Differences, however, become apparent during transitory stages. Under these circumstances, probably as a consequence of sudden changes of the stator voltage, the phase shift between the flux and the voltage might not be the same as during permanent operation regimes. As a matter of fact, as analyzed in Chapter 6, the flux amplitude and even its spinning speed notably vary during a voltage dip. Nevertheless,

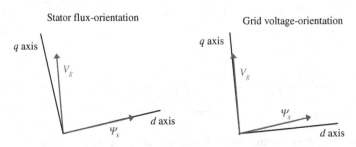

Figure 7.10 Reference frame orientation.

such exceptional circumstances are not necessary to have differences between both orientation techniques. As will be seen in Section 7.3.1, differences might appear even during stable cases when the grid voltage is perfectly steady. If vector control is implemented on a flux oriented reference, the control might become unstable depending on the d-axis component of the current. This phenomenon has been studied in depth by Heller and Schumacher [11] and Congwei et al. [12], where the phenomenon is analytically explained and the limiting set-point that makes the system unstable is found. A. Petterson et al., [13] compare both orientation techniques and prove that the voltage orientation is always stable, whereas the flux orientation might destabilize a vector control.

Section 7.3.1 applies the natural flux concept explained in Chapter 6 to intuitively explain why the flux orientation might destabilize the control. In the same way, we might explain the fact that the voltage orientation never presents stability difficulties.

7.2.5 Complete Control System

In previous sections the different parts of a classical vector control for a DFIM have been presented:

- Generation of the references: reference current calculations from the desired stator active and reactive powers
- Current control loops
- Reference frame transformations

Figure 7.11 shows a schematic block diagram of the interconnections between all these parts. The complete control structure must also include complementary issues as the start-up of the machine that will be discussed in Chapter 10.

Figure 7.11 depicts a control where the reference frame is oriented using the grid voltage instead of the stator flux. This choice was made in order to prevent stability problems that have been discussed in the previous section. The angle used in the transformations is thus obtained from the voltage measurements of the grid using a phase locked loop.

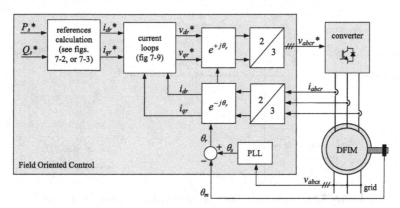

Figure 7.11   Schematic diagram of the vector control system.

7.3 SMALL SIGNAL STABILITY OF THE VECTOR CONTROL

In many closed-loop electrical systems, the stability is assured by the tuning the regulators. Grid connected converters, for example, usually include a current control loop. In such systems, the stability is guaranteed by designing the loop with a good margin phase.

In a DFIM with vector control, however, the system can be unstable even if the margin phase of the control loop is generous. Many factors can influence the system stability: for example, the choice of the reference frame orientation, the set-point of the currents, or the design of the regulators. But in all cases, the instability can be studied by means of the natural flux concept introduced in Chapter 6.

Previously, vector control has been developed based on the assumption that the phase shift between the voltage and the flux was 90°. This is fairly accurate in the steady state but in transient conditions, as previously mentioned, a natural flux appears in the stator that modifies the phase shift between the stator and the rotor. Under these circumstances the system may become unstable depending on the reference frame orientation.

7.3.1 Influence of the Reference Frame Orientation

As mentioned before, the difference between the two orientations becomes relevant under transient conditions. Figure 7.12 represents an example of the space vector trajectories of the stator flux and the grid voltage during a three-phase voltage dip. The voltage trajectory traces a circle, as it is a rotating vector of constant amplitude. The stator flux also traces a circle, but this circle is off-center due to the natural flux, as analyzed in Chapter 6.

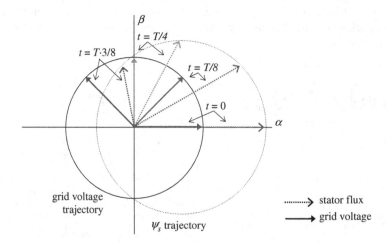

Figure 7.12   Reference frame orientation during a three-phase dip.

The solid arrows are the position of the d axis at different instants if a grid voltage orientation is used. The dotted arrows correspond to a stator-flux orientation. As can be seen, the position of the d axis is not the same in both orientations due to the displacement of the flux trajectory. It is important to remark that it is the presence of the natural flux that causes the difference between the two orientations.

We can examine the behavior of different variables of the machine using both reference frames.

Stator Flux Orientation

- The amplitude of the flux is variable; its phase is zero.
- The phase of the stator voltage is variable.
- The relationship between the stator powers and the rotor currents, Equation (7.5), is no longer valid, since $v_{ds} \neq 0$.

Grid Voltage Orientation

- The amplitude of the voltage is constant; its phase is 90°.
- The positive flux has a constant amplitude and its phase is zero.
- The natural flux is a rotating flux that turns inversely to the grid frequency.
- Relationship (7.5) is not valid but, as will seen in Section 7.4, it can be reinterpreted using the superposition principle.

However, the main difference appears in the rotational speed of the reference frame. In a grid orientation system, the reference frame turns synchronously with the grid voltage—thus at the grid frequency. In contrast, if the reference frame is orientated with the stator flux, the speed is no longer constant:

- At the beginning of the period ($t = 0$) both orientations are equivalent; the position of the d axis is zero.
- From this moment until $t = T/4$ the stator flux rotates more slowly than the voltage. The angle of the d axis is then lower in the stator-flux orientation, as can be observed in Figure 7.12.
- At instant $t = T/4$ the d axis remains behind the voltage, but its rotational speed increases so that the difference between the two vectors decreases.
- At $t = T/2$ the two orientations are again equivalent.

Figure 7.13 shows the evolution of the rotational angle using the two orientation techniques.

The difference between the two orientations is more manifest with the depth of the dip. When the dip is deeper than 50%, a curious circumstance happens: the natural flux is then greater than the positive flux. The trajectory of the flux is so displaced that it does not encircle the origin, as observed in Figure 7.14.

In that case, the rotational angle does not reach 180°; instead, it has a range as narrow as the depth of the dip increases, as shown in Figure 7.15. It is easy to deduce

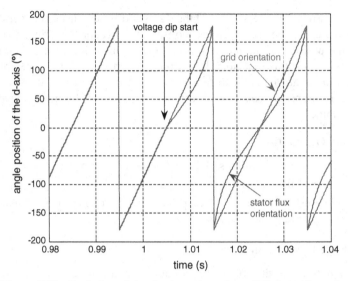

Figure 7.13 Evolution of the *d*-axis angle under a 40% three-phase dip.

that in the case of a total dip the angle is constant, since the only flux present in the stator is the natural flux that does not rotate.

As has been demonstrated, the natural flux causes the reference frame to rotate at variable speed when using a stator-flux orientation. This anomaly affects the stability of the machine as analytically demonstrated by Petersson et al. [13]. We will try to demonstrate it in a more intuitive way by examining the evolution of the natural flux. As will be seen, the instability is due to the fact that the natural flux increases incessantly until it induces an emf so high that it causes the machine to stop.

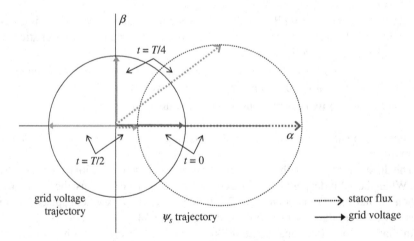

Figure 7.14 Reference frame orientation during a severe three-phase dip.

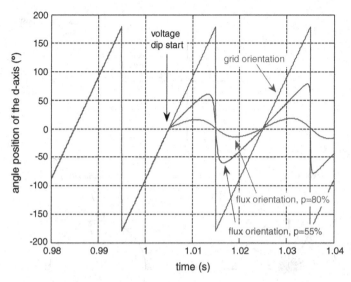

Figure 7.15 Evolution of the d-axis angle under different three-phase dips.

The time evolution of the natural flux was deduced in Section 6.6.1:

$$\frac{d}{dt}\vec{\psi}_{sn}^{s} = -\frac{R_s}{L_s}\vec{\psi}_{sn}^{s} + \frac{L_m}{L_s}R_s\vec{i}_r^{s} \qquad (7.15)$$

In the absence of rotor currents, $i_r = 0$, the stator resistance causes the natural flux to decay exponentially with the stator time constant. Generally, we can say that the natural flux disappears almost completely in a few seconds.

However, the rotor currents can modify the evolution of the natural flux:

- If the rotor current is in antiphase with the natural flux, the last term of expression (7.15) reinforces the action of the second term and therefore the flux damping would be accelerated.
- If the current is in phase, the flux damping would be slowed down, and might even be canceled out or become negative.

As observed, the angle between the rotor current and the stator flux space vectors turns out to be very relevant. This angle can be determined taking into account the following:

- The natural flux is fixed with the stator. Although its position depends on how it originated, we will assume that it is situated in the alpha axis.
- The rotor current is imposed by the vector control depending on the desired powers. If, for example, the active power is null, the rotor current will be aligned with the d axis. In general, the current can have components along the d and q axes. Anyhow, its space vector will rotate synchronously with the reference frame.

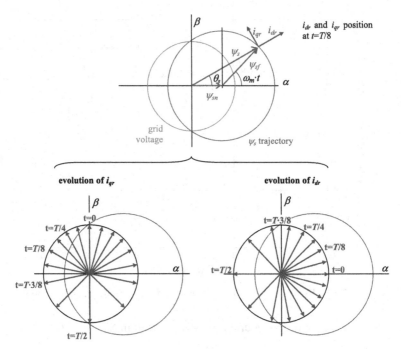

Figure 7.16   Evolution of the rotor current using a stator flux orientation.

If the reference frame is oriented with the grid voltage, the d and q axes rotate at constant speed. The angle of the rotor current goes from 0 to 2π uniformly and therefore its average action on the flux is null.

In contrast, if the reference frame is oriented using the stator flux, its rotational angle doesn't progress uniformly, as was shown in Figures 7.13 and 7.15, but it takes longer to evolve from 0° to 90° than from 90° to 180°. The same applies during the second half of the period: the angle takes longer from 270° to 360° than in the interval from 180° to 270°. This phenomenon can be observed graphically in Figure 7.16.

In the left half the evolution of the rotor current is shown if only a quadrature current is injected into the rotor. We can see that the average current during the period is upward; that is, the net current is perpendicular to the natural flux. According to Equation (7.15), the rotor current will then cause the flux to turn counterclockwise but it will not change its amplitude. In conclusion, a rotor current in the q axis causes the natural flux to turn, but it does not modify its decay—that will remain the same as in the absence of rotor currents.

The right half of the figure shows the evolution of the rotor current if only a component in the d axis exists. Since the prevailing sense of the rotor current is the same as the natural flux, its net action will be an increment of this flux. There will then be two terms influencing the natural flux in opposite senses: the second term of Equation (7.15) causes the flux to damp while the third term increases the flux. The final evolution of the flux depends on which of the two terms is bigger. If the current i_{dr} is small, the second term will be preponderant and the natural flux will decay.

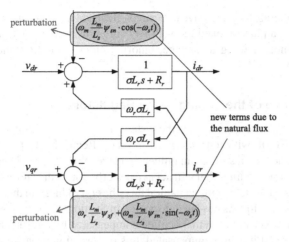

Figure 7.17   Plant of the current loop including the emf induced by the natural flux.

In contrast, if the current is large the third term will prevail and the natural flux will increase. If this happens, the anomaly in the orientation will augment, causing the average current to increase. That self-reinforcing process threaten control of the machine even if the initial natural flux were small. The system is said to be unstable.

In a flux oriented reference frame, there is a critical value for the d-axis current. Below this value any natural flux caused by a transient is damped, but above that value the system is unstable. This matter was analyzed by Congwei et al. [12], who found the condition for stability:

$$i_{dr} < \frac{2\hat{V}_g}{L_m \omega_s} \qquad (7.16)$$

This last expression gives the maximum value for the direct current that can be injected into the rotor when a flux orientation is used. Since the direct current is related to the reactive power delivered to the grid, this limit entails a maximum reactive power that cannot be surpassed without destabilizing the system:

$$Q_s < -\frac{\hat{V}_g^2}{L_s \omega_s} \qquad (7.17)$$

The minus sign indicates that the power is delivered to the grid. As a curious note, the maximum is equal (except for the sign) to the power consumed by the machine when there is no current in the rotor.

As noted before, if a grid orientation is used, the reference frame rotates at constant speed regardless of the presence or not of natural flux. Thus, there is no limitation to the rotor currents, at least regarding the system stability. This is why this orientation is preferred by most manufacturers. Other perturbations, such as voltage imbalances, can also cause anomalies in the orientation. Even if these anomalies don't cause instability problems, they can have undesired effects on the control. In order to avoid these problems, a phase looked loop (PLL) is often used to extract the angle of the positive sequence of the grid voltage, ignoring other harmonics and the negative sequence.

7.3.2 Influence of the Tuning of the Regulators

From now on, it will be assumed that the system is grid voltage oriented. The reference frame therefore turns with constant speed and is unaffected by the presence of natural flux. This flux, however, still has an influence on the DFIM control because it induces an emf in the rotor windings that will act as a perturbation in the current control loops.

During the steady state, without natural flux, the emf induced in the machine is low, proportional to the slip. As has been discussed in Section 7.2.3 (e.g., see Figure 7.7), this emf acts as a perturbation that is often compensated by a feed-forward signal in the control. Even if it is not compensated, this perturbation is easily rejected by the current regulators because it appears as DC in a synchronous reference frame. As a consequence, the emf has no effect on the current loops.

The situation is very different when a natural flux appears in the stator of the machine due to an event in the machine (a voltage dip, for example, or a sudden change of the set-points). As deduced in Chapter 6, this flux induces an additional emf, e_{rn}, fixed with the stator. In a synchronous reference frame that voltage is seen as an inversely rotational term at grid frequency. In that situation the perturbation in the control loops will be the addition of the two voltages: a DC voltage proportional to the slip and a rotational voltage caused by the natural flux.

Figure 7.17 shows the plant of the current control loops if there is a natural flux, ψ_{sn}, in the stator in the alpha axis as a result, for example, of a voltage dip. It is interesting to compare that diagram with the plant presented in Figure 7.7, which is only valid in steady state. Two new terms, related to the natural flux, now appear.

Unlike the perturbation due to the slip, the emf induced by the natural flux is more difficult to reject because it is seen as AC (at the grid frequency) by the control loops. As a consequence, a current arises in the rotor that superimposes to the current corresponding to normal operation. This new current modifies the evolution of the natural flux, accelerating or slowing its damping. It can even neutralize the damping depending on the characteristics of the current regulators. This latter case should be avoided: any perturbation, even if it is insignificant, will cause a natural flux that will increase nonstop until the system gets out of control. The influence of the regulators on the evolution of the natural flux will be analyzed by applying the superposition principle introduced in Section 6.6.2.

7.3.2.1 Influence of the Regulators on the Evolution of the Natural Flux As has been analyzed, all the variables of the machine have two components: one rotating at synchronous speed corresponding to normal operation and the other fixed with the stator and originating from the perturbation caused by the natural flux. In Chapter 6, the superposition principle was used to analyze independently the two components. The machine was analyzed as if it was the

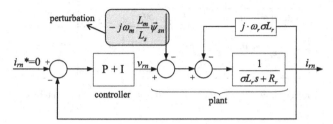

Figure 7.18 Rotor current loop for the natural machine.

addition of two isolated machines: the forced machine, where all the variables rotate synchronously, and the natural machine, where the variables are fixed to the stator.

The forced machine is the steady state machine. Its rotor current is perfectly regulated by the control loops and it is the machine that corresponds to normal operation.

The natural machine is a machine with no voltage at its stator. During normal operation, this machine has no flux—neither stator nor rotor current. During transient conditions, however, a natural flux appears. As we are interested in the evolution of the natural flux, we will focus our attention on that machine.

In this machine, the reference currents are zero, but depending on how the loop reacts to perturbations, some current might flow through the rotor. Figure 7.18 shows the flow chart of the rotor control loop with the perturbation caused by the natural flux highlighted.

If the bandwidth of the control loop is high and it firmly rejects the perturbation, the rotor current will be small. The situation of the machine will then be the same as if it were an open-circuited rotor. The natural flux will therefore decay with the stator time constant, $\tau_s = L_s/R_s$.

Otherwise, there will be a current circulating through the rotor. This current will affect the evolution of the natural flux. Restating the expression of the dynamics of the natural flux (7.15):

$$\frac{d}{dt}\vec{\psi}^s_{sn} = -\frac{R_s}{L_s}\vec{\psi}^s_{sn} + \frac{L_m}{L_s}R_s\vec{i}^s_r \qquad (7.18)$$

The expression clearly shows that the evolution has two opposing terms: the first term, the only one present when the rotor is open-circuited, damps the flux. The second term can increase or reduce the flux depending on the angle between the rotor current and the natural flux. This angle, referred to as α, becomes crucial to study the system stability.

If α is large, the rotor current will be almost in antiphase with the natural flux and therefore the flux damping will be accelerated. If α is small, the current will be approximately in-phase and the flux damping will be slowed down.

The angle α is determined by the PI regulator parameters. The influence of the PI control on the current can be analyzed by using the model of the natural machine that was deduced in Section 6.7 (Figure 7.19).

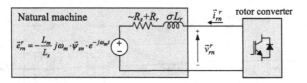

Figure 7.19   Rotor equivalent circuit for the natural machine.

Solving the rotor equivalent circuit leads to the following relationship between the natural flux, the converter voltage, and the rotor current:

$$\vec{v}_{rn}^{\,r} = -j\omega_m \frac{L_m}{L_s} \vec{\psi}_{rn}^{\,r} + R_r \vec{i}_{rn}^{\,r} + j\sigma L_r \frac{d}{dt} \vec{i}_{rn}^{\,r} \tag{7.19}$$

From that expression, we can deduce the relationship between the rotor phasors in steady state:

$$\underline{V}_{rn} = -j\omega_m \frac{L_m}{L_s} \underline{\psi}_{rn} + R_r \underline{I}_{rn} - jX_r \underline{I}_{rn} \tag{7.20}$$

where $X_r = \omega_m \sigma L_r$.

Note the negative sign of the third term. In the natural machine the rotor current is steady with the stator; hence, it rotates in the opposite sense with respect to the rotor windings. Consequently, the drop in the transitory inductance is 90° clockwise with respect to its current.

PI with a Predominant Proportional Action   If the PI proportional action is significantly higher than the integral, the converter voltage will be in antiphase with the rotor current:

$$\underline{V}_{rn} = -K'_P \underline{I}_{rn} \tag{7.21}$$

where K'_P is the proportional gain of the PI referred to the stator: $K'_P = K_P u^2$.

In those circumstances the rotor current would be almost in antiphase with the natural flux as shown on the left phasor diagram in Figure 7.20.

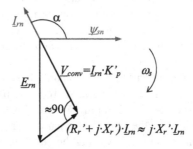

Figure 7.20   Phasor diagram when using a PI with predominant proportional action.

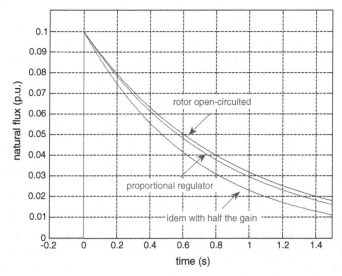

Figure 7.21 Natural flux evolution using a PI with predominant proportional action.

Such current accelerates the damping of the natural flux as predicted by Equation (7.18). To corroborate that assertion, the behavior of a DFIM has been simulated during a 10% three-phase voltage dip under three different situations:

- With its rotor open-circuited
- With its rotor connected to a converter with a current loop using a proportional regulator
- And finally using the same regulator but whose proportional gain is divided by 2

See Figure 7.21.

As predicted by the analysis carried out with the phasor diagrams, the damping of the natural flux is accelerated due to the effect of the rotor current. Besides, the lower the proportional gain, the bigger the rotor current and the quicker the damping of the flux.

PI with a Predominant Integral Action   The integration of an AC signal results in a signal that is phase shifted by 90°. Hence, the output voltage of a PI with predominant integral action is almost 90° from the rotor current. Figure 7.22 shows the corresponding phasor diagram.

From the phasor diagram we can see that the rotor current is in-phase with the natural flux and, consequently, it will slow the damping. That effect is shown in Figure 7.23, where the evolution of the natural flux is compared in three cases: without a rotor current, using a proportional regulator, and using a PI regulator.

General Case   We have deduced that the proportional action of the PI tends to accelerate the damping while its integral action slows down the decay. This second

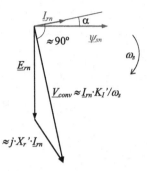

Figure 7.22 Phasor diagram when using a PI with predominant integral action.

effect is higher, so, in general, the net effect of the PI is a deceleration of the damping. It is interesting to detect when the damping is completely canceled, because if that happens the system becomes unstable. For that purpose we restate expression (7.18) in scalar notation:

$$\frac{d}{dt}\psi_{sn} = -\frac{R_s}{L_s}\psi_{sn} + \frac{L_m}{L_s}R_s i_r \cos(\alpha) \tag{7.22}$$

In order to have a stable system the natural flux must be damped; that is, its derivative should be negative:

$$-\frac{R_s}{L_s}\psi_{sn} + \frac{L_m}{L_s}R_s i_r \cos(\alpha) < 0 \tag{7.23}$$

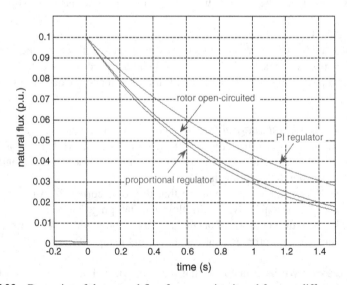

Figure 7.23 Dynamics of the natural flux for open-circuit and for two different regulators.

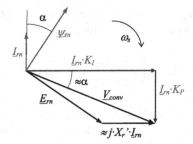

Figure 7.24 Phasor diagram when using a PI regulator.

which leads to

$$L_m i_r \cos(\alpha) < \psi_{sn} \tag{7.24}$$

Figure 7.24 shows the phasor diagram of the rotor circuit when using a general PI controller.

Habitually, the voltage drop in the leakage inductance is much smaller than the converter voltage, and therefore the phasor of the converter voltage is almost aligned with the emf induced by the natural flux. With this approximation, and neglecting the voltage drop in the rotor resistance, expression (7.20) can be simplified as follows:

$$\underline{V}_{conv} = \underline{E}_{rn} + X_r \underline{I}_{rn} = \frac{L_m}{L_s}\underline{\psi}_{sn}\omega_m + \sigma L_r \underline{I}_{rn} \tag{7.25}$$

On the other hand, the rotor converter is given by a PI controller, so it is the multiplication of the current by the regulator gain at the grid frequency:

$$\underline{V}_{conv} = PI\_gain(\omega_s)\underline{I}_{rn} = \sqrt{K_P'^2 + (K_I'/\omega_s)^2}\,\underline{I}_{rn} \tag{7.26}$$

From the last two expressions we can calculate the rotor current amplitude:

$$\underline{I}_{rn} = \frac{(L_m/L_s)\underline{\psi}_{sn}\omega_m}{\sqrt{K_P'^2 + (K_I'/\omega_s)^2} - \sigma L_r \omega_m} \tag{7.27}$$

Introducing this expression into the condition for stability, Equation (7.24), we have

$$\frac{(L_m^2/L_s)\omega_m \cos(\alpha)}{\sqrt{K_P'^2 + (K_I'/\omega_s)^2} - \sigma L_r \omega_m} < 1 \tag{7.28}$$

where the angle α can be approximated by taking into account that the phase of the emf is almost the same as the phase of the converter voltage, and thus

$$\alpha \approx \arctan\left(\frac{L_{rn}K_P}{L_{rn}K_I/\omega_s}\right) = \arctan(Tn\omega_s) \tag{7.29}$$

The importance of expression (7.28) is fundamental for the correct functioning of the vector control control. A system that doesn't comply with that condition will be unstable.

Figures 7.21 and 7.23 were obtained by simulating a machine that fulfilled the conditions of Equation (7.28) over its entire operation range. This machine, however, can become unstable if the gain of its regulators decreases. If, for example, the PI parameters are calculated in order to obtain a bandwidth of 250 Hz, the machine will be at the edge of instability: it will be stable for small rotational speeds but it will become unstable if the speed increases. According to the simplified expression (7.28), the critical speed is calculated to be 1.16 p.u. The accuracy of that value can be corroborated by means of Figure 7.25, where the evolution of the natural flux is shown during turbine acceleration.

The generator was operating at constant speed when at $t = 0$ there is a 10% voltage dip in the grid. At steady state the natural flux was zero, but the dip causes it to rise by 10% of its rated value. During the dip, the natural flux decreases, which is an indication that the machine is stable. Meanwhile, the machine is accelerated and at $t = 0.85$ s its speed exceeds the stability limit predicted by Equation (7.28). From that moment the machine is unstable and its natural flux increases. If no countermeasures are taken, it will continue to rise until the saturation of the rotor converter.

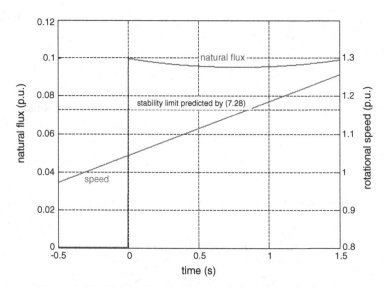

Figure 7.25   Natural flux evolution during a turbine acceleration.

7.4 VECTOR CONTROL BEHAVIOR UNDER UNBALANCED CONDITIONS

Throughout Section 7.2, when discussing the basis of vector control, we have assumed that the grid voltage was balanced with constant amplitude. Under those conditions, the direct component of the stator voltage was zero and its quadrature component was constant; meanwhile the stator flux was oriented with the d axis and was also constant. That is no longer true if the grid voltage is unbalanced, that is, if the voltages of the three phases have different amplitudes or their waveforms are not $120°$ offset in time.

As seen in Section 6.5, any unbalanced three-phase system can be seen as the addition of three balanced systems called sequences: the positive sequence, the negative sequence, and the zero sequence. If the stator of the DFIM is connected to an unbalanced grid, the positive sequence of the grid voltage creates a flux with constant amplitude, which rotates at the grid frequency. The negative sequence creates a stator flux with similar characteristics but rotating in the opposite sense; and finally, the zero sequence does not create any stator flux if the neutral point, as is usually the case, is isolated. The total stator flux is then the addition of two fluxes rotating in opposite senses: the positive flux and the negative flux. Its space vector in a stator reference frame can be written as

$$\vec{\psi}_s^s = \sqrt{2}\underline{\psi}_1 e^{j\omega_s t} + \sqrt{2}\underline{\psi}_2 e^{-j\omega_s t} \qquad (7.30)$$

where ψ_1 and ψ_2 are the phasors (in rms) of the positive and negative fluxes, respectively. These fluxes, as seen in Section 6.5, are proportional to the positive and negative sequence of the grid voltage, V_1 and V_2:

$$\begin{aligned} \underline{\psi}_1 &= \frac{V_1}{j\omega_s} \\ \underline{\psi}_2 &= \frac{V_2}{-j\omega_s} \end{aligned} \qquad (7.31)$$

In normal operation, the grid voltage is balanced and the negative sequence of the voltage, V_2, is zero. If V_2 is not zero, that means the grid is unbalanced and there is some negative flux in the stator of the machine.

The ratio between the amplitudes of the negative and the positive sequences is called the voltage unbalanced factor (VUF) and is a good way to measure the imbalance:

$$VUF = \frac{V_2}{V_1} \qquad (7.32)$$

The presence of negative flux marks the behavior of the vector control in three different ways: (1) it may affect the orientation of the reference frame, (2) it may saturate the rotor converter, and (3) it causes a second harmonic pulsation in the stator current and in the electromagnetic torque.

7.4.1 Reference Frame Orientation

The negative sequence, like any other harmonic, affects the orientation of the reference frame if no countermeasures are taken. In order to avoid problems, the rotational angle of the reference frame is calculated by means of a phase looked loop (PLL) that considers only the positive sequence of the grid voltage or the stator flux.

7.4.2 Saturation of the Rotor Converter

Since the negative sequence of the flux rotates in a countersense to the rotor, it can induce a very high emf in its windings depending on its amplitude. As deduced in Chapter 6, the amplitude of this emf is about twice the negative grid voltage (expressed in reference to the stator side):

$$\underline{E}_{r2} = \underline{V}_2 \frac{L_m}{L_s} (2 - s) \tag{7.33}$$

Introducing in that expression the concept of the voltage unbalance factor yields:

$$\underline{E}_{r2} = \underline{V}_1 \cdot VUF \cdot \frac{L_m}{L_s} (2 - s) \tag{7.34}$$

Under an asymmetrical voltage dip the VUF can reach high values: up to 50% for single-phase faults or up to 100% for phase-to-phase faults. In these cases the emf will probably exceed the maximum voltage of the rotor converter and thus the converter will saturate and lose control of the current. This situation will be thoroughly analyzed in Section 7.5.

Apart from voltage dips, VUF is normally below 1% in transmission lines and below 2% in distribution lines. It can be slightly higher in weak grids such as those of developing countries or small populated areas, but in any case it is not high enough to cause saturation of the converter of typical turbines.

7.4.3 Oscillations in the Stator Current and in the Electromagnetic Torque

From now on, it will be considered that a PLL is used in the orientation of the reference frame so that its orientation is immune to imbalance. Moreover, it will be assumed that the rotor converter does not saturate. In these circumstances, the control acts linearly and we can apply the superposition principle discussed in Section 6.6.2.

By using the superposition principle, the behavior of the machine can be analyzed as if we were dealing with two independent machines:

- The positive machine whose stator voltage is the positive sequence of the grid voltage. All the variables of this machine are synchronous with the grid: they turn counterclockwise.

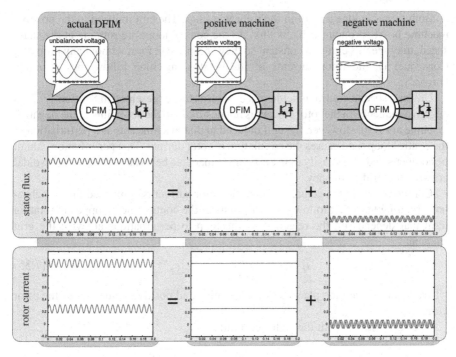

Figure 7.26 Superposition principle during three-phase dips.

- The negative machine whose stator voltage is the negative sequence of the grid voltage. The flux of that machine rotates in opposite sense to the rotor, that is, clockwise, and so do all the machine variables.

During normal operation, only the positive machine is relevant: all the variables of the negative machine are zero so they can be neglected. In case of imbalances, we will deal with a negative sequence in the stator voltage that will cause a negative flux to appear and the circulation of negative currents in the stator and rotor of the negative machine. In this latter case, the negative machine becomes relevant and any variable of the actual machine (the real one) will be the addition of that same variable of the positive and that of the negative machine.

Figure 7.26 represents the decomposition of the DFIM into these two machines under an unbalanced factor of 10%. Two variables of the machine, the stator flux and the rotor current, are represented in a synchronous reference frame. As can be seen, the stator flux in the actual machine is the addition of the flux in the positive and in the natural machine. The same can be said for the rotor current and for any other variable of the machine.

The stator flux in the positive machine is constant and synchronous with the grid. Consequently, it is seen as DC in the reference frame. The stator flux in the negative machine is also constant in amplitude, but it rotates in the opposite sense. Therefore, in the synchronous reference frame it is seen as sinusoidal at twice the grid frequency.

Both fluxes induce an emf in the rotor windings. The emf induced in the positive machine is synchronous, as is its flux. As the control loops are rotating at the same speed, they see this emf as a constant so they compensate for it without any problem. Consequently, the rotor currents of the positive machine follow the set-points correctly.

Unless a special control is set in place (such as those presented in Section 7.6.2), a negative sequence in the rotor current is not desirable so the set-points of the negative machine are zero. However, the emf induced in that machine acts as a perturbation in the control loops and causes the circulation of rotor current. That perturbation cannot be compensated for as easily as in the positive machine because it is seen as sinusoidal at twice the grid frequency.

Consequently, in the actual machine the rotor current is composed of a synchronous rotating term that follows the set-points of the control and an opposite rotating term that depends on the rejection of the loop to perturbations at twice the grid frequency:

$$\vec{i}_r^s = \sqrt{2}\underline{I}_{r1}e^{j\omega_s t} + \sqrt{2}\underline{I}_{r2}e^{-j\omega_s t} \tag{7.35}$$

where I_{r1} and I_{r2} are the phasors of the negative and positive sequences of the rotor current.

The same occurs with the stator current:

$$\vec{i}_s^s = \sqrt{2}\underline{I}_{s1}e^{j\omega_s t} + \sqrt{2}\underline{I}_{s2}e^{-j\omega_s t} \tag{7.36}$$

The stator current is then the addition of two space vectors that rotate in the opposite sense at the grid frequency. Twice per grid period the two vectors are aligned in the same sense, so they add constructively and the amplitude of the total stator current reaches its maximum. Also twice per period they are aligned but in the opposite sense. They then add destructively and the amplitude reaches its minimum. Figure 7.27 shows the amplitude of the stator current for an unbalanced factor of 10%.

The negative sequences cause fluctuations in the stator powers as well. Using the expression of these powers given in Equations (4.32) and (4.34) and splitting them into the positive and the negative sequence, we have

$$P_s = \tfrac{3}{2}\mathrm{Re}\left\{\vec{v}_s \cdot \vec{i}_s^{\,*}\right\} = 3\mathrm{Re}\left\{\left(\underline{V}_1 e^{j\omega_s t} + \underline{V}_2 e^{-j\omega_s t}\right) \cdot \left(\underline{I}_{s1} e^{j\omega_s t} + \underline{I}_{s2} e^{-j\omega_s t}\right)^*\right\}$$

$$Q_s = \tfrac{3}{2}\mathrm{Im}\left\{\vec{v}_s \cdot \vec{i}_s^{\,*}\right\} = 3\mathrm{Im}\left\{\left(\underline{V}_1 e^{j\omega_s t} + \underline{V}_2 e^{-j\omega_s t}\right) \cdot \left(\underline{I}_{s1} e^{j\omega_s t} + \underline{I}_{s2} e^{-j\omega_s t}\right)^*\right\} \tag{7.37}$$

which yields

$$P_s = 3\mathrm{Re}\left\{\underline{V}_1 \cdot \underline{I}_{s1}^* + \underline{V}_1 \cdot \underline{I}_{s2}^* e^{j2\omega_s t} + \underline{V}_2 \cdot \underline{I}_{s1}^* e^{-j2\omega_s t} + \underline{V}_2 \cdot \underline{I}_{s2}^*\right\}$$

$$Q_s = 3\mathrm{Im}\left\{\underline{V}_1 \cdot \underline{I}_{s1}^* + \underline{V}_1 \cdot \underline{I}_{s2}^* e^{j2\omega_s t} + \underline{V}_2 \cdot \underline{I}_{s1}^* \cdot e^{-j2\omega_s t} + \underline{V}_2 \cdot \underline{I}_{s2}^*\right\} \tag{7.38}$$

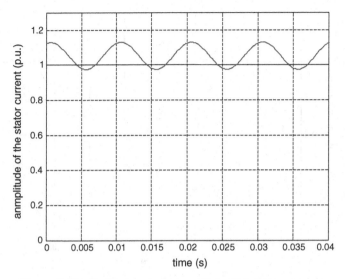

Figure 7.27 Oscillations in the amplitude of the stator current.

The first and the last terms of these equations give a constant transfer of power. The second and third terms cause pulsations at twice the grid frequency.

Similar fluctuations appear in the electromagnetic torque. If the shaft of the machine has a resonance close to double the grid frequency, these fluctuations can cause vibrations and mechanical stress.

The fluctuations in the torque can be calculated from Equation (4.36):

$$T_{em} = \tfrac{3}{2}p \, \text{Im}\left\{ \vec{\psi}_s \cdot \vec{i}_r^* \right\}$$
$$= 3p \, \text{Im}\left\{ \underline{\psi}_{s1} \cdot \underline{I}_{r1}^* + \underline{\psi}_{s1} \cdot \underline{I}_{r2}^* e^{j2\omega_s t} + \underline{\psi}_{s2} \cdot \underline{I}_{r1}^* e^{-j2\omega_s t} + \underline{\psi}_2 \cdot \underline{I}_{r2}^* \right\}$$

(7.39)

Again, the fluctuations are given by the second and third terms. As we can see, the amplitude of the torque ripple depends on the negative flux, ψ_{s2}, and the negative current, I_{r2}. The negative flux is due to the imbalance of the grid, and doesn't depend on the machine's control. On the contrary, the rotor current strongly depends on the control: if the control loops have a large bandwidth with a high rejection of perturbations, the negative current will be zero. In this case, the oscillations will be the product of the average torque by the unbalanced factor VUF. In any other case, the oscillations are usually bigger. As an example, in Figure 7.28a 15% of torque ripple can be observed in a commercial DFIM operating in a grid with 10% VUF.

7.5 VECTOR CONTROL BEHAVIOR UNDER VOLTAGE DIPS

All electrical drives connected to a grid are affected in a greater or lesser degree by voltage dips. The DFIM, however, is especially sensitive to grid disturbances and this

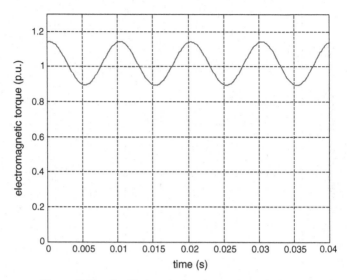

Figure 7.28 Oscillations in the electromagnetic torque.

is its main drawback. Vector control is based on the assumption that the voltage grid is stable and the stator flux is a space vector rotating at constant speed 90° behind the voltage. That hypothesis is true in steady state but not in the case of a voltage dip. During a voltage transient, a natural flux (negative, in the case of unbalanced dips) appears at the stator, which can affect the orientation of the reference frame. Additionally, this flux induces large voltages at the rotor terminals, which can saturate the rotor converter.

The choice of the orientation of the reference frame is very important for vector control under voltage dips. As analyzed in Section 7.3, the natural flux notably affects the orientation if a stator-flux-oriented frame is used. In this case, the behavior under voltage dips is complicated to study since the system it is not linear. Some authors have developed specific models that try to simplify the analysis [14].

From now on, it will be considered that the reference frame is aligned with the grid voltage, as is the case for most modern turbines. Additionally, the phase of the grid voltage will be calculated by means of a phase looked loop (PLL) in order to consider only the positive sequence of that voltage. By doing so, the orientation of the reference frame becomes immune to the voltage dips as its rotational speed remains constant and equal to the grid frequency.

Even so, the control continues to be affected by voltage dips as they induce large emfs in the rotor of the machine. Chapter 6 gives the amplitude of that voltage depending on the characteristics of the dip (type of dip, depth, instant of occurrence). If the induced emf exceeds the maximum voltage of the rotor converter, it will saturate. On the contrary, if the emf is low and doesn't surpass the limits of the converter, there won't be saturation and the control will act in a very different way. The analysis is then limited to small and severe voltage dips.

7.5.1 Small Dips

We will first examine the behavior of the DFIM under a "small" voltage dip, where "small" means that the induced emf does not saturate the converter. Since the reference frame is not affected by the dip and the converter doesn't saturate, the whole control system acts linearly and we can apply the superposition principle discussed in Section 6.6.2.

Using the superposition principle, the behavior of the machine can be analyzed as three independent machines:

- The positive machine whose stator voltage is the positive sequence of the grid voltage
- The natural flux whose stator is short-circuited but is still magnetized since there is a natural flux in its stator
- And, if the voltage dip is unbalanced, the negative machine whose stator voltage is the negative sequence of the grid voltage

7.5.1.1 Behavior During a Three-Phase Voltage Dip   In a three-phase voltage dip the grid voltage is balanced so its negative sequence is zero. Consequently, the generator can be analyzed as the addition of two machines: the positive and the natural as shown in Figure 7.29 for the case of a 30% voltage dip.

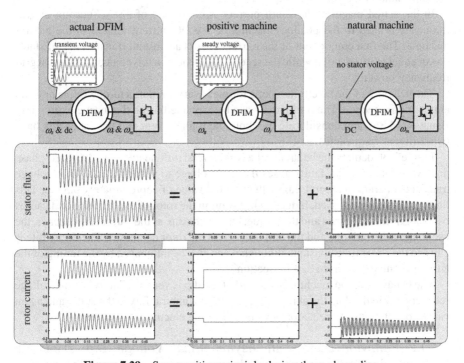

Figure 7.29   Superposition principle during three-phase dips.

The stator flux and the rotor current are depicted in the synchronous reference frame. The stator flux in the actual machine is the addition of the flux in the positive and in the natural machine and the same can be said for the rotor current or for any other variable of the machine.

The stator flux in the positive machine is constant and synchronous with the grid. Consequently, this machine is at steady state and all its variables are synchronous: in the stator windings the currents and voltages are sinusoidal at grid frequency, whereas in the rotor windings the currents and the voltages are sinusoidal at slip frequency. The emf induced in the rotor of the positive machine is seen as DC in the control loops so it is correctly compensated. Hence, that machine properly regulates the rotor current and the active and reactive powers. If the set-points of the active and reactive powers are unchanged, that current will vary in response to the lower voltage grid as seen in Figure 7.29: the quadrature current will increase, while the reactive power can increase or decrease depending on the reactive power generated.

The stator flux in the natural machine does not turn. It induces a large emf in the rotor windings that is seen as AC by the control loops and hence it is hard to compensate. An undesired current arises in the rotor that entails an almost similar stator current. During the dip, as the natural flux decays, the induced emf decreases and so does the rotor current, as shown in Figure 7.29. All the variables in this machine are fixed with the stator: the stator current is DC whereas the rotor current and voltage are sinusoidal at the rotating frequency.

In the actual machine the stator current is composed of an AC current at grid frequency related to the positive machine and a DC current related to the natural machine. The first component of the current entails a constant transference of stator power equal to its set-point while the second component involves an oscillation at grid frequency power.

The behavior of the electromagnetic torque is very similar: a torque ripple appears at the grid frequency. The oscillations can be explained using the torque expression, Equation (4.36), and considering that both the rotor current and the stator flux have two different frequencies.

Figure 7.30 depicts the behavior of a commercial turbine under a 30% three-phase dip. At instant $t = 0$ s a fault causes the grid voltage to suddenly fall. That voltage transient generates a natural flux with an initial value of approximately 0.3 p.u. This flux, as has been discussed, induces a high emf in the rotor of the machine and causes oscillations in the powers and the torque. During the dip, as the natural flux decays, the oscillations are reduced and the emf induced in the rotor decreases. The decay rate is close to the stator time constant although it can also be influenced by the tuning of the current regulator as discussed in Section 7.3.2.

The behavior of the machine at the end of the dip is very similar: when the voltage recovers its rated value, the abrupt change causes a natural flux in the stator and again causes oscillations in the stator powers and in the electromagnetic torque.

7.5.1.2 Behavior During an Asymmetrical Voltage Dip

During an asymmetrical dip the voltage is unbalanced and consequently the negative

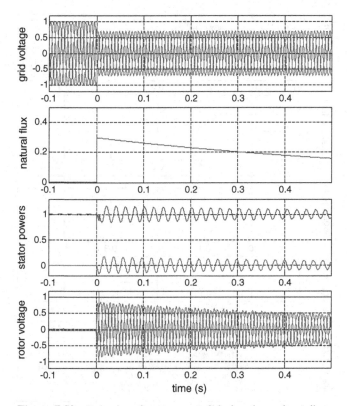

Figure 7.30 Behavior of vector control during three-phase dips.

sequence of the stator flux is nonzero. The machine can be analyzed as the addition of three machines: the positive, the natural, and the negative one.

In the last section, we have seen that the presence of the natural flux causes damped oscillations at the grid frequency in the stator powers and in the electromagnetic torque. On the other hand, when analyzing the behavior under unbalanced conditions in Section 7.4, it was deduced that the negative flux causes the powers and the torque to oscillate at twice the grid frequency. In an asymmetrical voltage dip, we found both effects simultaneously.

The oscillations at twice the grid period depend on the negative flux and therefore on the unbalanced factor—that is, on the type and depth of the dip. The oscillations at the grid frequency depend on the natural flux, and this flux can be higher or lower according to the instant of the timing of the fault.

As an example, Figure 7.31 shows the behavior of the machine during a phase-to-phase fault appearing at $t = 0$ s. This fault does not generate natural flux, so the machine is in the steady state from the beginning.

If the dip begins a quarter of the grid period later, the dip generates the maximum natural flux. In this case transient oscillations appear in the variables of the machine,

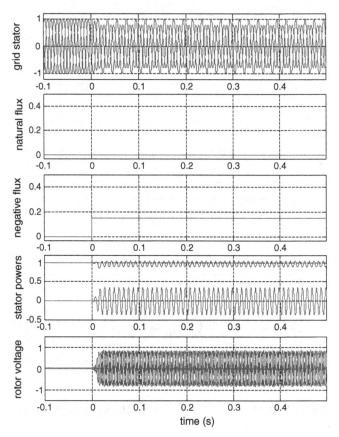

Figure 7.31   Behavior of vector control during a phase-to-phase dip beginning at $t = 0$ s.

as seen in Figure 7.32. Once the natural flux has completely decayed, the behavior is the same as in the previous case.

7.5.2 Severe Dips

We will consider "severe" dips—those that induce an emf that causes saturation of the rotor converter. Furthermore, since the converter doesn't control the current any more, this one can rise significantly, thus saturating the leakage inductances. Under these circumstances, the system is no longer linear and the superposition principle is not applicable. The study gets much more complicated than in the case of "small" dips. The analytical calculation of the behavior of the machine becomes very difficult and its study requires at least the simulation of the complete Park model. Moreover, some authors have proposed the use of more detailed models in order to take into account the magnetic saturation. Ekanayake et al. [10] propose a double cage model, whereas other authors use a model based on the finite-element method (FEM).

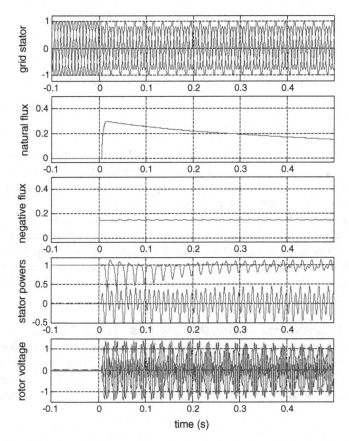

Figure 7.32 Behavior of vector control during a phase-to-phase dip beginning at $t = T/4$.

Broadly speaking, the saturation of the rotor converter has two effects on the system: (1) the loss of current control and (2) overvoltages in the DC-link bus.

7.5.2.1 Loss of the Current Control
When the rotor converter saturates the amplitude or its voltage reaches its maximum value, $V_{r,max}$,

$$V_{r,\text{max}} = \frac{V_{bus}}{\sqrt{3}} D_{\text{max}} \qquad (7.40)$$

where V_{bus} is the bus voltage of the DC link and D_{max} is the maximum duty cycle (around 97–99%).

The system can be simplified by the equivalent circuit of Figure 7.33: two voltage sources connected through the leakage inductance of the machine.

The amplitude of the rotor current is due to the difference between the two voltages and the value of the linking impedance. The deeper is the voltage dip, the

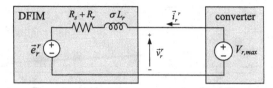

Figure 7.33 Equivalent circuit of the rotor during a three-phase voltage dip.

higher will be the induced emf, and the larger will be the rotor and stator currents. The rotor speed also has an influence on the current as the induced emf is proportional to it. Figure 7.34 shows the amplitude of the stator and rotor currents during an 80% three-phase voltage dip.

As can be seen, the highest currents occur at the beginning of the dip, when the natural flux is at its maximum. In most common turbines, the current can reach up to 3 p.u. for a full three-phase dip and even higher for a phase-to-phase dip. Unless a very oversized converter is used, that peak will damage the semiconductors of the converter. A protection technique, such as any of those presented in Chapter 8, becomes compulsory.

7.5.2.2 Overvoltages in the DC-Link Bus Figure 7.34 was obtained by simulating the machine behavior during an 80% voltage dip. For shallower dips, the peak current will be lower and may not exceed the rotor converter limits. In Figure 7.35, for instance, the currents during a 40% voltage dip are represented. In this case, the problem is not the rotor current, as it is easy to design a rotor converter able to

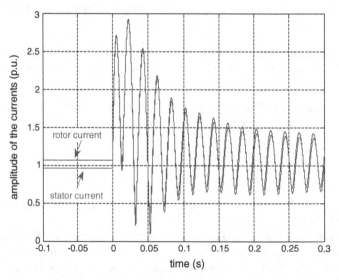

Figure 7.34 Stator and rotor currents during an 80% voltage dip.

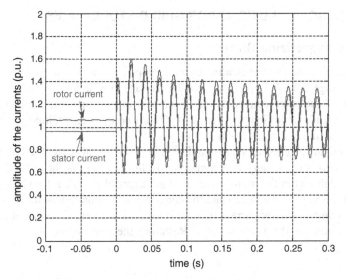

Figure 7.35 Stator and rotor currents during a 40% voltage dip.

manage it, but the DC-link voltage that reaches very high values, as depicted in the left half of Figure 7.36.

The DC-link bus voltage increases due to the large amount of power that the converter absorbs while saturated. During the voltage dip this power cannot be evacuated to the grid, because the power transference capability of the grid converter has dropped due to the low voltage of the grid. As a result, the power accumulates in the capacitors of the DC link and their voltage increases slowly but incessantly. The right half of Figure 7.36 depicts the powers managed by the two converters.

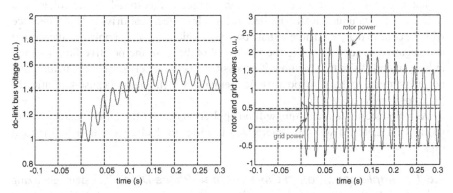

Figure 7.36 (Left) DC-link voltage during a 40% dip. (Right) Power transferred by the converters.

7.6 CONTROL SOLUTIONS FOR GRID DISTURBANCES

7.6.1 Demagnetizing Current

The demagnetizing current is a recently developed technique to protect the DFIM against voltage dips. As seen in Section 7.5, voltage dips, unless very shallow, cause saturation of the rotor converter and hence loss of the rotor current control. This technique avoids saturation of the converter by introducing an additional current into the rotor that minimizes its voltage.

Let's imagine, for example, a total three-phase dip. In this case, the stator voltage becomes zero and the only flux remaining in the machine is the natural flux that induces a high emf in the rotor converter. If nothing is done, the rotor voltage will be very similar to the emf induced and the converter will saturate as a result. However, by introducing a specific current into the rotor, it is possible to reduce the flux and hence to decrease the rotor voltage. Restating the expression of the rotor flux, Equation (6.1):

$$\vec{\psi}_r = \frac{L_m}{L_s}\vec{\psi}_s + \sigma L_r \vec{i}_r \qquad (7.41)$$

From that expression we can see that the rotor flux can be reduced by introducing into the rotor a current opposite to the stator flux:

$$\vec{i}_r = -K_d \vec{\psi}_s \qquad (7.42)$$

By doing so, the rotor flux will be

$$\vec{\psi}_r = \left(\frac{L_m}{L_s} - \sigma L_r K_d\right)\vec{\psi}_s \qquad (7.43)$$

That can even be zero by adjusting the gain, K_d.

The injection of a current opposite to the magnetic flux to reduce the voltage is a well-known technique mostly applied to brushless electrical drives. The difference with the control described here is that now the technique is solely applied to the natural and negative fluxes and not to the positive flux, because the positive flux does not induce an emf. With this method, it is possible to completely cancel the natural and negative fluxes, so that only positive flux remains, which does not cause saturation of the rotor converter.

As a first step, the technique will be applied to protect the DFIM against three-phase dips. Later, it will be generalized in order to deal also with asymmetrical dips.

7.6.1.1 *Implementation in Case of Three-Phase Dips*   The demagnetizing technique is better understood by applying the superposition principle to the machine. This principle has been used to analyze the behavior of the DFIM under three-phase

dips as the addition of the positive machine and the negative machine. It will now be used to calculate the required reference current in order to (1) demagnetize the machine and (2) generate the desired stator powers:

- In the *positive machine* all the variables are synchronous. The emf is relatively low, so the demagnetizing technique is not necessary. Still, we can introduce into the rotor a current in order to generate the desired stator powers. We will apply to this machine the basic vector control scheme depicted in Figure 7.3, so its rotor currents will be

$$
i^*_{qr1} = -\frac{2}{3}\frac{L_s}{L_m \hat{V}_{fault}} P_s
$$

$$
i^*_{dr1} = -\frac{2}{3}\frac{L_s}{L_m \hat{V}_{fault}} Q_s + \frac{\hat{V}_{fault}}{\omega_s L_m}
$$

(7.44)

where i^*_{dr1} and i^*_{qr1} are the two components of the rotor reference current, $\hat{V}_{fault}$ is the remaining grid voltage during the dip, and P_s and Q_s are the desired stator powers.

- In the stator of the *natural machine* there is only a fixed natural flux that induces a large emf in the rotor. In order to reduce the rotor voltage of this machine, we will demagnetize its rotor by introducing a current opposite to its flux:

$$
\vec{i}^*_{rn} = -K_d \vec{\psi}_{sn}
$$

(7.45)

Since the stator of this machine is zero, its power is zero. There is no electromagnetic torque either, as the phase shift between the rotor current and the stator flux is 180°.

In the actual machine, the reference current will be the addition of the two currents calculated above:

$$
\vec{i}^*_r = \vec{i}^*_{r1} + \vec{i}^*_{rn}
$$

(7.46)

By setting the reference of the rotor current to this value, we will simultaneously decrease the rotor voltage and obtain the desired powers. Actually, the interactions between the two machines cause additional powers, but these extra powers are oscillating so the average values of the total stator powers are not affected.

Figure 7.37 depicts the schematic block diagram of the vector control including the demagnetizing technique. It is interesting to compare that scheme with the one in Figure 7.11 corresponding to basic vector control. The two highlighted boxes are the reference currents for the two machines. The top one is related to the steady state and was already present in basic vector control. The bottom box is the calculation of the reference of the natural machine, that is, the demagnetizing current. During a voltage dip, this new term will help demagnetize the machine and hence reduce the rotor

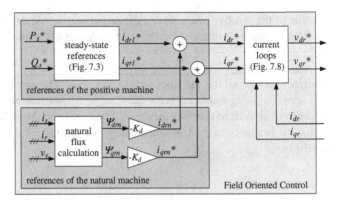

Figure 7.37 Control scheme with injection of demagnetizing current.

voltage. At steady state, however, this term will be zero and will have no effect on the behavior of the machine.

The natural flux can be calculated from the actual flux, ψ_s, and the positive flux, ψ_{s1}. The actual flux can be estimated by means of the stator and rotor currents (current model) or the stator voltage and current (voltage model) [15,16]. Considering that during three-phase dips the actual flux is the sum of the positive flux and the natural flux, the latter can be calculated as

$$\vec{\psi}_{sn} = \vec{\psi}_s - \vec{\psi}_{s1} = \vec{\psi}_s - \frac{\hat{V}_{fault}}{j\omega_s} \tag{7.47}$$

Figure 7.38 graphically shows the computation of the rotor current references when applying the demagnetizing technique during a 40% three-phase dip. The total current is the addition of the positive reference, calculated from the stator desired

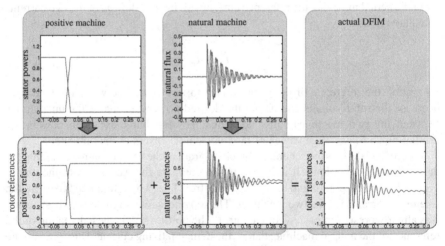

Figure 7.38 Demagnetizing technique applied to a 40% voltage dip.

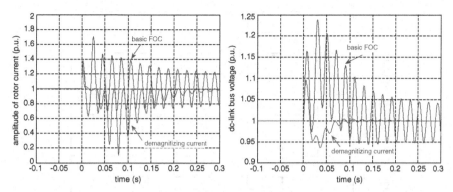

Figure 7.39 (Left) Rotor current during a 40% dip. (Right) DC-link voltage.

powers, and the natural reference, calculated in order to demagnetize the natural flux. Note that the set-points of the stator powers are modified during the dip in order to generate reactive power during the dip, as required in most grid codes.

Figure 7.38 shows also the main drawback of the technique: the current necessary to demagnetize is very high, 2.5 p.u. in the example of the figure. In some cases it is possible to overcome this problem by choosing a lower value for K_d or by limiting the natural references. In Figure 7.39, for instance, the behavior of the machine is compared using (1) the basic vector control and (2) a vector control with demagnetizing current that is limited so it does not exceed the rated current of the machine. As can be seen the results are fairly satisfactory even if the demagnetization is not complete: the current is limited to 1 p.u. except in the very beginning and the control of the DC-link voltage has improved as the power absorbed by the converter is smaller.

Meanwhile, the machine continues to generate powers. The demagnetizing current causes the powers to oscillate but their average value correctly tracks the set-points as it is observed in Figure 7.40.

Furthermore, the demagnetizing technique has an additional advantage: it accelerates the decaying of the natural flux so the machine recovers its steady state faster.

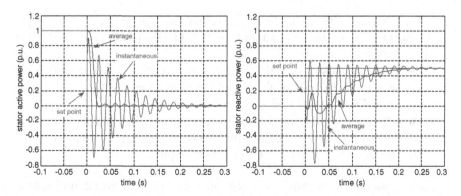

Figure 7.40 (Left) Active stator power during a 40% dip. (Left) Reactive power.

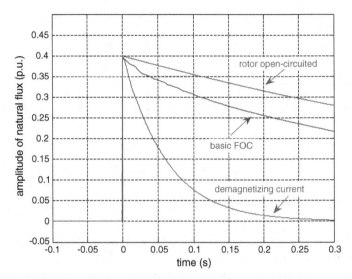

Figure 7.41 Natural flux evolution with and without demagnetizing current.

In Figure 7.41, the evolution of the natural flux is traced and compared with its decay, with the basic vector control and when there is no rotor current.

The behavior during the deepest dips is not so satisfactory: the current demand is very high. If it is limited, a large amount of power is transferred to the DC-link bus, which causes its voltage to rise unacceptably. The technique must be backed up by a hardware solution as will be seen in Chapter 9.

7.6.1.2 Implementation in a General Case
The demagnetizing current can be applied as well to asymmetrical voltage dips. In these dips there are two fluxes that cause overcurrents in the rotor: the natural flux and the negative flux. A demagnetizing current must be injected in the opposite sense to these two fluxes.

Using again the superposition principle, the rotor reference current can be calculated as the addition of three references:

- The reference of the *positive machine* determines the stator powers according to Equation (7.44), where V_{fault} must be replaced by the positive sequence of the grid voltage.
- Just as in three-phase dips, in the *natural machine* a current will be introduced in an opposite sense to its flux, the natural flux. This current will reduce the rotor voltage and will accelerate decay of the natural flux.
- The reference of the *negative machine* will also be in the opposite sense to its flux in order to avoid overvoltages in its rotor. This demagnetizing current will entail the generation of reactive power but not active power.

Grid operators usually require the generation of reactive power during voltage dips. It is a practical step to compute only the power transferred with the positive

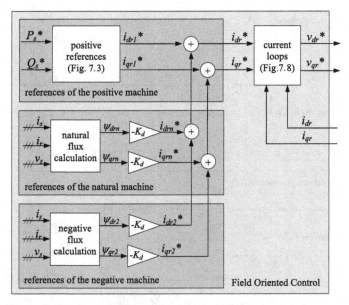

Figure 7.42 Control scheme with injection of demagnetizing current.

sequence, that is, the power generated by the positive machine. The reactive power of the negative machine will then be ignored.

The total current reference will then be the addition of three terms: the first one is used to match the stator power demanded by the grid operator; the aim of the other two terms is to reduce the rotor voltage. The whole scheme is represented in Figure 7.42.

In order to correctly calculate the current references, it is crucial to estimate the natural and negative fluxes. Figure 7.43 depicts two alternatives to calculate these two fluxes. In the top box a basic scheme is shown where the stator flux is calculated and referred to three reference frames. In the synchronous reference frame the natural and negative fluxes are seen as AC, while the positive flux is seen as DC and can be extracted by using lowpass filters. The same operation is performed in a static reference frame and in a reference rotating in the opposite sense in order to extract the natural and negative fluxes, respectively.

The use of filters introduces considerable time delays and adds amplitude and phase errors under transient conditions. Therefore, the performance of the demagnetizing technique is degraded. Better results are attained by using the scheme suggested by Xiang et al. [17] and shown in the bottom box of Figure 7.43.

Note that in the block diagrams of the previous figures the lines represent space vectors, with their two components. The superscript s denotes that the space vector is referred to a static reference frame ($\alpha\beta$ axis) whereas a denotes the use of a synchronous reference frame (dq axis) and $-a$ denotes an inversely rotating reference frame (i.e., a reference frame synchronous with the negative sequence).

In order to improve the transient response of the flux estimators, it is best to avoid the use of filters. However, it is mathematically impossible to independently estimate

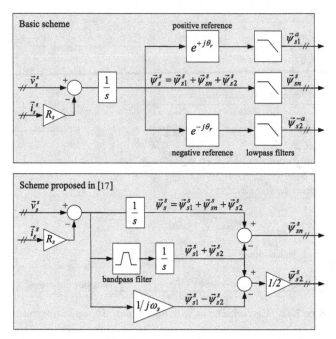

Figure 7.43 Identification of the natural and negative fluxes.

the natural and the negative fluxes without any kind of filter. Lopez et al. [18] propose a different approach that avoids filters. The scheme, represented in Figure 7.44, does not estimate the natural and negative fluxes, but a linear combination of these two fluxes. By using this scheme it is possible, for example, to cancel completely the negative flux and half of the natural flux.

The performance of this scheme can be observed in Figure 7.45, obtained for a 40% phase-to-phase voltage dip. The dip begins at $t = T/4$, so the natural flux is maximum.

7.6.2 Dual Control Techniques

Section 7.4 discussed how unbalanced conditions in the power network greatly affect the operation of the DFIM. An unbalanced voltage implies the presence of a negative

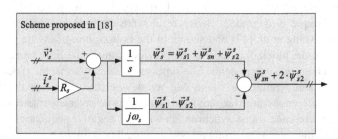

Figure 7.44 A different approach to estimate the natural and negative fluxes.

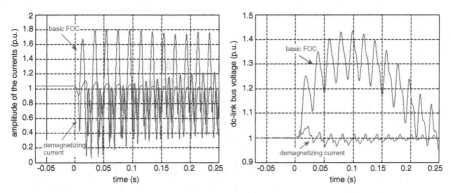

Figure 7.45   (Left) Rotor current during a 40% phase-to-phase dip. (Right) DC-link voltage.

sequence that causes oscillations in the torque, overcurrents in the stator and overvoltages in the rotor.

As will be demonstrated in the following, most of these problems can be overcome by introducing a precise amount of negative sequence in the current references. In this way, the current references are the addition of two sequences— one synchronized with the positive sequence of the grid voltage, and the other synchronized with the negative one.

In order to guarantee that the two sequences are well regulated, it is necessary to independently control each sequence. The original control loop is then substituted by two control loops, one working in a positive rotating reference frame and the other working in an inversely rotating frame [19]. This is why the technique is called dual control.

Figure 7.46 shows a typical schema of a dual control. The boxes in light gray are the two current controls: the top one regulates the positive sequence, while the bottom one regulates the negative sequence. As shown in the figure, the measured current must be

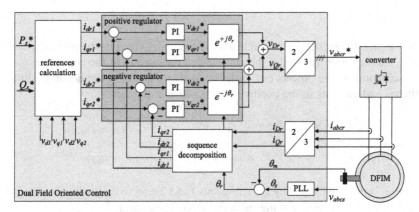

Figure 7.46   Schematic diagram of dual control.

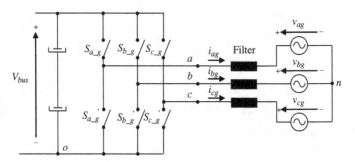

Figure 7.47 Grid side converter.

split into its two sequences before introducing it to the current regulators. The schema also includes the calculation of the positive and negative references from the desired powers and the grid voltages.

As an introduction, the dual techniques are first presented when applied to a grid connected converter. Later, the application of these techniques to a DFIM will be discussed.

7.6.2.1 *Grid Connected Converter*   Dual control for the DFIM stems from the dual control schemes developed for grid connected converters [20]. Although the techniques used in the DFIM are slightly different, the original techniques are presented first as an introduction because they can be used in the grid part of the back-to-back converter. Figure 7.47 shows the converter for which the dual control is applied.

The power transferred between the converter and the grid is given by

$$P_g = \tfrac{3}{2}\mathrm{Re}\left\{\vec{v}_g \cdot \vec{i}_g^*\right\}$$

$$Q_g = \tfrac{3}{2}\mathrm{Im}\left\{\vec{v}_g \cdot \vec{i}_g^*\right\} \tag{7.48}$$

where P_g and Q_g are the active and reactive powers delivered by the converter, and v_g and i_g are the space vectors of the voltage and the current of the converter.

Oscillations in the powers can easily be analyzed by using the decomposition of voltages and currents in the positive and negative sequences:

$$\vec{v}_g = V_1 e^{j\omega_g t} + V_2 e^{-j\omega_g t} = \left(v_{d1}+jv_{q1}\right)e^{j\omega_g t} + \left(v_{d2}+jv_{q2}\right)e^{-j\omega_g t}$$

$$\vec{i}_g = I_1 e^{j\omega_g t} + I_2 e^{-j\omega_g t} = \left(i_{d1}+ji_{q1}\right)e^{j\omega_g t} + \left(i_{d2}+ji_{q2}\right)e^{-j\omega_g t} \tag{7.49}$$

where ω_g is the grid frequency, the subscripts 1 and 2 refer to the positive and negative sequences, respectively, and the subscript g has been omitted.

Introducing the above decomposition in the expression of the grid powers leads to

$$P = P_0 + P_{\cos} \cos(2\omega_g t) + P_{\sin} \sin(2\omega_g t)$$
$$Q = Q_0 + Q_{\cos} \cos(2\omega_g t) + Q_{\sin} \sin(2\omega_g t) \tag{7.50}$$

where the coefficients P_0, $P_{\cos}$, $P_{\sin}$, Q_0, $Q_{\cos}$, $Q_{\sin}$ are given by the following matrix:

$$
\begin{bmatrix} P_0 \\ Q_0 \\ P_{\cos} \\ P_{\sin} \\ Q_{\cos} \\ Q_{\sin} \end{bmatrix} = \frac{3}{2}
\begin{bmatrix}
v_{d1} & v_{q1} & v_{d2} & v_{q2} \\
v_{q1} & -v_{d1} & v_{q2} & -v_{d2} \\
v_{d2} & v_{q2} & v_{d1} & v_{q1} \\
v_{q2} & -v_{d2} & -v_{q1} & v_{d1} \\
v_{q2} & -v_{q2} & v_{q1} & -v_{d1} \\
-v_{d2} & -v_{q2} & v_{d1} & v_{q1}
\end{bmatrix}
\cdot
\begin{bmatrix} i_{d1} \\ i_{q1} \\ i_{d2} \\ i_{q2} \end{bmatrix}
\tag{7.51}
$$

If the grid voltage is balanced, there will be no negative sequence; that is, v_{d2}, v_{q2}, i_{d2}, and i_{q2} are zero. Therefore, the coefficients $P_{\cos}$, $P_{\sin}$, $Q_{\cos}$, $Q_{\sin}$ are null and no oscillations appear in the powers.

With unbalanced input voltages, however, these coefficients are not equal to zero and the powers vary with time. Since the active power is delivered to the grid from the DC link, it determines the DC voltage level. Hence, if P_g varies then the DC-link voltage fluctuates, and a ripple appears at twice the grid frequency.

The coefficients $Q_{\cos}$, $Q_{\sin}$ express the variation of reactive power and have no effect on the voltage of DC link. This is why they are usually removed from the matrix:

$$
\begin{bmatrix} P_0 \\ Q_0 \\ P_{\cos} \\ P_{\sin} \end{bmatrix} = \frac{3}{2}
\begin{bmatrix}
v_{d1} & v_{q1} & v_{d2} & v_{q2} \\
v_{q1} & -v_{d1} & v_{q2} & -v_{d2} \\
v_{d2} & v_{q2} & v_{d1} & v_{q1} \\
v_{q2} & -v_{d2} & -v_{q1} & v_{d1}
\end{bmatrix}
\cdot
\begin{bmatrix} i_{d1} \\ i_{q1} \\ i_{d2} \\ i_{q2} \end{bmatrix}
\tag{7.52}
$$

To keep a constant DC level, the coefficients $P_{\cos}$, $P_{\sin}$ must be nullified. By inverting the matrix (7.52) the currents necessary to achieve that goal can be calculated:

$$
\begin{bmatrix} i_{d1} \\ i_{q1} \\ i_{d2} \\ i_{q2} \end{bmatrix} = \frac{P_0}{D} \cdot \frac{2}{3}
\begin{bmatrix} v_{d1} \\ v_{q1} \\ -v_{d2} \\ -v_{q2} \end{bmatrix}
+ \frac{Q_0}{R} \cdot \frac{2}{3}
\begin{bmatrix} v_{q1} \\ -v_{d1} \\ v_{q2} \\ -v_{d2} \end{bmatrix}
\tag{7.53}
$$

where

$$D = \left(v_{d1}^2 + v_{q1}^2\right) - \left(v_{d2}^2 + v_{q2}^2\right)$$
$$R = \left(v_{d1}^2 + v_{q1}^2\right) + \left(v_{d2}^2 + v_{q2}^2\right)$$
$$\tag{7.54}$$

If the grid converter regulates the current with the references calculated by Equation (7.53), the power delivered to the grid will be constant and equal to the value of P_0. The average value of the reactive power can also be controlled by adjusting the value of Q_0, although an alternating power will exist (as it is impossible to nullify all the alternating coefficients $P_{\cos}$, $P_{\sin}$, $Q_{\cos}$, $Q_{\sin}$).

It is interesting to note the presence of the term D in the denominator of Equation (7.53). This term becomes smaller as the imbalance gets greater, which means more current is necessary to transfer the same power. In the extreme of a total phase-to-phase voltage dip, the negative sequence equals the positive one. In this case D becomes null and therefore the machine cannot deliver any power. This entails a limitation for the dual control that does not exist in the basic vector control.

7.6.2.2 Setting of the References

In a grid connected converter the current references have been calculated with the aim of avoiding oscillations in the DC link while obtaining the desired average active and reactive powers. In the DFIM different strategies may be adopted in order to minimize:

- Stator power oscillations
- rms Stator current
- Torque ripple
- rms Rotor current
- Rotor voltage

Target 1: Constant Stator Power   One of the most common strategies is to obtain constant stator power. As in the case of a grid connected converter, it is impossible to avoid the ripple in the active power and in the reactive power at the same attempt, and therefore only the average value of the reactive power is controlled.

The easier way to calculate the reference of the rotor currents is to proceed in two steps: first, the stator currents that avoid the oscillations are calculated; second, the rotor currents are obtained from the stator currents.

The first step is identical to the previously analyzed case of the grid connected converter since the relations between the currents, the voltages, and the powers are the same. Therefore, the stator current that guarantees a constant active power is obtained by an expression very similar to Equation (7.53):

$$
\begin{bmatrix} i_{ds1} \\ i_{qs1} \\ i_{ds2} \\ i_{qs2} \end{bmatrix} = \frac{P_{s0}}{D} \cdot \frac{2}{3} \begin{bmatrix} v_{d1} \\ v_{q1} \\ -v_{d2} \\ -v_{q2} \end{bmatrix} + \frac{Q_{s0}}{R} \cdot \frac{2}{3} \begin{bmatrix} v_{q1} \\ -v_{d1} \\ v_{q2} \\ -v_{d2} \end{bmatrix} \tag{7.55}
$$

with

$$
\begin{aligned}
D &= \left(v_{d1}^2 + v_{q1}^2 \right) - \left(v_{d2}^2 + v_{q2}^2 \right) \\
R &= \left(v_{d1}^2 + v_{q1}^2 \right) + \left(v_{d2}^2 + v_{q2}^2 \right)
\end{aligned} \tag{7.56}
$$

where i_{xsx} and v_{xx} are the different components of the stator current and the grid voltage, respectively.

Once the stator currents have been determined, the second step involves the calculation of the references of the rotor currents. The relationship between both currents is given by

$$\vec{i}_r = \frac{\vec{\psi}_s}{L_m} - \frac{L_s}{L_m}\vec{i}_s$$

In basic vector control, discussed in Section 7.2, the grid voltage was assumed to be orientated along the q axis and the stator flux was aligned with the d axis since it was 90° behind. In unbalanced conditions this is no longer true, but we find something similar when reasoning with the symmetrical components. At steady state, and neglecting the small voltage drop in the stator resistance, the positive flux is 90° behind the positive sequence of the grid voltage and the negative flux is 90° ahead of the negative voltage:

$$\vec{\psi}_s = \psi_1 e^{j\omega_s t} + \psi_2 e^{-j\omega_s t} = \frac{V_1}{j\omega_s} e^{-j\omega_s t} + \frac{V_2}{-j\omega_s} e^{-j\omega_s t} \qquad (7.57)$$

Combining the two previous expressions in a matrix form yields

$$\begin{bmatrix} i_{dr1} \\ i_{qr1} \\ i_{dr2} \\ i_{qr2} \end{bmatrix} = \frac{1}{\omega_s L_m} \begin{bmatrix} v_{q1} \\ -v_{d1} \\ -v_{q2} \\ v_{d2} \end{bmatrix} - \frac{L_s}{L_m} \begin{bmatrix} i_{ds1} \\ i_{qs1} \\ i_{ds2} \\ i_{qs2} \end{bmatrix} \qquad (7.58)$$

Finally, from Equations (7.55) and (7.58) we finally obtain the reference of the rotor current:

$$\begin{bmatrix} i_{dr1} \\ i_{qr1} \\ i_{dr2} \\ i_{qr2} \end{bmatrix} = -\frac{2}{3}\frac{L_s}{L_m}\frac{P_{s0}}{D} \begin{bmatrix} v_{d1} \\ v_{q1} \\ -v_{d2} \\ -v_{q2} \end{bmatrix} - \frac{2}{3}\frac{L_s}{L_m}\frac{Q_{s0}}{R} \begin{bmatrix} v_{q1} \\ -v_{d1} \\ v_{q2} \\ -v_{d2} \end{bmatrix} + \frac{1}{\omega_s L_m} \begin{bmatrix} v_{q1} \\ -v_{d1} \\ -v_{q2} \\ v_{d2} \end{bmatrix} \qquad (7.59)$$

The first term of the expression is the rotor current that involves a transfer of a constant active power. The second term is related to the reactive current. The third term is the magnetizing current, that is, the current that we must inject into the rotor even if we don't want to generate any power.

Target 2: Balanced Stator Currents   The control system can also be addressed to avoid the negative sequence of the stator current and hence balance the current. This strategy is chosen sometimes with the objective to minimize the rms stator currents.

Without a negative sequence in the stator currents, it is inevitable that the active and reactive powers delivered to the grid would oscillate, but the system can still control their average values, P_{s0} and Q_{s0}.

As in previous cases, the active and reactive powers may be separated into their oscillating components:

$$P_s = P_{s0} + P_{s\,\cos}\cos(2\omega_g t) + P_{s\,\sin}\sin(2\omega_g t)$$
$$Q_s = Q_{s0} + Q_{s\,\cos}\cos(2\omega_g t) + Q_{s\,\sin}\sin(2\omega_g t)$$
(7.60)

The average powers P_0, Q_0, are given by

$$
\begin{bmatrix} P_{s0} \\ Q_{s0} \end{bmatrix} = \frac{3}{2}
\begin{bmatrix} v_{d1} & v_{q1} & v_{d2} & v_{q2} \\ v_{q1} & -v_{d1} & v_{q2} & -v_{d2} \end{bmatrix} \cdot
\begin{bmatrix} i_{ds1} \\ i_{qs1} \\ i_{ds2} \\ i_{qs2} \end{bmatrix}
$$
(7.61)

To obtain balanced stator currents, i_{ds2} and i_{qs2} must be null. The stator current must then be

$$
\begin{bmatrix} i_{ds1} \\ i_{qs1} \\ i_{ds2} \\ i_{qs2} \end{bmatrix} = \frac{2}{3} \frac{1}{v_{d1}^2 + v_{q1}^2}
\begin{bmatrix} v_{d1} & v_{q1} \\ v_{q1} & -v_{d1} \\ 0 & 0 \\ 0 & 0 \end{bmatrix} \cdot
\begin{bmatrix} P_{s0} \\ Q_{s0} \end{bmatrix}
$$
(7.62)

From this expression, and using the relationship between the stator and the rotor current, Equation (7.57), developed in the previous point, we obtain the references of the rotor current:

$$
\begin{bmatrix} i_{dr1} \\ i_{qr1} \\ i_{dr2} \\ i_{qr2} \end{bmatrix} = -\frac{2}{3} \frac{L_s}{L_m} \frac{1}{v_{d1}^2 + v_{q1}^2}
\begin{bmatrix} v_{d1} & v_{q1} \\ v_{q1} & -v_{d1} \\ 0 & 0 \\ 0 & 0 \end{bmatrix} \cdot
\begin{bmatrix} P_{s0} \\ Q_{s0} \end{bmatrix} + \frac{1}{\omega_s L_m}
\begin{bmatrix} v_{q1} \\ -v_{d1} \\ -v_{q2} \\ v_{d2} \end{bmatrix}
$$
(7.63)

Target 3: Constant Electromechanical Torque   Unbalanced voltages in the stator cause a ripple in the electromechanical torque. Obtaining constant electromechanical torque can be advantageous with regard to mechanical stresses since it avoids vibrations in the shaft.

Since the rotational speed can be considered unvarying, constant torque means constant power. The mechanical power delivered by the machine is

$$P_{mec} = \frac{3}{2}\frac{L_m}{L_s}\omega\mathrm{Im}\left\{\vec{\psi}_s \cdot \vec{i}_r^*\right\}$$
(7.64)

As for the stator power, the mechanical power also has a constant power and an oscillating term at double the grid frequency:

$$P_{mec} = P_{m0} + P_{m\,cos}\cos(2\omega_s t) + P_{m\,sin}\sin(2\omega_s t) \tag{7.65}$$

The objective then becomes to nullify the coefficients $P_{m\,cos}$ and $P_{m\,sin}$.

From Equation (7.64) and using the expression for the stator flux, Equation (7.57), we obtain

$$\begin{bmatrix} P_{m0} \\ P_{m\,cos} \\ P_{m\,sin} \end{bmatrix} = \frac{3}{2}\frac{L_m}{L_s}\frac{\omega}{\omega_s} \begin{bmatrix} -v_{d1} & -v_{q1} & v_{d2} & v_{q2} \\ v_{d2} & v_{q2} & -v_{d1} & -v_{q1} \\ v_{q2} & -v_{d2} & v_{q1} & -v_{d1} \end{bmatrix} \cdot \begin{bmatrix} i_{dr1} \\ i_{qr1} \\ i_{dr2} \\ i_{qr2} \end{bmatrix} \tag{7.66}$$

The rotor currents give 4 degrees of freedom, so to solve the above equation it is necessary to add one more condition. The most common one is to impose the reactive power expressed by Equation (7.61). From these equations the currents can be isolated, resulting in

$$\begin{bmatrix} i_{dr1} \\ i_{qr1} \\ i_{dr2} \\ i_{qr2} \end{bmatrix} = \frac{P_{m0}}{D}\frac{2}{3}\frac{L_s}{L_m}\frac{\omega_s}{\omega} \begin{bmatrix} v_{d1} \\ v_{q1} \\ v_{d2} \\ v_{q2} \end{bmatrix} + \left(\frac{Q_0}{D}\frac{2}{3}\frac{L_s}{L_m} + \frac{1}{L_m\omega_s}\right) \begin{bmatrix} v_{q1} \\ -v_{d1} \\ -v_{q2} \\ v_{d2} \end{bmatrix} \tag{7.67}$$

Target 4: Balanced Rotor Currents   It can also be interesting to obtain balanced rotor currents, imposing a negative sequence null for the rotor references. In fact, this strategy is very similar to the basic vector control, which operates only with a positive sequence. The only difference is that now the negative sequence is explicitly driven to zero by means of the negative current controller.

The references for this strategy are calculated as

$$\begin{bmatrix} i_{dr1} \\ i_{qr1} \\ i_{dr2} \\ i_{qr2} \end{bmatrix} = -\frac{2}{3}\frac{L_s}{L_m}\frac{1}{v_{d1}^2 + v_{q1}^2} \begin{bmatrix} v_{d1} & v_{q1} \\ v_{q1} & -v_{d1} \\ 0 & 0 \\ 0 & 0 \end{bmatrix} \cdot \begin{bmatrix} P_{s0} \\ Q_{s0} \end{bmatrix} + \frac{1}{\omega_s L_m} \begin{bmatrix} v_{q1} \\ -v_{d1} \\ 0 \\ 0 \end{bmatrix} \tag{7.68}$$

As can be observed, the expression is very similar to target 2 (balanced stator currents) but in this case the magnetizing current has no negative sequence.

Target 5: Minimum Rotor Voltage   With very unbalanced grid voltages, the main problem of the DFIM is not the oscillation of the stator power, the rms currents, or the torque ripple, but the rotor voltage. As analyzed in Section 7.5, the rotor voltage in unbalanced conditions can become very high. If this voltage exceeds its maximum, the rotor converter saturates and the regulators lose control of the currents. None of the previous strategies solve this problem. In order to minimize the rotor voltage the rotor current must be in the opposite sense of the negative sequence of the stator flux. By doing so, the total negative flux in the rotor is reduced or even canceled, as discussed in Section 7.6.1.

The following expression gives the reference currents calculated to cancel the negative rotor flux. The first term of the equation gives the currents involved in the transfer of the stator powers. The second term gives the necessary current to magnetize the machine. Finally, the third term is the demagnetizing current that has been calculated, taking into account that the negative flux is shifted 90° clockwise of the negative grid voltage (v_{d2}, v_{q2}).

$$
\begin{bmatrix} i_{dr1} \\ i_{qr1} \\ i_{dr2} \\ i_{qr2} \end{bmatrix} = -\frac{2L_s}{3L_m} \frac{1}{v_{d1}^2 + v_{q1}^2} \begin{bmatrix} v_{d1} & v_{q1} \\ v_{q1} & -v_{d1} \\ 0 & 0 \\ 0 & 0 \end{bmatrix} \cdot \begin{bmatrix} P_{s0} \\ Q_{s0} \end{bmatrix} + \frac{1}{\omega_s L_m} \begin{bmatrix} v_{q1} \\ -v_{d1} \\ 0 \\ 0 \end{bmatrix} + \frac{1}{\omega_s \sigma L_r} \frac{L_m}{L_s} \begin{bmatrix} 0 \\ 0 \\ v_{q2} \\ -v_{d2} \end{bmatrix}
$$

$$(7.69)$$

7.6.2.3 Comparison
The responses of the different alternatives of dual control are shown in Figures 7.48 to 7.50. The machine was rotating 20% above the synchronous speed (slip $= -20\%$) and generating the rated power when, at $t = 0.5$ s, a single-phase dip appears causing an unbalanced factor of 4%. The first result has been obtained using the basic vector control presented in Section 7.2. The others five pictures correspond to the five targets discussed.

As expected, the first strategy achieves elimination of the active power oscillation and slightly reduces the torque ripple. The results of the third strategy are somewhat opposite; the torque ripple is eliminated but only a small fraction of the power oscillations are damped. The second and fourth strategies (stator and rotor balanced

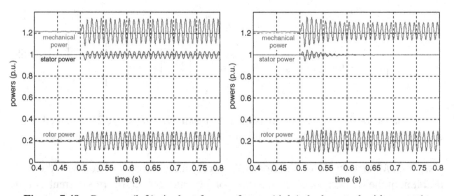

Figure 7.48 Powers: (left) single reference frame; (right) dual control with target 1.

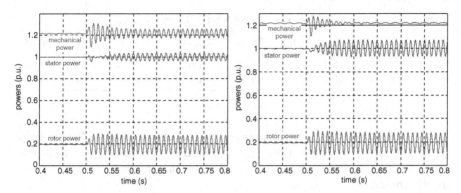

Figure 7.49 Powers using dual control: (left) target 2, (right) target 3.

currents) are very similar and seem to be an average solution since they damp at the same time the oscillations of both active power and torque.

Finally, the fifth strategy reduces the rotor voltage and, by doing so, avoids almost completely the oscillations of the rotor power. This last solution therefore helps to keep a constant level in the DC link of the back-to-back converter.

Table 7.1 shows the rms stator and rotor currents when the different strategies are used. As can be seen, all the strategies are very similar, and therefore, the minimization of the current does not seem to be a good criterion to choose the strategy.

The same results are obtained for different powers, slips, or unbalanced factors. Only the fifth strategy seems to be more current demanding when the generated power is low or the unbalanced factors increase.

7.6.2.4 *Current Controllers*
The first versions of dual control only used one current regulator (made up of two PIs, one for the d axis and the other for the q axis), typically implemented in a synchronous reference frame. The input of the controller is then the addition of the two sequences, as represented in Figure 7.51.

As the reference frame rotates in the same sense as the positive sequence, this sequence appears as DC and it is well regulated. The negative sequence, however,

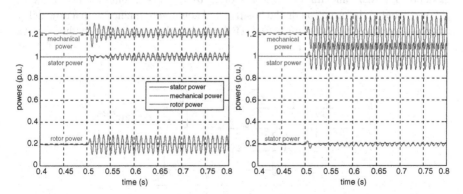

Figure 7.50 Powers using dual control: (left) target 4, (right) target 5.

TABLE 7.1   Stator and Rotor rms Currents for Different Strategies

rms Current	Basic Vector Control	Target 1	Target 2	Target 3	Target 4	Target 5
Stator	1.03	1.04	1.03	1.04	1.03	1.04
Rotor	1.17	1.17	1.16	1.17	1.16	1.17

appears as a double-frequency component (i.e., its components oscillate at twice the grid frequency, 100 Hz or 120 Hz) so it is not easy to regulate unless the control gain is increased. Since large gains usually cause instabilities, this option is not usually chosen.

In order to overcome this problem, typical schemes of dual control use independent current controllers for the positive and negative sequences. The positive sequence is regulated by a controller in a positive synchronous reference frame, which rotates counterclockwise, while the negative sequence is controlled in a negative frame that rotates clockwise. Since each sequence is seen as a DC value by its controller, this approach yields better performance without increasing the control gain.

The main idea of these schemes is that each controller regulates only one sequence. The feedback of the controllers must therefore be the sequence being controlled. That requires extracting the two sequences from the measured currents. In each reference frame one sequence appears as DC while the other appears at 100 or 120 Hz so the sequences can easily be extracted by the use of a lowpass filter or a notch filer tuned at that frequency (Figure 7.52).

As the process of extracting positive and negative sequence components is time consuming and adds amplitude and phase errors to the signals, the system is not properly decoupled under transient conditions. Therefore, system performance and stability are degraded. Furthermore, even when the network is perfectly balanced, the control system still has to split the current and flux and perform positive and negative sequence current controls. This unnecessarily affects the dynamic performance of the overall system.

In order to improve performance under transient conditions, a technique referred to as delayed signal cancellation (DSC) is often used. This technique allows one to achieve accurate information on the positive and negative sequence with a time delay of one-quarter of a period (5 ms at 50 Hz). A good explanation of the method can be found in Svensson et al. [21].

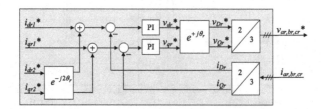

Figure 7.51   Dual control scheme using a single reference frame.

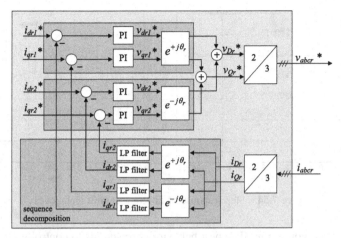

Figure 7.52 Dual control scheme using a sequence separator.

Others authors have proposed techniques that avoid some or all of the filters. Xu [22], for example, proposes using a main controller and an auxiliary controller. The main controller is designed in the same way as for conventional control without involving any sequence separation. Since it is implemented in a positive synchronous reference frame, it achieves good tracking for the positive sequence but not for the negative sequence. Regulation of this sequence is assured by the auxiliary controller that is implemented in a negative frame. Since this controller is specific for the negative sequence, it requires the extraction of that sequence and, hence, the use of filters (Figure 7.53).

A comparison of the different current controllers discussed is depicted in Figure 7.54. On the left half, the step response of the q-axis current is presented

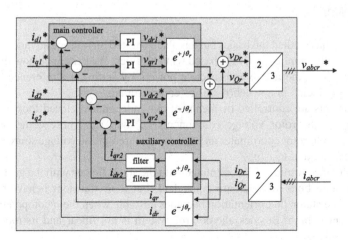

Figure 7.53 Dual control scheme with hierarchical controllers.

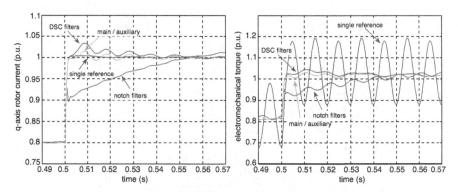

Figure 7.54   Step response of different control techniques.

under balanced conditions. The best behavior corresponds to control with only one single reference. The other alternatives behave very similar except for the notch filter that achieves a notably degraded response.

The comparison varies under unbalanced conditions where a negative sequence has to be controlled. The left picture represents the electromagnetic torque obtained when using target 3, that is, trying to avoid the torque ripple. At $t = 0.5$ s a step is made in the torque set-point in order to check the transient response of the different schemas. As can be observed, control with only one single reference is unable to track the negative sequence and thus it does not avoid the torque ripple. The other schemas control the negative sequence much better, reducing the ripple. The schema with notch filters again results in the worst transient behavior.

Other possible current controllers include the use of PI with resonant compensators tuned at twice the grid frequency to avoid the tracking error on the negative sequence [23].

7.7 SUMMARY

This chapter has studied the basics but has also ventured on to some advanced aspects of vector control applied to DFIM based wind turbines. We examined how the vector control regulates the active and reactive powers by setting the reference rotor dq current components in a synchronous reference frame.

The currents are controlled by means of two independent closed-loop regulators that calculate the rotor voltage components, v_{dr} and v_{qr}. These components are transformed into rotor coordinates and passed to a three-phase voltage source inverter by a PWM modulator.

The reference frame can be orientated with the stator flux or with the grid voltage. In steady state both methods are equivalent, but their transient behavior is very different. The stator-flux orientation becomes unstable when the d component of the rotor current is high. Besides, the control system is not linear and its modeling is complicated.

On the other hand, grid flux orientation stability is not influenced by the rotor current. With that orientation the control system is linear and easy to model.

In addition, tuning of the current regulators affects the system stability and may cancel the damping of the natural flux. If the perturbation rejection of the regulators is high, the flux decays with the stator time constant. In any other case, the flux damping is modified: slightly accelerated if the proportional action is predominant and reduced if the integral action is predominant. In the latter case, the damping can be canceled out so the system would become unstable. However, that problem is avoidable by introducing into the rotor a demagnetizing current.

Under unbalanced conditions, a negative sequence is added to the stator flux. As a consequence, a large amount of negative current flows through the rotor and stator. These currents are caused by the emf induced by the negative flux and their amplitudes depend on the current regulator rejection to a perturbation at twice the grid frequency. In addition, a second harmonic pulsation appears in the torque and stator powers.

Under a voltage dip, a transient natural flux appears in the stator. Moreover, if the dip is asymmetrical, an additional negative flux is added. These two fluxes can induce high emfs in the rotor. The vector control behavior under voltage dips largely depends on whether or not these overvoltages cause saturation of the rotor converter.

If the dip is deep enough to cause saturation, the converter loses control of the rotor current. In the case of very deep dips, the current rises to very high values that can damage the converter if no countermeasures are taken. Even for smaller dips, the converter absorbs a large amount of power that cannot be evacuated to the grid. The energy is then accumulated in the DC-link capacitors and can cause dangerous overvoltages. A commonly used solution is to provide the DC link with a braking chopper.

However, even if the converter does not saturate, the voltage dip affects the behavior of the machine. The natural flux causes transient oscillations at the grid frequency in the torque and stator powers. In the case of asymmetrical dips, the negative flux causes additional permanent oscillations at twice the grid frequency.

The emf induced by the natural and negative fluxes can be reduced by introducing into the rotor a demagnetizing current, that is, a current in the opposite sense to these two fluxes. This current avoids saturation of the rotor converter and minimizes the power transferred to the DC link. Furthermore, it accelerates the transient response so the machine reaches the steady state faster.

Several techniques of dual control have been presented that provide enhanced behavior during a network imbalance. Using these techniques, it is possible to reduce the torque ripple or the stator power oscillations. Another possibility is to provide balanced stator currents.

Once the vector control has been analyzed in depth in this chapter, the reader is ready to explore an alternative control solution detailed in the next chapter—the direct control technique.

REFERENCES

1. Longya Xu and Wei Cheng, "Torque and Reactive Power Control of a Doubly Fed Induction Machine by Position Sensorless Scheme," *IEEE Trans. Ind. Appl.* Vol. 31, No. 3, pp. 636–642, May/June 1995.

2. B. Hopfensperger, D. Atkinson, and R. A. Lakin, "Stator-Flux-Oriented Control of a Doubly-Fed Induction Machine with and Without Position Encoder," *IEE Proc. Electric Power Appl.*, Vol. 146, No. 6, pp. 597–605, November 1999.

3. R. Datta and V. T. Ranganathan, "Decoupled Control of Active and Reactive Power for a Grid-Connected Doubly-Fed Wound Rotor Induction Machine Without Position Sensors," in *Proceedings IEEE Industry Applications Conference*, Vol. 4, Phoenix, USA, October 1999, pp. 2623–2628.

4. S. Wang and Y. Ding, "Stability Analysis of Field Oriented Doubly-Fed Induction Machine Drive Based on Computer Simulation," *Electric Machines and Power Systems*, Vol. 21, pp. 11–24, January/February 1993.

5. W. Leonhard, *Control of Electrical Drives*. Springer-Verlag, 1996.

6. R. Pena, J. C. Clare, and G. M. Asher, "Doubly Fed Induction Generator Using Back-to-Back PWM Converters and Its Application to Variable-Speed Wind Energy Generation," *IEE Proc. Electric Power Appl.* Vol. 143, pp. 231–241, May 1996.

7. V. Kaura and V. Blasko, "Operation of a Phase Locked Loop System Under Distorted Utility Conditions," *IEEE Trans. Ind. Appl.*, Vol. 33, No. 1, pp. 58–63, January 1997.

8. R. Datta and V. T. Ranganathan, "Decoupled Control of Active and Reactive Power for a Grid-Connected Doubly-Fed Wound Rotor Induction Machine Without Position Sensors," in *Proceedings IEEE Industry Applications Conference*, Vol. 4, Phoenix, AZ, October 1999, pp. 2623–2628.

9. A. Perdana, O. Carlson, and J. Persson, "Dynamic Response of Grid-Connected Wind Turbine with Doubly Fed Induction Generator During Disturbances," in *Nordic Workshop on Power and Industrial Electronics*, Trondheim, Norway, 2004.

10. J. B. Ekanayake, L. Holdsworth, X. G. Wu, and N. Jenkins, "Dynamic Modeling of Doubly Fed Induction Generator Wind Turbines," *IEEE Trans. Power Systems*, Vol. 18, No. 2, pp. 803–809, May 2003.

11. M. Heller and W. Schumacher, "Stability Analysis of Doubly-Fed Induction Machines in Stator Flux Reference Frame," in *Proceedings 7th European Conference Power Electronics and Applications*, Vol. 2, Tronheim, Norway, 8–10 September 1997, pp. 707–710.

12. L. Congwei, W. Haiqing, S. Xudong, and L. Fahai, "Research of Stability of Doubly Fed Induction Motor Vector Control System," in *Proceedings 5th International Conference on Electrical Machines and Systems*, Vol. 2, Shenyang, China, 18–20 August 2001, pp. 1203–1206.

13. A. Petersson, L. Harnefors, and T. Thiringer, "Comparision Between Stator-Flux and Grid-Flux-Oriented Rotor Current Control of Doubly-Fed Induction Generators," in *Proceedings 35th Power Electronics Specialist Conference*, Vol. 1, Aachen, Germany, 20–25 June 2004, pp. 482–486.

14. A. Petersson, T. Thiringer, L. Harnefors, and T. Petru, "Modeling and Experimental Verification of Grid Interactions of a DFIG Wind Turbine," *Trans. Energy Conversion*, Vol. 20, No. 6, pp. 878–886, December 2005.

15. P. L. Jansen and R. D. Lorenz, "A Physically Insightful Approach to the Design and Accuracy Assessment of Flux Observers for Field Oriented Induction Machine Drives," *IEEE Trans. Ind. Appl.*, Vol. 30, pp. 101–110, January/February 1994.

16. M. Elbuluk, N. Langovsky, and M. D. Kankam, "Design and Implementation of a Closed-Loop Observer and Adaptive Controller for Induction Motor Drives," *IEEE Trans. Ind. Appl.*, Vol. 34, pp. 435–443, May/June 1998.

17. D. Xiang, L. Ran, P. Tavner, and S. Yang, "Control of a Doubly Fed Induction Generator in a Wind Turbine During Grid Fault Ride-Through," *IEEE Trans Energy Conversion*, Vol. 21, No. 3, pp. 652–662, September 2006.

18. J. Lopez, P. Sanchis, E. Gubia, A. Ursua, L. Marroyo, and X. Roboam, "Control of Doubly Fed Induction Generator Under Symmetrical Voltage Dips," Record of ISIE08, Cambridge, UK, July 2008.

19. L. Xu and Y. Wang, "Dynamic Modeling and Control of DFIG-Based Wind Turbines Under Unbalanced Network Conditions," *IEEE Trans. Power Systems*, Vol. 22, No. 1, pp. 314–323, February 2007.

20. H. Song and K. Nam, "Dual Current Control Scheme for PWM Converter Under Unbalanced Input Voltage Conditions," *IEEE Trans. Ind. Electron.*, Vol. 46, No. 5, pp. 953–959, October 1999.

21. J. Svensson, M. Bongiorno, and A. Sannino, "Practical Implementation of Delayed Signal Cancellation Method for Phase-Sequence Separation," *IEEE Trans. Power Delivery*, Vol. 22, No. 1, pp. 18–26, January 2007.

22. Lie Xu, "Enhanced Control and Operation of DFIG-Based Wind Farms During Network Disturbances," *IEEE Trans. Energy Conversion*, Vol. 23, No. 4, pp. 1073–1081, December 2008.

23. Y. Suh and T. A. Lipo, "Control Scheme in Hybrid Synchronous Stationary Frame for PWM AC/DC Converter Under Generalized Unbalanced Operating Conditions", *IEEE Trans. Ind. Appl.*, Vol. 42, No. 3, pp. 825–835, May/June 2006.

CHAPTER 8

Direct Control of the Doubly Fed Induction Machine

8.1 INTRODUCTION

In Chapter 7, probably the most extended control strategy for DFIM based wind turbines was analyzed, that is, the vector control (or field oriented control) technique. Apart from that, as these wind turbines are required to operate under a disturbed grid context, different adaptations of the classic and standard vector control for this faulty scenario were examined.

In this chapter, an equivalent exposition and analysis procedure is given along with other widely extended control techniques not only for DFIM based wind turbines, but also for drives in general: direct control techniques.

Direct control techniques are an alternative control solution for AC drives, in general, that present control principles and performance features, different from vector control techniques, as shown in this chapter. Many authors have worked in this field, developing many different versions of direct control techniques, suitable for application in DFIM based wind turbines. It is outside the scope of this chapter to describe all these variants of direct control techniques; only some of the most representative ones are examined. In fact, only the first developed versions of direct control techniques are studied in depth, accompanied by a newer variant known as *predictive direct control techniques.*

Once different versions of direct control techniques for DFIM based wind turbines are analyzed, as was done in Chapter 7, they are adapted to handle the difficulties demanded by a faulty grid scenario. Therefore, this chapter is structured as follows.

First, the basic principles of direct control techniques are presented, by analyzing in detail two of the original versions: *direct torque control* (*DTC*) and *direct power control* (*DPC*). Both controls share a common basic structure and philosophy, but they are oriented to directly control different magnitudes of the machine leading to slight differences. DTC seeks to control the torque and rotor flux

Doubly Fed Induction Machine: Modeling and Control for Wind Energy Generation,
First Edition. By G. Abad, J. López, M. A. Rodríguez, L. Marroyo, and G. Iwanski.

amplitude of the machine, while DPC controls the stator active and reactive powers. Both solutions guarantee control of the machine, achieving reasonably good control performances.

Second, an improved version of direct control techniques is studied, designed to avoid one of the most important drawbacks of direct control techniques as initially enunciated: the nonconstant switching frequency behavior. Thus, predictive direct control techniques, based on the same principles as direct control techniques, achieve operation at constant switching frequency by means of a slight increase in the complexity of the control. As done in the previous part, both versions of predictive direct torque control (P-DTC) and predictive direct power control (P-DPC) are examined. All of the direct control versions studied until this stage are applied, considering the use of a two-level voltage source converter (2L-VSC). One specific characteristic of these control techniques requires that the topology of the converter matches control design; that is, if a different converter topology is employed, the corresponding modifications to the control must be addressed.

Third, extension of the predictive direct control techniques to multilevel converter topologies is carried out. Again, P-DTC and P-DPC techniques are developed for the most commonly used multilevel converter topology: the three-level neutral point clamped (3L-NPC) converter.

Finally, adaptation of these direct control techniques is extended to a faulty environment of the grid voltage. The disturbances considered are voltage imbalance of the grid voltage and voltage dips. If special attention is not paid to these disturbances, that is, control of the wind turbine is unaltered, this situation would lead to undesired behavior (and often not permitted by the grid codes) of the wind turbine during the fault. Thus, the modifications required to handle these disturbances are studied, for DTC and DPC techniques.

In this way, the reader is provided with advanced control material applied to DFIM based wind turbines, enhancing knowledge of advanced control techniques and of wind turbine technology in particular.

8.2 DIRECT TORQUE CONTROL (DTC) OF THE DOUBLY FED INDUCTION MACHINE

The simplest version of DTC technique is studied in this section. This DTC version has also been reported in the literature as "classic DTC." This control technique has the following general characteristics:

- Fast dynamic response
- On-line implementation simplicity
- Reduced tuning and adjustment efforts on the part of controllers
- Robustness against model uncertainties
- Reliability

- Good perturbation rejection
- Nonconstant switching frequency behavior

In broad terms, these features are accepted as reasonably good; however, its nonconstant switching frequency behavior is its main drawback. This particular characteristic consists of a nonconstant switching behavior of the controlled switches of the converter used to supply the machine, leading in general to nonuniform semiconductor losses that require special attention. In addition, this nonuniform behavior also affects variables of the machine, such as the torque, currents, and fluxes, leading in an equivalent way to nonuniform behavior of the ripples of these variables. This nonuniform performance depends on the operation conditions of the machine; therefore, depending on, for instance, the operation speed, torque, or rotor flux amplitude, the working switching frequency of the converter would be different and consequently the same would occur for the ripple behavior of variables such as the torque, currents, and fluxes. However, this particular characteristic of direct control techniques, in general, can be partially mitigated or even eliminated, as will be shown later in subsequent sections.

On the other hand, the DTC is based on a direct control of two magnitudes of the DFIM, that is, the electromagnetic torque and the rotor flux amplitude of the machine. The background theory of DTC studied in this section embraces the basic principles of control strategies that will be further developed in later sections.

Finally, Figure 8.1 summarizes graphically a general classification of control techniques for AC machines. There is a wide range of control possibilities; however, within vector based control, field oriented control and direct control are the most representative. In this chapter, both versions of direct control are studied: direct torque control (DTC) and direct power control (DPC) under a circle flux trajectory basis. Added to this, the predictive direct control is also covered, satisfying a strong requirement—the constant switching frequency behavior.

8.2.1 Basic Control Principle

The DTC technique is based on a space vector representation of the achievable output AC voltages of the converter used for supplying, in this case, the two-level VSC. Figure 8.2 shows this space vector representation. The following can be remarked about DTC in an initial stage:

- There are two variables of the DFIM that are directly controlled: the torque and the rotor flux amplitude.
- The rotor flux and stator flux space vectors rotate clockwise (subsynchronism) or anticlockwise (hypersynchronism) to a distance noted by the angle δ.
- By modifying the distance between the space vectors, it is possible to control the torque.
- In order to influence the rotor flux trajectory and amplitude, different voltage vectors are injected into the rotor of the machine. These voltage vectors are

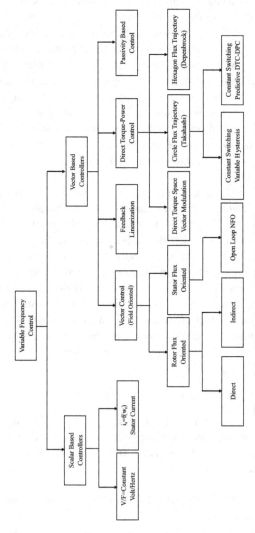

Figure 8.1 General classification of AC machine's controls.

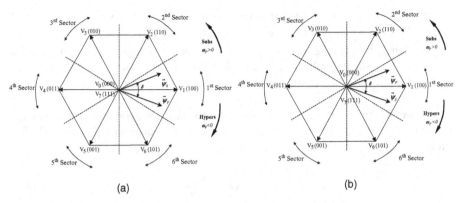

Figure 8.2   Flux space vectors in the rotor reference frame in (a) motor mode and (b) generator mode.

  provided by the voltage source converter that supplies the rotor, in this case, a
  two-level voltage source converter (Figure 8.3).

- Based on the space vector representation and analysis of all these magnitudes,
  the basic principles of the DTC can be studied.

- The DTC technique is very closely associated with the voltage source converter
  that is used. It creates directly the pulses for the controlled semiconductors of the
  converter, without using any modulation schema.

From Chapters 3 and 4, we know that the torque can be expressed according to the
following equation:

$$T_{em} = \frac{3}{2} p \frac{L_m}{\sigma L_r L_s} |\vec{\psi}_r| \cdot |\vec{\psi}_s| \sin\delta \tag{8.1}$$

  Notice that the torque depends on the stator and rotor flux amplitudes and the angle
δ between the space vectors. From this basic relation, we can deduce the following:

1. The direct connection of the stator to the grid creates a stator flux, as seen in
   Chapter 7. Hence, the stator flux space vector has constant amplitude and
   rotating speed, depending on the stator voltage applied.

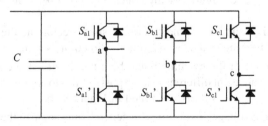

Figure 8.3   Two-level voltage source converter based on IGBTs.

2. By controlling the rotor flux amplitude and the angle δ, it is possible to control the torque magnitude to the desired value.

3. Hence, by creating a rotor flux space vector that rotates at the same angular speed as the stator flux to a distance δ, it is possible to control the machine, since the torque is also controlled.

Therefore, in order to create the desired rotor flux, the appropriate rotor voltage must be applied to the machine. We use the rotor voltage equation expressed in the DQ reference frame:

$$\vec{v}_r^r = R_r \vec{i}_r^r + \frac{d\vec{\psi}_r^r}{dt} \tag{8.2}$$

Mainly to avoid confusing nomenclature, this chapter assumes that the machine being studied has factor $u = 1$. Otherwise, if a general case is being studied ($u \neq 1$), the voltage created by the converter (denoted in space vector form) and the rotor voltage space vector would be related also by this factor u.

In addition, for simplicity in exposition, it is possible to neglect the voltage drop in the rotor resistance:

$$\vec{v}_r^r \cong \frac{d\vec{\psi}_r^r}{dt} \tag{8.3}$$

This last expression clearly shows that by applying the necessary voltage to the rotor, it is possible to directly influence the rotor flux. Assuming a constant time injection h of the rotor voltage, the approximated relation between the initial and final rotor flux amplitude yields

$$|\vec{\psi}_r^r|_{fin} \cong |\vec{\psi}_r^r|_{ini} + \int_0^h \vec{v}_r^r \, dt \tag{8.4}$$

Considering constant rotor voltage injection during h, we find

$$|\vec{\psi}_r^r|_{fin} \cong |\vec{\psi}_r^r|_{ini} + \vec{v}_r^r \, h \tag{8.5}$$

We can see how, by injecting different rotor voltage vectors, the rotor flux amplitude is controlled. Figure 8.4 illustrates the DTC operation principle:

- Initially, it is assumed that the stator flux space vector and the rotor flux space vectors are rotating in a subsynchronous direction as shown in Figure 8.4a.
- When the rotor of the machine is fed by a two-level voltage source converter (VSC), there are eight different voltage possibilities to apply in the rotor (V_0, V_1, V_2, V_3, V_4, V_5, V_6, V_7), as illustrated in Figure 8.4b.
- From all eight voltage possibilities, in general, only some of them are permitted, depending on where the rotor flux space vector is located. Figure 8.4c shows

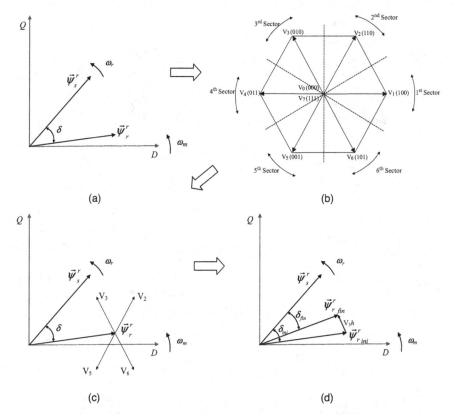

Figure 8.4 Space vector representation of stator and rotor fluxes, in DQ rotating reference frame: (a) initial stator and rotor flux locations, (b) hexagon voltages of the two-level converter, (c) four possible voltage vectors to modify the rotor flux location, and (d) rotor flux variation with V_3 injection.

four possible voltage vector injections (V_2, V_3, V_5, V_6) when the rotor flux space vector is located in sector 1.

- Figure 8.4d graphically illustrates the rotor flux space vector variation, after injection of voltage vector 3. Assuming that the stator flux rotates very slow, so that during the h time interval the stator flux has not moved, the consequence of the voltage injections is:

  o The rotor flux amplitude has been reduced.

  o The angle δ has also been reduced.

  o Both magnitude reductions yield a reduction in the torque, as deduced from expression (8.1).

Repeating a large sequence of different voltage injections (with different application time h_1, h_2, etc.) and now considering that the stator flux is rotating, it is possible to show how the rotor flux follows a circular trajectory behind the stator flux.

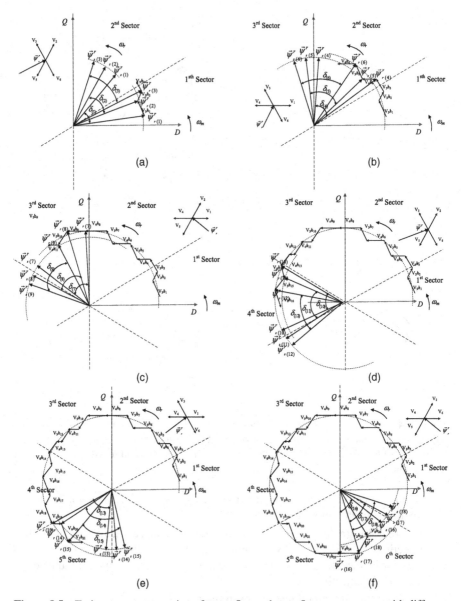

Figure 8.5 Trajectory representation of stator flux and rotor flux space vectors with different voltage vector injections having different durations (only active vectors): (a) 1st sector, (b) 2nd sector, (c) 3rd sector, (d) 4th sector, (e) 5th sector, and (f) 6th sector.

Figures 8.5 and 8.6 show that, by injecting different voltage vectors, the rotor flux describes a nearly circular trajectory. Note that when the smaller injection time h_i is taken, the appearance of the described trajectory by the rotor flux is more circular.

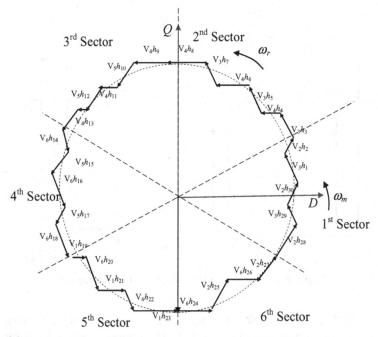

Figure 8.6 Trajectory representation of rotor flux space vector with different voltage vector injections having different durations (only active vectors) during one cycle.

8.2.2 Control Block Diagram

The control block diagram of the direct torque control (DTC) strategy is illustrated in Figure 8.7. The directly controlled variables are the electromagnetic torque and the rotor flux amplitude. From the torque and flux references, the control strategy calculates the pulses (S_a, S_b, S_c) for the controlled semiconductors of the two-level voltage converter.

By following the flux trajectory philosophy introduced in the previous subsection, this control technique selects the appropriate voltage vectors to keep the machine controlled.

The control strategy is divided into five different tasks, schematically represented in five different blocks:

1. Estimation block
2. Torque ON–OFF controllers
3. Flux ON–OFF controller
4. Voltage vector selection
5. Pulse generation block

Each block is described in subsequent sections in detail.

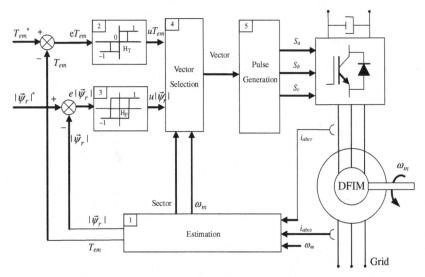

Figure 8.7 Direct torque control (DTC) block diagram.

8.2.2.1 Estimation Block

As mentioned before, the directly controlled variables are the torque and the rotor flux. Unfortunately, since these variables cannot be measured in an easy way, in general, they are estimated from measured magnitudes of the machine. In this case, the torque and the rotor flux amplitude can be estimated as shown in Figure 8.8. First, from the rotor and stator current measurements together with the rotor angular position θ_m, the stator and rotor flux $\alpha\beta$ components are estimated. Second, the torque and rotor flux amplitudes are immediately calculated. As will be studied in Chapter 10, this is just a simple example of how these variables can be estimated; many alternative estimation methods are possible.

In addition, once the rotor flux $\alpha\beta$ components are estimated, it is also necessary to calculate the position where the rotor flux vector lies, that is, the rotor flux angle. This angle provides the rotor flux location; with this angle, the sector is derived according to Table 8.1.

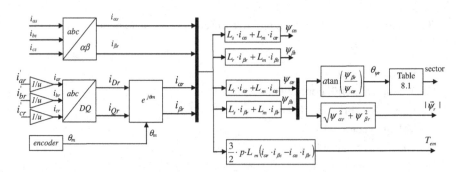

Figure 8.8 Estimation block diagram.

TABLE 8.1 Sector Versus Angle Look-up Table

Rotor Flux Angle	Sector
(270°, 30°)	1
(30°, 90°)	2
(90°, 150°)	3
(150°, 210°)	4
(210°, 270°)	5
(270°, 30°)	6

8.2.2.2 Voltage Vector Selection and ON–OFF Hysteresis Controllers

The DTC technique selects the required rotor voltage vector directly from the rotor flux and electromagnetic torque errors, using hysteresis ON–OFF controllers (blocks 2 and 3 from Figure 8.7). It chooses the needed voltage vector to correct the errors in the controlled variables.

The flux controller is based on a two-level hysteresis comparator with H_F hysteresis band, while the torque controller uses a three-level hysteresis comparator with H_T hysteresis band. The schematic representation of these two controllers is shown in Figure 8.9.

Depending on the value of the torque error, the output uT_{em} can take a value of -1, 0, or 1. $uT_{em} = 0$ means zero rotor voltage vector requirement, as will be shown later. Similarly, depending on the value of the flux error, the output $u|\vec{\psi}_r|$ can take only a value of -1 or 1 (not 0). Note that these tasks can be very simply implemented in a digital based control board.

As mentioned in the introduction to this chapter, the reader can find different alternative hysteresis philosophies or variants in the specialized literature. However, this section seeks to concentrate exclusively on a representative one.

Once the uT_{em} and $u|\vec{\psi}_r|$ signals are defined by means of the ON–OFF controllers, together with the information about the rotor flux space vector position (i.e., the sector), it is possible to select the rotor voltage vector. It is chosen from Table 8.2.

The voltage vectors are selected from the ON–OFF controllers' outputs and the sector where the rotor flux space vector is located. As an example, if the rotor flux

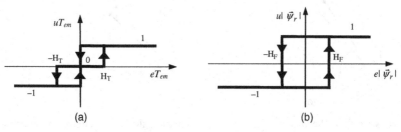

Figure 8.9 (a) ON–OFF electromagnetic torque controller with hysteresis band. (b) ON–OFF rotor flux controller with hysteresis band.

TABLE 8.2 Vector Selection as a Function of the Torque and Flux ON–OFF Controller Outputs (k = Sector)

		uT_{em}		
		1	0	−1
$u\lvert\vec{\psi}_r\rvert$	1	$V_{(k-1)}$	V_0, V_7	$V_{(k+1)}$
	−1	$V_{(k-2)}$	V_0, V_7	$V_{(k+2)}$

space vector is located in sector 5, $uT_{em}=1$ and $u\lvert\vec{\psi}_r\rvert = -1$, the selected voltage vector is V_3. For each sector, only four active vectors are permitted ($V_{(k-2)}$, $V_{(k-1)}$, $V_{(k+1)}$, $V_{(k+2)}$,) and the zero vector (V_0, V_7).

The reader can deduce how this table is created, by the analysis of Section 8.2.1. Note that the zero vector produces nearly zero rotor flux variation (Equation (8.5)) and also very small positive or negative torque variation, depending on the subsynchronous or hypersynchronous operation of the machine. This fact is illustrated graphically in Figure 8.10. Due to the fact that the zero vector maintains the rotor flux unaltered, the stator flux movement produces an angle δ increase (subsynchronism) or decrease (hypersynchronism), provoking a torque increase or decrease, respectively (Equation (8.1)).

When the output of the ON–OFF controller is set to 1, it means a positive variation requirement. On the contrary, when the output is set to −1, a negative variation is required. According to these requirements, the adequate active voltage vector is injected. Alternatively, when the output of the torque ON–OFF controller is set to zero, and in order to minimize the torque and the flux ripples, the zero vector is injected.

No matter how large the flux error is, if the torque controller output is equal to zero, the voltage vector injected will be a zero vector. It must be highlighted that the zero vector is not really needed to keep the torque and flux controlled; however, it is used to

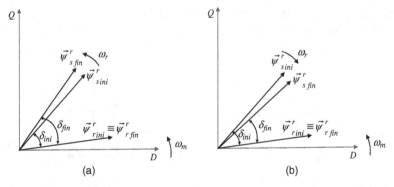

Figure 8.10 (a) The δ angle increase (T_{em} increase), due to a zero vector injection at subsynchronism. (b) The δ angle decrease (T_{em} decrease), due to a zero vector injection at hypersynchronism.

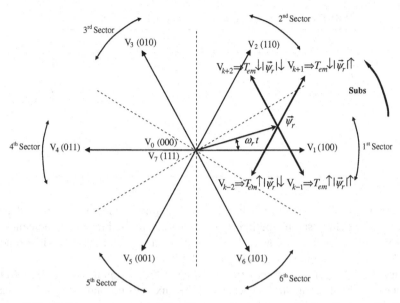

Figure 8.11 Voltage vectors and their effect on the torque and rotor flux amplitudes at subsynchronous speed in the motor mode.

reduce the torque and flux ripples at steady state operation. For almost every application of DTC, it is advantageous if the torque and flux ripples are minimized as much as possible. We can say that zero voltage vector injection is commonly used in many DTC schemas.

Deeper analysis showing the exact torque and flux variation, depending on the voltage vector injected and the operating point of the machine, is carried out in subsequent sections.

Finally, Table 8.2 (not considering the zero vectors effect) can be expressed graphically as illustrated in Figure 8.11.

8.2.2.3 Pulse Generation
Once the vector that will correct the torque and flux errors is selected, the next task is to create the pulses for the controlled semiconductors of the two-level converter. For the analyzed DTC technique, this task is simple. We use Table 8.3 to determine the voltage vector injected. More sophisticated DTC schemas require a more complex pulse generation, as studied in subsequent sections.

8.2.2.4 Torque and Flux Waveforms
Here, the typical steady state behavior of the directly controlled variables is shown under DTC. Figure 8.12 illustrates the torque and rotor flux amplitude waveforms when the machine operates at hypersynchronous speed.

If a reduction of the torque and flux ripples is required to improve accuracy in the controlled variables, the hysteresis band widths of both controllers (H_T and H_F) should

TABLE 8.3 Pulse Generation Look-up Table

Vector	S_a	S_b	S_c
V_0	0	0	0
V_1	1	0	0
V_2	1	1	0
V_3	0	1	0
V_4	0	1	1
V_5	0	0	1
V_6	1	0	1
V_7	1	1	1

be set to small values. Unfortunately, there exists a limit for those values, related to the minimum switching sample period (h) of the hardware used for the implementation and minimum conduction time of semiconductors. Hence, an overshoot in torque or flux beyond the hysteresis bands is unavoidable.

Figure 8.12 displays a short period of time at the beginning of the third sector. Torque overshoot occurs three times; when the flux goes beyond the hysteresis band limit, the inevitable torque and flux ripples are produced.

In any case, if reduced torque and flux ripples are required, in general, a small hysteresis band should be used together with a small sample time period (h). Nevertheless, the choice of the hysteresis bands is not so obvious even for a given sample time (h), since they will produce different torque and flux ripples depending on the operation conditions of the machine.

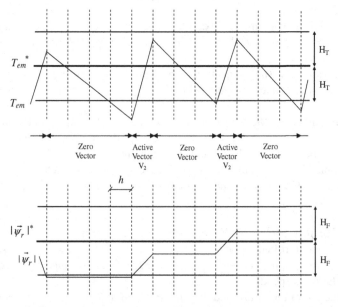

Figure 8.12 Steady state torque and flux waveforms at hypersynchronism.

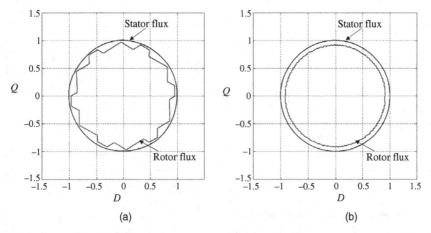

Figure 8.13 Stator and rotor fluxes DQ plot: (a) $H_T = 17\%$, $H_F = 5.5\%$ and (b) $H_T = 17\%$, $H_F = 0.55\%$.

Figure 8.13 shows the DQ plot of the stator and rotor fluxes of the machine. It can be noticed that since the stator is supplied directly from a sinusoidal voltage from the grid, the stator flux describes a circular trajectory. However, due to the hysteresis band selected for the rotor flux in DTC, there is a ripple superimposed on its circular trajectory.

Under these circumstances, the slopes of the electromagnetic torque and the rotor flux variations, due to an active vector or zero vector selection, take on different values depending on the operating conditions; the converter that supplies the rotor operates at variable switching frequency.

8.2.3 Example 8.1: Direct Torque Control of a 2 MW DFIM

This example shows a 2 MW DFIM, controlled by the DTC technique described in this section. The hysteresis band of the torque ON–OFF controller is set to 5% of the rated torque, while the flux hysteresis is set to 1.5% of the rated flux. The machine operates at constant subsynchronous speed $\omega_{mpu} = 0.75$. The DC bus voltage of the bidirectional converter is controlled to 1000 V by the grid side converter.

At the beginning of the experiment the machine operates as a generator at nominal torque. In the middle of the experiment, the torque is reversed to positive, so the machine begins to operate as a motor. Hence, this experiment shows the steady state and transient performances with DTC. Figure 8.14 displays the most characteristic variables of the machine.

Figures 8.14a and 8.14b illustrate the torque and flux behaviors, when the DFIM is directly connected to the grid through the stator. In the middle of the experiment, the torque is reversed from nominal negative (generator) to nominal positive (motor). A very quick and safe transient response is seen, since the torque reaches the new

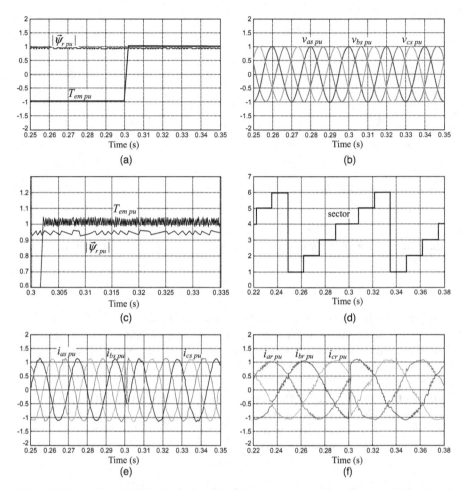

Figure 8.14 A 2 MW DFIM controlled by DTC at subsynchronous speed: steady state and transient response. (a) Torque and rotor flux amplitude, (b) *abc* stator voltages, (c) zoom of the torque and rotor flux amplitude, (d) sector, (e) *abc* stator currents, and (f) *abc* rotor currents.

reference value without further overshoot than that defined by the hysteresis band of the ON–OFF controller, as shown in Figure 8.14c. In addition, the flux is also not affected by the transient. On the other hand, the sector where the rotor flux space vector is located is presented in Figure 8.14d.

Finally, the stator and rotor currents behaviors are shown in Figure 8.14e and 8.14f, demonstrating that despite a very severe change of torque, the currents do not overshoot. Note that at steady state, the ripple present in the stator and rotor currents is an image of the torque and flux ripple.

In a similar way, Figure 8.15 illustrates the rotor voltages and the pulses of the controlled semiconductors for the same experiment of the previous figure.

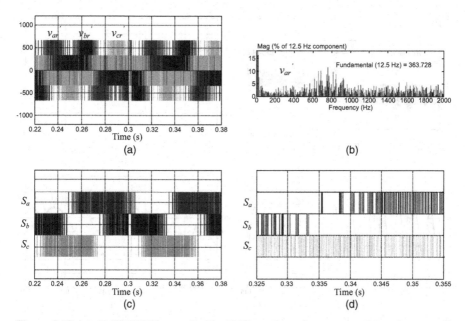

Figure 8.15 A 2 MW DFIM controlled by DTC at subsynchronous speed: steady state and transient response. (a) *abc* Rotor real voltages, (b) spectrum of a phase rotor voltage, (c) pulses of the controlled semiconductors, and (d) zoom of the pulses of the controlled semiconductors.

Figures 8.15a and 8.15b, show the rotor voltages and its spectrum. Notice how there is a group of higher amplitude harmonics around 800 Hz. This shape of spectra is typical for DTC driven systems. Instead of being more localized around one specific frequency, as in sinusoidal pulse width modulation techniques, for instance, the higher amplitude harmonics appear grouped in a wider area. Note also that the real rotor voltages are shown, rather than the stator referred voltages.

Figures 8.15c and 8.15d show the pulses for the controlled semiconductors, after choosing the voltage vectors required to correct the torque and rotor flux errors.

On the other hand, Figure 8.16 shows a variable speed experiment. The speed of the machine is modified from hypersynchronism to subsynchronism. The torque and the flux references are unaltered. In this case the hysteresis of the torque is set to 8% while the flux hysteresis is set to 1%.

Figures 8.16a and 8.16b show how the torque and flux ripple behaviors are different at hypersynchronism and subsynchronism. Figures 8.16c and 8.16d show the stator and rotor current behaviors.

8.2.4 Study of Rotor Voltage Vector Effect in the DFIM

As studied in the previous subsection, the DTC strategy, as it was initially developed, determines rotor voltage vectors in order to control directly the rotor flux and

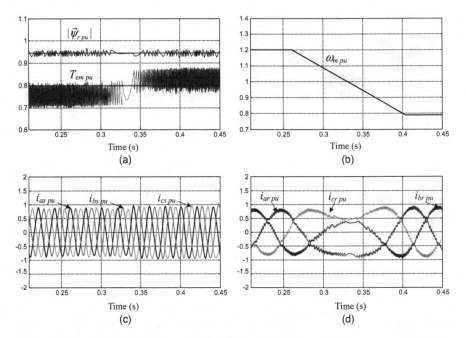

Figure 8.16 A 2 MW DFIM controlled by DTC at variable speed: (a) torque and rotor flux amplitude, (b) speed, (c) *abc* stator currents, and (d) *abc* rotor currents.

the electromagnetic torque of the machine. Depending on the operation conditions of the machine and the position of the rotor and stator flux space vectors expressed in the rotor reference frame (*DQ*), the rotor voltage vectors affect the torque and flux in different ways. Until now, only Table 8.2 showed this dependence. However, it is possible to analyze this important fact in much more detail.

Consequently, in this section, we study the voltage vector effect on the DFIM. For that purpose, *the torque and the flux derivative expressions* are derived next, deduced from basic model equations of the machine presented in Chapter 4, in the stator reference frame (note that quite long manipulation of equations is necessary to produce these expressions):

$$\frac{d|\vec{\psi}_r|}{dt} = \frac{1}{|\vec{\psi}_r|}\left[\left(\frac{R_r L_m}{\sigma L_s L_r}\right)\mathrm{Re}\{\vec{\psi}_r \cdot \vec{\psi}_s^*\} - \left(\frac{R_r}{\sigma L_r}\right)|\vec{\psi}_r|^2 + \mathrm{Re}\{\vec{\psi}_r \cdot \vec{v}_r^*\}\right] \qquad (8.6)$$

$$\frac{dT_{em}}{dt} = \frac{3}{2}p\frac{L_m}{\sigma L_s L_r}\left[\left(\frac{R_s}{\sigma L_s} + \frac{R_r}{\sigma L_r}\right)\mathrm{Im}\{\vec{\psi}_r \cdot \vec{\psi}_s^*\}\right.$$

$$\left. -\omega_m\mathrm{Re}\{\vec{\psi}_r \cdot \vec{\psi}_s^*\} + \mathrm{Im}\{\vec{v}_s \cdot \vec{\psi}_r^*\} + \mathrm{Im}\{\vec{\psi}_s \cdot \vec{v}_r^*\}\right] \qquad (8.7)$$

In order to further develop these expressions, it is necessary to consider the following space vector representations, in the stator reference frame:

$$\vec{\psi}_r = |\vec{\psi}_r| e^{j\omega_s t} \tag{8.8}$$

$$\vec{\psi}_s = |\vec{\psi}_s| e^{j(\omega_s t + \delta)} \tag{8.9}$$

$$\vec{v}_s \cong |\vec{v}_s| e^{j(\omega_s t + \delta + \pi/2)} \tag{8.10}$$

$$\vec{v}_r = \tfrac{2}{3} V_{bus} e^{j[\omega_m t + (\pi/3)(n-1)]} \tag{8.11}$$

where n is the subindex of the rotor voltage vector and takes values from 1 to 6. For zero vectors V_0 and V_7, $\vec{v}_r = 0$. By substituting expressions (8.8)–(8.11) into Equation (8.6), the rotor flux derivative yields

$$\frac{d|\vec{\psi}_r|}{dt} = \left(\frac{R_r L_m}{\sigma L_s L_r}\right)|\vec{\psi}_s|\cos\delta - \left(\frac{R_r}{\sigma L_r}\right)|\vec{\psi}_r| + \tfrac{2}{3} V_{bus} \cos\left(\omega_r t - \frac{\pi}{3}(n-1)\right) \tag{8.12}$$

The rotor flux derivative is composed of two constant terms and one cosine term, with the ω_r pulsation and amplitude dependent only on the DC bus voltage. The zero vector produces a constant rotor flux variation; this means that the cosine term is valid only for the active vectors. So it can clearly be seen that, depending on the operation condition of the machine (i.e. rotor and stator flux, torque, etc.), the flux variation created by a voltage vector is different.

Similarly, the torque derivative expression (8.7) at steady state may be simplified, considering that the stator flux vector module is nearly constant

$$|\vec{\psi}_s| \cong \frac{|\vec{v}_s|}{\omega_s}$$

$$\frac{dT_{em}}{dt} = T_{em}\left(\frac{\omega_r}{\tan\delta} - \left(\frac{R_s}{\sigma L_s} + \frac{R_r}{\sigma L_r}\right)\right) + p\frac{L_m}{\sigma L_s L_r}V_{bus}|\vec{\psi}_s|\sin\left(\omega_r t + \delta - \frac{\pi}{3}(n-1)\right) \tag{8.13}$$

As similarly occurs with the rotor flux, in this case again the torque derivative expression is composed of one constant term and one sine term with ω_r pulsation. Its amplitude is only dependent on the DC bus voltage, since the stator flux module is considered constant. Again, the zero vectors produce constant torque variation. Figure 8.17 graphically shows the torque and flux derivative evolutions of a DFIM (Equations (8.12) and (8.13)), at steady state and for each rotor voltage vector.

In general, the constant terms of the rotor flux derivative expression (8.12) are very small, so the zero vectors do not produce rotor flux variation. On the other hand, the constant term of expression (8.13) is dependent on the electromagnetic torque and the speed of the machine. So depending on the machine's operating conditions, it can be positive or negative as shown in Figure 8.17.

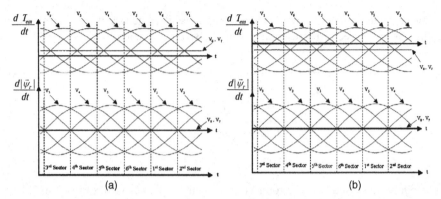

Figure 8.17 Torque and flux slopes as a function of the time in motor or generator mode at hypersynchronism and subsynchronism, for each rotor voltage vector: (a) subsynchronism and (b) hypersynchronism.

For the same reason, the phase shift order of each active vector varies from subsynchronous to hypersynchronous operation, since ω_r varies from positive to negative values.

Finally, note that the larger is the DC bus voltage, the greater are the torque and flux variations that can produce the active voltage vectors. It must be highlighted that the DTC technique does not employ these torque and flux variation expressions; they are only studied for comprehensive purposes. However, as will be shown later, there are several versions of direct control techniques that can use these expressions. In fact, expressions (8.12) and (8.13) can be real time implemented within the control strategy, as illustrated in Figure 8.18. Considering the DC bus voltage, the stator

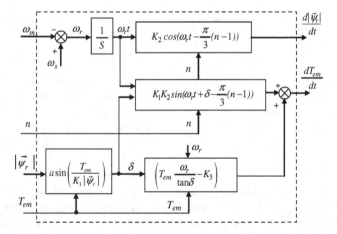

Figure 8.18 Torque and flux derivative expressions calculation block diagram.

voltage, and the stator flux as having constant magnitudes, the torque and rotor flux derivatives will only depend on four magnitudes:

$$K_1 = \tfrac{3}{2}p\frac{L_m}{\sigma L_s L_r}|\vec{\psi}_s| \tag{8.14}$$

$$K_2 = \tfrac{2}{3}V_{bus} \tag{8.15}$$

$$K_3 = \frac{R_s}{\sigma L_s} + \frac{R_r}{\sigma L_r} \tag{8.16}$$

Added to this, from Equations (8.12) and (8.13), it can be considered that for small switching sample periods (h), the torque and flux evolutions can be approximated by straight lines as illustrated, for instance, in Figure 8.12. From the numerical value of the slopes, it can be seen that the zero voltage vector is the vector that will produce the smallest torque variation, since at steady state in each sector k, only k and k + 3 vectors produce smaller slopes, and, in general, they are not allowed to be used in DTC strategies (see Table 8.2).

Further analysis of the torque variation produced by the zero vectors can be deduced by considering the first term of Equation (8.13). At fixed torque and rotor flux operation conditions, the slope varies proportionally to the slip speed. More specifically, as shown in Figure 8.19, at speeds near to the synchronous speed, the slope of the zero vectors becomes smaller. This fact implies that, at speeds near synchronism, the required rotor voltage vector amplitude is very small; the torque ripple will also be small since the most demanded vector will be the zero vector. Finally, note that the transition from positive to negative torque slope, in general, is not exactly at the synchronous speed.

Depending on the speed of operation, for instance, the torque and flux variations produced by the voltage vectors are different. This fact, added to the constant hysteresis bands (H_T and H_F) used by the DTC technique, yields a nonconstant switching frequency behavior for the system. When the DFIM is driven by DTC,

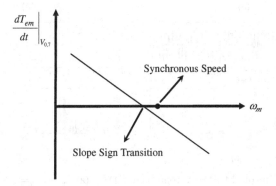

Figure 8.19   Torque variation produced by the zero vector, at constant torque and rotor flux.

the switching frequency of the converters, the stator and rotor currents, and so on have nonconstant behaviors.

This issue is one of the most important drawbacks of the DTC technique presented in this section. In general, torque, controlled semiconductors of the converter, stator currents exchanged with the grid, and so on should operate at a constant switching frequency. For this reason, many authors have tried to improve this behavior, modifying the original basic DTC schema presented in this subsection. In subsequent sections, one possibility for achieving constant switching behavior by DTC is studied.

8.2.5 Example 8.2: Spectrum Analysis in Direct Torque Control of a 2 MW DFIM

This example shows the switching behavior of a 2 MW DFIM driven by DTC. Figures 8.20a and 8.20b shows the torque and stator current spectra at rated torque at $\omega_m = 1.4$ p.u. (deliberately taken high) speed. Alternatively, Figures 8.20c and 8.20d show the same variables under the same control but at $\omega_m = 0.8$ p.u. speed. These spectrum shapes are typical with DTC. In all the cases, the higher amplitude harmonic content is grouped around a frequency. At $\omega_m = 1.4$ p.u. speed, the group of most significant harmonics is around 1500 Hz, while at $\omega_m = 0.8$ p.u. speed, it is around 1250 Hz.

In general, the torque and the rotor flux amplitude do not modify the frequency of the first group of harmonics; however, the hysteresis bands and the speed do.

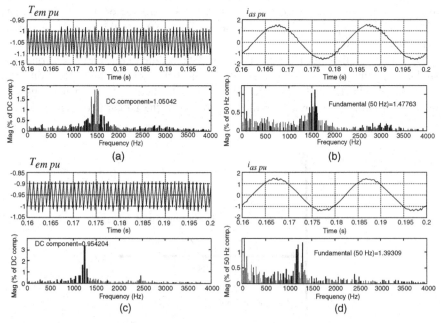

Figure 8.20 Spectra of 2 MW DFIM controlled by DTC: (a) T_{em} at $\omega_m = 1.4$ p.u. speed, (b) i_{as} at $\omega_m = 1.4$ p.u. speed, (c) T_{em} at $\omega_m = 0.8$ p.u. speed, and (d) i_{as} at $\omega_m = 0.8$ p.u. speed.

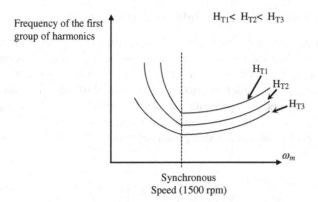

Figure 8.21 First group of harmonics as a function of time at different H_T values.

For instance, Figure 8.21 shows how the frequency of the first group of harmonics is modified in function by the speed and the torque hysteresis band (the flux hysteresis band is left constant for simplicity).

Notice, that if a constant first harmonic is required at a different speed, that is, constant switching frequency behavior, it is possible to implement a variable torque hysteresis band ON–OFF controller as illustrated in Figure 8.22. Having a constant hysteresis band of the flux controller, it is possible to modify the torque hysteresis by means of a look-up table that modifies H_T as a function of ω_m. The look-up table can be derived from Figure 8.21, interpolating values previously calculated by simulation or by off-line tests.

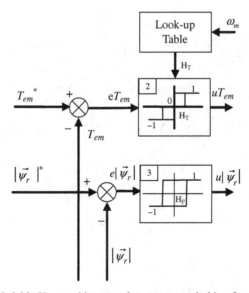

Figure 8.22 Variable H_T to achieve nearly constant switching frequency behavior.

8.2.6 Rotor Flux Amplitude Reference Generation

As studied before in Chapters 1 and 7, the wind turbine in general operates with a reactive power reference, instead of rotor flux amplitude references as does the DTC. This fact obliges us to introduce an extra block, for adapting the reactive power reference into the rotor flux reference. This issue is illustrated in Figure 8.23.

By considering the *dq* representation of the DFIM developed in Chapter 7, it is possible to have to a single expression to calculate the rotor flux amplitude reference, by knowing the torque and reactive power references [30]:

$$\left|\vec{\psi}_r\right|^* = \sqrt{\left(k_1|\vec{\psi}_s| + k_2 \frac{Q_s^*}{|\vec{\psi}_s|}\right)^2 + \left(k_3 \frac{T_{em}^*}{|\vec{\psi}_s|}\right)^2} \tag{8.17}$$

with constants

$$k_1 = \sigma \frac{L_r}{L_m} + \frac{L_m}{L_s} \tag{8.18}$$

$$k_2 = \frac{-\sigma \cdot L_r \cdot L_s}{1.5 \cdot \omega_s \cdot L_m} \tag{8.19}$$

$$k_3 = \frac{-\sigma \cdot L_r \cdot L_s}{1.5 \cdot p \cdot L_m} \tag{8.20}$$

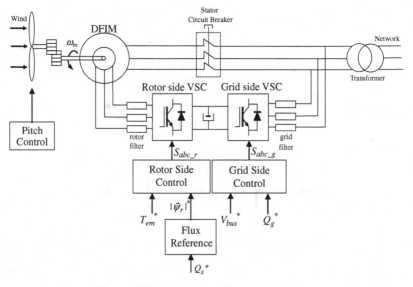

Figure 8.23 Rotor flux reference generation.

The stator flux amplitude can be calculated by assuming zero voltage drop in the stator resistance:

$$|\vec{\psi}_s| \cong \frac{|\vec{v}_s|}{\omega_s} \tag{8.21}$$

8.3 DIRECT POWER CONTROL (DPC) OF THE DOUBLY FED INDUCTION MACHINE

Direct power control is based on the same control principles as the direct torque control technique. The unique difference is the directly controlled variables. In the case of DTC, the electromagnetic torque and the rotor flux are directly controlled, while in DPC, the stator active and reactive powers are directly controlled.

First, a conceptual study of the basic control principle of DPC is carried out. This conceptual analysis is essential to understand the fundaments of the DPC strategy and provides the necessary clearance to understand the parallelism between DTC and DPC principles.

Second, as done in the previous section, the theory related to DPC is developed accompanied by illustrative examples. From the performance characteristics point of view, since DPC is based on the same fundamentals as DTC, in broad terms it is possible to affirm that both controls present very similar behaviors.

8.3.1 Basic Control Principle

As mentioned before, the DPC technique is based on direct control of the stator active and reactive powers of the DFIM. As presented in previous chapters, the stator active and reactive powers of the machine can be calculated directly from the stator voltage and currents as follows:

$$P_s = \tfrac{3}{2}\mathrm{Re}\left\{ \vec{v}_s \cdot \vec{i}_s^{\,*} \right\} \tag{8.22}$$

$$Q_s = \tfrac{3}{2}\mathrm{Im}\left\{ \vec{v}_s \cdot \vec{i}_s^{\,*} \right\} \tag{8.23}$$

As done in DTC, by injecting directly rotor voltage vectors, the stator active and reactive powers are controlled. This objective is achieved by DPC, together with the necessary creation of a rotatory rotor flux space vector. However, from study of Equations (8.22) and (8.23), it is not possible to know how the injection of different rotor voltage vectors can influence the creation of desired P_s and Q_s. Note that the stator voltage is fixed by the grid, while the established stator current depends on how the rotor voltage vectors have been chosen. Hence, in order to clarify this study, by substitution of DFIM model equations in Equations (8.22) and (8.23) we obtain

$$P_s = \frac{3}{2}\frac{L_m}{\sigma L_s L_r}\omega_s |\vec{\psi}_s||\vec{\psi}_r|\sin\delta \tag{8.24}$$

$$Q_s = \frac{3}{2}\frac{\omega_s}{\sigma L_s}|\vec{\psi}_s|\left[\frac{L_m}{L_r}|\vec{\psi}_s|-|\vec{\psi}_r|\cos\delta\right] \tag{8.25}$$

where δ is the phase shift between the stator and the rotor flux space vectors. The voltage drop in the stator resistance has been neglected. These last two expressions show:

1. The dependence between the stator active and reactive powers and the stator and rotor flux space vectors.
2. The stator active and reactive powers can be controlled by modifying the relative angle between the rotor and stator flux space vectors (δ) and their amplitudes.
3. Assuming that the stator voltage is fixed, there are many constant terms that can be grouped to simplify the expressions:

$$P_s = K_1 |\vec{\psi}_r|\sin\delta \tag{8.26}$$

$$Q_s = K_2\left[K_3-|\vec{\psi}_r|\cos\delta\right] \tag{8.27}$$

Consequently, according to the basic control principle of DTC studied in Section 8.2.1, since it is possible to know how the injected rotor voltage vectors influence the rotor flux and its relative distance to the stator flux, it is possible to know their influence on the stator active and reactive powers as well. More specifically, by checking the terms $|\vec{\psi}_r|\sin\delta$ and $|\vec{\psi}_r|\cos\delta$, it is possible to modify P_s and Q_s. Thus, Figure 8.24 illustrates the terms $|\vec{\psi}_r|\sin\delta$ and $|\vec{\psi}_r|\cos\delta$, for given stator and rotor flux space vector locations.

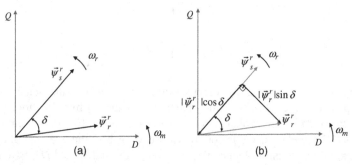

Figure 8.24 (a) Space vector representation of stator and rotor fluxes, in DQ rotating reference frame. (b) The $|\vec{\psi}_r|\sin\delta$ and $|\vec{\psi}_r|\cos\delta$ projections.

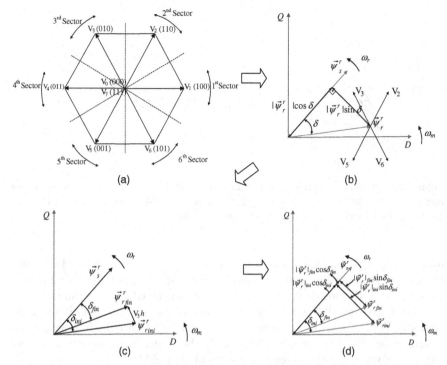

Figure 8.25 Space vector representation of stator and rotor fluxes, in DQ rotating reference frame (connected to Figure 8.24): (a) hexagon voltages of the two-level VSC, (b) four possible voltage vectors to modify the rotor flux location, (c) rotor flux variation with V_3 injection, and (d) $|\vec{\psi}_r|\sin\delta$ term and $|\vec{\psi}_r|\cos\delta$ term variations with V_3 injection.

Then, considering that a two-level VSC feeds the rotor of the DFIM, there are several rotor voltage injection possibilities, as Figures 8.25a and 8.25b illustrate. If voltage vector V_3 is applied during a time interval h, Figure 8.25c shows the rotor flux space vector movement from its initial position to the final position. Note that during the time interval h, it is assumed that the stator flux space vector position is unaltered. Hence, Figure 8.25d graphically illustrates the $|\vec{\psi}_r|\sin\delta$ term and the $|\vec{\psi}_r|\cos\delta$ term variations due to V_3 voltage vector injection. Notice that $|\vec{\psi}_r|\sin\delta$ is decreased while $|\vec{\psi}_r|\cos\delta$ is slightly increased, producing according to Equations (8.26) and (8.27), a P_s decrease and a Q_s decrease as well:

$$V_3 \Rightarrow \begin{cases} |\vec{\psi}_r|\cos\delta \uparrow & \Rightarrow \quad Q_s \downarrow \\ |\vec{\psi}_r|\sin\delta \downarrow & \Rightarrow \quad P_s \downarrow \end{cases}$$

Consequently, following the DTC philosophy, repeating a large sequence of different voltage injections (with different application times h_1, h_2, etc.) and now considering that the stator flux is rotating, we know how the rotor flux follows a circular trajectory behind the stator flux (Section 8.2.1). Figure 8.26 shows that,

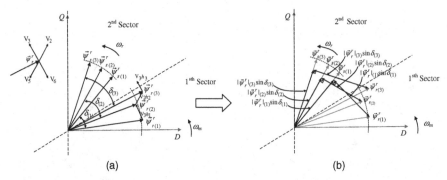

Figure 8.26 Trajectory representation of stator and rotor fluxes with different voltage vector injections having different durations (only active vectors): (a) flux representations in 1st sector and (b) $|\vec{\psi}_r|\sin\delta$ and $|\vec{\psi}_r|\cos\delta$ representations in 1st sector.

by injecting different voltage vectors, the rotor flux describes a nearly circular trajectory and, accordingly, the terms $|\vec{\psi}_r|\sin\delta$ and $|\vec{\psi}_r|\cos\delta$ maintain nearly a constant value. This means that the P_s and Q_s are maintained nearly constant as well. Note in this case as well that the smaller the injection time h_i is, the more circular is the appearance of the described trajectory by the rotor flux. Hence, by following direct control basic principles, this graphical study shows that it is possible to control directly the stator active and reactive powers of the DFIM.

Similarly, from a different perspective, it can be said that although the direct power control technique does not impose direct rotor flux control, indirectly by controlling the stator active and reactive powers, it imposes an amplitude for the rotor flux space vector and its rotation to a distance δ from the stator flux space vector. This is a necessary condition to ensure that the DFIM reaches a stable steady state operating point, because as shown in Chapters 3 and 4, proper creation and rotation of the stator flux space vector is already guaranteed by feeding the stator of the DFIM from the grid.

8.3.2 Control Block Diagram

The control block diagram of the direct torque control (DPC) strategy is illustrated in Figure 8.27. It follows exactly the same philosophy of the control bock diagram of DTC presented in Section 8.2.2. The directly controlled variables are the stator active and reactive powers. From the P_s and Q_s references, the control strategy calculates the pulses (S_a, S_b, S_c) for the controlled semiconductors of the two-level VSC.

This control technique selects the appropriate voltage vectors to keep the machine under control. The control strategy is divided into five different tasks, schematically represented in five different blocks:

1. Estimation block
2. Stator active power ON–OFF controllers
3. Stator reactive power ON–OFF controller

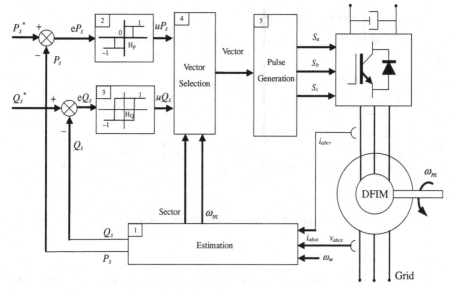

Figure 8.27   Direct power control (DPC) block diagram.

4. Voltage vector selection
5. Pulse generation block

Since each block is very similar to the DTC technique presented previously, only the most important differences are studied.

8.3.2.1 Estimation Block   Due to the fact that that the directly controlled variables P_s and Q_s can be directly calculated from the measured stator voltage and currents, in DPC, the estimation structure is slightly different from DTC. As shown in Figure 8.28, from the rotor and stator current measurements together with the rotor angular position θ_m, the rotor flux $\alpha\beta$ components are estimated for the sector calculation (same look-up Table 8.1). On the other hand, from measurements of stator voltage and currents, the actual values of P_s and Q_s are calculated.

As mentioned in Section 8.2.1, this is just a simple example of how these variables can be estimated; many alternative estimation methods are possible. Note also that compared to the estimator structure of DTC, the stator voltage measurement is required. In addition, the sector location is deduced from the rotor flux space vector position as in DTC.

8.3.2.2 Voltage   Vector   Selection   and   ON–OFF   Hysteresis Controllers   The DPC technique also selects the required rotor voltage vector directly from P_s and Q_s errors, using hysteresis ON–OFF controllers (blocks 2 and 3 from Figure 8.27). It chooses the needed voltage vector to correct the errors in the controlled variables.

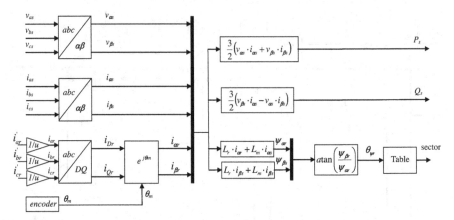

Figure 8.28   Estimation block diagram.

The ON–OFF controller structure is equivalent to DTC. The Q_s controller is based on a two-level hysteresis comparator with H_Q hysteresis band, while the P_s controller uses a three-level hysteresis comparator with H_P hysteresis band. The schematic representation of these two controllers is shown in Figure 8.29.

Once the uP_s and uQ_s signals are defined by means of the ON–OFF controllers, together with the information about the rotor flux space vector position (i.e., the sector), it is possible to select the rotor voltage vector. It is chosen from the look-up table of Table 8.4.

- The voltage vectors are selected from the ON–OFF controllers' outputs and the sector where the rotor flux space vector is located, as in DTC.
- As in DTC, for each sector only four active vectors are permitted ($V_{(k-2)}$, $V_{(k-1)}$, $V_{(k+1)}$, $V_{(k+2)}$,) and the zero vector (V_0, V_7).
- The reader can deduce how this table is created, by the graphical analysis of Section 8.3.1.

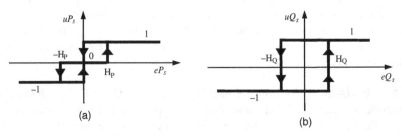

Figure 8.29   (a) ON–OFF P_s controller with hysteresis band. (b) ON–OFF Q_s controller with hysteresis band.

TABLE 8.4 **Vector Selection as a Function of the Torque and Flux ON–OFF Controller Outputs (k = Sector)**

		uP_s		
		1	0	−1
uQ_s	1	$V_{(k-2)}$	V_0, V_7	$V_{(k+2)}$
	−1	$V_{(k-1)}$	V_0, V_7	$V_{(k+1)}$

- Note that the zero vector in DPC produces both P_s and Q_s variation. This fact is illustrated graphically in Figure 8.30.
- Deeper analysis, showing the exact torque and flux variations depending on the voltage vector injected and the operating point of the machine, is carried out in subsequent sections.

Finally, not considering the zero vectors effect, Table 8.4 can be expressed graphically as illustrated in Figure 8.31.

8.3.2.3 Pulse Generation As occurred in DTC, once the vector that corrects the stator power errors is selected, the next task is to create the pulses for the controlled semiconductors of the two-level VSC. In DPC again, this task is simply done by using Table 8.3.

8.3.2.4 Stator Power Waveforms Figure 8.32 illustrates the stator active and reactive power waveforms when the machine operates at hypersynchronous speed at steady state under DPC. Very similar behavior of the directly controlled variables to DTC is observed.

Noticed that, in DPC, when the zero vector is injected, the reactive power varies. This is one of the main differences in the behavior compared to DTC.

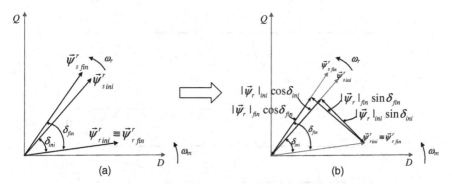

Figure 8.30 (a) The δ angle increase due to a zero vector injection at subsynchronism. (b) Term $|\vec{\psi}_r|\sin\delta$ increases (P_s increase) and term $|\vec{\psi}_r|\cos\delta$ decreases (Q_s increase) due to a zero vector injection at subsynchronism.

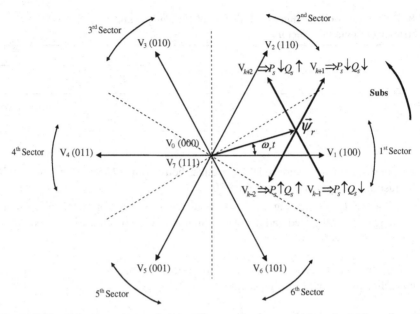

Figure 8.31   Voltage vectors and their effect on the stator active and reactive powers at subsynchronous speed in the motor mode.

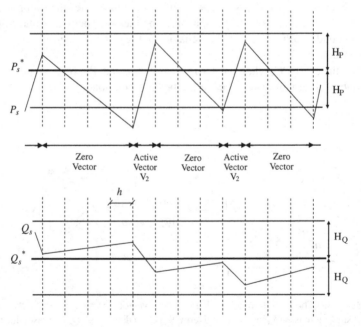

Figure 8.32   Steady state P_s and Q_s waveforms at hypersynchronism in sector 3.

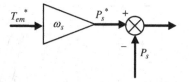

Figure 8.33 Reference adaptation.

8.3.2.5 Active Power Reference Generation Strategy As studied in Chapter 1, if we assume that the higher level wind turbine control generates a torque reference, then the DFIM is controlled in torque rather than in P_s, as DPC is designed. This fact can simply be coordinated by means of the reference adaptation of Figure 8.33.

8.3.3 Example 8.3: Direct Power Control of a 2 MW DFIM

This example shows a 2 MW DFIM, controlled by the DPC technique described in this section. Both hysteresis bands of the ON–OFF controllers are set to 2.5% of the rated power. The machine operates at constant hypersynchronous speed, $\omega_{m\,pu} = 1.25$. The DC bus voltage of the back-to-back converter is controlled to 1000 V by the grid side converter.

At the beginning of the experiment the machine operates as a generator at nominal active power. The reactive power reference is set to zero. In the middle of the experiment, the active power is reversed to positive, so the machine begins to operate as a motor. Hence, this experiment shows the steady state and transient performances with DPC. Figure 8.34 shows the most characteristic variables of the machine.

Figure 8.34a illustrates the P_s and Q_s behavior, when the DFIM is connected directly to the grid through the stator. In the middle of the experiment, P_s is reversed from nominal negative (generator) to nominal positive (motor). A very quick and safe transient response is seen, since P_s reaches the new reference value without further overshoot than that defined by the hysteresis band of the ON–OFF controller, as shown in Figure 8.34c. During the transient, Q_s is also properly controlled.

The sector where the rotor flux space vector is located is presented in Figure 8.34d. Finally, the stator and rotor currents behaviors, shown in Figures 8.34e and 8.34f, demonstrate that despite a very severe change of power demand, the currents do not overshoot.

8.3.4 Study of Rotor Voltage Vector Effect in the DFIM

As for the DTC studied in previous sections, depending on the operating conditions of the machine and the position of the rotor and stator flux space vectors expressed in the rotor reference frame, the rotor voltage vectors affect the stator active and reactive powers in a different manner. Until now, only Table 8.4 showed this dependence.

However, in this subsection, the mathematical expressions that relate the stator active and reactive power variations with the rotor voltage vectors will be derived. For

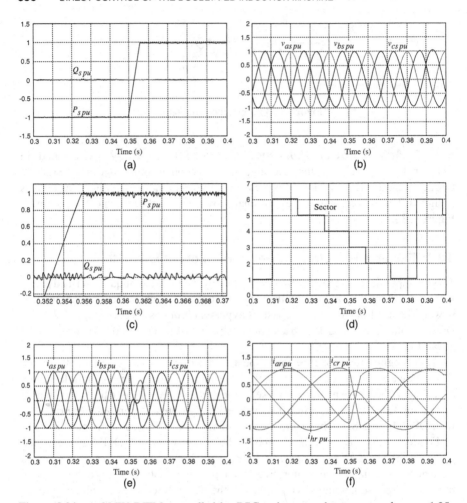

Figure 8.34 A 2 MW DFIM controlled by DPC at hypersynchronous speed, $\omega_m = 1.25$: steady state and transient response. (a) Stator active and reactive power, (b) abc stator voltages, (c) zoom of the stator active and reactive power, (d) sector, (e) abc stator currents, and (f) abc rotor currents.

that purpose, it is necessary to consider the following space vectors represented in the rotor reference frame.

$$\vec{\psi}_r^r = |\vec{\psi}_r|e^{j\omega_r t} \tag{8.28}$$

$$\vec{\psi}_s^r = |\vec{\psi}_s|e^{j(\omega_r t + \delta)} \tag{8.29}$$

$$\vec{v}_s^r \cong |\vec{v}_s|e^{j(\omega_r t + \delta + \pi/2)} \tag{8.30}$$

$$\vec{v}_r^r = \tfrac{2}{3}V_{bus}e^{j(\pi/3)(n-1)} \tag{8.31}$$

In addition, from Equations (8.22) and (8.23), the stator active and reactive power variations can be calculated in coordinates referenced to the rotor reference frame:

$$\frac{dP_s}{dt} = \frac{3}{2}\left(\frac{dv_{Ds}}{dt}i_{Ds} + \frac{di_{Qs}}{dt}v_{Qs} + \frac{dv_{Qs}}{dt}i_{Qs} + v_{Ds}\frac{di_{Ds}}{dt}\right) \tag{8.32}$$

$$\frac{dQ_s}{dt} = \frac{3}{2}\left(\frac{dv_{Qs}}{dt}i_{Ds} + \frac{di_{Ds}}{dt}v_{Qs} - \frac{dv_{Ds}}{dt}i_{Qs} - v_{Ds}\frac{di_{Qs}}{dt}\right) \tag{8.33}$$

By manipulating Equations (8.32) and (8.33), and considering that the amplitude of the stator flux space vector is constant, the stator active and reactive power variations yield

$$\frac{dP_s}{dt} = -P_s\left(\frac{R_s}{\sigma L_s} + \frac{R_r}{\sigma L_r}\right) - \omega_r Q_s + \frac{3}{2}\omega_r \frac{|\vec{v}_s|^2}{\sigma L_s \omega_s}$$
$$+ \frac{L_m}{\sigma L_s L_r}V_{bus}|\vec{v}_s|\sin\left(\omega_r t + \delta - \frac{\pi}{3}(n-1)\right) \tag{8.34}$$

$$\frac{dQ_s}{dt} = -Q_s\left(\frac{R_s}{\sigma L_s} + \frac{R_r}{\sigma L_r}\right) + \omega_r P_s + \frac{3}{2}\frac{R_r}{L_r}\frac{|\vec{v}_s|^2}{\sigma L_s \omega_s}$$
$$- \frac{L_m}{\sigma L_s L_r}V_{bus}|\vec{v}_s|\cos\left(\omega_r t + \delta - \frac{\pi}{3}(n-1)\right) \tag{8.35}$$

The power derivatives are composed of three constant terms and a cosine or a sine term, with ω_r pulsation and amplitude dependent on the DC bus voltage. The zero vector produces constant active and reactive power variation, since the cosine and sine terms of expressions (8.34) and (8.35) are only valid for the active vectors.

On the other hand, by combining expressions (8.24) and (8.25), the angle δ between the rotor and the stator flux space vectors can be expressed as a function of the active and reactive powers:

$$\delta = a\tan\left(\frac{L_r}{L_m}\frac{P_s}{\frac{3}{2}\frac{L_m}{\sigma L_s L_r}\frac{|\vec{v}_s|^2}{\omega_s} - Q_s}\right) \tag{8.36}$$

Figure 8.35 graphically illustrates the stator active and reactive power derivative evolutions of a specific DFIM at steady state and for each rotor voltage vector.

The whole period of the power derivative waveforms can be divided into six different sectors, with the limit between sectors being the zero crossing of the reactive

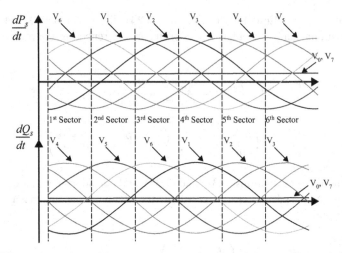

Figure 8.35 Active and reactive power slopes as a function of the rotor flux angle in motor or generator mode and subsynchronism, for each rotor voltage vector.

power derivatives. In that way, the reactive power derivatives will have a uniform sign in every sector. Considering the DC bus voltage and the stator voltage as constant magnitudes, the stator active and reactive power derivatives will depend only on four magnitudes, as shown in the simplified block diagram of Figure 8.36.

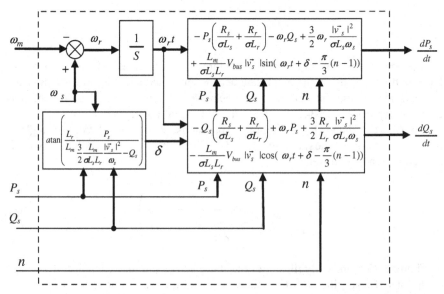

Figure 8.36 Simplified prediction schema: power derivative expression calculation.

8.4 PREDICTIVE DIRECT TORQUE CONTROL (P-DTC) OF THE DOUBLY FED INDUCTION MACHINE AT CONSTANT SWITCHING FREQUENCY

In this section, the predictive direct torque control (P-DTC) strategy of the doubly fed induction machine (DFIM) is presented. It is especially designed to operate at a considerably low constant switching frequency, reducing the electromagnetic torque and rotor flux ripples, in order to provide good steady state and fast dynamic performances. This control is convenient for high power drive and generator applications, with restricted switching frequency. The DFIM is connected to the grid by the stator and the rotor is fed by a two-level voltage source converter. In addition, this control method allows one to implement a technique that reduces the switching power losses of the converter. Finally, two illustrative examples show that the proposed DTC method effectively reduces the torque and flux ripples at low and higher switching frequencies.

Thus, the P-DTC strategy presented in this section is based on direct control of the electromagnetic torque and the rotor flux of the machine. A sequence of three voltage vectors will be introduced at a constant switching period, in order to reduce the ripples of both directly controlled variables, that is, the torque and the rotor flux. For that purpose, a ripple reduction criterion based on a prediction of the torque and flux evolution over time is implemented.

Therefore, the predictive control technique presented maintains the main features of direct control techniques:

- Constant switching frequency behavior
- Fast dynamic response
- On-line implementation simplicity
- Reduced tuning and adjustment efforts on the part of the controllers
- Robustness against model uncertainties
- Reliability

Obviously, the improvement achieved by means of this control is accomplished by slightly increasing the complexity of the control and, consequently, degrading its implementation simplicity.

8.4.1 Basic Control Principle

The predictive DTC (P-DTC) technique is based on the same control principles of classic DTC techniques. In the study of the space vector diagram, it seeks to create a circular trajectory of the rotor flux space vector, as depicted in Figure 8.37.

By injecting the available rotor voltage vectors, the amplitude of the rotor flux space vector and its movement are controlled, in order to locate it to a specific distance from the stator flux space vector. Thus, with this simple control schema, it is possible

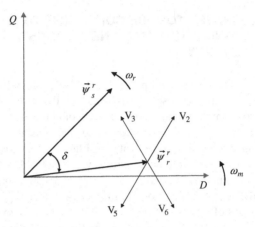

Figure 8.37   Basic DTC principle.

to modify the electromagnetic torque of the machine according to the well-known formula

$$T_{em} = \frac{3}{2} p \frac{L_m}{\sigma L_r L_s} |\vec{\psi}_r| \cdot |\vec{\psi}_s| \sin \delta \qquad (8.37)$$

However, as seen in previous sections, if no special treatment is applied to the rotor voltage injection, classic DTC techniques operate at a nonconstant switching frequency. Hence, the predictive DTC technique presented in this section introduces the required modifications to the basic control DTC structure, in order to achieve constant switching frequency behavior.

The basic principles of the predictive DTC strategy can be summarized as follows:

1. A constant switching period is defined. The duration of this constant period is denoted as h. As in classic DTC, all the actions taken by the control technique are in the time domain (Figure 8.38), but in P-DTC the control procedure is discretized.

2. During this period h, at steady state operation, three different rotor voltage vectors are injected, with the objective of minimizing a cost function related to the torque and rotor flux amplitude errors. Hence, the torque and rotor flux evolutions within the switching period can take the shape shown in Figure 8.39.

h_{c1}, $h_{c2} - h_{c1}$ are the time intervals dedicated to injected vectors 1 and 2, respectively. Note that the third injected vector is not an active vector, but is a zero vector. The injection time for the third vector is $h - h_{c2}$.

Note that according to the chosen three vector sequence together with the time intervals dedicated for each vector, the torque and rotor flux evolutions within the switching period can be different.

3. This procedure is repeated again at constant period h as illustrated in Figure 8.40. In general, the vector sequence as well as the time intervals dedicated for each vector (h_{c1}, h_{c2}, $h - h_{c2}$) are modified for every sample time.

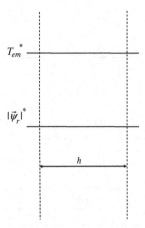

Figure 8.38 Constant switching period.

Therefore, by following this philosophy, the P-DTC is in charge of performing the following tasks every sample period h:

1. Before the beginning of the switching period, from the information given by the torque and rotor flux errors, select the appropriate three voltage vector sequence.
2. Before the beginning of the switching period as well, calculate the corresponding time intervals (h_{c1}, h_{c2}, $h - h_{c2}$) for the selected three voltage vector sequence, with the objective of minimizing a cost function related to the torque and rotor flux errors. In this book, the torque and rotor flux amplitude ripples are

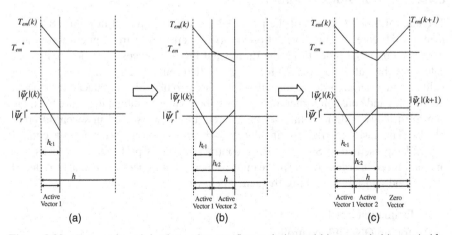

Figure 8.39 Rotor voltage injection and torque-flux evolutions within one switching period h: (a) first vector injection, (b) second vector injection, and (c) third vector injection.

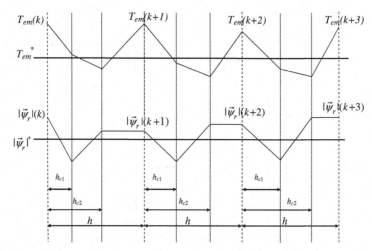

FIGURE 8.40   Torque and rotor flux evolutions after three switching periods according to P-DTC.

minimized; however, the reader can find that there exist several alternative minimization criteria in related specialized literature.

3. During the switching period, apply the vector sequence calculated at the beginning of the period.

Hence, the presented general procedure is suitable to be applied within a DTC context, and can easily be implemented in standard control hardware boards, as shown in subsequent sections.

8.4.2 Control Block Diagram

The control block diagram of the proposed strategy is depicted in Figure 8.41. As mentioned before, the directly controlled variables are the electromagnetic torque and the rotor flux amplitude. From the torque and flux references, the control strategy calculates the pulses (S_a, S_b, S_c) for the controlled semiconductors of the two-level voltage converter, in order to meet the following two objectives: reduced torque and flux ripples and constant switching frequency behavior. As shown in Figure 8.41, the control strategy is divided into seven different tasks represented in seven different blocks. The tasks carried out in blocks 1–4 are the basic direct torque control principles presented in Section 8.2. In contrast, the tasks performed in blocks 5–7 are introduced in order to achieve the improvements mentioned above, that is, reduced ripples and constant switching frequency behavior:

1. Estimation block
2. Torque ON–OFF controller
3. Flux ON–OFF controller

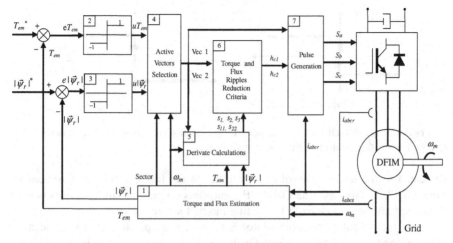

Figure 8.41 Predictive DTC block diagram with reduced torque and flux ripples and constant switching frequency.

4. Voltage vector sequence selection
5. Slope calculations
6. Torque and flux ripple reduction criteria
7. Pulse generation

Each block is described in subsequent sections in detail.

8.4.2.1 Estimation Block The estimation block for this P-DTC strategy is exactly the same as for the classic DTC. Since it is based on the same control principles, the torque, rotor flux amplitude, and sector where the rotor flux space vector is located can be estimated in the same manner as presented in Section 8.2.2.1.

8.4.2.2 First Voltage Vector Selection and ON–OFF Controllers Once the estimated flux and torque values are obtained, they are compared with their corresponding torque and flux references, so the torque and flux errors are obtained. The basic principle of direct control techniques again determines the choice of one specific active voltage vector of the converter, in order to correct the torque and flux errors. The choice of this vector is directly calculated from the torque and flux errors themselves and the sector where the rotor flux lies.

The proposed predictive DTC strategy adopts this principle to select the first active voltage vector of the three-vector sequence employed to minimize the torque and flux ripples. Hence, the first active voltage vector selection is based on a look-up table mapped by the output of two ON–OFF comparators without hysteresis bands. These tasks are accomplished in blocks 2, 3, and 4 of Figure 8.41. One comparator is

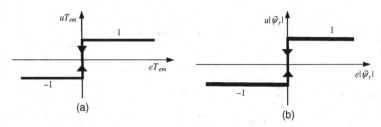

Figure 8.42 (a) ON–OFF electromagnetic torque controller without hysteresis band. (b) ON–OFF rotor flux controller without hysteresis band.

dedicated to the torque control and the other to the flux control, as illustrated in Figure 8.42. In this case, the ON–OFF controllers do not present hysteresis bands, since they are only in charge of selecting the first active vector.

When the output of the comparator is set to 1 (i.e., positive error), it means a positive slope variation requirement. On the contrary, when the output is set to –1 (i.e., negative error), a negative slope variation is required.

Once the outputs of the ON–OFF controllers are calculated, the first active vector of the three vector sequence can be selected. This selection is simply made by a look-up table, Table 8.5.

Note that this table is equivalent to Table 8.2 of classic DTC, but the column corresponding to the zero vector output has been omitted.

8.4.2.3 Second and Third Voltage Vector Selection and Torque and Flux Waveforms
As advanced before, once the first active vector is selected from the DTC table, in order to correct the electromagnetic torque and the rotor flux errors, the predictive DTC control strategy employs a sequence of three different voltage vectors in a constant switching period h, under steady state operating conditions. Hence, a second active vector is injected followed always by a zero vector with the objective of minimizing the torque and the rotor flux ripples, from their reference values. The typical torque and flux waveforms for this control strategy are illustrated in Figure 8.43.

In the adopted notation, the slopes of the torque variation are called s_1, s_2, and s_3 for the active and zero vectors at each switching period. Similarly, s_{11}, s_{22}, and s_{33} are the slopes of the rotor flux variation, for the active and the zero vectors. The slope of the torque s_3 at hypersynchronous speed is negative, while at subsynchronous speed it is positive.

TABLE 8.5 First Active Vector Selection as a Function of the Torque and Flux ON–OFF Controllers' Outputs (k = Sector)

		uT_{em}	
		1	−1
$u|\vec{\psi}_r|$	1	$V_{(k-1)}$	$V_{(k+1)}$
	−1	$V_{(k-2)}$	$V_{(k+2)}$

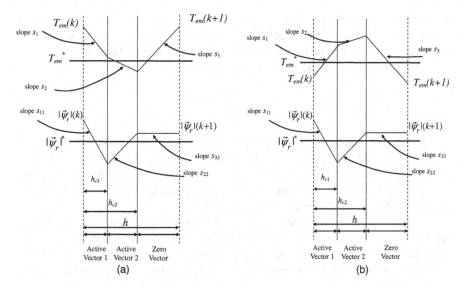

Figure 8.43 Steady state torque and flux waveforms at subsynchronous and hypersynchronous speeds with a three vector based DTC strategy: (a) subsynchronism and (b) hypersynchronism.

As defined before, h_{c1} and $(h_{c2}-h_{c1})$ are the time intervals dedicated for the first and second active vectors. The second active vector produces the same sign variation as the first active vector in the torque, while it produces the opposite sign variation in the rotor flux.

On the other hand, the zero vector produces as opposite sign torque variation to the first two active vectors, while the rotor flux is maintained nearly constant, as can be inferred from Equations (8.12) and (8.13). These slopes are calculated in block 5 of Figure 8.41. The presence of the zero vector in the three-vector sequence is very useful at steady state operating conditions, since it provides "small" torque variations and almost zero flux variations, yielding a reduction in the torque and flux ripples.

Therefore, once the first active vector has been chosen, if the torque and the flux ripples can be reduced, the choice of the second active vector is made according to Table 8.6. It defines the correct choice of the active vectors, depending on the speed of the machine and the sector in which the rotor flux is located.

For instance, when the machine is rotating at subsynchronous speed, if the rotor flux is in the first sector, and the actual torque and flux values are close enough to their

TABLE 8.6 Active Vector Selections at Steady State (k = Sector)

	Active Vectors
Subsynchronism	$V_{(k+1)}, V_{(k+2)}$
Hypersynchronism	$V_{(k-1)}, V_{(k-2)}$

references, the combination of active vectors 2 and 3 produces the torque and flux waveforms according to Figure 8.43. Hence, if Table 8.5 defines V_3 as the first vector, for instance, the second active vector will be V_2 according to Table 8.6.

On the other hand, taking into account Equation (8.3):

$$\vec{v}_r^r \cong \frac{d\vec{\psi}_r^r}{dt} \tag{8.38}$$

It is deduced that the rotor space flux vector and the rotor voltage space vector at steady state are always going to be shifted approximately 90°. So from a different point of view, the proposed choice of the three vectors guarantees the generation of the rotor voltage space vector $\vec{v}_r^r$, using the nearest three voltage vectors provided by the two-level VSC, as can be observed in Figure 8.44.

When the machine operates at hypersynchronism, the zero vector produces negative torque variation as revealed in Equation (8.13). As a consequence, at steady state, the injected two vectors must present positive torque variation, in accordance with Figure 8.43b. It means (from Table 8.5) that the possible first two active vectors are V_{k-1} and V_{k-2}.

For the example of Figure 8.44, since the sector is 1, the possible first active vectors are V_5 and V_6; together with the zero vector, they are the nearest three voltage vectors of the two-level VSC that guarantee a good quality rotor voltage vector creation. Note that if the first active vector selection was V_{k+1} or V_{k+2}, the steady state would not have been reached.

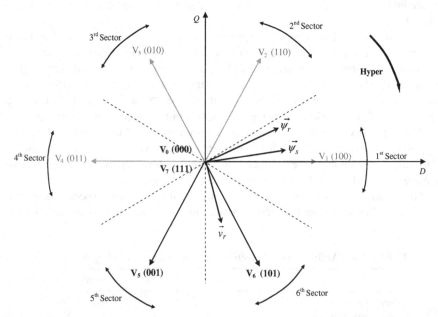

Figure 8.44 Choice of the nearest three voltage vectors at steady state in hypersynchronism.

8.4.2.4 Torque and Flux Ripple Reduction Criteria at Constant Switching Frequency

The switching instants h_{c1} and h_{c2}, as mentioned before, are calculated in order to reduce the rotor flux and the torque ripples. Hence, the minimization criteria presented in this section seek to minimize both the directly controlled variables' ripples. So, once the active vectors are identified by using Tables 8.5 and 8.6, their corresponding time intervals are calculated based on a prediction of the torque and the rotor flux. This task is graphically represented in block 6 of Figure 8.41.

First, taking advantage of the fact that the slope of the flux when the zero vector is applied is nearly zero ($s_{33} = 0$), the expression that relates h_{c1} and h_{c2} is derived by minimizing the flux ripple. Therefore, the square of the rms flux ripple is calculated as follows:

$$|\vec{\psi}_r|^2_{ripple} = \frac{1}{h_{c2}} \int_0^{h_{c1}} \left(s_{11}t + |\vec{\psi}_r|(k) - |\vec{\psi}_r|^* \right)^2 dt$$

$$+ \frac{1}{h_{c2}} \int_{h_{c1}}^{h_{c2}} \left(s_{22}t - s_{22}h_{c1} + s_{11}h_{c1} + |\vec{\psi}_r|(k) - |\vec{\psi}_r|^* \right)^2 dt \qquad (8.39)$$

In order to find the minimum of this ripple, we differentiate this expression with respect to the switching instant of the first active vector h_{c1}:

$$\frac{d|\vec{\psi}_r|^2_{ripple}}{dh_{c1}} = 0 \qquad (8.40)$$

By solving this equation, the optimal switching instant h_{c1} is obtained:

$$h_{c1} = \frac{2\left(|\vec{\psi}_r|^* - |\vec{\psi}_r|(k) \right) - s_{22}h_{c2}}{2s_{11} - s_{22}} \qquad (8.41)$$

Second, once the relation $h_{c1} = f(h_{c2})$ is obtained, it is possible to calculate the equivalent slope s_{12} of the first two active vectors for the torque. As illustrated in Figure 8.45, the equivalent slope is calculated by means of the expression

$$s_{12} = \frac{s_1 h_{c1} + s_2(h_{c2} - h_{c1})}{h_{c2}} \qquad (8.42)$$

After that, as done with the flux ripple, the square of the rms torque ripple is calculated with the integral

$$T^2_{em\_ripple} = \frac{1}{h} \int_0^{h_{c2}} (s_{12}t + T_{em}(k) - T_{em}^*)^2 dt$$

$$+ \frac{1}{h} \int_{h_{c2}}^{h} (s_3 t - s_3 h_{c2} + s_{12}h_{c_2} + T_{em}(k) - T_{em}^*)^2 dt \qquad (8.43)$$

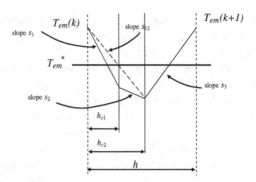

Figure 8.45   Equivalent slope s_{12} for the torque, of the first two active vectors.

Considering that the equivalent slope s_{12} is constant, the square of the rms torque ripple can be minimized by the expression

$$\frac{dT^2_{em\_ripple}}{dh_{c2}} = 0 \tag{8.44}$$

That leads to an expression equivalent to Equation (8.41):

$$h_{c2} = \frac{2(T^*_{em} - T_{em}(k)) - s_3 h}{2s_{12} - s_3} \tag{8.45}$$

Finally, by combining this last expression and expressions (8.41) and (8.42)), both switching instants are calculated by means of the following:

$$h_{c1} = \frac{2s_{22}T^*_{em} - 2s_{22}T_{em}(k) - s_{22}s_3 h + (2s_3 - 4s_2)|\vec{\psi}_r|^* + (4s_2 - 2s_3)|\vec{\psi}_r|(k)}{2s_{22}s_1 - 4s_{11}s_2 + 2s_{11}s_3 - s_{22}s_3} \tag{8.46}$$

$$h_{c2} = \frac{(2s_{22} - 4s_{11})T^*_{em} + (4s_{11} - 2s_{22})T_{em}(k)}{2s_{22}s_1 - 4s_{11}s_2 + 2s_{11}s_3 - s_{22}s_3}$$
$$+ \frac{(2s_{11} - s_{22})s_3 h + (4s_1 - 4s_2)|\vec{\psi}_r|^* + (4s_2 - 4s_1)|\vec{\psi}_r|(k)}{2s_{22}s_1 - 4s_{11}s_2 + 2s_{11}s_3 - s_{22}s_3} \tag{8.47}$$

Notice that the obtained expressions for h_{c1} and h_{c2} depend on the actual and reference values of the torque and rotor flux and on the slopes of the three-vector sequence chosen. For that purpose it is supposed that the slopes are already known (block 5 of Figure 8.41).

If the computed values for h_{c1} and h_{c2} are greater than h or smaller than zero, or if h_{c2} is smaller than h_{c1}, it means that the torque and the flux cannot reach their references yet; hence, the first active vector is maintained during the whole switching period h.

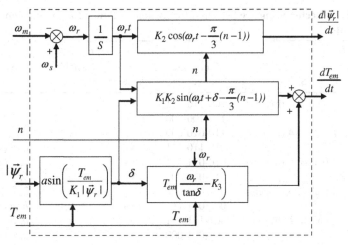

Figure 8.46 Torque and flux derivative expressions calculation block diagram.

As an alternative to these approximated expressions, more complex torque reduction criteria could be adopted, but, in general, it would lead to very complex and computationally expensive expressions. For simplicity, and because further torque reduction would be almost insignificant, the presented solution has been considered as the simplest solution.

8.4.2.5 *Slope Calculations* The prediction based control schema presented in the previous subsection requires us to know the torque and flux variations produced by each rotor voltage vector. This task can be performed with the block diagram studied in Section 8.2.4. The simplified block diagram that can be used is depicted again in Figure 8.46. The task is accomplished in block 5 of the general control block diagram of Figure 8.41.

8.4.2.6 *Pulse Generation* To conclude with the predictive DTC strategy description, after choosing the voltage vector sequences together with the switching instants h_{c1} and h_{c2}, the pulse generation task must be performed. This last task is represented in block 7 of the whole control block diagram of Figure 8.41.

Before analyzing how the pulse generation can be performed, it is necessary to study in detail the switching behavior of the two-level VSC operating with P-DTC. Under steady state operating conditions, focusing the analysis within the switching period h, if three vectors are used only two legs of the two-level VSC are switched at different instants. This is due to the fact that the sequence of vectors will always be composed of two consecutive active vectors and one zero vector, as discussed in previous sections. The right choice of the zero vector between V_0 and V_7 provides another degree of freedom—the reduction of the switching power losses of the converter.

TABLE 8.7 Vector Sequence Selection for Switching Power Loss Reduction at Steady State

Active Vectors	Sector	Candidate Sequences	Zero Vector	Inevitable Switching Leg	Switching Legs
V_1–V_2	6 (subs), 3 (hypers)	100-110-111	V_7	b	b, c ($|i_{ar}|>|i_{cr}|$)
		110-100-000	V_0		a, b ($|i_{ar}|<|i_{cr}|$)
V_2–V_3	1 (subs), 4 (hypers)	110-010-000	V_0	a	a, b ($|i_{br}|<|i_{cr}|$)
		010-110-111	V_7		a, c ($|i_{br}|>|i_{cr}|$)
V_3–V_4	2 (subs), 5 (hypers)	010-011-111	V_7	c	a, c ($|i_{ar}|<|i_{br}|$)
		011-010-000	V_0		b, c ($|i_{ar}|>|i_{br}|$)
V_4–V_5	3 (subs), 6 (hypers)	011-001-000	V_0	b	b, c ($|i_{cr}|<|i_{ar}|$)
		001-011-111	V_7		b, a ($|i_{cr}|>|i_{ar}|$)
V_5–V_6	4 (subs), 1 (hypers)	001-101-111	V_7	a	a, b ($|i_{br}|<|i_{cr}|$)
		101-001-000	V_0		a, c ($|i_{br}|>|i_{cr}|$)
V_6–V_1	5 (subs), 2 (hypers)	101-100-000	V_0	c	a, c ($|i_{ar}|<|i_{br}|$)
		100-101-111	V_7		b, c ($|i_{ar}|>|i_{br}|$)

For each pair of required active vectors, two different sequences exist that allow the commutations of the converter to be reduced. Paying attention to both candidate sequences, the sequence that requires commutation of the legs that are transmitting the smallest currents will be employed, in order to reduce the switching power losses. Table 8.7 summarizes the information necessary to select the correct sequence of vectors, guaranteeing only four commutations per switching period h.

For instance, when the machine is operating at steady state, if active vectors V_2 and V_3 are going to be used because the rotor flux is located in sector 1 at subsynchronous speed (Table 8.7), leg a of the VSC will always switch and the decision to choose one sequence from the two candidates will be carried out by checking the highest current between i_{br} and i_{cr}.

Consequently, once the voltage vectors are selected together with the appropriate candidate sequence, in order to reduce the switching losses of the VSC, the pulses for the controlled semiconductors (S_a, S_b, S_c) must be generated. This task can simply be performed by a sawtooth comparison based schema. Comparing the switching instants h_{c1} and h_{c2}, with a sawtooth as depicted in Figure 8.47, the pulses S_a, S_b, S_c can be created with their corresponding time duration.

- Figure 8.47 shows the pulse creation at two switching periods, in sector 1 and at subsynchronous speed.
- Since $|i_{br}|<|i_{cr}|$, the corresponding vector sequence is V_2-V_3-V_0. In this case, this example shows V_2 as the first vector selected from Table 8.5 of P-DTC. Note that the selected first active vector from Table 8.5, is not always the first vector injected of the three-vector sequence, if a reduced switching schema is required

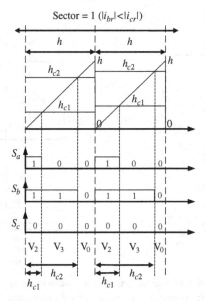

Figure 8.47 Sawtooth based pulse generation schema.

according to Table 8.7. However, this fact does not influence the P-DTC ripple performance; both candidate sequences ensure the same minimized ripples in torque and flux, as illustrated in Figure 8.48.

- Within each switching period, the commutation of only two arms is necessary— arm a and arm b.
- However, at the beginning of each switching period, the same arms of the converter commutate simultaneously.
- Depending on the sector and the required voltage vectors, the pulse generation from the comparison could change.
- The reader can find many alternative pulse generation schemas that are beyond the scope of this book. The presented schema is thought to be implemented in a digital based control hardware.

To conclude, Figure 8.49 illustrates the simplified block diagram of the pulse generation.

8.4.3 Example 8.4: Predictive Direct Torque Control of 15 kW and 2 MW DFIMs at 800 Hz Constant Switching Frequency

In this simulation experiment, the capacity of the DTC control technique to operate at "low" switching frequencies is explored. Hence, two different machines of 2 MW and 15 kW are driven by this control strategy with a considerably low switching frequency

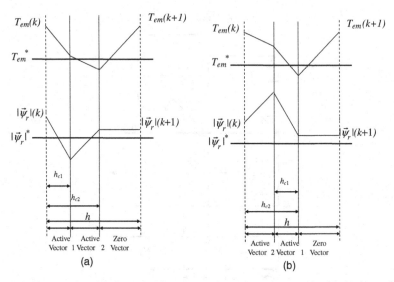

Figure 8.48 Torque and flux waveforms at subsynchronous speed with both candidate sequences of three vectors: (a) first active vector is the first injected vector and (b) first active vector is the second injected vector.

of 800 Hz. The experiment is carried out according to the following operating conditions:

- Switching frequency: 800 Hz
- Speed: 1150 rev/min ($s = 0.25$)
- $V_{bus} = 500$ V
- Torque step: from + nominal (motor mode) to −nominal (generator mode).

The results are shown in Figure 8.50.

By this simulation experiment, it is observed how the quality of the steady state behavior is reduced due to the reduction in the switching frequency, but not due to a degradation of the control strategy performance.

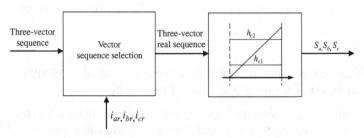

Figure 8.49 Pulse generation simplified block diagram.

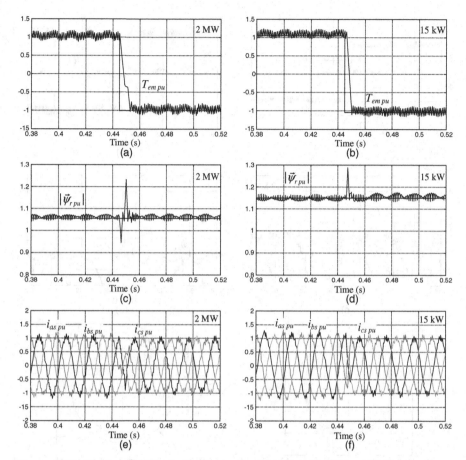

Figure 8.50 Predictive DTC simulation results at 800 Hz of 2 MW and 15 kW machines: transient and steady state behavior. (a) Electromagnetic torque of 2 MW, (b) electromagnetic torque of 15 kW, (c) rotor flux of 2 MW, (d) rotor flux of 15 kW, (e) stator currents of 2 MW, and (f) stator currents of 15 kW.

The DTC control technique is still able to reduce the torque and flux ripples, but at the imposed switching frequency. The system can handle considerably high torque and flux ripples, as shown in Figures 8.50a to 8.50d.

The transient response capacity is reasonably fast and safe as Figure 8.50 illustrates. It is able to operate with an absolute absence of overcurrents due to the strong change in the torque demand.

It can be concluded that both 2 MW and 15 kW machine performances are very similar.

Continuing with the same experiment, Figure 8.51 shows more magnitudes.

- Figures 8.51a and 8.51b show that the stator active and reactive powers take values according to the set torque and rotor flux references. In both machines an

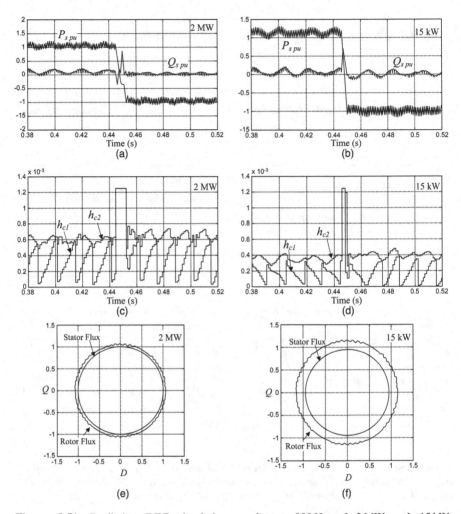

Figure 8.51 Predictive DTC simulation results at 800 Hz of 2 MW and 15 kW machines: transient and steady state behavior. (a) P_s and Q_s of 2 MW, (b) P_s and Q_s of 15 kW (c) h_{c1} and h_{c2} of 2 MW, (d) h_{c1} and h_{c2} of 15 kW, (e) fluxes DQ plots of 2 MW, and (f) fluxes DQ plots of 15 kW.

oscillation in the powers is observed due to nonextinguished transients. As torque and flux are the directly controlled variables, it is typical that some other noncontrolled variables, in this case the stator powers, present oscillations that are damped after several cycles.

- Figures 8.51c and 8.51d present the switching time instants (h_{c1} and h_{c2}) of both machines. There are significant differences between both cases. It is due to the fact that they are working at different rotor voltage since they have a different stator–rotor turn ratio. Note that the experiment is the same as the nominal active

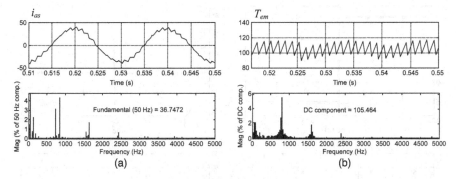

Figure 8.52 Predictive DTC simulation results at 800 Hz of a 15 kW machine: (a) stator current spectrum and (d) torque spectrum.

power step and has approximately zero reactive power exchange with the grid, for both machines (Figures 8.51a and 8.51b).

- Finally, Figures 8.51e and 8.51f show the *DQ* plots of the stator and rotor fluxes. In both cases circular trajectories are achieved, with a ripple in the rotor flux due to the switching frequency used and the characteristics of each machine.

To conclude, the constant switching behavior of the predictive DTC strategy is shown in Figure 8.52.

Figure 8.52a illustrates the constant switching frequency behavior of the stator current. The higher amplitude harmonics appear at multiples of the switching frequency (800 Hz). Apart from these groups of harmonics, as stated before, there are also some low frequency harmonics due to nondamped transients.

Finally, Figure 8.52b reveals that the torque switching behavior is also constant. The properties of the spectrum again are the typical ones for a constant switching frequency. Although not shown, this behavior can be extrapolated to all the other magnitudes of the machine driven by the P-DTC. For a real application, the currents and torque have ripples that are too high. This fact is due to the selected low switching frequency. If better ripple performance is desired, obtaining better quality of currents and torque, the switching frequency should be increased (if the VSC allows doing it) or passive filters could be added on the rotor side of the DFIM. In this example, no importance has been given to the poor quality of the currents, in order to show how it is possible to achieve good control performances with the P-DTC technique, even under low frequency scenarios, and how it is possible to avoid control degradation.

8.4.4 Example 8.5: Predictive Direct Torque Control of a 15 kW DFIM at 4 kHz Constant Switching Frequency

In this example, the behavior of a 15 kW DFIM at steady state and the transient response to a torque step reference is presented, under constant speed operating conditions. The predictive DTC with reduced torque and flux ripple control strategy is

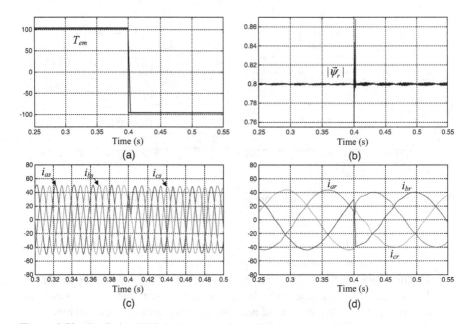

Figure 8.53 Predictive DTC simulation results at 4 kHz: transient and steady state behavior. (a) Electromagnetic torque, (b) rotor flux, (c) stator currents, and (d) rotor currents.

in charge of controlling the machine. The main characteristics of the experiment are summarized as follows:

- Switching frequency: 4 kHz (considerably high)
- Speed: 1350 rev/min ($s = 0.1$); $V_{bus} = 500$ V
- Torque step: from 100 N·m (motor mode), to -100 N·m (generator mode)

The results are shown in Figure 8.53.

- Basically, by analyzing the torque and the flux magnitudes in Figures 8.53a and 8.53b, a very reduced and uniform ripple is observed. Note the absence of rotor flux overshoot. This implies good quality and very uniform current waveform.
- Added to this, a good quality transient response is achieved by this control technique, that is, the currents are quickly controlled within their safety limits— Figures 8.53c and 8.53d.
- Compared with the previous example at 800 Hz, we can see that the quality of the steady state as well as the transient response has been improved.

8.5 PREDICTIVE DIRECT POWER CONTROL (P-DPC) OF THE DOUBLY FED INDUCTION MACHINE AT CONSTANT SWITCHING FREQUENCY

In this section, the predictive direct power control (P-DPC) technique is presented for the doubly fed induction machine (DFIM). Based on this predictive control, the strategy is able to operate at considerably low constant switching frequencies and performs a power ripple minimization technique, in order to improve the steady state and transient response behaviors of the machine.

A similar exposition procedure is followed as in the previous section when we studied the P-DTC technique. Added to this, due to their very similar control philosophy, very similar features and performances are achieved with both P-DTC and P-DPC techniques.

8.5.1 Basic Control Principle

The predictive DPC (P-DPC) technique is based on the same control principles of predictive DTC, so in consequence, it is also based on the same principles as classic DTC and DPC techniques. As studied in Section 8.3, the directly controlled variables are the stator active and reactive powers:

$$P_s = \frac{3}{2} \frac{L_m}{\sigma L_s L_r} \omega_s |\vec{\psi}_s||\vec{\psi}_r| \sin \delta \qquad (8.48)$$

$$Q_s = \frac{3}{2} \frac{\omega_s}{\sigma L_s} |\vec{\psi}_s| \left[\frac{L_m}{L_r} |\vec{\psi}_s| - |\vec{\psi}_r| \cos \delta \right] \qquad (8.49)$$

The P-DPC seeks to create a circular trajectory of the rotor flux space vector, injecting different rotor voltage vectors of different durations. Hence, by changing the values of the terms $|\vec{\psi}_r| \sin \delta$ and $|\vec{\psi}_r| \cos \delta$, it locates the rotor flux space vector, with a given amplitude, to a distance from the stator flux space vector (Figure 8.54).

Hence, the predictive DPC technique presented in this section introduces the required modifications to the basic control DPC structure, in order to achieve constant switching frequency behavior as achieved in predictive DTC. Exactly the same

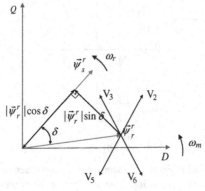

Figure 8.54 Terms $|\vec{\psi}_r| \sin \delta$ and $|\vec{\psi}_r| \cos \delta$ in P-DPC.

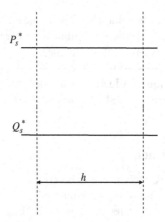

Figure 8.55   Constant switching period.

control philosophy is followed in this case again. The basic principle of the predictive DPC strategy can be summarized as follows:

1. A constant switching period is defined. The duration of this constant period is denoted as h (Figure 8.55).

2. During this period h, at steady state operation, three different rotor voltage vectors are injected, with the objective of minimizing a cost function related to the stator active and reactive power errors. Hence, the stator active and reactive power evolutions within the switching period can take the shape shown in Figure 8.56.

h_{c1}, $h_{c2} - h_{c1}$ are the time intervals dedicated to injected vectors 1 and 2, respectively. Note that the third injected vector is not an active vector, but is a zero

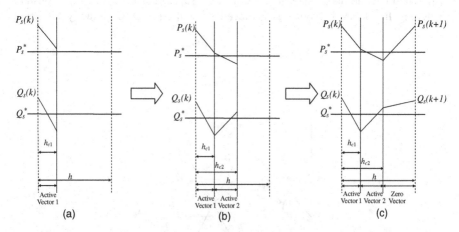

Figure 8.56   Rotor voltage injection and stator active and reactive power evolutions within one switching period h: (a) first vector injection, (b) second vector injection, and (c) third vector injection.

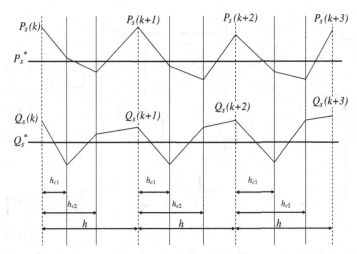

Figure 8.57 P_s and Q_s evolutions after three switching periods according to P-DPC.

vector. The injection time for the third vector is $h - h_{c2}$. In DPC, the zero vector produces a small Q_s variation.

3. This procedure is repeated again at constant period h as illustrated in Figure 8.57. In general, the vector sequence as well as the time intervals dedicated for each vector (h_{c1}, h_{c2}, $h - h_{c2}$), are modified for every sample time.

Therefore, by following the same philosophy as for P-DTC, the P-DPC is in charge of performing the following tasks every sample period h:

1. At the beginning of the switching period, from the information given by the P_s and Q_s errors, select the appropriate three voltage vector sequence.
2. At the beginning of the switching period as well, calculate the corresponding time intervals (h_{c1}, h_{c2}, $h - h_{c2}$) for the selected three voltage vector sequence, with the objective of minimizing a cost function related to the P_s and Q_s errors.
3. During the switching period, apply the vector sequence calculated at the beginning of the period.

Consequently, with this control philosophy, the P-DPC block diagram structure is studied in detail in subsequent sections.

8.5.2 Control Block Diagram

The control block diagram of the proposed strategy is given in Figure 8.58. As mentioned before, the directly controlled variables are P_s and Q_s. From the P_s and Q_s references, the control strategy calculates the pulses (S_a, S_b, S_c) for the controlled semiconductors of the two-level VSC, in order to meet the following two objectives:

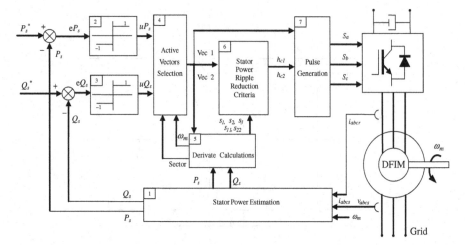

Figure 8.58 Predictive DPC block diagram with constant switching frequency.

reduced P_s and Q_s ripples and constant switching frequency behavior. As shown in Figure 8.58, the control strategy is divided into seven different tasks represented in seven different blocks, as we did in P-DTC. The tasks carried out in blocks 1–4 are the basic direct power control principles. In contrast, the tasks performed in blocks 5–7 are introduced in order to achieve the improvements mentioned above—reduced ripples and constant switching frequency behavior:

1. Estimation block
2. P_s ON–OFF controller
3. Q_s ON–OFF controller
4. Voltage vector sequence selection
5. Slope calculations
6. P_s and Q_s reduction criteria
7. Pulse generation

Each block is described in subsequent sections in detail.

8.5.2.1 Estimation Block The estimation block for this P-DPC strategy is exactly the same as for the classic DPC. Since it is based on the same control principles, P_s, Q_s, and the sector where the rotor flux space vector is located can be estimated in the same manner as presented in Section 8.3.

8.5.2.2 First Voltage Vector Selection and ON–OFF Controllers In the first block of Figure 8.58, the stator active and reactive power values are calculated. After that, the stator active and reactive power references are compared with their estimated values, so the stator power errors are obtained.

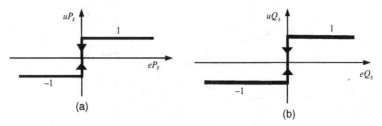

Figure 8.59 (a) ON–OFF stator active power controller without hysteresis band. (b) ON–OFF stator reactive power controller without hysteresis band.

The basic principle of the direct control techniques determines the choice of one specific active voltage vector of the converter, in order to correct the stator power errors. The choice of this vector is calculated directly from the stator active and reactive power errors themselves and the sector where the rotor flux lies.

The proposed predictive DPC strategy adopts this principle to select the first active voltage vector of the three-vector sequence employed to minimize the stator power ripple. Hence, the first active voltage vector selection is based on a look-up table mapped by the output of two ON–OFF comparators without hysteresis bands. These tasks are accomplished in blocks 2, 3, and 4 of Figure 8.58. One comparator is dedicated to the P_s control and the other to the Q_s control, as illustrated in Figure 8.59. In this case, as occurred in P-DTC, the ON–OFF controllers do not present hysteresis bands, since they are only in charge of selecting the first active vector.

When the output of the comparator is set to 1 (i.e., positive error), it means a positive slope variation requirement. In contrast, when the output is set to -1 (i.e., negative error), a negative slope variation is required.

Once the outputs of the ON–OFF controllers are calculated, the first active vector of the three-vector sequence can be selected. This selection is simply made by the look-up table of Table 8.8.

8.5.2.3 Second and Third Voltage Vector Selection and Power Waveforms

Once the first active vector is chosen from the classical DPC switching table, the predictive DPC control strategy employs a sequence of three different voltage vectors in a constant switching period h, under steady state operating conditions. Hence, a second active vector is injected followed always by a zero vector

TABLE 8.8 First Active Vector Selection as a Function of the Stator Active and Reactive Power ON-OFF Controllers' Outputs (k = Sector)

		uP_s	
		1	-1
uQ_s	1	$V_{(k-2)}$	$V_{(k+2)}$
	-1	$V_{(k-1)}$	$V_{(k+1)}$

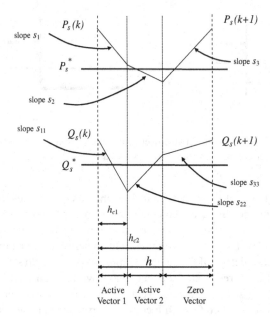

Figure 8.60   Steady state active and reactive power waveforms at motor and generator modes in subsynchronism.

with the objective of minimizing the stator active and reactive power ripples. The typical power waveforms for this control strategy are shown in Figure 8.60.

The time intervals dedicated to each active vector, h_{c1} and $h_{c2} - h_{c1}$, are calculated according to the minimization criteria that will be presented in the next section.

So, after selecting the first active vector, if the stator power ripples can be reduced, the choice of the second active vector is made according to Table 8.9. It defines the correct choice of the active vectors, depending on the speed of the machine and the sector in which the rotor flux is located.

As can be noticed, the equivalence to the P-DTC technique is present in every task of the control block diagram. Note also that the notations for the active power slope variations (s_1, s_2, s_3) and the reactive power slope variations (s_{11}, s_{22}, s_{33}) are also equivalent to those in P-DTC.

8.5.2.4 Power Ripple Reduction Criteria at Constant Switching Frequency   The switching instants h_{c1} and h_{c2}, as mentioned before, are

TABLE 8.9   Second Active Vector Selection at Steady State (k = Sector)

	Active Vectors
Subsynchronism	$V_{(k+1)}$, $V_{(k+2)}$
Hypersynchronism	$V_{(k-1)}$, $V_{(k-2)}$

calculated in order to reduce the active power ripple and the reactive power ripple. So, once the active vectors are identified by using Tables 8.8 and 8.9, their corresponding time intervals are calculated based on a prediction of the stator active and reactive powers. This task is graphically represented in block 6 of Figure 8.58.

The same procedure as for P-DTC is followed in this task as well. Hence, taking advantage of the fact that the slope of the reactive power when the zero vector is applied is very small ($s_{33} = 0$ is approximated), and considering the equivalent slope s_{12} of the first two active vectors for the stator active power can be calculated by means of the expression

$$s_{12} = \frac{s_1 h_{c1} + s_2(h_{c2} - h_{c1})}{h_{c2}} \tag{8.50}$$

the expression for h_{c1} and h_{c2} is derived by minimizing the ripples:

$$Q_{s\,ripple}^2 = \frac{1}{h_{c2}} \int_0^{h_{c1}} (s_{11}t + Q_s(k) - Q_s^*)^2 dt$$
$$+ \frac{1}{h_{c2}} \int_{h_{c1}}^{h_{c2}} (s_{22}t - s_{22}h_{c1} + s_{11}h_{c1} + Q_s(k) - Q_s^*)^2 dt \tag{8.51}$$

$$P_{s\,ripple}^2 = \frac{1}{h} \int_0^{h_{c2}} (s_{12}t + P_s(k) - P_s^*)^2 dt$$
$$+ \frac{1}{h} \int_{h_{c2}}^{h} (s_3 t - s_3 h_{c2} + s_{12}h_{c2} + P_s(k) - P_s^*)^2 dt \tag{8.52}$$

Hence, the switching time instants yield

$$h_{c1} = \frac{2s_{22}P_s^* - 2s_{22}P_s(k) - s_{22}s_3 h}{2s_{22}s_1 - 4s_{11}s_2 + 2s_{11}s_3 - s_{22}s_3}$$
$$+ \frac{(2s_3 - 4s_2)Q_s^* + (4s_2 - 2s_3)Q_s(k)}{2s_{22}s_1 - 4s_{11}s_2 + 2s_{11}s_3 - s_{22}s_3} \tag{8.53}$$

$$h_{c2} = \frac{(2s_{22} - 4s_{11})P_s^* + (4s_{11} - 2s_{22})P_s(k) + (2s_{11} - s_{22})s_3 h}{2s_{22}s_1 - 4s_{11}s_2 + 2s_{11}s_3 - s_{22}s_3}$$
$$+ \frac{(4s_1 - 4s_2)Q_s^* + (4s_2 - 4s_1)Q_s(k)}{2s_{22}s_1 - 4s_{11}s_2 + 2s_{11}s_3 - s_{22}s_3} \tag{8.54}$$

8.5.2.5 Slope Calculations   The prediction based control schema presented in the previous subsection requires us to know the torque and flux variations produced by each rotor voltage vector. This task can be performed with the block diagram studied in Section 8.3. The simplified block diagram that can be used is depicted again in

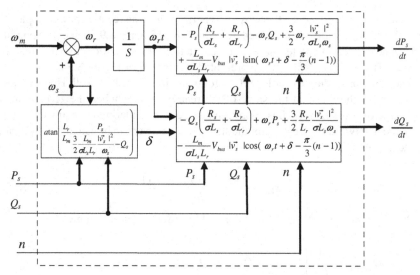

Figure 8.61 Stator active and reactive power derivative expressions calculation block diagram.

Figure 8.61. The task is accomplished in block 5 of the general control block diagram of Figure 8.58.

8.5.2.6 Pulse Generation To conclude with the predictive DPC strategy description, after choosing the voltage vector sequence together with the switching instants h_{c1} and h_{c2}, the pulse generation task must be performed. This last task is represented in block 7 of the whole control block diagram of Figure 8.58.

The pulse generation philosophy, in this case, is exactly the same as in the P-DTC strategy. So the procedure presented in Section 8.4 can be applied for pulse generation in P-DPC also.

8.5.3 Example 8.6: Predictive Direct Power Control of a 15 kW DFIM at 1 kHz Constant Switching Frequency

In this example, the behavior of a 15 kW DFIM at steady state and the transient response to a torque step reference is studied, under constant speed operating conditions. The predictive DPC at constant switching frequency is in charge of controlling the machine. The main characteristics of the experiment are summarized as follows:

- Switching frequency: 1 kHz (considerably low)
- Speed: 1200 rev/min ($s = 0.2$); $V_{bus} = 500$ V
- Stator active and reactive power steps in motor mode.

The results are shown in Figures 8.62 and 8.63. Figures 8.62a and 8.62c show a simultaneous P_s and Q_s step change at constant speed. It can be noticed that, at the transient as well as at steady state, the P-DPC is able to keep the variables controlled safely at reduced power ripples.

Figures 8.62b and 8.62d also show noncontrolled variables such as torque and rotor flux amplitude. It can be noticed that although they are not being directly controlled, their ripple behavior is equivalent to that obtained with the P-DTC strategy.

Figures 8.62e and 8.62f show the stator and rotor currents during the experiment. It can be noticed that, at all times, the current behavior is safe, with an absolute absence of overcurrents.

Perhaps the quality of the currents would not be good enough for a real application due to the adopted low switching frequency. Hence, if a more realistic scenario needs to be simulated, with lower ripples in variables such as currents and torque, the switching frequency should be increased (if the VSC allows it) or passive filters should be included in the system, as studied in next section. In the same way, Figures 8.62g and 8.62h present the switching instant as well as the sector behavior.

Finally, Figures 8.63a and 8.63b show the spectra of the torque and the stator current of the experiment. Notice that the behavior is typical of constant switching frequency schemas. The most representative groups of harmonics appear at frequencies that are a multiple of the switching frequency (1 kHz). In addition, the short nondamped oscillation present in the torque also produces some low amplitude and frequency harmonics as well.

8.6 MULTILEVEL CONVERTER BASED PREDICTIVE DIRECT POWER AND DIRECT TORQUE CONTROL OF THE DOUBLY FED INDUCTION MACHINE AT CONSTANT SWITCHING FREQUENCY

8.6.1 Introduction

When it comes to implement DTC or DPC control techniques in a system fed by a multilevel converter topology, there are several aspects that must be taken into account:

- The basic control principles of direct techniques studied in previous sections of this chapter are oriented to DFIM fed, by a two-level VSC.
- If the converter is changed by a more complex multilevel VSC topology, the main difference is that the multilevel converter allows operation with a higher number of voltage vectors, rather than with the eight voltage vectors of the two-level VSC.
- This additional voltage vector availability, in general, from the control strategy designer's point of view, maintains the basic control structure of direct control techniques, but leads to a modification requirement of elements, such as a table for vector selection, slopes of the voltage vector calculation, and pulse generation.

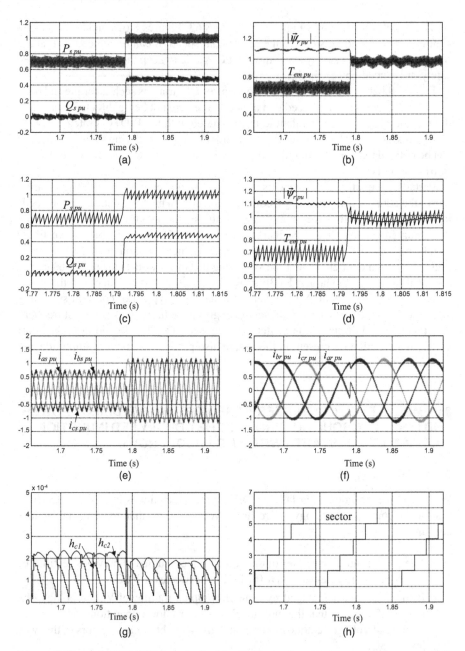

Figure 8.62 A 15 kW DFIM controlled by DPC at subsynchronous speed, $\omega_{mpu} = 0.8$: steady state and transient response. (a) Stator active and reactive powers, (b) torque and rotor flux amplitudes, (c) stator active and reactive power zoom, (d) torque and rotor flux amplitudes zoom, (e) *abc* stator currents, (f) *abc* rotor currents, (g) switching instants, and (h) sector.

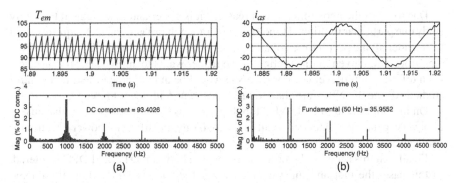

Figure 8.63 Predictive DPC simulation results at 1 kHz of 15 kW machine: (a) torque spectrum and (b) stator current spectrum.

- This fact is an unavoidable consequence of direct control philosophy, which directly selects the voltage vector without the need of an independent modulator. Thus, in vector control technique, for instance, the adaptation of the modulator to the requirements of the multilevel converter is enough; it is not necessary to adapt the vector control itself (Figure 8.64).
- In addition, if the employed multilevel VSC topology requires inclusion of an algorithm for balancing the DC bus voltage capacitors, the direct control technique must also perform this task.
- On the contrary, obviously, the inclusion of the multilevel converter, as mentioned in Chapter 2, implies benefits such as higher output voltage levels, with better quality of magnitudes like torque or stator currents, and the possibility of reaching higher voltage with the given semiconductors rather than with two-level VSC.

Hence, in this section, DTC and DPC techniques are developed for only one multilevel converter topology: that is, *three-level neutral point clamped VSC topology* (3L-NPC). Details such as the voltage vector table, pulse generation, and voltage balancing are presented and analyzed in depth for this particular multilevel topology.

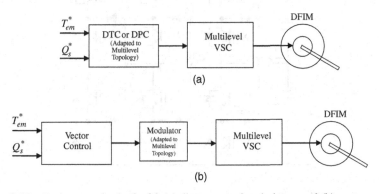

Figure 8.64 Basic control schema for (a) direct control techniques and (b) vector control.

Due to all of the above-mentioned reasons, it is easier and more logical to simply study one particular multilevel topology case, rather than a general technique oriented to consider all the peculiarities of all existing multilevel topologies. Hence, thanks to the information given in this section, if the reader wants to apply DTC, for instance, for a different multilevel topology, he/she would be able to adapt the direct control studied for the 3L-NPC into the required, different multilevel topology.

On the other hand, the 3L-NPC topology is chosen for the exposition because it is a very generally extended multilevel topology in different power electronics applications.

Finally, in this section, the order of exposition between DTC and DPC is altered. In this case, the first and most detailed control studied is DPC; after introducing all the general aspects of multilevel direct control into DPC, the study is extended to DTC.

8.6.2 Three-Level NPC VSC Based DPC of the DFIM

As advanced before, the 3L-NPC VSC is used to feed the rotor of the DFIM, as illustrated in Figure 8.65. The most important characteristics affecting the control strategy can be summarized as follows:

- The DC bus is composed of two capacitors C_1 and C_2, allowing one to obtain three different voltage levels (0, $V_{bus}/2$, V_{bus}) between phase output and the o point of the DC bus.
- These three voltage levels, at each phase of the VSC, can be generated by changing the semiconductors' commands (S_{a1}, S_{a2}, S_{b1}, S_{b2}, S_{c1}, S_{c2},) as studied in Chapter 2.
- As discussed in Chapter 2, many authors perform the synthesis of the output voltage space vector according to a space vector modulation (SVM) schema. Hence, depending on where the desired output voltage space vector is located in

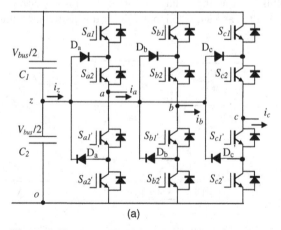

(a)

Figure 8.65 3L-NPC VSC with IGBTs: (a) topology.

the space vector diagram, the nearest three voltage vectors of the VSC are employed to synthesize it, in a constant switching period. In this section, the direct control techniques studied perform the pulse generation in a similar way as in a SVM schema, as presented later.

8.6.2.1 Model of the Doubly Fed Induction Machine and the Rotor Side Filter

In this section, an additional modification is considered in the system. A passive filter is located between the rotor and the rotor side converter. This filter is called the rotor side filter, and among many possible variants of passive filters, an inductive filter is employed. The inclusion of the filter can be necessary, for instance, in order to improve the generated power quality (simultaneous better THDs of currents and less electromagnetic torque ripples) of the DFIM and meet power generation standards and grid codes. This is a typical challenge of high power DFIM, where the restrictions in commutations of the semiconductors cause them to operate at "low" switching frequencies. Hence, despite the fact that the use of multilevel topologies allows one to achieve better power quality than standard two-level topologies under the same switching conditions, in general, the necessity of a further improvement of the generated power quality, for instance, by passive filter inclusion, is also required.

Perhaps in some cases the necessity of the rotor side filter can be avoided; however, if it is used, it influences the control strategy. Thus, the general case using the filter is studied in this section, considering and studying the effect of this filter in the control strategy.

Hence, the simplified equivalent electrical circuit of the doubly fed induction machine and the considered inductive filter can be represented as depicted in Figure 8.66. The inductance of the filter is L_f, while a small R_f also accompanies the predominant inductance.

Thus, considering the rotor side inductive filter, the rotor voltage equation (4.10) is transformed into the following converter voltage equation:

$$\vec{v}_f^r = R_r \vec{i}_r^r + \frac{d\vec{\psi}_r^r}{dt} + R_f \vec{i}_r^r + L_f \frac{d\vec{i}_r^r}{dt} \qquad (8.55)$$

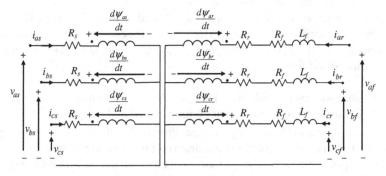

Figure 8.66 Electrical model of the DFIM and the rotor side filter.

It is important to highlight that, under these circumstances, the injected voltage vector by the VSC is not the same as the rotor voltage vector. The rotor side filter makes these two voltages become different and the control strategy must take it into account, as detailed in subsequent subsections.

On the other hand, from this last equation and Equations (4.9), (4.11), and (4.12), the state space equations of the machine and the filter model, selecting the stator and rotor currents as state variables yields

$$
\frac{d}{dt}\begin{bmatrix} \vec{i}_s^r \\ \vec{i}_r^r \end{bmatrix} = \frac{1}{\sigma_f L_s L_{rf}} \left(\begin{bmatrix} -R_s L_{rf} - j\omega_m L_s L_{rf} & R_{rf} L_m - j\omega_m L_m L_{rf} \\ R_s L_m + j\omega_m L_s L_m & -R_{rf} L_s + j\omega_m L_m^2 \end{bmatrix} \begin{bmatrix} \vec{i}_s^r \\ \vec{i}_r^r \end{bmatrix} \right.
$$
$$
\left. + \begin{bmatrix} L_{rf} & -L_m \\ -L_m & L_s \end{bmatrix} \begin{bmatrix} \vec{v}_s^r \\ \vec{v}_f^r \end{bmatrix} \right)
$$

(8.56)

where $L_{rf} = L_r + L_f$, $R_{rf} = R_r + R_f$, and $\sigma_f = 1 - L_m/L_s L_{rf}$.

8.6.2.2 Basic Control Principle The multilevel predictive DPC technique presented in this section is based on exactly same control principles as two-level VSC oriented P-DTC and P-DPC. The directly controlled variables are the stator active and reactive powers, based on the following known expressions:

$$
P_s = \tfrac{3}{2}\frac{L_m}{\sigma L_s L_r}\omega_s|\vec{\psi}_s||\vec{\psi}_r|\sin\delta
$$

(8.57)

$$
Q_s = \tfrac{3}{2}\frac{\omega_s}{\sigma L_s}|\vec{\psi}_s|\left[\frac{L_m}{L_r}|\vec{\psi}_s| - |\vec{\psi}_r|\cos\delta\right]
$$

(8.58)

The multilevel P-DPC as usual seeks to create a circular trajectory of the rotor flux space vector, injecting different rotor voltage vectors of different durations. The main difference compared to two-level VSC P-DPC is that, in this case, there are a larger number of voltage vectors available. Hence, by changing the values of the terms $|\vec{\psi}_r|\sin\delta$ and $|\vec{\psi}_r|\cos\delta$, it locates the rotor flux space vector, with a given amplitude, at a distance from the stator flux space vector.

Exactly the same control philosophy is followed in this case again. One of the important characteristics of this control technique is the constant switching frequency achievement as well. The basic principle of the multilevel predictive DPC strategy can be summarized as follows:

1. As done with two-level VSC P-DTC and P-DPC, a constant switching period is defined. The duration of this constant period is noted as h (Figure 8.67).

2. During this period h, at steady state operation, three different rotor voltage vectors are injected, with the objective of minimizing a cost function related to the stator active and reactive power errors. Hence, the stator active and reactive power evolutions within the switching period can take the shape shown in Figure 8.68.

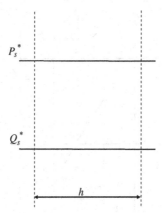

Figure 8.67 Constant switching period.

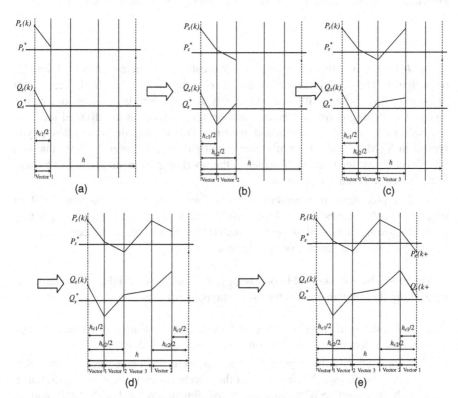

Figure 8.68 Rotor voltage injection and stator active and reactive power evolutions within one switching period h, for multilevel P-DPC: (a) first vector injection half of the interval, (b) second vector injection half of the interval, (c) third vector injection, (d) second vector injection, remaining half of the interval, and (e) first vector injection, remaining half of the interval.

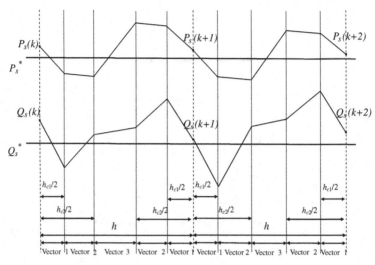

Figure 8.69 P_s and Q_s evolutions after two switching periods, according to multilevel P-DPC.

h_{c1}, $h_{c2} - h_{c1}$ are the time intervals dedicated to injected vectors 1 and 2, respectively. The injection time for the third vector is $h - h_{c2}$, in this case, due to the multilevel schema; it is not necessarily a zero vector. In addition, during the constant switching period h, vector 1 and vector 2 injections are divided into two intervals. Only vector 3 is injected at one time. This vector injection philosophy is typical in SVM schemas for multilevel topologies (see Chapter 2). Note that if a multilevel converter is used, in general, the operating voltages are high so these issues must be taken into account.

3. This procedure is repeated again at constant period h, as illustrated in Figure 8.69. As occurs in two-level VSC based P-DTC and P-DPC, in general, the vector sequence as well as the time intervals dedicated for each vector (h_{c1}, h_{c2}, $h - h_{c2}$) are modified for every sample time.

Therefore, by following this philosophy, the multilevel P-DPC is in charge of performing the following tasks for every sample period h:

1. Before the switching period begins, from the information given by the P_s and Q_s errors, select the appropriate three voltage vector sequence.
2. Before the switching period begins as well, calculate the corresponding time intervals (h_{c1}, h_{c2}, $h - h_{c2}$) for the selected three voltage vector sequence, with the objective of minimizing a cost function related with the P_s and Q_s errors.
3. During the switching period, apply the vector sequence (splitting vectors 1 and 2 into two sequences) calculated at the beginning of the period.

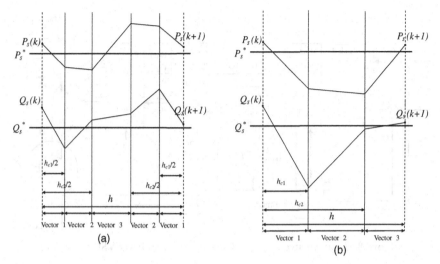

Figure 8.70 P_s and Q_s evolutions with two different commutation schemas, according to multilevel P-DPC: (a) three voltage vectors dividing the sequence into five intervals and (b) three voltages vectors continuously applied.

Note that, as illustrated in Figure 8.70, there is a significant difference in the ripple performance of P_s and Q_s, when the three voltage vectors are applied in five intervals (Figure 8.70a) or in three intervals (Figure 8.70b) per switching period h, with equal time h_{c1} and h_{c2}. Although the final values of P_s and Q_s are exactly the same, in an injection schema of three intervals the ripple is much higher. This operation mode, as advanced before, implies more commutations of switches within the switching period h, but reduces the simultaneous commutations that the schema in Figure 8.70b requires, at the moment of changing the switching period, and a new sequence must be injected.

Consequently, with this control philosophy and considering these voltage injection issues, the multilevel P-DPC block diagram structure is studied in detail in subsequent sections.

8.6.2.3 Control Block Diagram The block diagram of the proposed control strategy is illustrated in Figure 8.71. As mentioned before, the directly controlled magnitudes are P_s and Q_s. From the references, the control strategy calculates the pulses (S_{1a}, S_{2a}, S_{1b}, S_{2b}, S_{1c}, S_{2c}) for the IGBTs of the three-level NPC voltage source converter (VSC), in order to meet the following two objectives: reduced power ripples and constant switching frequency behavior. Figure 8.71 shows that the control strategy is divided into eight different tasks represented in eight different blocks.

1. Estimation block.
2. P_s ON–OFF controller
3. Q_s ON–OFF controller

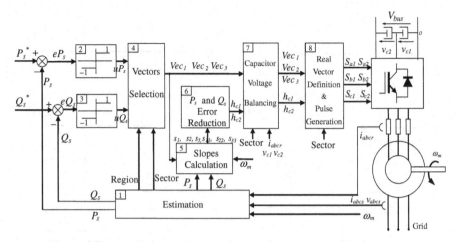

Figure 8.71 Predictive DPC block diagram for multilevel VSC fed DFIM.

4. Voltage vector sequence selection
5. Slope calculations
6. Stator active and reactive power errors reduction criteria
7. DC bus capacitor's voltage balancing
8. Real vector definition and pulse generation

Each block is described in subsequent sections in detail.

Estimation Block Sector and Region Calculation   Compared with previously studied predictive DTC and DPC strategies, there are two new considerations: rotor side filter and three-level NPC VSC inclusions. Thus, the estimation block is changed slightly from previous versions and adapted to these two new considerations.

The selection of the voltage vectors of the three-level NPC converter is made according to the nearest three vector choice principle [45]. The same voltage vector choice procedure as done in P-DTC and P-DPC is carried out. As can be derived from Equation (8.55), the relation between the rotor magnitudes space vectors at steady state is

$$\vec{v}_f^r = R_{rf}\vec{i}_r^r + j\omega_r(\vec{\psi}_r^r + L_f\vec{i}_r^r) \tag{8.59}$$

Note that since the rotor side filter has been located between the 3L-NPC VSC and the rotor, now the converter's voltage in not directly applied to the rotor of the machine. Hence, modifying this last expression, it is possible to define a new space vector, $\vec{x}^r$, as follows:

$$\vec{x}^r = -j\vec{v}_f^r = -jR_{rf}\vec{i}_r^r + \omega_r(\vec{\psi}_r^r + L_f\vec{i}_r^r) \tag{8.60}$$

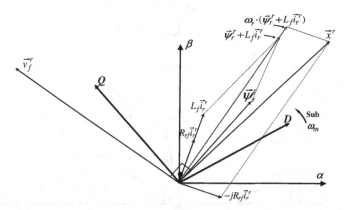

Figure 8.72   Relation between the VSC output voltage, the rotor flux, and the rotor current space vectors at steady state.

As graphically shown in Figure 8.72, the inverter voltage space vector will always be 90° shifted to the $\vec{x}^r$ space vector. This variable, which maintains the 90° constant phase shift with the converter's output voltage vector, is very useful, as shown next.

In addition, the three-level voltage vector diagram may be divided into 12 different sectors, as illustrated in Figure 8.73. Note that, in this case, the sectors present 30° length.

- In the example represented, the space vector $\vec{x}^r$ is lying in the 4th sector.
- At steady state operating conditions, the converter voltage space vector lies in the 7th sector (90° shift) and can be generated by using the following nearest vectors: 021, 022, 122, and 011, guaranteeing that this choice of voltage vectors produces reduced P_s and Q_s variations from the reference values.
- In addition, taking advantage of the symmetry present in the three-level NPC converter and in order to reduce the implementation complexity of the proposed control strategy, the task of knowing which are the nearest vectors, for a given converter voltage space vector position, will be carried out by generating an equivalent voltage space vector $\vec{v}_{f\_eq}$ in sectors 2 and 3 of the vector diagram.
- Moreover, by dividing sectors 2 and 3 into four different regions, the equivalent voltage vector of the example illustrated in Figure 8.73 lies in region 1. Sufficient information to calculate the angle θ_{eq} of the equivalent inverter voltage space vector is provided in Table 8.10.
- Once θ_{eq} has been calculated, the region may be deduced by normalizing the voltage vector to its maximum value ($V_{bus}/\sqrt{3}$ not considering overmodulation), and checking the DQ components of the normalized voltage vector according to the information provided in Table 8.11.

All the required calculations performed in block 1, of Figure 8.71, are represented in the simplified block diagram of Figure 8.74.

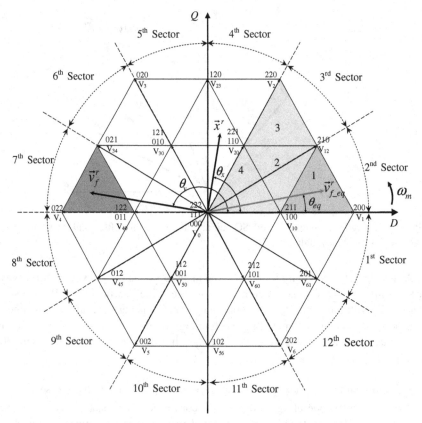

Figure 8.73 Three-level space vector diagram at subsynchronism.

Note that, in this case, instead of calculating the stator active and reactive powers in $\alpha\beta$ coordinates, DQ coordinates have been used.

First Voltage Vector Selection and ON–OFF Controllers   As with previously studied predictive direct controls, the basic principle of these direct control techniques determines the choice of one specific active voltage vector of the converter, in order to

TABLE 8.10   Angle of the Equivalent Voltage Space Vector as a Function of the Sector Where the $\vec{x}$ Space Vector Lies at Subsynchronism

θ_x	θ_v	Sector	θ_{eq}
[330°, 30°]	[60°, 120°]	1, 2	$-\theta_v + 120°$
[30°, 90°]	[120°, 180°]	3, 4	$\theta_v - 120°$
[90°, 150°]	[180°, 240°]	5, 6	$-\theta_v - 120°$
[150°, 210°]	[240°, 300°]	7, 8	$\theta_v + 120°$
[210°, 270°]	[300°, 0°]	9, 10	$-\theta_v$
[270°, 330°]	[0°, 60°]	11, 12	θ_v

TABLE 8.11   Region Calculation as a Function of v_D and v_Q

Case	Region
$v_D > 1,\ v_Q < \sqrt{3}(v_D - 1)$	1
$v_Q < \sqrt{3}/2,\ v_D > 1 - v_Q/\sqrt{3},\ v_D < 1 + v_Q/\sqrt{3}$	2
$v_Q > \sqrt{3}/2$	3
$v_D < 1 - v_Q/\sqrt{3},\ v_Q < \sqrt{3}/2$	4

correct the stator power errors. The choice of this vector is made directly from the P_s and Q_s errors themselves, the sector where $\vec{x}^r$ lies, and the region of the equivalent voltage vector $\vec{v}_{f\_eq}$.

In this case again, the first active voltage vector selection is based on a look-up table mapped by the output of two ON–OFF comparators without hysteresis bands. These tasks are accomplished in blocks 2, 3, and 4 of Figure 8.71. One comparator is dedicated to the P_s control and the other to the Q_s control, as illustrated in Figure 8.75.

Depending on the sector and the region, only four vectors are permitted in order to apply the four different combinations to the P_s and Q_s variation, as depicted in Figure 8.76 (note that only one of the two double vectors will be used).

Hence, if the voltage vector lies in region 1, 2, or 3, the four permitted vectors can be calculated from the four equivalent vectors V_{10}, V_1, V_{20}, V_2, and V_{12} of sectors 2 and 3. As shown in Figure 8.76, these permitted vectors are the closest vectors to the voltage reference vector when it lies in region 1, 2, or 3. The look-up table is made up according to Table 8.12. It defines the first voltage vector as a function of the sector and the stator power errors.

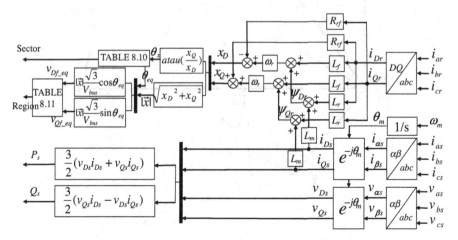

Figure 8.74   Stator powers, sector, and region estimations.

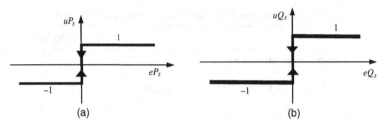

Figure 8.75 (a) ON–OFF stator active power controller without hysteresis band. (b) ON–OFF stator reactive power controller without hysteresis band.

Three of the permitted voltage vectors belong to the region where the equivalent converter voltage vector is located; these vectors are the closest vectors that provide the three-level NPC converter with the predicted required converter voltage vector. In contrast, the fourth permitted vector requires a jump to the subsequent region and in general it is only used in dynamic operation requirements, for instance, when a change in P_s and Q_s is demanded.

There is no difference when we use different redundant vectors, since they produce the same line-to-line voltage at the output of the converter. This means that they

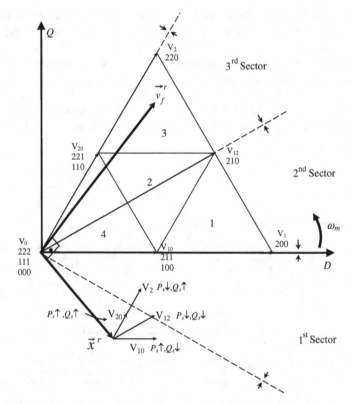

Figure 8.76 Vector selection for regions 1, 2, and 3 at subsynchronism.

TABLE 8.12 First Vector Selection for Regions 1, 2, and 3 at Subsynchronism

Sectors	eP_s, eQ_s			
	(1, 1)	(1, −1)	(−1, 1)	(−1, −1)
4, 8, 12	V_{20}	V_{10}	V_2	V_{12}
3, 7, 11	V_{20}	V_{10}	V_{12}	V_1
2, 6, 10	V_{10}	V_{20}	V_1	V_{12}
1, 5, 9	V_{10}	V_{20}	V_{12}	V_2

produce the same directly controlled magnitudes (P_s and Q_s) variations. However, the redundant vectors are used to balance the DC bus capacitor voltages according to the criteria shown in Section 2.3.1.

On the other hand, if the rotor voltage vector lies in region 4, the first vector selection is made according a slightly different procedure. Since, in region 4, the closest four vectors to the converter voltage vector do not produce equal stator power variations along one sector, the choice is made as in the two-level converter based DTC or DPC strategies. Omitting regions 1, 2, and 3 from the three-level space vector diagram, as structured in Figure 8.77, the four permitted vectors are selected from six

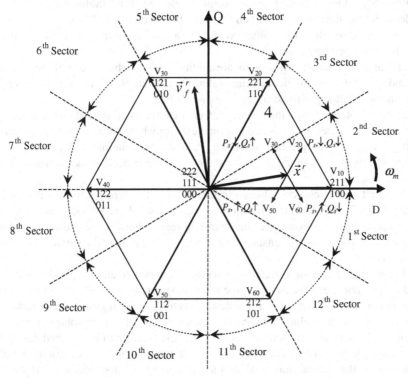

Figure 8.77 Three-level space vector diagram omitting regions 1, 2, and 3.

TABLE 8.13 First Vector Selection for All the Sectors and Region 4 at Subsynchronism

Sectors	eP_s, eQ_s			
	$(1, -1)$	$(1, 1)$	$(-1, -1)$	$(-1, 1)$
1, 2	V_{60}	V_{50}	V_{20}	V_{30}
3, 4	V_{10}	V_{60}	V_{30}	V_{40}
5, 6	V_{20}	V_{10}	V_{40}	V_{50}
7, 8	V_{30}	V_{20}	V_{50}	V_{60}
9, 10	V_{40}	V_{30}	V_{60}	V_{10}
11, 12	V_{50}	V_{40}	V_{10}	V_{20}

vector candidates—V_{10}, V_{20}, V_{30}, V_{40}, V_{50}, and V_{60}—according to the information given in Table 8.13.

It must be stressed that Table 8.13 is equivalent to the table of vectors defined for DPC based on the two-level VSC.

Second and Third Vector Selection Strategy at Constant Switching Frequency Once the first active vector is selected from Tables 8.13 or 8.12, in order to correct the controlled P_s and Q_s errors, the predictive DPC strategy employs a sequence of three different voltage vectors in a constant switching period h, under steady state operating conditions.

Hence, a second and a third vector are applied, with the objective of minimizing the P_s and Q_s errors, from their reference values. One example of the typical waveform evolutions for this control strategy is shown in Figure 8.78.

In the adopted notation the slopes of P_s are denoted as s_1, s_2, and s_3, while s_{11}, s_{22}, and s_{33} are the slopes of Q_s. Depending on the sector where the $\vec{x}^r$ vector lies, different voltage vectors will produce different P_s and Q_s variations. These slopes are calculated in block 5 of Figure 8.71; the calculation procedure is described in the next section.

The first vector choice defines the second and the third vectors as mapped in Table 8.14. The three vectors injected during one switching period h, under steady state operating conditions, are the nearest three vectors to the converter voltage reference vector. Thus, it is ensured that these vectors produce the smallest P_s and Q_s variation.

It could occur that the selected first vector (Figures 8.76 and 8.77 and Tables 8.13 and 8.12) is not one of the nearest three vectors to the required voltage vector, so there is a region jump requirement: for instance, vector V_{20} for region 1. In general, this situation is due to a change in the demanded steady state, provoking a transient response. Under these circumstances, only the first vector will be injected during the whole sample period h, in order to achieve fast dynamic operation capacity. Finally, selection of the second and third vectors as well as the first vector is graphically accomplished in block 4 of Figure 8.71.

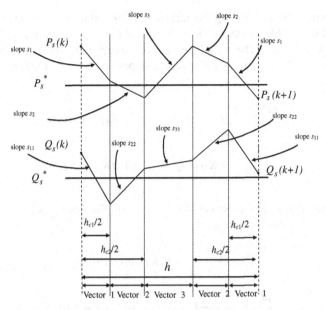

Figure 8.78 Steady state active and reactive power waveforms at motor and generator modes in subsynchronism for 3L-NPC VSC based P-DPC.

Derivative Expressions Calculation In this subsection, the mathematical expressions that relate the stator active and reactive power variations with the converter voltage vectors are derived. For predictive purposes, it is necessary to know the amount of modification for P_s and Q_s, for each converter voltage vector. First, the stator power derivative expressions are considered in DQ coordinates:

$$\frac{dP_s}{dt} = \frac{3}{2}\left(\frac{dv_{Ds}}{dt}i_{Ds} + \frac{di_{Qs}}{dt}v_{Qs} + \frac{dv_{Qs}}{dt}i_{Qs} + v_{Ds}\frac{di_{Ds}}{dt}\right) \qquad (8.61)$$

$$\frac{dQ_s}{dt} = \frac{3}{2}\left(\frac{dv_{Qs}}{dt}i_{Ds} + \frac{di_{Ds}}{dt}v_{Qs} - \frac{dv_{Ds}}{dt}i_{Qs} - v_{Ds}\frac{di_{Qs}}{dt}\right) \qquad (8.62)$$

TABLE 8.14 Second and Third Vector Selections According to the Nearest Three Vectors Criteria in Equivalent Sectors 2 and 3

Region	Vectors
1	V_{10}, V_1, V_{12}
2	V_{10}, V_{12}, V_{20}
3	V_{20}, V_2, V_{12}
4	V_{20}, V_{10}, V_0

Consequently, by substituting the state space representation of the machine and the filter, Equation (8.56), into expressions (8.61) and (8.62), it is possible to deduce the derivative expressions of the stator active and reactive powers, for each voltage vector of the three-level NPC converter. Hence, the slopes produced by the zero vectors are calculated as

$$\left.\frac{dP_s}{dt}\right|_{V_0} = K_{Lf}\left(-P_s K_1 - \omega_r Q_s / K_{Lf} + \omega_r K_2\right) \tag{8.63}$$

$$\left.\frac{dQ_s}{dt}\right|_{V_0} = K_{Lf}\left(-Q_s K_1 + \omega_r P_s / K_{Lf} + K_3\right) \tag{8.64}$$

After that, the slopes produced by the rest of the vectors can be calculated according to the following equations:

$$\left.\frac{dP_s}{dt}\right|_{V} = \left.\frac{dP_s}{dt}\right|_{V_0} + K_v K_4 \sin(\omega_r t + \delta + \theta) \tag{8.65}$$

$$\left.\frac{dQ_s}{dt}\right|_{V} = \left.\frac{dQ_s}{dt}\right|_{V_0} - K_v K_4 \cos(\omega_r t + \delta + \theta) \tag{8.66}$$

with δ being the phase shift between the rotor and stator flux space vectors and the constant values served in Table 8.15 and:

$$K_{Lf} = \frac{1}{\sigma_f L_s L_{rf}} \tag{8.67}$$

$$K_1 = R_s L_{rf} + R_{rf} L_s \tag{8.68}$$

$$K_2 = \frac{3}{2} L_{rf} \frac{|\vec{v}_s|^2}{\omega_s} \tag{8.69}$$

$$K_3 = \frac{3}{2} R_{rf} \frac{|\vec{v}_s|^2}{\omega_s} \tag{8.70}$$

$$K_4 = K_{Lf} L_m V_{bus} |\vec{v}_s| \tag{8.71}$$

Therefore, in Figure 8.79, the stator active power and reactive power derivative expressions are graphically represented, when the machine operates at subsynchronous speed with an inverter voltage vector that lies in regions 1, 2, and 3. Analyzing the resulting sinusoidal waveforms, for instance, in sector 12, under these operating conditions regions 2 and 3 are required. In that sector, according to Figure 8.79, vectors V_{20}, V_{10}, V_2, and V_{12} are the corresponding voltage vectors. As can be noted from Figure 8.79, these four vectors produce the four required combinations of P_s and Q_s variations $(1, 1), (1, -1), (-1, 1),$ and $(-1, -1)$, with the smallest slope magnitude. Consequently, by choosing these four vectors, control of the stator active reactive power, with reduced error ripples is ensured.

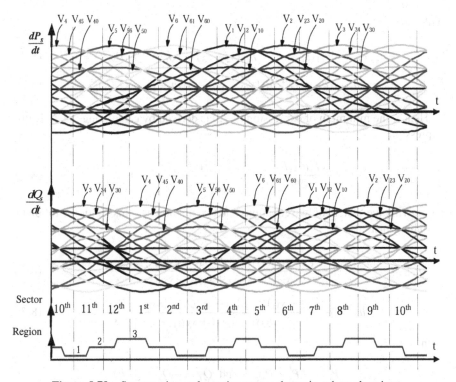

Figure 8.79 Stator active and reactive power slopes in subsynchronism.

Power Error Reduction Criteria at Constant Switching Frequency The switching instants h_{c1} and h_{c2} shown in Figure 8.78 are calculated in order to obtain the objective of reducing the squared P_s and Q_s errors at the end of the switching period. So, once the three vectors are chosen, their corresponding time intervals are calculated based on a prediction of P_s and Q_s. This task is graphically represented in block 6 of Figure 8.71.

Thus, since the objective is to minimize the sum of the squared P_s and Q_s errors at the end of the switching period h, it can be mathematically expressed as

$$f = e_{P_s}^2 + e_{Q_s}^2 \tag{8.72}$$

TABLE 8.15 Amplitude and Phase Shift Constants for Vectors of the Three-Level NPC Converter with $N = 1$–6

	V	K_v	θ
Large vectors	$V_N \Rightarrow V_1, V_2, V_3, V_4, V_5, V_6$	1	$-\frac{\pi}{3}(N-1)$
Medium vectors	$V_{N,N+1} \Rightarrow V_{12}, V_{23}, V_{34}, V_{45}, V_{56}, V_{61}$	$\frac{\sqrt{3}}{2}$	$-\frac{\pi}{3}(N-1)-\frac{\pi}{6}$
Small vectors	$V_{N0} \Rightarrow V_{10}, V_{20}, V_{30}, V_{40}, V_{50}, V_{60}$	$\frac{1}{2}$	$-\frac{\pi}{3}(N-1)$

with

$$e_{P_s}^2 = [h_{c1}(s_1 - s_2) + h_{c2}(s_2 - s_3) + s_3 h - P_s^* + P_s(k)]^2 \tag{8.73}$$

$$e_{Q_s}^2 = [h_{c1}(s_{11} - s_{22}) + h_{c2}(s_{22} - s_{33}) + s_{33} h - Q_s^* + Q_s(k)]^2 \tag{8.74}$$

Hence, the minimum of expression (8.72) is derived by calculating the derivative expressions

$$\frac{\partial f(h_{c1}, h_{c2})}{\partial h_{c1}} = 0, \qquad \frac{\partial f(h_{c1}, h_{c2})}{\partial h_{c2}} = 0 \tag{8.75}$$

and solving the system of two equations and two unknowns. This yields the following switching instants:

$$h_{c1} = -\frac{(s_{22} - s_{11})\bar{P}_s + (s_1 - s_2)\bar{Q}_s + (s_{22}s_3 - s_2 s_{33} + s_{33}s_1 - s_{11}s_3)h}{s_{22}s_1 - s_{33}s_1 - s_{22}s_3 + s_3 s_{11} - s_2 s_{11} + s_{33}s_2} \tag{8.76}$$

$$h_{c2} = -\frac{(s_{22} - s_{33})\bar{P}_s + (s_3 - s_2)\bar{Q}_s + (s_{22}s_3 - s_2 s_{33})h}{s_{22}s_1 - s_{33}s_1 - s_{22}s_3 + s_3 s_{11} - s_2 s_{11} + s_{33}s_2} \tag{8.77}$$

with

$$\bar{P}_s = -P_s^* + P_s(k) \quad \text{and} \quad \bar{Q}_s = -Q_s^* + Q_s(k)$$

The pair of equations (8.76) and (8.77) will minimize the P_s and Q_s errors in the same proportion. These expressions are valid for magnitudes of different scales and do not require a special weighting treatment.

Note that in previously studied predictive direct controls, the switching instants were calculated by focusing on the power ripple minimization. However, in this case, for simplicity and just to show a different solution, power errors at the end of the switching period are minimized.

Balancing of DC Bus Capacitors Voltages   The medium vectors of the three-level NPC converter must be selected appropriately to ensure the DC bus capacitor voltage balance. That choice is simply made by considering only the medium vectors V_{10} and V_{20} of equivalent sectors 2 and 3 and the output currents of the converter (i_{ar}, i_{br}, i_{cr}). This task is accomplished in block 7 of Figure 8.71.

Hence, from the equivalent sectors 2 and 3, Tables 2.8 and 2.9 from Chapter 2 provide sufficient information to perform the voltage balancing, according to the typical criteria for 3L-NPC converters. The reader should note the equivalence between the sextants (Chapter 2) and sectors used in this chapter, to appropriately be able to interpret these tables.

Vector Sequence Definition: Reduction of the Commutations   Once the three vectors and their duty cycles (h_{c1} and h_{c2}) are selected from the equivalent sectors 2 and 3, the next task is to define the right implementation sequence, in order to reduce the commutations of the semiconductors. As can be seen in Figure 8.78, the vectors are generated by a triangular based comparison, since the first and second vectors of the sequence are split into two different application time intervals. This strategy is different from the one employed in two-level converter based P-DTC and P-DPC strategies. In this case, for an equal sample period h, the switching frequency of the devices is increased compared to the two-level converter based control strategies. In contrast, the P_s and Q_s ripples are decreased for the triangular based vector sequence implementation. In addition, many authors adopt this implementation procedure for three-level NPC converter applications, since, in general, it produces more symmetric commutations of the switching devices, especially considering the transition to subsequent sample periods.

Hence, depending on the medium vector choice, defined by the DC bus capacitor voltage balancing algorithm that has been described in the previous section, Table 2.10 from Chapter 2 provides the vector sequence order that reduces the commutations in a switching period. According to that, it is found that the zero vector employed will always be 111 since it produces the lowest commutations.

Real Vector Definition   The last task of the proposed algorithm is to generate the real voltage vectors, from the calculated equivalent vectors in sectors 2 and 3. The real vector definition can be performed by simply interchanging the phases according to Table 2.11 presented in Chapter 2, considering the sector where the $\vec{x}^r$ space vector lies. The reader again must note the equivalence between sextants and sectors. Finally, these two last tasks are accomplished in block 8 of Figure 8.71.

Hence, Figure 8.80 illustrates how the pulses for the controlled semiconductors can be created, according to a triangular based comparison. A sequence of vectors, 100-200-210–210-200-100, is generated, with two specific switching lengths, h_{c1} and h_{c2}.

Therefore, thanks to the triangular based comparison of h_{c1} and h_{c2} values, the pulses can easily be created. The sequence of three different vectors is divided into five different parts. In this particular case, only semiconductors S_{a1} and S_{b2} commutate within the switching period h.

On the other hand, when Table 2.10 is used for a reduced commutation schema within the switching period, the first vector defined by Tables 8.13 or 8.12, can actually be injected as the first, the second, or the third vector, as illustrated in Figure 8.81.

8.6.2.4 *Example 8.7: 3L-NPC VSC Based P-DPC of a 15 kW DFIM*   In this example, an experimental result of the control strategy presented in the previous section is carried out. Despite the fact that this control technique is oriented to high power applications, for practical reasons, the results have been obtained in a 15 kW test bench, composed of a 380 V–15 kW–1500 rev/min doubly fed induction machine.

The selection of the rotor side inductive filter is made according to a simulation and experimental comparison procedure. Table 8.16 shows the obtained stator current THD with several inductance values, at nominal power conditions.

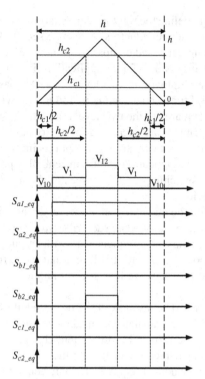

Figure 8.80   Triangular based pulse generation schema: 100-200-210–210-200-100 sequence of vectors injection.

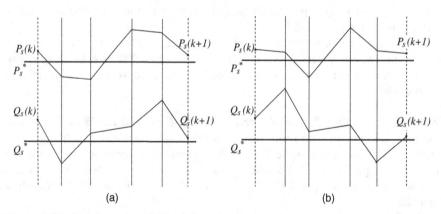

Figure 8.81   Two examples of P_s and Q_s evolutions, according to multilevel P-DPC: (a) first vector defined by Table 8.12 is finally the first vector injected; (b) first vector defined by Table 8.12 is finally the second vector injected.

TABLE 8.16   Generated Stator Current Quality for Different Rotor Side Filter Values (15 kW, 11 kVAR, 1250 rev/min and 500 Hz Switching Frequency)

L_f (mH)	Stator Current THD	
	Simulation	Experiment
0	7.1%	7.8%
4	4%	4.7%
10	2.6%	3.1%
14	2.5%	2.8%

In order to produce the stator current THD that reasonably meets the IEEE Standard 519-1992 recommendation, the selected inductive filter will be 14 mH, for subsequent experimental trials.

Hence, the steady state behavior of the presented predictive DPC strategy will be shown in two different scenarios: with a 14 mH inductive rotor filter and without a rotor filter. It is obvious that, with the filter, the achieved current and power qualities will be better than without the filter.

Hence, the machine is magnetized by the stator, interchanging a stator reactive power of 11 kVAR with the grid. The DC bus voltage of the three-level NPC converter has been set to 320 V. The switching frequency is imposed to a considerably low value of 0.5 kHz. In both cases, the stator active power reference is set to the nominal power of 15 kW, implying that the machine will operate at the nominal torque. The speed of the machine is controlled externally to 1250 rev/min first and then to 750 rev/min. Under these operating conditions, the three-level NPC converter will operate in the fourth region (1250 rev/min) and in the first, second, and third regions (750 rev/min) of the space vector diagram.

As can be seen in Figure 8.82, the quality of the generated stator and rotor currents is better when the machine is fed with the rotor filter; on the other hand, while at higher modulation indexes (Figures 8.82c and 8.82d), all the voltage levels of the three-level NPC converter are used.

8.6.3   Three-Level NPC VSC Based DTC of the DFIM

In this next section, the predictive DTC strategy is depicted for the three-level NPC converter. Since this control strategy is equivalent to the DPC strategy presented in this chapter, only the most significant features of the predictive DTC will be analyzed. First, the control block diagram will presented; second, only the new different blocks will be analyzed, that is, the first vector selection blocks, the slope calculation block, and the error reduction block.

8.6.3.1 *Control Block Diagram*   The block diagram of the proposed control strategy is shown in Figure 8.83. The directly controlled magnitudes are the electromagnetic torque and the rotor flux amplitude. From the references, the

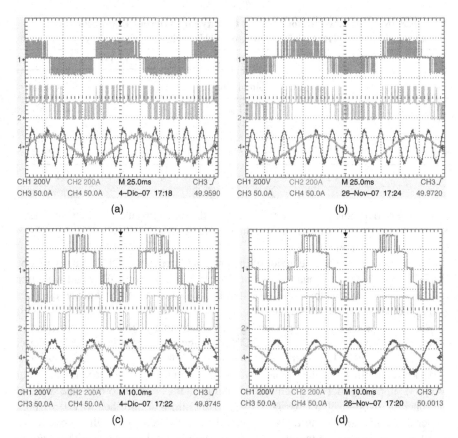

Figure 8.82 Voltage (CH1-v_{ab} and CH2-v_{ao}) and current (CH3-i_{as} and CH4-i_{ar}) waveforms of predictive DPC control strategy, with and without rotor side inductive filter, at steady state with $P_s = 15$ kW, $Q_s = 11$ kVAR, and 0.5 kHz switching frequency: (a) 1250 rev/min without filter, (b) 1250 rev/min with 14 mH inductive filter, (c) 750 rev/min without filter, and (d) 750 rev/min with 14 mH inductive filter.

control strategy calculates the pulses (S_{1a}, S_{2a}, S_{1b}, S_{2b}, S_{1c}, S_{2c}) for the IGBTs of the three-level NPC voltage source converter (VSC), in order to meet the following two objectives: reduced ripples and constant switching frequency behavior. Figure 8.83 shows that the control strategy is divided into eight different tasks represented in eight different blocks. Note that this control block diagram is equivalent to the block diagram of Figure 8.71 for predictive DPC strategy. In fact, blocks 1, 2, 3, 7, and 8 are exactly the same, so for simplicity they will not be analyzed again in this section. On the other hand, blocks 4, 5, and 6 present some differences that will be analyzed in subsequent subsections.

8.6.3.2 First Vector Selection Strategy
The basic principle of the direct control techniques determines the choice of one specific voltage vector of the

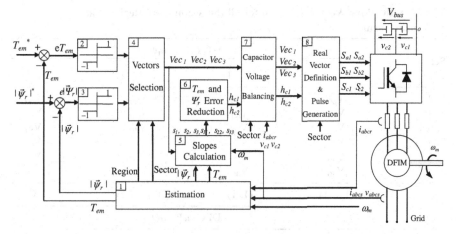

Figure 8.83 Predictive DTC block diagram.

converter, in order to correct the directly controlled magnitude's errors. Equivalently, as occurred with the predictive DPC, the choice of this vector is directly calculated to the electromagnetic torque and the rotor flux amplitude errors themselves, the sector where the $\vec{x}^r$ lies, and the region of the equivalent voltage vector $\vec{v}_{f\_eq}$.

Once again, the proposed predictive DTC strategy adopts this principle to select the first voltage vector of the three-vector sequence employed to minimize the T_{em} and $|\vec{\psi}|_r$ errors. Hence, the first voltage vector selection is based on a look-up table mapped by the output of two ON–OFF comparators without hysteresis bands. These tasks are accomplished in blocks 2, 3, and 4 of Figure 8.83. When the estimated inverter voltage vector lies in regions 1, 2, or 3, the look-up table is that shown in Table 8.17.

On the contrary, when the inverter space vector is located in region 4, the look-up table is that shown in Table 8.18. Note that, in both cases, the required information to select the first vector for the predictive DTC and DPC strategies has been included.

After that, the second and third vector selection procedure of DTC strategy employs the same table as in DPC strategy (Table 8.14).

TABLE 8.17   First Vector Selection for Regions 1, 2, and 3 at Subsynchronism

Sectors		$(eT_{em}, e|\vec{\psi}_r|)$ or (eP_s, eQ_s)			
DTC	DPC	$(1, 1)$	$(1, -1)$	$(-1, 1)$	$(-1, -1)$
1, 5, 9	4, 8, 12	V_{20}	V_{10}	V_2	V_{12}
2, 6, 10	3, 7, 11	V_{20}	V_{10}	V_{12}	V_1
3, 7, 11	2, 6, 10	V_{10}	V_{20}	V_1	V_{12}
4, 8, 12	1, 5, 9	V_{10}	V_{20}	V_{12}	V_2

TABLE 8.18   First Vector Selection for all the Sectors and Region 4 at Subsynchronism

Sector	DTC $(uT_{em}, u|\vec{\psi}_r|)$			
	$(1, 1)$	$(1, -1)$	$(-1, 1)$	$(-1, -1)$
1, 2	V_{60}	V_{50}	V_{20}	V_{30}
3, 4	V_{10}	V_{60}	V_{30}	V_{40}
5, 6	V_{20}	V_{10}	V_{40}	V_{50}
7, 8	V_{30}	V_{20}	V_{50}	V_{60}
9, 10	V_{40}	V_{30}	V_{60}	V_{10}
11, 12	V_{50}	V_{40}	V_{10}	V_{20}
	$(1, -1)$	$(1, 1)$	$(-1, -1)$	$(-1, 1)$
Sector	DPC (uP_s, uQ_s)			

8.6.3.3 Derivative Expressions Calculation

In this subsection, the mathematical expressions that relate the electromagnetic torque and the rotor flux amplitude variations with the inverter voltage vectors are presented. From the machine's model equations the torque and flux derivative expressions yield:

$$
\frac{d|\vec{\psi}_r|}{dt} = \left(\frac{R_{rf}L_m}{\sigma L_s L_r}\right)|\vec{\psi}_s|\cos\delta - \left(\frac{R_{rf}}{\sigma L_r}\right)|\vec{\psi}_r|
$$
$$
- \left(\frac{L_f L_m}{\sigma L_s L_r}\right)\omega_r|\vec{\psi}_s|\sin\delta + \frac{2}{3}V_{bus}\cos\left(\omega_r t - \frac{\pi}{3}(n-1)\right) \tag{8.78}
$$

$$
\frac{dT_{em}}{dt} = T_{em}\left(\left(\frac{L_f}{\sigma L_r}+1\right)\frac{\omega_r}{\tan\delta} - \left(\frac{R_s}{\sigma L_s}+\frac{R_r}{\sigma L_r}\right)\right) - \frac{3}{2}pL_f\left(\frac{L_m}{\sigma L_s L_r}\right)^2\omega_r|\vec{\psi}_s|^2
$$
$$
+ p\frac{L_m}{\sigma L_s L_r}V_{bus}|\vec{\psi}_s|\sin\left(\omega_r t + \delta - \frac{\pi}{3}(n-1)\right) \tag{8.79}
$$

8.6.3.4 Torque and Flux Error Reduction Criteria at Constant Switching Frequency

The expected electromagnetic torque and rotor flux evolutions in one switching period are shown in Figure 8.84. Since the same error reduction procedure as predictive DPC is followed, the expressions for the duty cycles are

$$
h_{c1} = -\frac{(s_{22}-s_{11})\bar{T}_{em}+(s_1-s_2)\bar{F}_r+(s_{22}s_3-s_2s_{33}+s_{33}s_1-s_{11}s_3)h}{s_{22}s_1-s_{33}s_1-s_{22}s_3+s_3s_{11}-s_2s_{11}+s_{33}s_2} \tag{8.80}
$$

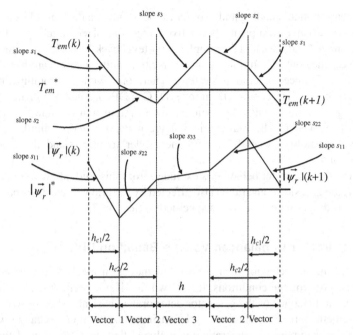

Figure 8.84 Steady state electromagnetic torque and rotor flux amplitude waveforms at motor and generator modes in subsynchronism.

$$h_{c2} = -\frac{(s_{22} - s_{33})\bar{T}_{em} + (s_3 - s_2)\bar{F}_r + (s_{22}s_3 - s_2s_{33})h}{s_{22}s_1 - s_{33}s_1 - s_{22}s_3 + s_3s_{11} - s_2s_{11} + s_{33}s_2} \qquad (8.81)$$

with

$$\bar{T}_{em} = -T_{em}^* + T_{em}(k) \quad \text{and} \quad \bar{F}_r = -|\vec{\psi}_r|^* + |\vec{\psi}_r|(k)$$

8.7 CONTROL SOLUTIONS FOR GRID VOLTAGE DISTURBANCES, BASED ON DIRECT CONTROL TECHNIQUES

8.7.1 Introduction

In this section, the behavior of the wind turbine generation system is studied under abnormal grid voltage situations. Mainly an unbalanced grid voltage scenario is addressed, but the presence of voltage dips in the grid is also considered. It will be shown that if no special control efforts are employed, the wind turbine system behavior, controlled by any of the direct control strategies presented in this chapter, can deteriorate.

Therefore, in order to tackle these problems, different control modifications to the direct control techniques studied in this chapter will be provided for rotor side and converter control of the wind turbine.

For an easier understanding and in order to focus more on the control solutions for these two disturbances, simple nonpredictive direct control techniques are employed (see Sections 8.2 and 8.3), and the simplest two-level back-to-back converter is used as the VSC. The control solutions proposed for this faulty scenario are based on the appropriate reference generation strategy. Therefore, with the combination of a simple but efficient direct control technique (DTC or DPC), together with the required reference generation strategy, the wind turbine system will be able to tackle the problems provoked by the faulty grid voltage context. Finally, simulation based illustrative examples successfully help the reader to understand the theoretical considerations.

The grid side converter behavior is excluded from this study; however, as seen in Chapter 7, the grid side converter is also affected by the grid voltage disturbances and should accordingly be controlled as specified in Section 7.6.

8.7.2 Control for Unbalanced Voltage Based on DPC

In this section, the behavior of a doubly fed induction machine is studied under unbalanced grid voltage conditions. It's shown that if no special control efforts are employed, the behavior of the generator deteriorates, basically due to two reasons: electromagnetic torque oscillations and nonsinusoidal current exchange with the grid. Those phenomena are theoretically analyzed first, as a function of the stator active and reactive instantaneous power exchange, by the stator of the DFIM. This analysis provides the main ideas for generation of the active and reactive power references, for the rotor side converter (RSC) controlled by means of direct power control (DPC) techniques. Therefore, this section proposes the algorithm that generates the RSC power references, without the necessity of a sequence component extraction, in order to eliminate torque oscillations and achieve sinusoidal stator current exchanges through the stator. Finally, a simulation based example successfully validates the studied power reference generation methods.

8.7.2.1 Description of the Scenario The analyzed system considers the imbalance created in the grid voltage by the presence of an unbalanced load, as illustrated in Figure 8.85. In a general situation, not all the impedances (R_{abc}) of the load considered are equal. The unbalanced consumption of currents of the load (i_{abcL}), together with the impedance of the lines (Z_{N1}, Z_{N2}), induces an unbalanced voltage in the stator and in the grid side converter of the wind turbine.

For simplicity in the exposition, the transformer of the wind turbine system has been omitted in the analysis. It must be highlighted that there could be other different situations where the voltage seen by the wind turbine is unbalanced, but in almost all the cases, the permanent imbalance is created by a permanent unbalanced load consuming unbalanced currents. Thus, the presentation in this section focuses on the scenario illustrated in Figure 8.85. The VSC employed in the study is the classic 2L-VSC.

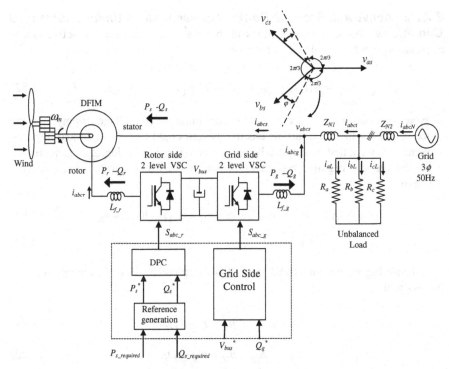

Figure 8.85   Wind energy generation system under unbalanced voltage grid conditions.

In this study, the DPC technique is adopted for RSC as enunciated in Section 8.3; note that any other DPC technique could also be applied. Through the rotor side converter, the stator active and reactive powers (P_s and Q_s) exchanged with the grid are directly controlled. Hence, the pulses for the controlled semiconductors of the two-level back-to-back converter are calculated from the stator active and reactive power references. Thus, by means of these direct control techniques, it is possible to achieve quick dynamic performances, without using current regulators and with reasonably good implementation simplicity. This strategy can be understood as an alternative solution to control methods based on current loops, as studied in Chapter 7. In this case, an easy and standard version of DPC has been utilized for rotor side converter control. One of the most important reasons for this choice is that, with this work, we want to show how the unbalanced problem can be tackled, by simply adding the appropriate power reference generation strategy to a classic DPC technique; however, as mentioned before, it is also possible to use any other direct power control. The proper reference generations are studied in depth in subsequent sections.

Prior to the design of the control strategies, to address this unbalanced situation, the following subsection presents the theoretical bases for analyzing the effect of unbalanced voltage in the active and reactive powers.

8.7.2.2 Active and Reactive Power Representations Under Unbalanced Conditions
As seen several times in this book, the active and reactive power expressions can be calculated as follows:

$$P = \tfrac{3}{2}\text{Re}\{\vec{v}\cdot\vec{i}^{*}\} = \tfrac{3}{2}(v_{\alpha}i_{\alpha} + v_{\beta}i_{\beta}) \quad Q = \tfrac{3}{2}\text{Im}\{\vec{v}\cdot\vec{i}^{*}\} = \tfrac{3}{2}(v_{\beta}i_{\alpha} - v_{\alpha}i_{\beta}) \quad (8.82)$$

Note that since this section studies an unbalanced voltage supply, all the space vectors of the wind turbine electric system could have positive and negative sequences as well; that is, they can be unbalanced. Hence, considering a general case with positive and negative sequences in both voltage and current space vectors; we have

$$\vec{v} = (\vec{v}_1 + \vec{v}_2) = v_{\alpha} + jv_{\beta} = (v_{\alpha1} + v_{\alpha2}) + j(v_{\beta1} + v_{\beta2}) \qquad (8.83)$$

$$\vec{i} = (\vec{i}_1 + \vec{i}_2) = i_{\alpha} + ji_{\beta} = (i_{\alpha1} + i_{\alpha2}) + j(i_{\beta1} + i_{\beta2}) \qquad (8.84)$$

Substituting expressions (8.83) and (8.84) into equations of active and reactive powers yields

$$P = A_P + B_P + C_P + D_P \qquad (8.85)$$

$$Q = A_Q + B_Q + C_Q + D_Q \qquad (8.86)$$

with

$$A_P = \tfrac{3}{2}\text{Re}\{\vec{v}_1\cdot\vec{i}_1^{*}\} = \tfrac{3}{2}(v_{\alpha1}i_{\alpha1} + v_{\beta1}i_{\beta1}) \qquad (8.87)$$

$$B_P = \tfrac{3}{2}\text{Re}\{\vec{v}_2\cdot\vec{i}_2^{*}\} = \tfrac{3}{2}(v_{\alpha2}i_{\alpha2} + v_{\beta2}i_{\beta2}) \qquad (8.88)$$

$$C_P = \tfrac{3}{2}\text{Re}\{\vec{v}_1\cdot\vec{i}_2^{*}\} = \tfrac{3}{2}(v_{\alpha1}i_{\alpha2} + v_{\beta1}i_{\beta2}) \qquad (8.89)$$

$$D_P = \tfrac{3}{2}\text{Re}\{\vec{v}_2\cdot\vec{i}_1^{*}\} = \tfrac{3}{2}(v_{\alpha2}i_{\alpha1} + v_{\beta2}i_{\beta1}) \qquad (8.90)$$

$$A_Q = \tfrac{3}{2}\text{Im}\{\vec{v}_1\cdot\vec{i}_1^{*}\} = \tfrac{3}{2}(v_{\beta1}i_{\alpha1} - v_{\alpha1}i_{\beta1}) \qquad (8.91)$$

$$B_Q = \tfrac{3}{2}\text{Im}\{\vec{v}_2\cdot\vec{i}_2^{*}\} = \tfrac{3}{2}(v_{\beta2}i_{\alpha2} - v_{\alpha2}i_{\beta2}) \qquad (8.92)$$

$$C_Q = \tfrac{3}{2}\text{Im}\{\vec{v}_1\cdot\vec{i}_2^{*}\} = \tfrac{3}{2}(v_{\beta1}i_{\alpha2} - v_{\alpha1}i_{\beta2}) \qquad (8.93)$$

$$D_Q = \tfrac{3}{2}\text{Im}\{\vec{v}_2\cdot\vec{i}_1^{*}\} = \tfrac{3}{2}(v_{\beta2}i_{\alpha1} - v_{\alpha2}i_{\beta1}) \qquad (8.94)$$

Note that, for simplicity, the time dependence of each term has been omitted. On the other hand, the terms A_P, A_Q and B_P, B_Q are constant at steady state, since they are composed of same sequence product. However, terms C_P, C_Q and D_P, D_Q oscillate at 2ω angular speed, since they are composed of positive and negative sequence products. In a subsequent subsection, this basic analysis is applied to the DFIM.

8.7.2.3 DFIM Analysis and Control Under Unbalanced Voltage This section finds the required modifications for the control, in order to make the wind turbine operative under unbalanced voltage situations. First, the system under unbalanced voltage is studied and then the control strategy is designed.

DFIM Analysis   The DFIM can be modeled with the following voltage and flux equations, in the stator reference frame as seen in previous sections:

$$\vec{v}_s = R_s\vec{i}_s + \frac{d\vec{\psi}_s}{dt} \tag{8.95}$$

$$\vec{\psi}_s = L_s\vec{i}_s + L_m\vec{i}_r \tag{8.96}$$

On the other hand, the electromagnetic torque can be expressed using the following equation:

$$T_{em} = \tfrac{3}{2}p\,\mathrm{Im}\big\{\vec{\psi}_s^{*}\cdot\vec{i}_s\big\} = \tfrac{3}{2}p(\psi_{\alpha s}i_{\beta s} - \psi_{\beta s}i_{\alpha s}) \tag{8.97}$$

It must be highlighted that as the stator voltage is unbalanced, all the space vectors of Equations (8.95)–(8.97), may present positive and negative sequences.

The influence of the unbalanced voltage will be analyzed in the stator flux. From expressions (8.83) and (8.95), the relationship between stator flux and the stator voltage results:

$$\vec{v}_{s1} = R_s\vec{i}_{s1} + j\omega_s\vec{\psi}_{s1} \tag{8.98}$$

$$\vec{v}_{s2} = R_s\vec{i}_{s2} - j\omega_s\vec{\psi}_{s2} \tag{8.99}$$

The stator flux evolves as the superposition of the positive and negative sequence fluxes—the first one rotating clockwise and the second one anticlockwise, as illustrated in Figure 8.86:

Substituting the positive and negative sequence components of the stator current and stator flux into the torque expression (8.97), we have

$$T_{em} = \tfrac{3}{2}p\,\mathrm{Im}\big\{\vec{\psi}_{s1}^{*}\vec{i}_{s1} + \vec{\psi}_{s1}^{*}\vec{i}_{s2} + \vec{\psi}_{s2}^{*}\vec{i}_{s1} + \vec{\psi}_{s2}^{*}\vec{i}_{s2}\big\} \tag{8.100}$$

It is deduced that the torque is composed of two constant terms, associated with equal sequence products and two 2ω pulsating terms associated with nonequal sequence products.

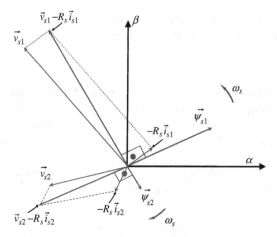

Figure 8.86   Relationship between stator voltage, current, and flux space vectors.

Substituting the flux from expressions (8.98) and (8.99) in torque equation (8.100), we find

$$T_{em} = \tfrac{3}{2}p\frac{1}{\omega_s}\mathrm{Re}\left\{\vec{v}_{s1}\vec{i}_{s1}^{*} - \vec{v}_{s2}\vec{i}_{s2}^{*} - \vec{v}_{s2}\vec{i}_{s1}^{*} + \vec{v}_{s1}\vec{i}_{s2}^{*} - R_s(|\vec{i}_{s1}|^2 - |\vec{i}_{s2}|^2)\right\} \quad (8.101)$$

By similitude with the active power expression, Equation (8.85) and defining the constant term E_{s_T}, we have

$$E_{s\_T} = R_s(|\vec{i}_{s1}|^2 - |\vec{i}_{s2}|^2) \quad (8.102)$$

The torque can be expressed as

$$T_{em} = \frac{p}{\omega_s}(A_{Ps} - B_{Ps} + C_{Ps} - D_{Ps} - E_{Ts}) \quad (8.103)$$

where the subscript s indicates the stator variables.

On the other hand, the stator active and reactive power expressions, as a function of the positive and negative sequences of the voltage and currents, yield

$$P_s = A_{Ps} + B_{Ps} + C_{Ps} + D_{Ps} \quad (8.104)$$

$$Q_s = A_{Qs} + B_{Qs} + C_{Qs} + D_{Qs} \quad (8.105)$$

As can be observed from Equations (8.103) and (8.104), the stator active power and the electromagnetic torque are closely related since they share four common terms: A_{Ps}, B_{Ps}, C_{Ps}, and D_{Ps}. Hence, the relationship between the torque and the active power can be derived:

$$P_s = \frac{T_{em}\omega_s}{p} + 2B_{Ps} + 2D_{Ps} + E_{Ts} \quad (8.106)$$

This expression is exploited later, in order to improve the control performance under unbalanced grid voltage conditions.

Active Power Oscillations Cancellation Strategy (APOC)   As depicted in Figure 8.85, the control strategy for the DFIM is divided into two different general blocks. The first block is a DPC technique, directly controlling the stator active and reactive powers. The second block calculates the required stator power references, to deal with the problems generated by the unbalanced grid voltage, that is, electromagnetic torque oscillations and nonsinusoidal stator currents.

However, in this subsection, the control strategy that produces constant active power is analyzed. From the stator active power expression, Equation (8.104), we notice that addition of the two oscillating terms is zero:

$$C_{Ps} + D_{Ps} = 0 \tag{8.107}$$

The active power will remain constant, and since C_{Ps} and D_{Ps} depend on both positive and negative stator currents sequences, it is impossible to make both oscillating terms zero at the same time ($C_{Ps} = 0$ and $D_{Ps} = 0$), because it will require the imposition of zero stator currents. The required power ($P_{s\_required}$) is equal to the reference active power (P_s^*):

$$P_s^* = P_{s\_required} = A_{Ps} + B_{Ps} \tag{8.108}$$

but the electromagnetic torque, Equation (8.103), will have oscillating terms,

$$
\begin{aligned}
T_{em} &= \frac{p}{\omega_s}(A_{Ps} - B_{Ps} - 2D_{Ps} - E_{Ts}) \\
&= \frac{p}{\omega_s}(A_{Ps} - B_{Ps} + 2C_{Ps} - E_{Ts})
\end{aligned} \tag{8.109}
$$

Torque Oscillations Cancellation Strategy (TOC)   As occurs with the active power, the analysis of expression (8.103) shows that the only way to achieve constant electromagnetic torque, under unbalanced grid voltage conditions, is by imposing

$$C_{Ps} - D_{Ps} = 0 \tag{8.110}$$

The resulting electromagnetic torque is

$$T_{em} = \frac{p}{\omega_s}(A_{Ps} - B_{Ps} - E_{Ts}) \tag{8.111}$$

This imposition, for a given voltage imbalance, in general does not make either the positive or negative sequences of the stator currents zero.

Under this situation, the required stator power and the reference stator active power are related by the following expression:

$$P_s^* = P_{s\_required} + 2C_{Ps} = P_{s\_required} + 2D_{Ps} \qquad (8.112)$$

Note that by equating the oscillating terms of this last expression and expression (8.104), we have expression (8.110). On the other hand, as a function of electromagnetic torque from Equation (8.106), the active power reference results:

$$
\begin{aligned}
P_s^* &= \frac{T_{em}\omega_s}{p} + 2B_{Ps} + 2C_{Ps} + E_{Ps} \\
&= \frac{T_{em}\omega_s}{p} + 2B_{Ps} + 2D_{Ps} + E_{Ps}
\end{aligned}
\qquad (8.113)
$$

In addition, as shown later in a simulation validation, with this power reference generation strategy, the stator currents exchanged with the grid are unbalanced as well, but they are sinusoidal, ensuring that there is no further quality deterioration.

Torque Oscillations Cancellation Without Sequence Calculation (TOC-WSC)   In the previous two power reference generation strategies, positive and negative sequence calculations of voltage and currents were needed, to avoid power–torque oscillations. In this subsection, an improved method to avoid these sequence calculations is presented.

For a given active power reference ($P_{s\_required}$), an oscillating term obtained from the electromagnetic torque and stator power is added as shown in Figure 8.87:

$$P_s^* = P_{s\_required} + \left(P_s - \frac{\omega_s}{p}T_{em}\right) \qquad (8.114)$$

Substituting into expression (8.114) Equations (8.103) and (8.104), the stator active power reference expression yields

$$P_s^* = P_{s\_required} + (2B_{Ps} + 2D_{Ps} + E_{Ts}) \qquad (8.115)$$

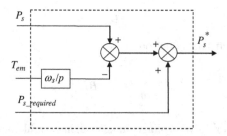

Figure 8.87   Stator active power generation strategy.

Again, this last expression must be equal to expression (8.104); thus, the oscillating terms yields

$$C_{Ps} + D_{Ps} = 2D_{Ps} \qquad (8.116)$$

This means that, with this last equation, the condition to cancel the torque oscillations has been obtained.

Note that with this reference generation strategy, as mentioned before, since the positive and negative sequence calculations are not needed to avoid the electromagnetic torque oscillations, implementation of the strategy is simpler.

8.7.2.4 Example 8.8: System Behavior Controlled by DPC Under Unbalanced Voltage with Oscillatory Power References

This example shows the behavior of a 690 V and 2 MW DFIM supplied by an unbalanced grid voltage, when it is controlled by a classic DPC but with the studied torque oscillations elimination strategy (TOC-WSC). It seeks to show the improvements derived from the proper power references generation strategy. The programmed unbalanced voltage in p.u. is $v_{as} = 1 \angle 0°$, $v_{bs} = 1.17 \angle -115.3°$, $v_{cs} = 1.17 \angle +115.3°$. Figure 8.88 covers the simulation results. The speed of the machine is controlled to constant 0.9 p.u. subsynchronous speed. During the first half of the experiment, oscillating power references are commanded, while in the middle of the experiment, the power references are constant.

Thus, Figure 8.88a illustrates the unbalanced stator voltage. During the first half of the experiment, the stator currents are unbalanced but sinusoidal. Both the stator voltage and currents present the same imbalance, as can be noticed from Figures 8.88a and 8.88b. However, once compensation of the imbalance is disabled, the stator currents further deteriorate, becoming nonsinusoidal. This deterioration of currents is typical when direct control techniques are employed with constant references during permanent voltage imbalance of the grid.

In order to achieve the shown improvement in the stator currents, the stator active power performance can be seen in Figure 8.88c. The active power oscillations commanded have a frequency of 100 Hz, as studied theoretically. Note that in the middle of the experiment, both references are set to a constant value.

In addition, as exhibited in Figure 8.88d, the torque oscillations are totally eliminated in the first half of the experiment thanks to the control strategy employed. Again, as studied analytically, the torque oscillations are at 100 Hz when they are not disabled. Note that this oscillatory torque would create oscillatory behavior of the speed, disturbing the performance of the wind turbine operation.

Finally, Figures 8.88e and 8.88f illustrate how the rotor voltages (shown filtered for an easier analysis) and currents under both situations, have 100 Hz oscillations superimposed on the fundamental component. Note that an additional rotor voltage is required, to keep the stator currents sinusoidal in the absence of torque oscillations.

Hence, thanks to the programmed power reference strategy, it is possible to avoid two of the major problems derived from an unbalanced stator voltage scenario:

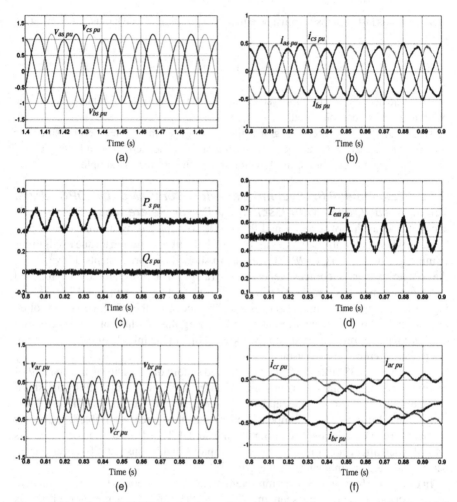

Figure 8.88   The 2 MW DFIM behavior under unbalanced voltage and controlled by DPC with oscillatory references: (a) unbalanced stator voltage, (b) stator currents, (c) stator active and reactive powers, (d) electromagnetic torque, (e) filtered rotor voltage, and (f) rotor currents.

nonsinusoidal exchange of currents through the stator and oscillatory torque behavior. However, the stator currents must still be unbalanced to eliminate the torque oscillations.

8.7.3   Control for Unbalanced Voltage Based on DTC

This section explores the issue of controlling the doubly fed induction machine (DFIM) under an unbalanced grid voltage situation and by using a direct torque control (DTC) technique. The effect of the imbalance on the machine is studied first,

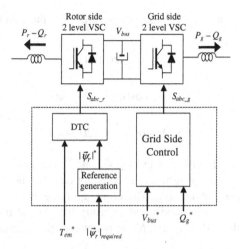

Figure 8.89   General control block diagram under unbalanced voltage grid conditions.

based on an amplitude space vector analysis. An alternative analysis method for studying the unbalanced situations is to use the symmetric decomposition method seen in the previous section. Next, a special control strategy is proposed in order to address the two main difficulties generated by the imbalance; torque oscillations and nonsinusoidal current exchange with the grid through the stator of the machine. This control strategy is composed of DTC control (Section 8.2) and a special flux reference generation schema (Figure 8.89). As similarly occurs with the DPC, if constant torque and rotor flux amplitude are required, the DFIM will exchange nonsinusoidal currents with the grid through the stator in the presence of unbalanced voltage. This fact in general is not accepted in energy generation applications, since it can unstabilize the electric network among other reasons.

Consequently, the same philosophy as in the previous section is followed to tackle the unbalanced situation. By simply adding a special rotor flux reference generation strategy, to the widely used DTC for the DFIM, it is possible to solve the unbalanced voltage problem, without modifying either the control philosophy or the hardware-system structures.

Prior to the design of the control strategies, to address this unbalanced situation, the following subsections present the theoretical principle for analyzing the effect of unbalanced, voltage based on a different methodology than the previous section. In this case now, the method is based on an amplitude analysis of the space vector magnitudes.

8.7.3.1 Analysis of the Unbalanced Voltage Distortion Based on the Amplitude of Space Vectors   In this case again, the unbalanced voltage considered is as represented in Figure 8.85. There is no zero sequence consideration and, for simplicity, two of the phases have equal amplitude and equal angle displacement φ from the 120° (*b* and *c* phases in Figure 8.85). Hence,

the voltage is mathematically represented according to the following three voltage equations:

$$v_a = \hat{V}_a \cos(\omega t) \tag{8.117}$$

$$v_b = \hat{V}_b \cos(\omega t - 120° - \varphi) \tag{8.118}$$

$$v_c = \hat{V}_c \cos(\omega t + 120° + \varphi) \tag{8.119}$$

where v_a, v_b, and v_c are the voltages and $\hat{V}_a$, $\hat{V}_b$, and $\hat{V}_c$ are the constant amplitudes of the voltages. In this case the zero sequence is not present; the imbalance produces $\hat{V}_b = \hat{V}_c$ and constant φ phase shift. Therefore, applying the space vector definition (see the Appendix for more detail) and by using the general trigonometric equivalences, it is possible to derive the $\alpha\beta$ components of the voltage space vector:

$$v_\alpha = \hat{V}_\alpha \cos(\omega t) \tag{8.120}$$

$$v_\beta = \hat{V}_\beta \sin(\omega t) \tag{8.121}$$

with the following constant amplitudes of the $\alpha\beta$ components:

$$\hat{V}_\alpha = \tfrac{1}{3}\left(2\hat{V}_a + \hat{V}_b\left(\cos\varphi + \sqrt{3}\sin + \varphi\right)\right) \tag{8.122}$$

$$\hat{V}_\beta = \hat{V}_b\left(\cos\varphi - \frac{\sin\varphi}{\sqrt{3}}\right) \tag{8.123}$$

Once again by using general trigonometric relations, the amplitude of the voltage space vector $|\vec{v}|$ can be expressed as follows:

$$|\vec{v}|^2 = \frac{\hat{V}_\alpha^2 + \hat{V}_\beta^2}{2} + \frac{\hat{V}_\alpha^2 - \hat{V}_\beta^2}{2}\cos(2\omega t) \tag{8.124}$$

This last equation shows the oscillating behavior of the voltage space vector amplitude when it is represented as a function of time. On the other hand, the space vector representation in the $\alpha\beta$ complex plane would yield an elliptic trajectory when unbalanced, as depicted in Figure 8.90 and seen in previous chapters.

8.7.3.2 Analysis of the DFIM at Steady State

Once the unbalanced voltage has been studied, we turn to how it affects the DFIM behavior. Since the unbalanced grid voltage is directly connected to the stator, the relation between the stator voltage and the stator flux space vectors is required in the $\alpha\beta$ reference frame—based on the well-known relation

$$\vec{v}_s = R_s \vec{i}_s + \frac{d\vec{\psi}_s}{dt} \tag{8.125}$$

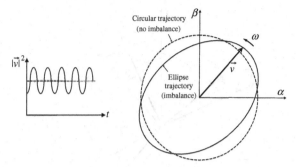

Figure 8.90 Unbalanced effect on the space vector amplitude and trajectory.

Similarly, the electromagnetic torque is calculated according to

$$T_{em} = \tfrac{3}{2}p \frac{L_m}{\sigma L_s L_r} \mathrm{Im}\{\vec{\psi}_r^{\,*} \cdot \vec{\psi}_s\} \qquad (8.126)$$

Hence, the unbalanced grid voltage connected directly to the stator will also directly provoke equal behavior in the stator flux space vector, that is, nonconstant amplitude and elliptic trajectory in the $\alpha\beta$ complex plane. This fact can be understood by checking expression (8.125) and neglecting the voltage drop in the stator resistance.

Consequently, the stator and rotor flux space vector can be mathematically represented as follows:

$$\vec{\psi}_s = \psi_{\alpha s} + j\psi_{\beta s} = \hat{\Psi}_{\alpha s}\cos(\omega t + \delta) + j\hat{\Psi}_{\beta s}\sin(\omega t + \delta) \qquad (8.127)$$

$$\vec{\psi}_r = \psi_{\alpha r} + j\psi_{\beta r} = \hat{\Psi}_{\alpha r}\cos(\omega t) + j\hat{\Psi}_{\beta r}\sin(\omega t) \qquad (8.128)$$

Note that, for the DFIM model, the rotor flux has been selected as the space vector at 0°. Hence, the stator voltage will be phase shifted approximately 90° to the stator flux space vector, as illustrated in Figure 8.91.

In addition, depending on the rotor flux control performed, the rotor flux space vector can describe a circular or an elliptic trajectory in the $\alpha\beta$ complex plane. The general unbalanced case has been considered in Equation (8.128).

8.7.3.3 Conditions for Constant Electromagnetic Torque
By using the general electromagnetic torque expression (8.126) and substituting the flux expressions (8.127) and (8.128), the electromagnetic torque can be represented as follows at steady state:

$$T_{em} = \frac{3}{4}p\frac{L_m}{\sigma L_s L_r}\Big[(\hat{\Psi}_{\beta s}\hat{\Psi}_{\alpha r} + \hat{\Psi}_{\alpha s}\hat{\Psi}_{\beta r})\sin\delta $$
$$ + (\hat{\Psi}_{\beta s}\hat{\Psi}_{\alpha r} - \hat{\Psi}_{\alpha s}\hat{\Psi}_{\beta r})\sin(2\omega t + \delta)\Big] \qquad (8.129)$$

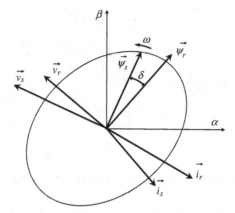

Figure 8.91   Space vector magnitudes of the DFIM under unbalanced conditions.

For an unbalanced voltage situation, the electromagnetic torque is composed of a constant term and an oscillating term. In general, a real application will require constant electromagnetic torque behavior at steady state as discussed before, demanding a cancellation in the oscillatory term of Equation (8.129). This means that the following equation must be true:

$$\frac{\hat{\Psi}_{\alpha r}}{\hat{\Psi}_{\beta r}} = \frac{\hat{\Psi}_{\alpha s}}{\hat{\Psi}_{\beta s}} \tag{8.130}$$

In fact, if this relation between the flux amplitudes is not maintained under an unbalanced situation, and at the same time, the control strategy imposes a constant electromagnetic torque, the DFIM will reach a steady state that imposes nonsinusoidal $\alpha\beta$ stator and rotor fluxes and currents.

For this reason, the rotor flux reference generation strategy needs to create a rotor flux, in accordance with expression (8.130), in order to exchange sinusoidal currents with the grid together with a generation of constant electromagnetic torque.

8.7.3.4 *Rotor Flux Reference Generation Strategy*   Since the stator flux behavior is mainly imposed by the grid stator voltage, as deduced from expression (8.125), the stator and rotor flux space vector amplitudes can be expressed in a similar manner to Equation (8.124), at steady state:

$$|\vec{\psi}_s|^2 = \frac{\hat{\Psi}_{\alpha s}^2 + \hat{\Psi}_{\beta s}^2}{2} + \frac{\hat{\Psi}_{\alpha s}^2 - \hat{\Psi}_{\beta s}^2}{2}\cos(2\omega t + \delta) \tag{8.131}$$

$$|\vec{\psi}_r|^2 = \frac{\hat{\Psi}_{\alpha r}^2 + \hat{\Psi}_{\beta r}^2}{2} + \frac{\hat{\Psi}_{\alpha r}^2 - \hat{\Psi}_{\beta r}^2}{2}\cos(2\omega t) \tag{8.132}$$

Note that since the stator flux is unbalanced, the rotor flux will be unbalanced as well, in order to fulfill Equation (8.130), that is, the condition for sinusoidal current exchange. Consequently, the rotor flux space vector amplitude needs to be oscillatory, as represented by expression (8.132).

The next task is to define how this oscillatory rotor flux space vector amplitude can be created. For implementation simplicity, a closed-loop reference generation strategy has been adopted, as illustrated in Figure 8.92.

Note that the oscillating term is simply derived from the rotor flux α squared component itself:

$$(\psi_{\alpha r})^2 = \frac{\hat{\Psi}_{\alpha r}^2}{2} + \frac{\hat{\Psi}_{\alpha r}^2}{2} \cos(2\omega t) \qquad (8.133)$$

for expression (8.130), the $\alpha\beta$ components of the fluxes are calculated by using the same procedure, but, in this case, we remove the oscillating terms with a lowpass filter. The speed variation of the oscillating amplitude is adjusted by the constant gain K and G is an attenuation gain that can be positive or negative depending on the unbalanced voltage.

Hence, with this rotor flux reference generation strategy, it is possible to maintain the quality of the current exchanges, under an unbalanced situation, and maintain constant electromagnetic torque.

It must be highlighted that there are many alternative solutions for generating the required oscillating flux reference, to that proposed in Figure 8.92.

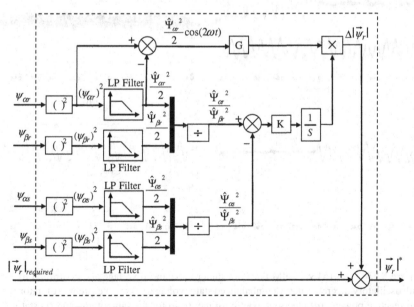

Figure 8.92 Rotor flux amplitude generation strategy.

8.7.3.5 *Example 8.9: System Behavior Controlled by DTC Under Unbalanced Voltage with Oscillatory Rotor Flux Reference* This example shows the behavior of a 380 V and 15 kW DFIM supplied by unbalanced grid voltage, when it is controlled by a classic DTC with the presented flux reference generation strategy. It seeks to show the improvements derived from the proper flux reference generation strategy. The programmed unbalanced voltage in p.u. is $v_{as} = 1\angle 0°$, $v_{bs} = 1.25\angle{-113.6°}$, $v_{cs} = 1.25\angle{+113.6°}$. Figure 8.93 covers the simulation results. The speed of the machine is controlled to 0.8 p.u. subsynchronous speed. During the first half of the experiment, an oscillating flux reference is commanded, while after the middle of the experiment it is maintained constant.

Therefore, Figure 8.93a illustrates the unbalanced stator voltage. Compared to the previously seen example related to imbalance with DPC, a slightly more unbalanced

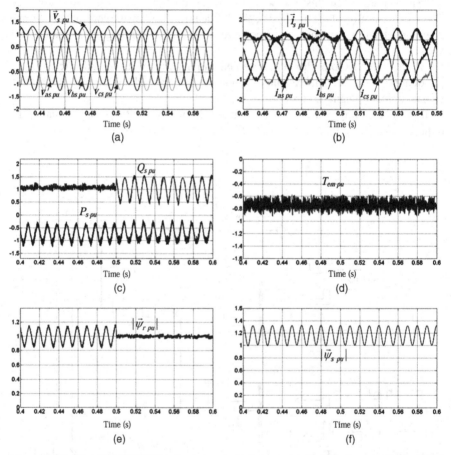

Figure 8.93 The 15 kW DFIM behavior under unbalanced voltage and controlled by DTC with oscillatory flux reference: (a) unbalanced stator voltage, (b) stator currents, (c) stator active and reactive powers, (d) electromagnetic torque, (e) rotor flux amplitude, and (f) stator flux amplitude.

grid has been simulated. During the first half of the experiment, the stator currents are unbalanced but sinusoidal. Both the stator voltage and currents have the same imbalance as can be noticed from Figures 8.93a and 8.93b. However, once the controlled compensation of the imbalance is disabled, the stator currents deteriorate further, becoming nonsinusoidal. The amplitude of the space vector also shown in the figures oscillates at 100 Hz, as studied theoretically in previous sections.

In order to achieve the shown improvement in the stator currents, the stator active power performance can be seen in Figure 8.93c. The active power oscillations commanded are at 100 Hz, due to the effect of the DTC, but note that these power are not being directly controlled. In the middle of the experiment, both references are set to oscillatory values, in a different way from what occurred in DPC. This behavior is due to the fact that the DTC is inherently maintaining the torque constant, since it controls directly the torque itself. In this case, as exhibited in Figure 8.93d, the torque is maintained constant all the time by the DTC.

Therefore, Figure 8.93e shows the rotor flux amplitude control of the simulation experiment. Once the oscillations are disabled in the middle of the experiment, the stator currents deteriorate further, becoming nonsinusoidal.

Finally, Figure 8.93f illustrates the continuous oscillating behavior of the stator flux amplitude. This occurs due to a direct connection of the stator to the grid. Thanks to the programmed oscillating rotor flux reference strategy, it is possible to avoid the nonsinusoidal exchange of currents with the grid.

8.7.4 Control for Voltage Dips Based on DTC

This final section shows a special rotor flux amplitude reference generation strategy for doubly fed induction machine (DFIM) based wind turbines. It is specially designed to address perturbations such as voltage dips, keeping the torque of the wind turbine controlled and considerably reducing the stator and rotor overcurrents during faults. A direct torque control (DTC) strategy that provides fast dynamic response accompanies the overall rotor side control of the wind turbine. Despite the fact that the proposed control does not totally eliminate the necessity of the typical crowbar protection for this kind of turbine (see Chapters 7 and 9), it eliminates activation of this protection during some low depth voltage dips.

8.7.4.1 *Description of the Scenario* This section focuses on analysis of the control of DFIM based high power wind turbines when they operate in the presence of voltage dips. As seen in previous chapters, most of the wind turbine manufacturers build these wind turbines with a back-to-back converter sized to approximately 30% of the nominal power. When the machine is affected by voltage dips, the reduced converter design needs special crowbar protection in order to avoid damage in the wind turbine and to meet the grid code requirements.

The main objective of the control strategy proposed in this section is to eliminate the necessity of the activation of the crowbar protection when low depth voltage dips occurs. Hence, by using DTC, with a proper rotor flux generation strategy, during the fault it will be possible to maintain the machine connected to the grid, generating power from the wind, reducing overcurrents, and eliminating the torque oscillations that normally produce such voltage dips.

The proposed control block diagram is exactly the same as proposed in a previous section in Figure 8.89, however, in this case, the flux reference generation is done according to different criteria. The grid side converter controls the DC bus voltage and the reactive power exchange to the grid through this converter; however, as done in the previous sections, its behavior is not studied under this situation.

Thus, the voltage dip is directly seen by the stator of the machine and by the grid side converter. Once again, the transformer and its effect have been neglected in order to simplify the exposition. Therefore, when the wind turbine is affected by a voltage dip, we will need to address three major problems:

1. From the control strategy point of view, the dip produces control difficulties, since it is a perturbation in the winding of the machine that is not being directly controlled (the stator).

2. The dip generates a disturbance in the stator flux, making a higher rotor voltage necessary to control the machine currents.

3. If special improvements are not adopted, the power delivered through the rotor by the back-to-back converter will be increased due to the increase of voltage and currents in the rotor of the machine, provoking finally an increase in the DC bus voltage.

Taking this into account, depending on the dip depth and asymmetry, together with the machine operating conditions at the moment of the dip (speed, torque, mechanical power, etc.), we can see that crowbar protection is inevitable in most of the faulty situations. However, in this section a control strategy that eliminates the necessity of crowbar activation in some low depth voltage dips is presented.

8.7.4.2 Analytical Study of the DFIM Behavior During the Voltage Dip   The stator flux evolution of the machine is imposed by the stator voltage equation:

$$\vec{v}_s = R_s \vec{i}_s + \frac{d\vec{\psi}_s}{dt} \tag{8.134}$$

In general, since very high stator currents are not allowed, the stator flux evolution can be approximated by the addition of a sinusoidal term and an exponential term (neglecting R_s):

$$\psi_{\alpha s} = K_1 e^{-K_2 t} + K_3 \cos(\omega_s t + K_4) \tag{8.135}$$

$$\psi_{\beta s} = K_5 e^{-K_2 t} + K_3 \sin(\omega_s t + K_4) \tag{8.136}$$

where K_1, K_2, K_3, K_4, and K_5 are constants that depend on the nature of the voltage dip and the moment when it occurs. On the other hand, sinusoidal current exchanges with the grid are always preferred by the application during the fault. This means that the stator and rotor currents should be sinusoidal. However, by checking the expressions that relate the stator and rotor currents as a function of the fluxes, we find

$$\vec{i}_s = \frac{L_m}{\sigma L_r L_s}\left(\frac{L_r}{L_m}\vec{\psi}_s - \vec{\psi}_r\right) \tag{8.137}$$

$$\vec{i}_r = \frac{L_m}{\sigma L_r L_s}\left(\frac{L_s}{L_m}\vec{\psi}_r - \vec{\psi}_s\right) \tag{8.138}$$

It is very hard to achieve sinusoidal current exchanges, since only the rotor flux amplitude is controlled by a DTC technique. So if, due to the dip, the stator flux presents an unavoidable exponential term, it will probably be seen also in the stator and rotor currents.

Consequently, as proposed in the next section, a solution that reasonably cancels the exponential terms from expressions (8.137) and (8.138) is to generate equal oscillation in the rotor flux amplitude as in the stator flux amplitude. Thus, as shown is subsequent sections, the quality of the currents is substantially improved with this oscillatory rotor flux, rather than with constant flux.

8.7.4.3 Rotor Flux Reference Generation Strategy for Voltage Dip Effect Mitigation   As depicted in Figure 8.94, the proposed rotor flux amplitude reference generation strategy adds a term $(\Delta|\vec{\psi}_r|)$ to the required reference rotor flux amplitude according to the following expression:

$$\Delta|\vec{\psi}_r| = |\vec{\psi}_s| - \frac{|\vec{v}_s|}{\omega_s} \tag{8.139}$$

where $|\vec{\psi}_s|$ is the estimated stator flux amplitude and $|\vec{v}_s|$ is the voltage of the grid not affected by the dip. This voltage can be calculated by several methods, for instance, using a simple small bandwidth lowpass filter, as illustrated in Figure 8.94. It must be highlighted that constants K_1–K_5 from expressions (8.135) and (8.136), are not needed in the rotor flux reference generation, thus reducing its complexity.

Note that at steady state, without the presence of a dip, the term $\Delta|\vec{\psi}_r|$ will be zero. On the other hand, when a dip occurs, the added term to the rotor flux reference will be

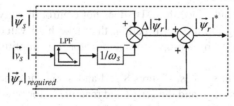

Figure 8.94 Rotor flux reference generation strategy for voltage dip effect mitigation.

approximately equal to the oscillations provoked by the dip in the stator flux amplitude. For simpler exposition, the voltage drop in the stator resistance has been neglected.

8.7.4.4 Example 8.10: System Behavior Controlled by DTC Under Voltage Dip
This example shows the behavior of a 690 V and 2 MW DFIM affected by a three-phase balanced voltage dip of 30% depth. The performance is shown when it is controlled by a classic DTC with the presented flux reference generation strategy and without it. It seeks to show the improvements derived from the proper flux references generation strategy as well as to show the general behavior of the wind turbine during the dip when it is controlled by DTC. Figure 8.95 covers the simulation results when the rotor flux amplitude reference is left constant. The speed of the machine is controlled to 0.9 p.u. subsynchronous.

Figure 8.95a illustrates the stator voltage when the voltage dip occurs. The dip occurs suddenly and it lasts 0.1 second. Figure 8.95b shows the torque behavior. Before the dip, the wind turbine was generating energy with -0.2 p.u. torque. During the dip, no action is taken by the torque reference, so it tries to keep constant at the same value. However, as seen in the figure, the torque cannot be maintained to the desired value and several peaks appear.

As shown in Figure 8.95c, the rotor flux amplitude is kept constant and there is no problem for the system to achieve this objective. On the other hand, due to the direct connection of the stator to the grid, the dip provokes a strong oscillation in the stator flux amplitude.

Unfortunately, the behavior of the currents is not acceptable in a realistic situation. As noted in Figures 8.95f and 8.95h, during the dip, they take too high values so the converter would need to be disconnected to avoid damage to it. However, in this simulation, for a comprehensive analysis, the current flow has been allowed.

Figure 8.95g, showing the rotor voltages (filtered), reveals a very important fact. During the dip, the DTC tries to control the torque and flux to the required values. When the required voltage is higher than that available by the converter ($V_{bus} = 1250$ V), it is not possible to control the torque, provoking the torque peaks shown in Figure 8.95b. Note that the torque peaks occur when the rotor voltage exceeds the rotor voltage limit for controllability ($V_{bus}/\sqrt{3}$). Consequently:

- The torque in this case cannot be controlled all the time, since the perturbation provoked in the stator flux by the dip requires more rotor voltage than the back-to-back converter can provide.
- On the other hand, since the DTC does not control directly the rotor or stator currents, during the dip perturbation, they take high values that can only be reduced indirectly by a proper torque or rotor flux amplitude command.

Under these circumstances, Figures 8.95d and 8.95e show how the perturbation affects the DC bus voltage and the active and reactive stator powers. In this case, the oscillatory behavior of the rotor voltage and currents provokes power transmission

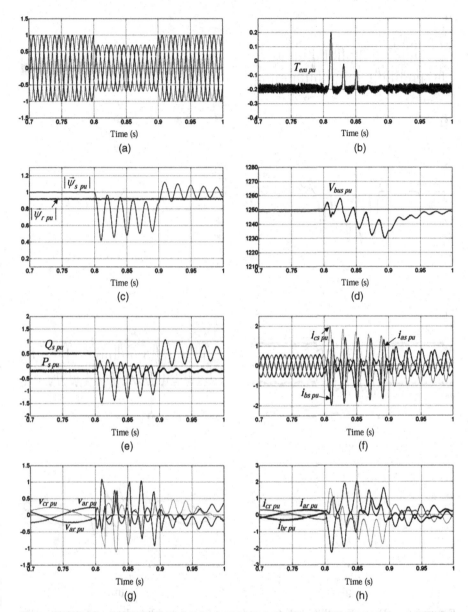

Figure 8.95   The 2 MW DFIM behavior under a voltage dip of 30% depth and controlled by DTC with constant rotor flux reference: (a) unbalanced stator voltage, (b) electromagnetic torque, (c) rotor and stator flux amplitudes, (d) DC bus voltage, (e) stator active and reactive powers, (f) stator currents, (g) rotor voltages (filtered), and (h) rotor currents.

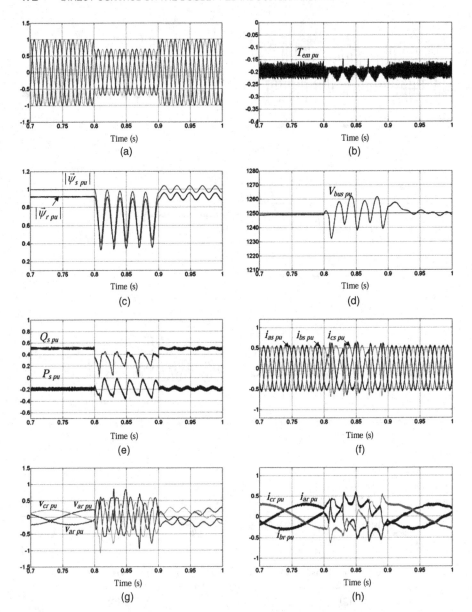

Figure 8.96 The 2 MW DFIM behavior under a voltage dip of 30% depth and controlled by DTC with oscillatory rotor flux reference: (a) unbalanced stator voltage, (b) electromagnetic torque, (c) rotor and stator flux amplitudes, (d) DC bus voltage, (e) stator active and reactive powers, (f) stator currents, (g) filtered rotor voltage, and (h) rotor currents.

through the back-to-back converter, producing a nonsignificant perturbation in the DC bus.

Finally, in a realistic application, due to the high currents, the rotor side converter would inevitably be disabled during this dip, enabling the crowbar for energy dissipation through the rotor. However, this situation would end up with the machine in a different state from that presented in this simulation.

Thus, the next simulation experiment shows how, by means of the proposed rotor flux reference generation, the disturbing effect of the dip can be mitigated, avoiding in this case the necessity to disable the converter and the crowbar action. Figure 8.96 shows the simulation experiment, under the same operating conditions of the wind turbine, affected by the same dip but with a rotor flux reference generated as presented in Figure 8.94.

Figure 8.96a illustrates a stator voltage dip equal to the previous experiment. Thanks to the employed rotor flux reference generation, we see in Figure 8.96b how the torque in this case is kept reasonably close to the desired value of -0.2 p.u.

Figure 8.96c shows the stator and rotor flux amplitudes. Notice that the rotor flux amplitude presents the same oscillation as the stator flux amplitude. In this case, also, the stator and rotor currents are kept reasonably safe (Figures 8.96f and 8.96h), eliminating the strong overcurrents of the previous example and therefore avoiding also the necessity to disable the converter. Unfortunately, during the dip, the stator and rotor currents are not maintained sinusoidal, implying a nonsinusoidal exchange of currents with the grid through the stator.

It is also seen that, under these circumstances, the required rotor voltage by DTC exceeds the limit but less severely than in the previous example, allowing the torque and the rotor flux amplitude to be kept reasonably controlled (Figure 8.96g).

In addition, Figures 8.96d and 8.96e show how the perturbation affects the DC bus voltage and the active and reactive stator powers. In this case, also, the oscillatory behavior of the rotor voltage and currents provokes power transmission through the back-to-back converter, producing a nonsignificant perturbation in the DC bus.

Consequently, this example demonstrates how, by the proper choice of the rotor flux reference, during a small dip, it is possible to control the torque of the wind turbine. There is no necessity to disconnect the converter (also avoiding crowbar activation) to avoid damage to it; the rotor and stator currents and DC bus voltage are maintained at their safe values, and the wind turbine can continue operating as if the dip did not exist.

8.8 SUMMARY

This chapter has examined and discussed a wide range of direct control solutions, arising as alternative solutions to vector control techniques for DFIM based wind turbines. The performance characteristics achieved by both vector and direct control philosophies, in broad terms, can be categorized as very similar, making them two valid solutions for wind energy generation contexts.

In addition, both control solutions can be adapted to deal with faulty conditions of the grid; therefore, they are both valid controls for a realistic wind energy generation environment.

Finally, thanks to the advanced control knowledge acquired in Chapters 7 and 8, the reader is ready to address the last control aspects treated in this book for grid connected wind turbines. The next two chapters deal with some more control issues related to the practical operation of DFIM based wind turbines.

REFERENCES

1. W. Leonhard, *Control of Electrical Drives*. Springer-Verlag, 1985.

2. M. P. Kazmierkowski, R. Krishnan, and F. Blaabjerg, *Control in Power Electronics: Selected Problems*. Academic Press, 2002.

3. I. Boldea and S. A. Nasar, *Vector Control of AC Drives*. CRC Press, 1992.

4. P. Vas, *Sensorless Vector and Direct Torque Control*. Oxford University Press, 1998.

5. N. Mohan, T. Undeland and R. Robbins, *Power Electronics*. John Wiley & Sons. Inc. 1989.

6. A. Veltman, D. W. J. Pulle, and R. W. De Doncker, *Fundamentals of Electric Drives*. Springer, 2007.

7. B. K. Bose, *Power Electronics and Motor Drives*. Elsevier, 2006.

8. A. M. Tryznadlowski, *Control of Induction Motors*. Academic Press, 2001.

9. A. Hughes, *Electric Motors and Drives*. Elsevier, 1990.

10. M. Barnes, *Variable Speed Drives and Power Electronics*. Elsevier, 2003.

11. I. Boldea, *Variable Speed Generators*. CRC Press/Taylor & Francis, 2006.

12. I. Boldea and S. A. Nasar, *The Induction Machine Handbook*. CRC Press, 2006.

13. F. Blaschke, "A New Method for the Structural Decoupling of A.C. Induction Machines," in *Conf. Rec. IFAC*, Dusseldorf, Germany, October 1971.

14. I. Takahashi and Y. Ohmori, "High-Performance Direct Torque Control of an Induction Motor," *IEEE Trans. Ind. Appl.*, Vol. 25, pp. 257–264, March/April 1989.

15. I. Takahashi and T. Noguchi, "A New Quick Response and High Efficiency Control Strategy of an Induction Motor," *IEEE Trans. Ind. Appl.*, Vol. IA-22 pp. 820–827, September/October 1986.

16. M. Depenbrock, "Direct Self-control (DSC) of Inverter-Fed Induction Machine," *IEEE Trans. Power Electron.*, Vol. 3, pp. 420–429, October 1988.

17. D. Casadei, F. Profumo, G. Serra, and A. Tani, "FOC and DTC: Two Viable Schemes for Induction Motors Torque Control," *IEEE Trans. Power Electron.*, Vol. 17, No. 5, pp. 779–787, September 2002.

18. E. Flach, R. Hoffmann, and P. Mutschler, "Direct Mean Torque Control of an Induction Motor," in *Proceedings EPE'97 Conference*, 1997.

19. D. Casadei, G. Serra, and K. Tani, "Implementation of a Direct Control Algorithm for Induction Motors Based on Discrete Space Vector Modulation," *IEEE Trans. Power Electron.*, Vol. 15, No. 4, pp. 769–777, July 2000.

20. J. K. Kang and S.K. Sul, "New Direct Torque Control of Induction Motor for Minimum Torque Ripple and Constant Switching Frequency," *IEEE Trans. Ind. Appl.*, Vol. 35, No.5, pp. 1076–1086, September/October 1999.

21. K. B. Lee, J.-H. Song, I. Choy, and J.-Y. Yoo, "Improvement of Low-Speed Operation Performance of DTC for Three-Level Inverter-Fed Induction Motors," *IEEE Trans. Ind. Electron.*, Vol. 48, No. 5, pp. 255–264, October 2001.

22. C. A. Martins, X. Roboam, T. A. Meynard, and A. S. Carvalho, "Switching Frequency Imposition and Ripple Reduction in DTC Drives by Using a Multilevel Converter," *IEEE Trans. Power Electron.*, Vol. 17, No. 2, pp. 286–297, March 2002.

23. J. Rodríguez, J. Pontt, S. Kouro, and P. Correa, "Direct Torque Control with Imposed Switching Frequency in an 11-Level Cascaded Inverter," *IEEE Trans. Ind. Electron.*, Vol. 51, No. 4, pp. 827–833, August 2004.

24. S. Kouro, R. Bernal, H. Miranda, C. A. Silva, and J. Rodríguez "High-Performance Torque and Flux Control for Multilevel Inverter Fed Induction Motors," *IEEE Trans. Power Electron.*, Vol. 22, No. 6, pp. 2116–2123, November 2007.

25. K. Hatua and V. T. Ranganathan, "Direct Torque Control Schemes for Split-Phase Induction Machine," *IEEE Trans. Ind. Appl.*, Vol. 41, No. 5, pp. 1243–1254, September/October 2005.

26. F. Morel, J. M. Retif, X. Lin-Shi, and A. M. Llor, "Fixed Switching Frequency Hybrid Control for a Permanent Magnet Synchronous Machine," in *Proceedings IEEE ICIT'04 Conference*, 2004.

27. M. Pacas and J. Weber, "Predictive Direct Torque Control for the PM Synchronous Machine," *IEEE Trans. Ind. Electron.*, Vol. 52, No. 5, pp. 1350–1356, October 2005.

28. Z. Xu and M. F. Rahman, "Direct Torque and Flux Regulation of an IPM Synchronous Motor Drive Using Variable Structure Control Approach," *IEEE Trans. Power Electron.*, Vol. 22, No. 6, pp. 2487–2498, November 2007.

29. I. Sarasola, J. Poza, M. A. Rodríguez, and G. Abad, "Predictive Direct Torque Control of Brushless Doubly Fed Machine with Reduced Torque Ripple at Constant Switching Frequency," in *Proceedings IEEE ISIE'07 Conference*, 2007.

30. K. P. Gokhale, et al., "Controller for a Wound Rotor Induction Machine," U.S. Patent 1999.

31. S. Aurtenechea, M. A. Rodríguez, E. Oyarbide, and J. R. Torrealday, "Predictive Control Strategy for DC/AC Converters Based on Direct Power Control," *IEEE Trans. Ind. Electron.*, Vol. 54, No. 3, pp. 1261–1271, June 2007.

32. R. Vargas, P. Cortés, U. Ammann, J. Rodríguez, and J. Pontt, "Predictive Control of a Three-Phase Neutral-Point-Clamped Inverter," *IEEE Trans. Ind. Electron.*, Vol. 54, No. 5, pp. 2697–2705, October 2007.

33. R. Pena, J. C. Clare, and G. M. Asher, "Doubly Fed Induction Generator Using Back-to-Back PWM Converters and Its Application to Variable-Speed Wind-Energy Generation," *Proc. IEE Electric Power Appl.*, Vol. 143, No. 3, pp. 231–241, May 1996.

34. S. A. Gomez and J. L. R Amenedo, "Grid Synchronization of Doubly Fed Induction Generators Using Direct Torque Control," in *Proceedings IEEE IECON'02 Conference*, 2002.

35. R. Datta and V. T. Ranganathan, "Direct Power Control of Grid-Connected Wound Rotor Induction Machine Without Rotor Position Sensors," *IEEE Trans. Power Electron.*, Vol. 16, No. 3, pp. 390–399, May 2001.

36. L. Xu and P. Cartwright, "Direct Active and Reactive Power Control of DFIG for Wind Energy Generation," *IEEE Trans. Energy Conversion*, Vol. 21, No. 3, pp. 750–758, September 2006.

37. G. Abad, M. A. Rodríguez, and J. Poza, "Predictive Direct Power Control of the Doubly Fed Induction Machine with Reduced Power Ripple at Low Constant Switching Frequency," in *Proceedings IEEE ISIE'07 Conference*, 2007.

38. A. Linder and R. Kennel, "Direct Model Predictive Control–A New Direct Predictive Control Strategy for Electrical Drives," in *Proceedings EPE'05 Conference*, 2005.

39. Z. Sorchini and P. T. Krein, "Formal Derivation of Direct Torque Control for Induction Machines," *IEEE Trans. Power Electron.*, Vol. 21, No. 5, pp. 1428–1436, September 2006.

40. A. K. Jain, S. Mathapati, V. T. Ranganathan, and V. Narayanan, "Integrated Starter Generator for 42-V Power Net Using Induction Machine and Direct Torque Control Technique," *IEEE Trans. Power Electron.*, Vol. 21, No. 3, pp. 701–710, May 2006.

41. G. Abad, M. A. Rodríguez, and J. Poza, "Two Level VSC Based Predictive Direct Torque Control of the Doubly Fed Induction Machine with Reduced Torque and Flux Ripples at Low Constant Switching Frequency," *IEEE Trans. Power Electron.*, Vol. 23, No. 3, May 2008.

42. G. Abad, M. A. Rodríguez, and J. Poza, "Two Level VSC Based Predictive Direct Power Control of the Doubly Fed Induction Machine with Reduced Power Ripple at Low Constant Switching Frequency," *IEEE Trans. Energy Conversion*, Vol. 23, No. 2, pp. 570–580, June 2008.

43. G. Abad, M. A. Rodríguez, and J. Poza, "Three Level NPC Converter Based Predictive Direct Power Control of the Doubly Fed Induction Machine at Low Constant Switching Frequency," *IEEE Trans. Ind. Electron.*, Vol. 55, No. 12, pp. 4417–4429, 2008.

44. G. Abad, M. A. Rodriguez, G. Iwanski, and J. Poza, "Direct Power Control of Doubly-Fed-Induction-Generator-Based Wind Turbines Under Unbalanced Grid Voltage," *IEEE Trans. Power Electron.*, Vol. 25, No. 2, pp. 442–452, February 2010.

45. J. Pou, D. Boroyevich, and R. Pindado, "New Feedforward Space-Vector PWM Method to Obtain Balanced AC Output Voltages in a Three-Level Neutral-Point-Clamped Converter," *IEEE Trans. Ind. Electron.*, Vol. 49, No. 5, pp. 1026–1034, October 2002.

46. G. Abad, M. A. Rodriguez, J. Poza, and J. M. Canales, "Direct Torque Control for Doubly Fed Induction Machine-Based Wind Turbines Under Voltage Dips and Without Crowbar Protection," *IEEE Trans. Energy Conversion*, Vol. 24, No. 3, pp. 586–588, 2010.

47. M. Bobrowska-Rafal, K. Rafal, G. Abad, and M. Jasinski, "Control of PWM Rectifier Under Grid Voltage Dips," *Bull. Polish Acad. Sci.* Vol. 57, No. 4, 2009.

48. R. Peña, R. Cardenas, E. Escobar, J. Clare, and P. Wheeler, "Control System for Unbalanced Operation of Stand-Alone Doubly Fed Induction Generators," *IEEE Trans. Energy Conversion*, Vol. 22, No. 2, pp. 544–545, June 2007.

49. G. Iwański and W. Koczara, "Sensorless Direct Voltage Control of the Stand-Alone Slip-Ring Induction Generator," *IEEE Trans. Ind. Electron.*, Vol. 54, No. 2. pp. 1234–1236, April 2007.

50. T. K. A. Brekken and N. Mohan, "Control of a Doubly Fed Induction Wind Generator Under Unbalanced Grid Voltage Conditions," *IEEE Trans. Energy Conversion*, Vol. 22, No. 1, pp. 129–135, March 2007.

51. L. Xu and Y. Wang, "Dynamic Modeling and Control of DFIG-Based Wind Turbines Under Unbalanced Network Conditions," *IEEE Trans. Power Syst.*, Vol. 22, No. 1, pp. 314–323, February 2007.

52. C. J. Ramos, A. P. Martins, and A. S. Carvalho, "Rotor Current Controller with Voltage Harmonics Compensation for a DFIG Operating Under Unbalanced and Distorted Stator Voltage," in *Conference IECON*, 2007.

53. D. Santos-Martin, J. L. Rodriguez-Amenedo, and S. Arnalte, "Direct Power Control Applied to Doubly Fed Induction Generator Under Unbalanced Grid Voltage Conditions," *IEEE Trans. Power Electron.*, Vol. 23, No. 5, pp. 2328–2336, September 2008.

54. D. Santos-Martin, J. L. Rodriguez-Amenedo, and S. Arnalte, "Dynamic Programming Power Control for Doubly Fed Induction Generators," *IEEE Trans. Power Electron*, Vol. 23, No. 5, pp. 2337–2345, September 2008.

55. S. Seman, J. Niiranen, S. Kanerva, A. Arkkio, and J. Saitz, "Performance Study of a Doubly Fed Wind-Power Induction Generator Under Network Disturbances," *IEEE Trans. Energy Conversion*, Vol. 21, No. 4, pp. 883–890, December 2006.

56. S. Dominguez Rubira and M. D. McCulloch, "Control Method Comparison of Doubly Fed Wind Generators Connected to the Grid by Asymmetric Transmission Lines," *IEEE Trans. Ind. Appl.*, Vol. 36, No. 4, pp. 986–991, July/August 2000.

57. M. Chomat, J. Bendl, and L. Schreier, "Extended Vector Control of Doubly Fed Machine Under Unbalanced Power Network Conditions," in *International Conference on Power Electronics Machines and Drives*, 2002, pp. 329–334.

58. J. Lopez, E. Gubia, P. Sanchis, X. Roboam, and L. Marroyo, "Wind Turbines Based on Doubly Fed Induction Generator Under Asymmetrical Voltage Dips," *IEEE Trans. Energy Conversion*, Vol. 23, No. 1, March 2008.

59. L. Xu, "Coordinated Control of DFIG's Rotor and Grid Side Converters During Network Unbalance," *IEEE Trans. Power Electron*, Vol. 23, No. 3, May 2008.

60. J. Yao, H. Li, Y. Liao, and Z. Chen, "An Improved Control Strategy of Limiting the DC-Link Voltage Fluctuation for a Doubly Fed Induction Wind Generator," *IEEE Trans. Power Electron*, Vol. 23, No. 3, May 2008.

61. P. Rioual, H. Pouliquen, and J. P. Louis, "Regulation of a PWM Rectifier in the Unbalanced Network State Using a Generalized Model," *IEEE Trans. Power Electron*, Vol. 11, No. 3, pp. 495–502, May 1996.

62. H. S. Song and K. Nam, "Dual Current Control Scheme for PWM Converter Under Unbalanced Input Voltage Conditions," *IEEE Trans. Ind. Electron*, Vol. 46, No. 5, pp. 953–959, October 1999.

63. I. Etxeberria-Otadui, U. Viscarret, M. Caballero, A. Rufer, and S. Bacha, "New Optimized PWM VSC Control Structures and Strategies Under Unbalanced Voltage Transients," *IEEE Trans. Ind. Electron*, Vol. 54, No. 5, pp. 2902–2914, October 2007.

64. J. Eloy-García, S. Arnaltes, and J. L. Rodríguez-Amenedo, "Direct Power Control of Voltage Source Inverters with Unbalanced Grid Voltages," *IET Power Electron.*, Vol. 1, No. 3, pp. 395–407, September 2008.

65. T. Noguchi, H. Tomiki, S. Kondo, and I. Takahashi, "Direct Power Control of PWM Converter Without Power-Source Voltage Sensors," *IEEE Trans. Ind. Appl.*, Vol. 34, pp. 473, May/June 1998.

66. C. Fortescue, "Method of Symmetrical Coordinates Applied to the Solution of Polyphase Networks," *Trans. Am. Inst. Elect. Eng.*, Vol. 37, No. Part II, pp. 1027–1140, June 1918.

CHAPTER 9

Hardware Solutions for LVRT

9.1 INTRODUCTION

The preceding two chapters have studied the control of DFIM wind turbines. The effect of voltage dips has been analyzed and control solutions to deal with the problematic situations that result from these faulty scenarios have been provided. It has been proved how control by itself is incapable of handling any kind of voltage dip. It is often the case, during the most severe voltage dips, that DFIM based wind turbines require additional hardware protection to avoid disconnection from the grid.

Additionally, the latest grid code requirements specify the behavior of wind turbines when voltage dips occur, defining mainly:

- Under what type of dips the wind turbine must remain connected to the grid
- How much reactive power the wind turbine must provide to the grid in order to contribute to the fault clearance

This performance of the wind turbines is known as the *low voltage ride-through* (*LVRT*). This chapter examines some of the most commonly used hardware solutions in DFIM based wind turbines, in order to fulfill the LVRT requirements. First, we inform the reader about the specific grid code requirements related to this type of faulty scenarios; and second, we present and analyze several hardware solutions for different voltage dips.

A review of DFIM based wind turbines in Chapters 6 to 9 would give the reader advanced understanding of the technology, its advances, and its constraints.

9.2 GRID CODES RELATED TO LVRT

The grid code requirements regarding the behavior of wind turbines under voltage dips has rapidly changed in the last years.

Doubly Fed Induction Machine: Modeling and Control for Wind Energy Generation,
First Edition. By G. Abad, J. López, M. A. Rodríguez, L. Marroyo, and G. Iwanski.
© 2011 the Institute of Electrical and Electronic Engineers, Inc. Published 2011 by John Wiley & Sons, Inc.

Some decades ago wind turbines were considered to behave similarly to large industrial loads and the same quality standards applied to both. In general, the requirements were focused on protection of the turbines themselves and on prevention of the wind turbines from operating under islanding conditions. Accordingly, grid codes demanded disconnection of the turbines in the case of over currents and in the case of abnormal frequency or voltage in the grid. This behavior contributes to the dip as the wind turbines stop generating electric power [1].

This loss of power was tolerable in the past, when the share of wind turbines for power generation was insignificant. In recent years, however, the increasing contribution of wind turbines to the grid has displaced conventional power generation and therefore the loss of a considerable part of the wind generators cannot be accepted anymore. As an example, the European outage on 4 November 2006, caused the tripping of 4892 MW of wind-origin power in Western Europe, exacerbating the imbalance between demand and supply in this area and risking a continental outage [2].

In order to guarantee the grid stability, many countries have recently enforced grid codes that demand wind farms perform similarly to power plants. A core element of these requirements is the ability of wind turbines to remain connected to the grid during and after faults, even with extensive frequency and voltage variations. This performance is commonly known as *low voltage ride through* ability.

Figure 9.1 depicts the requirements for LVRT for different grid codes. Wind turbines must be designed, built, and operated so as to remain in service without tripping if the voltage stays in the zone above a corresponding line. The figure refers to three-phase voltage dips and requirements can be different for asymmetrical dips.

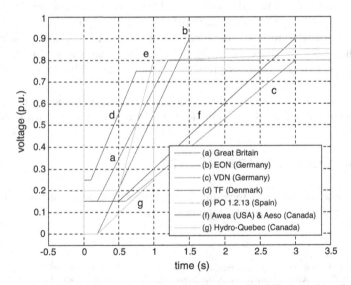

Figure 9.1   Low voltage ride-through requirements for different grid codes.

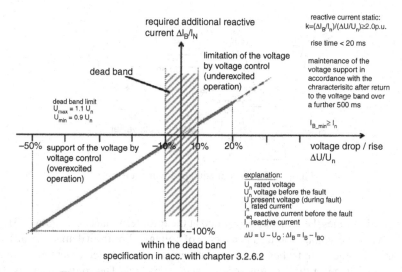

Figure 9.2 Reactive current required during a voltage dip.

It can be observed that most countries set the lowest voltage that wind farms must withstand at 15% or 20%. Some countries (Canada, Germany) require supporting zero voltage, a capability often referred to as *zero voltage ride through* (*ZVRT*).

Moreover, most grid codes (Spain [3], Germany [4], United Kingdom [5], and Canada [6]) demand the injection of reactive current during the dip. Figure 9.2 shows, for example, the required additional reactive current according to the latest E.ON guidelines [7]. As can be seen, if the voltage drops more than 10%, the turbine must provide a reactive current. The deeper the dip, the higher must be the reactive current injected.

Although the oldest grid codes didn't specify the timing of the reactive injection, it is common for most recent ones to establish a maximum time. The current Spanish operation procedure, for example, stipulates that the reactive current must attain the demanded value before 150 ms after the beginning of the fault [3]; meanwhile, the latest German code requires a settled time equal to 80 ms [4]. This is an important point because, as we will see later, some hardware solutions (such as crowbars) hamper the injection of reactive current during the first instants of the dip.

9.3 CROWBAR

A crowbar is a device widely used to protect electronic circuits against an overvoltage condition in their power supply. It operates by putting a short-circuit or low resistance path between the terminals. The circuit is therefore protected; meanwhile, the short-circuit current can cause, in general, blow-out of an upstream fuse.

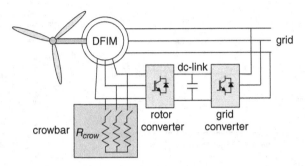

Figure 9.3   System equipped with a crowbar.

In wind turbines the crowbar is installed at the rotor terminals (see Figure 9.3) and prevents the overvoltage induced in the case of voltages dips from damaging the rotor converter. It is activated when an anomalous situation is detected (overcurrent in the rotor, overvoltage in the DC link, or low stator voltage). The rotor current is then diverted to the crowbar and the rotor converter is switched off.

Figure 9.4 shows the equivalent circuit of the system of Figure 9.3 using the rotor model deduced in Chapter 6. As can be seen, when the crowbar is activated the circuit becomes an impedance divider: the converter voltage is then a fraction of the emf induced in the rotor windings. Another way to understand how it works is the following: when the crowbar is connected, a large current circulates across the rotor, causing a big voltage drop in the rotor internal impedance, R_r' and L_r'; this reduces the remaining voltage at the rotor terminals.

In Figure 9.3 the crowbar was made up of three resistors and three switches. Since the rotor current is AC, the switches must be bidirectional. In order to reduce the complexity and cost of the circuit, many manufacturers use the alternative circuit of Figure 9.5.

This alternative schema rectifies the rotor currents, in general, by means of a diode bridge, and hence it requires only one unidirectional switch. Neglecting the harmonics, both schemas are equivalent if their resistances fulfill the following relationship:

$$R_{crowDC} = \frac{\pi^2}{6} R_{crow} \qquad (9.1)$$

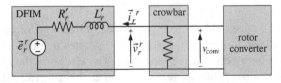

Figure 9.4   Equivalent circuit of the system when the crowbar is activated.

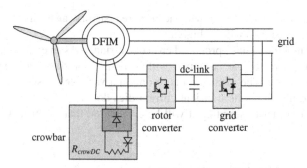

Figure 9.5   Alternative implementation of a crowbar.

Earlier versions of the crowbar used thyristors (or SCRs) as switches. However, the problem with thyristors was that their cut-off is not controlled: once the crowbar is triggered, it remains connected until the circuit breaker of the generator stops the short-circuit current. As a result, the wind turbine is disconnected from the grid and stops generating electric power even if the grid recovers its normal operation.

In order to provide the LVRT capability, the crowbar short-circuit has to be eliminated without disconnecting the turbine from the grid. Today most manufacturers use the "active crowbar" [8] in which the activation and also the deactivation can be actively controlled. Modern versions of active crowbars are usually based on the scheme of Figure 9.5 and include at least one switch with cut-off capability, such as a GTO or IGBTs. This design allows direct disconnection of the crowbar and instant rotor converter reactivation, enabling the resumption of normal operation in the turbine.

The active crowbar has an additional advantage: Since the resistor can be connected and disconnected, it is possible to perform a pulse width modulation (PWM) in order to simulate a variable resistor [9]. It can be interesting, for example, to simulate a low resistance at the beginning of the dip, when the rotor current is very large, and to increase the resistance as the current decreases.

Another possibility to obtain a variable resistance is to use different resistors that can be switched independently, as Figure 9.6 depicts. At the beginning of the dip, the resistors can be connected simultaneously in order to provide a low resistance path to the rotor currents and to avoid exceeding the maximum voltage of the rotor converter.

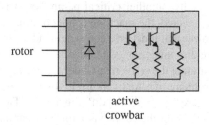

Figure 9.6   Active crowbar with a set of various resistors.

Once the rotor current has died down, one or more resistors can be disconnected so the resulting total resistance will be higher.

Many manufacturers have proposed crowbar variants. As an example, in Llorente and Visiers [10] a crowbar equipped with varistors (or some voltage-dependent resistor) is proposed. The idea again is to obtain a variable resistance, lower at the beginning and higher as the transient current dies down.

Next, and in order to simplify, it will be considered that a constant resistance is used.

9.3.1 Design of an Active Crowbar

In an active crowbar there are two parameters that must be carefully chosen: its resistance R_{crow} and its activation time.

Regarding the resistance R_{crow}:

- If a very low value is chosen, the short-circuit current will be very large. The crowbar switch should then be oversized and, as will be seen, the electromagnetic torque will have a big peak.
- The rotor current can be reduced by using a higher resistance. However, if the resistance is too big the crowbar won't pull the rotor voltage low enough. In this case, the rotor current will circulate across the rotor converter via its freewheeling diodes even if it is inactive. In order to properly protect the rotor converter, the voltage should be kept below the following value:

$$\hat{V}_r < \frac{V_{bus}}{\sqrt{3}} u \qquad (9.2)$$

where V_r is the peak value of the rotor voltage, V_{bus} is the DC-link bus voltage, and u is the turn ratio.

The crowbar resistance also has an influence on the decay of the natural flux. As an example, Figure 9.7 depicts the evolution of the natural flux during a total three-phase voltage dip for three different situations: with a low resistance, with a high resistance, and with the rotor open-circuited, that is, with an infinite resistance.

As can be observed, the crowbar assists the demagnetizing of the machine and this effect is accentuated when, as is usual, its resistance is low.

The activation time is also another critical parameter of the crowbar operation. While the crowbar is connected, the machine is not controlled, so it is impossible to generate the reactive power that most grid codes demand during voltage dips. On the other hand, a premature disconnection, when the natural flux is still too high, might cause the rotor emf to saturate the converter. Under those circumstances, control over the current is lost and problems of rotor overcurrents or overvoltage in the DC-link bus appear. For instance, Figure 9.8 shows the difference in the DC-link voltage during a three-phase 80% voltage dip for the cases when the crowbar is connected for 100 ms and 50 ms.

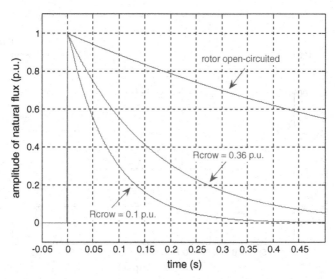

Figure 9.7 Dynamics of the natural flux in a full voltage dip with and without a crowbar.

The connection time is then a compromise between safety and fulfillment of grid codes. Over the past years grid codes have toughened their requirements, demanding each year quicker injection of reactive current. The manufacturers have reacted by shortening the time during which the crowbar is connected but there is a minimum time that must be respected at the risk of losing control of the turbine.

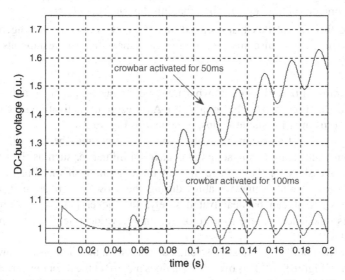

Figure 9.8 DC-link bus voltage connecting the crowbar for 50 and 100 ms.

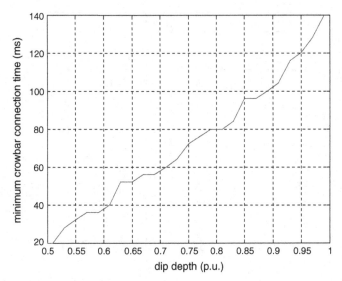

Figure 9.9   Minimum connection time under three-phase dips.

9.3.2 Behavior Under Three-Phase Dips

The main difficulty for the DFIM in three-phase dips is the large emf induced in the rotor by the natural flux. Since this emf is transient the crowbar can be an effective solution for these voltage dips:

- The crowbar is connected for the first stages of the dip, when the emf is higher, thus preventing that emf from damaging the rotor converter.
- Once the natural flux has decayed and the emf is no longer dangerous, the crowbar can be disconnected so that the machine can resume its normal operation.

The minimum time that the crowbar must be connected depends mainly on the natural flux that is caused by the voltage dip. As a result, the activation time is longer for severe dips than for shallow dips, as shown in Figure 9.9.

The figure is just an example, as the connection time may be very different for each turbine and many other factors also have an influence, such as the rotational speed and the slope of the voltage drop at the beginning of the dip. In general, manufacturers monitor some variables during the crowbar activation in order to determine when it can be deactivated safely. In Bücker [11], for example, the crowbar is disconnected when the rotor currents have declined below a predetermined value.

The crowbar activation causes a strong current rise in the rotor, as is shown in the upper part of Figure 9.10. This over-current does not affect the rotor converter since it is disconnected and its current is zero; see the lower part of the Figure 9.10.

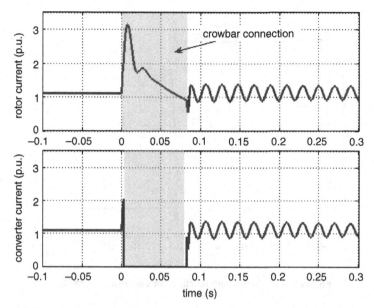

Figure 9.10 Rotor and converter currents using a crowbar during a three-phase dip.

The rotor over current, however, causes a sudden increase of the electromagnetic torque, typically reaching 2 to 3 times its rated value, as represented in Figure 9.11. Manufacturers usually install a slip clutch in the shaft in order to prevent the torque peak from damaging the costly gearbox.

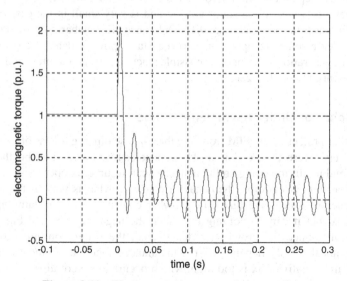

Figure 9.11 Electromagnetic torque using a crowbar.

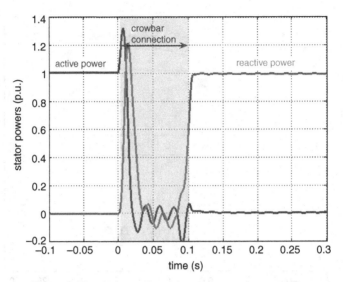

Figure 9.12   Active and reactive currents using a crowbar.

Finally, Figure 9.12 shows the active and reactive currents in the stator of the generator. The currents have been calculated taking into account only the positive sequence of currents and voltages, as recommended by the IEC 61400-21 standard [12] that is used in most grid codes [3,4]. During the crowbar connection, the rotor converter is deactivated so the generator remains uncontrolled and acts as a squirrel cage machine with a rotor resistance slightly higher than normal. Consequently, the generator absorbs reactive power while it can generate or absorb active power depending on its rotational speed. The amount of both powers, nevertheless, depends on the stator voltage and is very small in the case of severe voltage dips, as is the case in Figure 9.12. Only once the crowbar is disconnected can the converter resume its operation and the stator powers can be controlled. In Figure 9.12 generation of as much as possible reactive current was preferred, keeping the active current equal to zero.

9.3.3 Behavior Under Asymmetrical Dips

The crowbar protects the DFIM from any three-phase dip, regardless of its depth. If the dip is very severe, the crowbar must be connected for a longer time, but the natural flux will end up disappearing so the converter can resume its operation.

Under asymmetrical dips, the behavior of the crowbar is very different due to the presence of a negative flux. The crowbar contributes to the damping of the natural flux, but it can do nothing to reduce the negative flux; see Figure 9.13. Consequently, once the crowbar is deactivated, the stator continues to have a negative flux that only depends on the unbalanced factor as stated by Equation (6.35). If the negative flux is too high, which occurs in severe dips, the converter saturates and loses control of the current, resulting in rotor overcurrents or

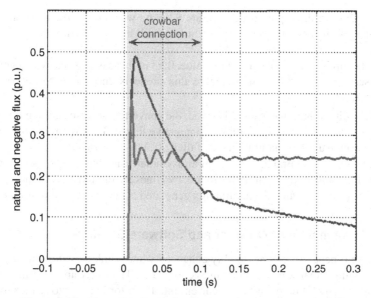

Figure 9.13 Natural and negative fluxes under an asymmetrical dip.

overvoltage in the DC-link bus. Unlike in the case of three-phase dips, a longer connection time is now completely useless.

Figure 9.14 shows the minimum time that the crowbar must be connected in the case of a phase-to-phase dip. It is interesting to compare this figure with the one

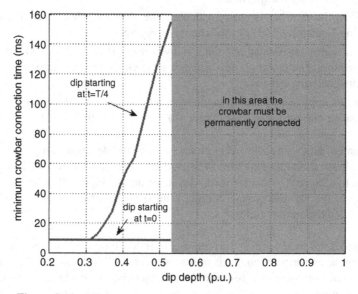

Figure 9.14 Minimum connection time under phase-to-phase dips.

corresponding to three-phase dips (Figure 9.9). As we can see, the behavior under asymmetrical dips is completely different for small and severe dips:

- If the dip is deeper than a certain value, 0.53 in this case, the crowbar cannot be disconnected because the negative flux alone is enough to cause converter saturation.
- If the dip is shallower than this value, the converter can also saturate due to the combined action of the natural and negative fluxes. The crowbar must then be connected until the natural flux has fallen off. The necessary connection time can vary since the natural flux caused by an asymmetrical voltage dip depends heavily on the timing of the fault. In the figure, two lines are plotted corresponding to the extreme cases: when the natural flux generated is maximum and when it is zero.

9.3.4 Combination of Crowbar and Software Solutions

The crowbar behavior can be notably enhanced when combined with the demagnetizing technique that was presented in Section 7.6.1. That software solution is based on the injection of a demagnetizing current into the rotor in order to reduce the emf induced by the natural and negative fluxes. When used alone, that solution demands a large capacity of rotor current, but this drawback can be overcome when using the active crowbar.

In López et al. [13], this combined solution is presented and the following operational sequence is suggested:

1. Immediately upon detection of a grid fault connect the crowbar, quickly demagnetizing the machine.
2. While the crowbar is active, calculate the necessary demagnetizing current using Equation (7.45) or the schema of Figure 7.42.
3. If the current estimated at point 2 is lower than the converter maximum value, disconnect the crowbar, thus activating the converter and starting to inject a demagnetizing current.
4. As the amplitude of the demagnetization decreases, introduce progressively a positive current to the grid in order to generate reactive power.

Compared with using a crowbar alone (without the demagnetizing current) this solution allows shortening the time of activation of the crowbar; hence, the injection of a reactive current is accelerated. As an example, Figure 9.15 compares the evolution of the DC-link voltage during a three-phase 80% voltage dip with and without a demagnetizing current:

- Without a demagnetizing current, a connection time of 50 ms is not enough: when the converter resumes its operation there is still too much natural flux and the converter saturates, resulting in an unacceptable increase of the DC-link voltage bus.

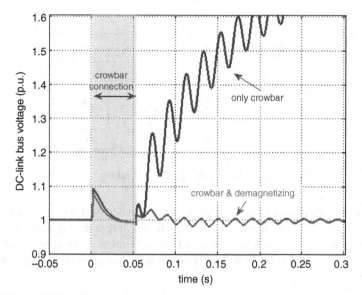

Figure 9.15 DC-link voltage using a crowbar and demagnetizing current.

- With demagnetizing current the DC-link voltage is controlled because it reduces the induced emf, therefore avoiding saturation of the rotor converter.

The combined use of the crowbar and demagnetizing current enhances the behavior not only under three-phase dips but also under asymmetrical voltage dips. As an example, Figure 9.16 shows the necessary connection time for the crowbar using this technique. Note the reduction in time compared with the results presented in Figures 9.9 and 9.14, traced without a demagnetizing current.

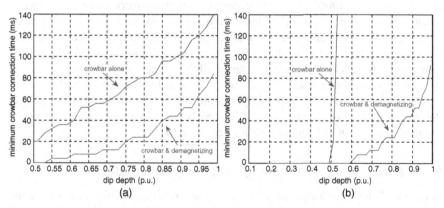

Figure 9.16 Minimum crowbar connection time using a demagnetizing current: (a) three-phase dips and (b) asymmetrical dips.

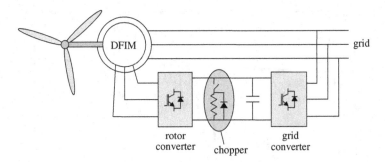

Figure 9.17 System equipped with braking chopper.

Obviously, the enhancement is more noticeable when the current capacity of the converter is higher and, hence, it can introduce a larger demagnetizing current. The figure was traced for a converter whose maximum current was established at 1.5 times its rated value. For bigger converters shorter times are obtained and vice versa.

9.4 BRAKING CHOPPER

A braking chopper is an electric device connected to the DC-link bus of the back-to-back converters to prevent an uncontrolled increase of their voltage. They are used in those drives, such as locomotives or streetcars, where the electrical machine may function as a generator when the drive is braking. In this kind of situation, the machine feeds energy back to the DC-link bus. As very frequently the bus cannot evacuate this energy to the power supply, this energy is accumulated in the DC link and may cause overvoltages in the DC link if it is not dissipated.

Figure 9.17 shows a system equipped with a braking chopper. The braking chopper is made up of a resistor that can be connected or disconnected by means of a switch. A freewheeling diode is also necessary to prevent overvoltages in the switch when it is turned off. Control of the switch is often made by an ON–OFF (also known as hysteresis) controller: when the actual DC bus voltage exceeds a specified level, for example, 1.2 p.u., the resistor is connected and the surplus energy is dissipated. The resistor is kept connected until the voltage drops below a minimum specified level, for example, 1.1 p.u., when the resistor is disconnected.

The installation of a braking chopper in modern commercial turbines to protect the converters from overvoltages in the DC link is more and more common. It can be installed alone or in combination with an active crowbar.

9.4.1 Performance of a Braking Chopper Installed Alone

In Chapter 7 the behavior of the DFIM during severe voltage dips was analyzed and two associated problems were discussed: (1) large rotor currents and (2) overvoltages in the DC-link bus. A braking chopper is a very effective solution for the second

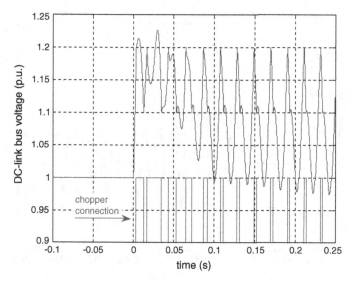

Figure 9.18   DC-link bus voltage using a braking chopper.

problem if it is sized to withstand all the power absorbed by the rotor converter during the dip. However, the chopper does not affect the dynamics of the rotor currents and therefore does not solve the first problem.

As an example, Figure 9.18 shows the evolution of the DC-link bus voltage during a three-phase 80% dip. The figure includes also the connection function of the chopper.

Although the bus voltage is completely controlled, during the earliest moments of the dip the rotor converter saturates and hence it loses control of the rotor current. Depending on the machine and dip characteristics, the current must reach up to 3 p.u. for three-phase dips, as observed in Figure 9.19, and even higher for asymmetrical voltage dips. Unlike with the crowbar operation, the current flows across the converter so it must be oversized.

9.4.2 Combination of Crowbar and Braking Chopper

The braking chopper is often associated with an active crowbar, as shown in Figure 9.20. This arrangement creates a synergy since both solutions complement each other:

- The active crowbar is activated at the beginning of the dip and prevents the initial overcurrents from damaging the rotor converter.
- The braking chopper makes it possible to disconnect the crowbar earlier and hence to accelerate injection of reactive power in to the grid.

Figure 9.20 shows the schema of the whole system.

The addition of the crowbar avoids the need to oversize the rotor converter that the chopper alone requires. On the other hand and compared with the crowbar alone, the

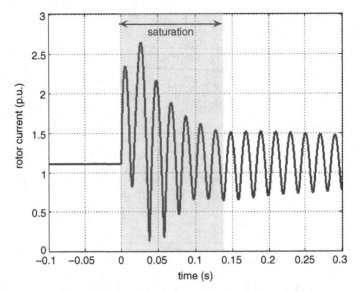

Figure 9.19   Rotor current using a braking chopper.

chopper allows the shortening of the crowbar activation and, hence, it accelerates the injection of a reactive current. This can be observed in Figure 9.21, where the dynamics of the DC-link voltage are compared for three cases:

- Using only a crowbar that is connected for 50 ms
- Using only a crowbar that is connected for 80 ms
- Combining a chopper and a crowbar activated for 50 ms

As can be observed, when the crowbar is used alone, its activation time must be longer than if the crowbar is combined with a chopper.

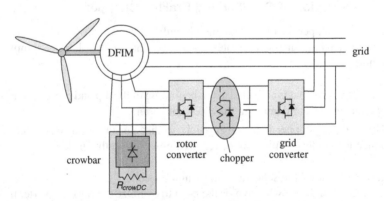

Figure 9.20   System combining crowbar and braking chopper.

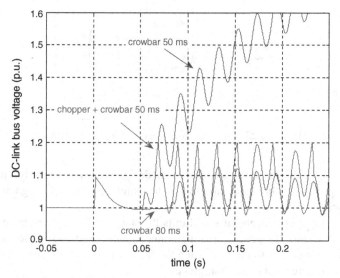

Figure 9.21   DC-link bus voltage using a crowbar and a combination of crowbar + chopper.

9.5 OTHER PROTECTION TECHNIQUES

9.5.1 Replacement Loads

This solution provides the wind turbine with a standby load, as Figure 9.22 shows. When a network fault occurs, the generator is disconnected from the grid and is connected to the standby load by means of a special switch placed in the stator of the machine [14]. After removal of the grid fault condition, the standby load is disconnected and the stator windings are reconnected to the power grid.

In one embodiment of the solution, the standby is sized so it can dissipate all the rated power of the turbine. The load is then controlled in such a way that the voltage of the generator is kept at the same level as the voltage that was present prior to the fault. The wind turbine can therefore continue to generate power so speed overshoots are avoided.

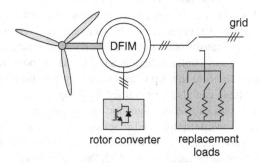

Figure 9.22   Different alternatives to connect the series resistances.

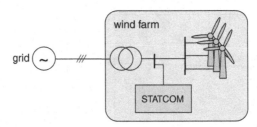

Figure 9.23 Wind farm equipped with a STATCOM.

Another possibility is to use a reduced load designed to dissipate only a part of the generated power. In this case, the goal is to maintain a certain magnetization in the generator.

In any case, the replacement load limits the stator and rotor currents during the fault. Besides, since the generator is maintained magnetized during the fault, it can quickly be reconnected to the grid when the fault is removed without performing a time-consuming resynchronization process.

The main drawback of the solution is that during the fault the turbine does not generate reactive power. In the strict sense, this solution does not provide LVRT capability and therefore would only be used when the grid codes require fast reconnection times but do not demand a reactive current. Another possibility, if LVRT capability is required, is to use the grid converter of the back-to-back converter to generate reactive power, as suggested by Nielsen [15].

9.5.2 Wind Farm Solutions

The solutions presented in the previous sections are addressed to improve the performance of individual turbines. Sometimes, however, wind farm operators prefer to install a general solution directed at the whole farm. This avoids retrofitting the already installed turbines and, hence, can be the most economical solution in the case of old farms.

One of the most common options is the so-called STATCOM [16]. The STATCOM (STATic COMpensator) uses a voltage source converter interfaced in parallel to the transmission line of the wind park, as shown in Figure 9.23.

The STATCOM does not have any power source so the active power exchanged with the line must be zero. However, it can absorb or generate reactive power and, by doing so, it can vary the transmission line voltage. By introducing a large amount of reactive power in the case of a voltage dip, the STATCOM can boost the voltage inside the park, attenuating therefore the dip consequences in the turbines individually.

The wind farm voltage can also be modified by using a static synchronous series compensator (SSSC), also known as a dynamic voltage restorer (DVR). This device uses a voltage source converter interfaced in series with the transmission line. In the case of a dip, the converter generates between its terminals the missing voltage so that the farm voltage is kept at its rated value. With fast and accurate control, the wind turbines only experience a very short transient that does not affect their behavior.

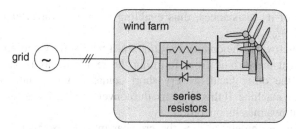

Figure 9.24   Series resistors installed in a wind farm.

Another option, cheaper and more reliable than the SSSC, is the use of series bank resistors [17], as shown in Figure 9.24. This solution includes a series resistor between the grid and the turbines. The resistor is bypassed in normal operation by means of a switch. In the case of a voltage dip, the switch is opened so that the current is diverted to the resistor. The aim of the resistor is to increase the short-circuit impedance, therefore reducing the overcurrents in the stator and in the rotor. Moreover, the addition of the resistor accelerates the damping of the natural flux, as if the stator resistance was higher. The resistor therefore assists the demagnetizing of the machine and the resumption of its normal operation.

9.6 SUMMARY

Modern grid codes require wind turbines to remain connected to the grid even in the case of severe dips, that is, to have a low voltage ride through (LVRT) capability. Moreover, the latest grid codes demand the injection of reactive current during voltage dips with the aim of contributing to the fault clearance and assisting the recovery of the grid voltage.

This chapter has presented some of the most commonly used hardware solutions to provide a LVRT capability. Their advantages and drawbacks have been examined and some examples of their behavior have been given.

The most wide spread hardware solution, the crowbar, was analyzed. The crowbar short-circuits the rotor by connecting low impedance resistances at its terminals. When a wind turbine detects a voltage dip, the rotor converter is switched off and the rotor current is diverted to the crowbar.

A side effect of the crowbar is the large short-circuit current that flows across the rotor and stator of the machine during its connection, especially at the beginning. This overcurrent can be reduced by increasing the impedance of the resistances, but if a too large impedance is used, the rotor voltage can be too high and the crowbar does not protect the converter any more.

The oldest variants of crowbars couldn't be autonomously disconnected, so they forced the turbine to be separated from the grid. In order to provide the LVRT capability, the crowbar short-circuit has to be eliminated without disconnecting the turbine from the grid. This can be done by using the "active crowbar," which allows disconnection of its resistances, thus enabling the rotor converter to resume its normal operation.

In an active crowbar the connection time must be carefully chosen. A very long time can prevent the turbine from fulfilling the requirements of reactive current injection. On the other hand, a very short connection may not be enough to demagnetize the machine. If this happens, the converter will saturate and lose current control when it resumes its operation.

Under severe asymmetrical dips the crowbar must be permanently connected, impeding the turbine from injecting reactive current and, even worse, causing the turbine to consume reactive power. This occurs because the negative flux induces in the rotor enough emf to saturate the rotor converter.

These problems can be overcome if a demagnetizing current is injected into the rotor when the crowbar is disconnected. Thus, the use of the demagnetizing technique notably enhances the behavior of the crowbar under three-phase and asymmetrical voltage dips.

Another hardware solution that is increasingly used is the braking chopper. This protection system uses a resistor installed in the intermediate circuit (DC-link bus) of the rotor converter. The resistance is connected in order to discharge the DC link when its bus voltage is too high, and it is disconnected when it is no longer necessary.

When used alone, the braking chopper prevents overvoltages in the DC link during voltage dips but it does not reduce the rotor currents. Thus, the rotor converter must be current-oversized. This drawback is often overcome by combining the chopper with an active crowbar. In this case, the crowbar is connected at the very beginning of the dip, when the rotor converter is higher. After a few milliseconds, the crowbar can be disconnected, and the rotor converter resumes its operation. The addition of the braking chopper allows a shorter disconnection of the crowbar and therefore a quicker injection of reactive current.

Other solutions have briefly been presented: replacement loads (or standby loads) whose purpose is to supplant the grid during the dip or other wind farm solutions such as STATCOMs, DVRs, or series resistors installed between the grid and the turbines.

REFERENCES

1. J. M. Rodríguez et al., "Incidence on Power System Dynamics of High Penetration of Fixed Speed and Doubly Fed Wind Energy Systems. Study of the Spanish Case," *IEEE Trans. Power Syst.*, Vol. 17, No. 4, pp. 1089–1095, November 2002.
2. Union for the Co-ordination of Transmission of Electricity, "Final Report on the Disturbances on 4 November 2006," available at: http://www.ucte.org/publications/otherreports.
3. "PO 12.3: Requisitos de respuesta frente a huecos de tensión de las instalaciones eólicas," Red Eléctrica Española, Spain, January 2000, available at http://www.aeeolica.es/doc/privado/procedimiento_verificacion_v3.pdf.
4. Technical guidelines (Technische Richtlinien), Fördergesellschaft Windenergie e.V, Germany, available at http://www.wind-fgw.de/.

5. The Grid Code, Connection Conditions (CC), National Grid, UK, available at http://www. nationalgrid.com/uk/Electricity/Codes/gridcode/gridcodedocs/.

6. Requirements for the Interconnection of Distribution Generation to the Hydro-Québec Medium-Voltage Distribution System, Hydro-Québec, Canada, available at http://www. hydroquebec.com/transenergie/fr/commerce/producteurs_prives.html.

7. "Grid Code, High and Extra High Voltage," E.ON Netz GmbH Bayreuth, Germany, April 2006, available at http://www.eon-netz.com.

8. Lorenz Feddersen(Vestas Wind Systems), "Circuit to be Used in a Wind Power Plant," U.S. Patent No. US7102247 September 2006.

9. M. Rodriquez, G. Abad, I. Sarasola, and A. Gilabert, "Crowbar Control Algorithms for Doubly Fed Induction Generator During Voltage Dips," presented at *European Conference on Power Electronics and Applications*, 2005.

10. J. I. Llorente and M. Visiers(Gamesa Innovation & Technology), "Control and Protection of a Doubly-Fed Induction Generator System," European Patent No. EP1499009, October 2007. Patent also granted in US, Patent No. US7518256, April 2009.

11. Andreas Bücker(General Electric Company), "Method for Operating a Wind Power Plant and Method for Operating It," U.S. Patent No. US7321221, February 2008.

12. IEC 61400-21: *Wind Turbine Generator Systems, Part 21: Measurement and Assessment of Power Quality Characteristics of Grid Connected Wind Turbines*. 2nd edition. International Electrotechnical Commission, Geneva, Switzerland, 2008.

13. J. López, L. Marroyo, E. Gubia, E. Olea, and I. Ruiz, "Ride-through of Wind Turbines with Doubly-Fed Induction Generator Under Symmetrical Voltage Dips," *IEEE Trans. Ind. Electron.*, Vol. 56, pp. 4246–4254, October 2009.

14. Roho Reinhard (Asea Brown Boveri, U.S.A.), "Power Station Having a Generator Which Is Driven by a Turbine, as Well as a Method for Operating Such a Power Station," U.S. Patent No. US6239511, May 2005. Patent also granted in Europe, Patent No. EP0984552, October 2005.

15. John Godsk Nielsen(Vestas Wind Systems), "Method for Controlling a Power-Grid Connected Wind Turbine Generator During Grid Faults and Apparatus for Implementing Said Method," U.S. Patent No. US7332827, February 2008. Patent also granted in Europe, Patent No. EP1595328, September 2009.

16. J. I. Llorente and M. Visiers(Gamesa Innovation & Technology), "Method and Device for Injecting Reactive Current During a Mains Supply Voltage Dip," European Patent No. EP1803932 (A1), July 2007.

17. Janos Rajda, Anthony William Galbraith, and Colin David Schauder(Satcon Technology Corp U.S.A), "Device, System and Method for Providing a Low-Voltage Fault Ride-Through for a Wind Generator Farm," U.S. Patent No. US7514907, April 2009.

Complementary Control Issues: Estimator Structures and Start-Up of Grid-Connected DFIM

10.1 INTRODUCTION

After having accomplished, in a broad manner, control of the DFIM as well as the solutions to address a faulty scenario of the grid with Chapters 6 to 9, this chapter looks ahead to real applications, covering some of the most important aspects not addressed yet, but necessary for a practical operation of DFIM based wind turbines.

Therefore, as discussed several times in previous chapters, the reduced size of the converter employed to supply the rotor of the DFIM makes this concept a cost effective solution for wind turbines. However, it also presents some major drawbacks or particular limitations with which the reader must be familiar.

For instance, the start-up (acceleration from zero speed until normal operation) of the doubly fed induction machine with a reduced size back-to-back converter implies a specific procedure that requires encoder calibration followed by synchronization with the grid voltage, prior to connection of the machine with the grid. These two actions are necessary, due to the fact that the limited voltage of the back-to-back converter reduces the speed range (around $\pm 30\%$ from synchronous speed) in which the DFIM can be controlled, so low speeds, including zero speed, are out of the controllable range.

Consequently, this chapter describes these two encoder calibration and grid synchronization processes, contextualized in an example of a general sequential control procedure that is normally utilized in DFIM based wind turbines. Also, various illustrative examples show the most representative steps and stages that the machine commonly follows in a real application: start-up of grid side converter, encoder calibration, wait for minimum wind speed, grid synchronization, connection with the grid, normal operation, and so on.

Added to this, the first part of the chapter provides the reader with a wider view of estimators and observer structures for measuring or estimating normally

---

Doubly Fed Induction Machine: Modeling and Control for Wind Energy Generation,
First Edition. By G. Abad, J. López, M. A. Rodríguez, L. Marroyo, and G. Iwanski.

immeasurable magnitudes of the machine (torque, fluxes, etc.), necessary for implementation of control.

10.2   ESTIMATOR AND OBSERVER STRUCTURES

This first section covers estimation techniques of immeasurable variables of the DFIM. In preceding chapters, we have seen how magnitudes such as torque and fluxes are necessary for the proper control performance of the machine. Unfortunately, these variables are not commonly directly available by sensors; hence, in this section we provide various estimation possibilities. Consequently, different versions of estimators are presented, with different grades of complexity and therefore of accuracy, with the objective of narrowing down the huge amount of estimation techniques available in the specialized literature.

Depending on the control technique employed (vector control, direct power control, etc.), not all the required estimated magnitudes are equal; therefore, some of the studied estimators will be useful for only its corresponding control technique.

10.2.1   General Considerations

Doubly fed induction machine based wind turbines require several current, voltage, and speed sensors, as illustrated in Figure 10.1. Depending on the employed control strategies, some of the sensors can be used for different purposes, but in general, unless sensorless control techniques are adopted, the following magnitudes must be sensorized: i_s, i_r, i_g, v_s, v_g, and speed-position.

The following sections introduce how these sensor measurements can be used to perform estimations of nonavailable magnitudes of the machine such as torque and fluxes.

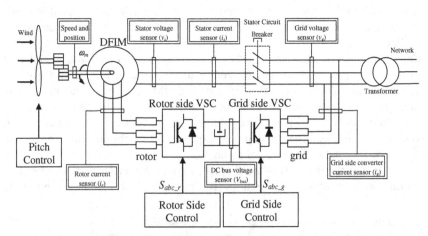

Figure 10.1   DFIM based wind turbine sensorization schematic.

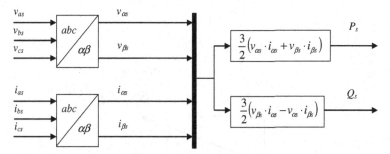

Figure 10.2   Stator active and reactive power estimation.

10.2.2 Stator Active and Reactive Power Estimation for Rotor Side DPC

DPC of the DFIM requires stator active and reactive power control. Since these magnitudes can not be directly measured in general, they are estimated from the stator voltage and current measurements, according the following well-known expressions:

$$P_s = \tfrac{3}{2}\mathrm{Re}\left\{\vec{v}_s \cdot \vec{i}_s^{\,*}\right\} = \tfrac{3}{2}\left(v_{\alpha s}i_{\alpha s} + v_{\beta s}i_{\beta s}\right) \tag{10.1}$$

$$Q_s = \tfrac{3}{2}\mathrm{Im}\left\{\vec{v}_s \cdot \vec{i}_s^{\,*}\right\} = \tfrac{3}{2}\left(v_{\beta s}\,i_{\alpha s} - v_{\alpha s}\,i_{\beta s}\right) \tag{10.2}$$

The $\alpha\beta$ components of voltage and currents from the measured values are calculated with the Clarke transformation (see Appendix for more details):

$$\begin{bmatrix} x_\alpha \\ x_\beta \end{bmatrix} = \frac{2}{3}\begin{bmatrix} 1 & -\dfrac{1}{2} & -\dfrac{1}{2} \\ 0 & \dfrac{\sqrt{3}}{2} & -\dfrac{\sqrt{3}}{2} \end{bmatrix} \cdot \begin{bmatrix} x_a \\ x_b \\ x_c \end{bmatrix} \tag{10.3}$$

Finally, the block diagram of the estimator is depicted in Figure 10.2.

10.2.3 Stator Flux Estimator from Stator Voltage for Rotor Side Vector Control

As discussed in Chapter 7, the angle of rotation, θ_s, of the stator flux space vector is necessary for vector control of the DFIM. There are several alternative methods to estimate the stator flux. For instance, it is possible to exploit the fact that the DFIM is directly connected to the grid, only taking into consideration measurements from the stator. Hence, the stator flux can be calculated according to the next expression:

$$\vec{\psi}_s = \int \left(\vec{v}_s - R_s\vec{i}_s\right)dt \tag{10.4}$$

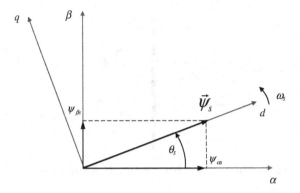

Figure 10.3 *dq* Reference frame alignment with stator flux space vector.

We break this down into $\alpha\beta$ components, yielding

$$\psi_{\alpha s} = \int (v_{\alpha s} - R_s i_{\alpha s})dt \qquad (10.5)$$

$$\psi_{\beta s} = \int (v_{\beta s} - R_s i_{\beta s})dt \qquad (10.6)$$

This means that the stator flux is estimated from only stator voltage and current measurements. This fact is graphically illustrated in Figure 10.3. The amplitude and the angle are calculated according to the expressions

$$|\vec{\psi}_s| = \sqrt{\psi_{\alpha s}^2 + \psi_{\beta s}^2} \qquad (10.7)$$

$$\theta_s = a\tan\left(\frac{\psi_{\beta s}}{\psi_{\alpha s}}\right) \qquad (10.8)$$

However, this estimator presents a pure integration requirement that brings many problems if it is implemented in real systems, where the measured stator voltage or current can be polluted by small offsets or drifts.

In order to avoid these offsets, which accumulate increasing offset error in the stator flux $\alpha\beta$ components, there are several solutions proposed in the literature. A simple one consists of substituting the integrator ($1/s$) with a lowpass filter $1/(s + \omega_c)$.

As can be deduced from the Bode diagrams of Figure 10.4, the unavoidable DC offsets of the stator current and voltage measurements are not integrated as occurs with a pure integrator. However, the lowpass filter modifies the amplitude and the phase shift of the flux estimation and this modification depends on the frequency of the signal.

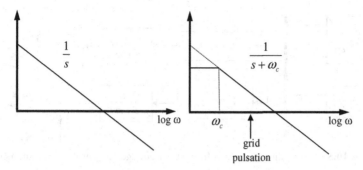

Figure 10.4 Bode diagrams of a pure integrator and a lowpass filter.

Consequently, these errors must be corrected, for instance, by following the next procedure: defining the transfer functions of the integrator and lowpass filter.

$$F(s) = \frac{1}{s}$$

$$F_c(s) = \frac{1}{s + \omega_c}$$

(10.9)

Thus, taking advantage of the fact that the frequency of the stator magnitudes is a constant fixed by the grid, it is possible to quantify the amplitude and angle errors:

$$\left.\begin{array}{l} |F(j\omega)| = \dfrac{1}{\omega} \\[2mm] |F_c(j\omega)| = \dfrac{1}{\sqrt{\omega^2 + \omega_c^2}} \end{array}\right\} \rightarrow \dfrac{|F(j\omega)|}{|F_c(j\omega)|} = \sqrt{\dfrac{\omega^2 + \omega_c^2}{\omega^2}} = \sqrt{1 + \dfrac{\omega_c^2}{\omega^2}} = \varepsilon_{mod}$$

(10.10)

$$\left.\begin{array}{l} \angle F(j\omega) = -90° \\[2mm] \angle F_c(j\omega) = -atan\dfrac{\omega}{\omega_c} \end{array}\right\} \rightarrow \angle F(j\omega) - \angle F_c(j\omega) = -90° + a\tan\dfrac{\omega}{\omega_c} = \varepsilon_{ang} \quad (10.11)$$

Consequently, the simplified block diagram of the estimator is depicted in Figure 10.5.

10.2.3.1 Example 10.1: Error Correction in a 50 Hz Network   The DFIM

is connected to a grid of 50 Hz and the lowpass filter is tuned at 10 Hz. The angular frequencies are

$$\omega = 2\pi \cdot 50 = 314.16 \, \text{rd/s}$$

$$\omega_c = 2\pi \cdot 10 = 62.83 \, \text{rd/s}$$

(10.12)

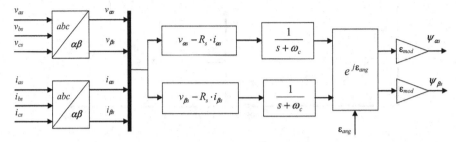

Figure 10.5   Stator flux estimation from stator voltage and currents measurements.

Thus, the errors yield

$$\varepsilon_{\text{mod}} = \sqrt{1 + \frac{\omega_c^2}{\omega^2}} = \sqrt{1 + \frac{62.83^2}{314.16^2}} = 1.02 \tag{10.13}$$

$$\varepsilon_{ang} = -90° + a\tan\frac{\omega}{\omega_c} = -90° + a\tan\frac{314.16}{62.83} = -90° + 78.69° = 11.31° \tag{10.14}$$

10.2.4 Stator Flux Synchronization from Stator Voltage for Rotor Side Vector Control

A very similar method to the previous one consists of using the measured stator voltages and a phase locked loop (PLL) to derive only the angle of the stator flux space vector, θ_s. Since the stator flux and the stator voltage space vectors are approximately 90° phase shifted (Figure 10.6), this method synchronizes with the stator voltage.

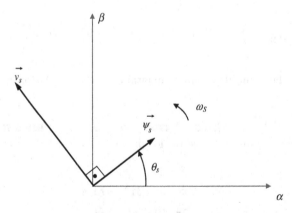

Figure 10.6   The 90° phase shift between the stator voltage and flux space vectors.

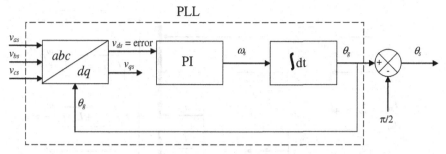

Figure 10.7 Estimation of stator flux space vector angle θ_s.

Thus, by neglecting the stator resistance, the relation between the stator voltage and flux at steady state yields

$$\vec{v}_s = R_s\vec{i}_s + \frac{d\vec{\psi}_s}{dt} \rightarrow \vec{v}_s \cong j\omega_s\vec{\psi}_s \qquad (10.15)$$

This means that the stator flux angle can be estimated as illustrated in the block diagram of Figure 10.7. This estimation is based on a steady state relation, so it is only accurate if the stator voltage is not affected by sudden variations. In that case, the PLL can reject these voltage variations, because θ_s is not affected.

Accuracy can be improved if the stator resistance is also considered in the expression. The PLL needs both rotational and Clarke transformations, tuning of a PI controller, and a pure integrator. Due to its closed-loop philosophy, the problems related to the offsets do not have as dramatic affect as in the previous estimator structure.

Finally, once the PLL achieves synchronization with the stator voltage, the phase angle of the stator flux is simply obtained by subtracting 90° from the grid angle θ_g.

10.2.5 Stator and Rotor Fluxes Estimation for Rotor Side DPC, DTC, and Vector Control

As studied in Chapters 7 and 8, DPC and DTC need knowledge of the rotor flux space vector for the sector calculation. On the other hand, FOC also needs the stator flux space vector knowledge for alignment with the rotatory dq frame, as seen in previous sections. Hence, from the stator and rotor currents, both fluxes can be calculated according to the model based expressions:

$$\vec{\psi}_s = L_s\vec{i}_s + L_m\vec{i}_r \qquad (10.16)$$

$$\vec{\psi}_r = L_m\vec{i}_s + L_r\vec{i}_r \qquad (10.17)$$

Note that, in this case, stator and rotor current measurements are needed in addition to two previously off-line estimated L_s, L_r, and L_m inductance values. Both fluxes are

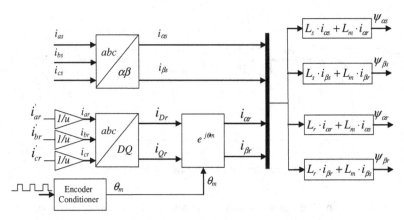

Figure 10.8 Stator and rotor fluxes estimation from currents measurements.

calculated in the $\alpha\beta$ stationary frame. In addition, as illustrated in Figure 10.8, the angular position of the rotor, θ_m, is also needed for stationary frame conversion of DQ rotor current components.

Due to their similarity and for simplicity in the exposition, both flux estimators are represented by only one estimator; however, they are not strictly needed at the same time by any control in general.

10.2.6 Stator and Rotor Flux Full Order Observer

There also exists a more sophisticated estimation procedure based on state observers. Compared with the previously presented estimators, the observers are based on closed-loop estimation. In general, they are more complex than simple estimators, but observers allow estimating directly immeasurable magnitudes with reasonably good accuracy, when the estimated parameters can have uncertainties.

In this section, an example of a full order observer is studied from a wide range of existing alternative possibilities. The general block diagram of the observer is shown in Figure 10.9. The most relevant characteristics can be summarized as follows:

- Due to its full order nature, it needs both stator and rotor abc voltages as inputs, together with the stator abc currents.
- The objective of the observer is to estimate the stator and rotor fluxes' $\alpha\beta$ components.
- The structure of the observer is based on the state-space representation of the system that is to be observed, that is, the DFIM.
- From the voltage inputs, together with the A, B, and C matrices of the state-space representation of the DFIM, the observer calculates an estimated state vector $\hat{x}$ and an estimated output $\hat{i}_s$.

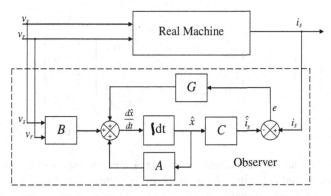

Figure 10.9 Full order observer for stator and rotor fluxes.

- In order to improve the estimation, a closed-loop structure is provided to the observer, by adding a proportional part of the error between the real (measured) stator current i_s and the estimated stator current $\hat{i}_s$, to the first derivative of the state vector $d\hat{x}/dt$.
- Note that the observer tries to be robust to uncertainty in parameters from matrices A, B, and C, by minimizing the error between i_s and $\hat{i}_s$.
- The feedback matrix G must be tuned, in order to obtain a quicker dynamic observation than the DFIM dynamic.

Hence, as seen in Chapter 4, the state-space representation of the DFIM, by choosing as the state vector the stator and rotor fluxes, is

$$\frac{dx}{dt} = Ax + Bu \tag{10.18}$$

$$\frac{d}{dt}\begin{bmatrix} \psi_{\alpha s} \\ \psi_{\beta s} \\ \psi_{\alpha r} \\ \psi_{\beta r} \end{bmatrix} = \begin{bmatrix} \dfrac{-R_s}{\sigma L_s} & 0 & \dfrac{R_s L_m}{\sigma L_s L_r} & 0 \\ 0 & \dfrac{-R_s}{\sigma L_s} & 0 & \dfrac{R_s L_m}{\sigma L_s L_r} \\ \dfrac{R_r L_m}{\sigma L_s L_r} & 0 & \dfrac{-R_r}{\sigma L_r} & -\omega_m \\ 0 & \dfrac{R_r L_m}{\sigma L_s L_r} & \omega_m & \dfrac{-R_r}{\sigma L_r} \end{bmatrix} \cdot \begin{bmatrix} \psi_{\alpha s} \\ \psi_{\beta s} \\ \psi_{\alpha r} \\ \psi_{\beta r} \end{bmatrix} + \begin{bmatrix} v_{\alpha s} \\ v_{\beta s} \\ v_{\alpha r} \\ v_{\beta r} \end{bmatrix} \tag{10.19}$$

$$y = Cx \tag{10.20}$$

$$
\begin{bmatrix} i_{\alpha s} \\ i_{\beta s} \end{bmatrix} = \begin{bmatrix} \dfrac{1}{\sigma L_s} & 0 & \dfrac{-L_m}{\sigma L_s L_r} & 0 \\ 0 & \dfrac{1}{\sigma L_s} & 0 & \dfrac{-L_m}{\sigma L_s L_r} \end{bmatrix} \cdot \begin{bmatrix} \psi_{\alpha s} \\ \psi_{\beta s} \\ \psi_{\alpha r} \\ \psi_{\beta r} \end{bmatrix}
\tag{10.21}
$$

Note that not all the coefficients of matrix A are constant, since some of them depend on the speed ω_m. Thus, the dynamic of the state-space system (the DFIM) is given by the eigenvals of matrix A:

$$
\text{eig}(A) \tag{10.22}
$$

The resulting expressions for the eigenvals are significantly complex so they have been omitted.

On the other hand, the observer is defined as follows (assuming no uncertainty in matrices A, B, and C):

$$
\frac{d\hat{x}}{dt} = A\hat{x} + Bu + G(y - \hat{y}) \tag{10.23}
$$

$$
\hat{y} = C\hat{x} \tag{10.24}
$$

with the estimated vector being

$$
\hat{x} = \begin{bmatrix} \hat{\psi}_{\alpha s} \\ \hat{\psi}_{\beta s} \\ \hat{\psi}_{\alpha r} \\ \hat{\psi}_{\beta r} \end{bmatrix} \tag{10.25}
$$

and

$$
\hat{y} = \begin{bmatrix} \hat{i}_{\alpha s} \\ \hat{i}_{\beta s} \end{bmatrix} \tag{10.26}
$$

In this case, the dynamic of the observer is given by the eigenvals of matrix $A - GC$:

$$
\text{eig}(A - GC) \tag{10.27}
$$

Again, the resulting expressions for the eigenvals are significantly complex, so they are not shown.

Therefore, matrix G is chosen with the next criterion: the dynamic of the observer is k times (with $k > 1$) faster than the dynamic of the DFIM. This means that

$$\text{eig}(A - GC) = k\text{eig}(A) \tag{10.28}$$

Consequently, by using the G matrix, we have

$$G = \begin{bmatrix} \dfrac{ar_{12}}{c_{12}}(1-k^2) & 0 \\[2ex] 0 & \dfrac{ar_{12}}{c_{12}}(1-k^2) \\[2ex] \dfrac{ar_{22}}{c_{12}}(1-k) - \dfrac{ar_{11}}{c_{12}}k(1-k) & \dfrac{ai_{22}}{c_{12}}(k-1) \\[2ex] \dfrac{ai_{22}}{c_{12}}(1-k) & \dfrac{ar_{22}}{c_{12}}(1-k) - \dfrac{ar_{11}}{c_{12}}k(1-k) \end{bmatrix} \tag{10.29}$$

with the terms

$$ar_{11} = \frac{-R_s}{\sigma L_s} \tag{10.30}$$

$$ar_{22} = \frac{-R_r}{\sigma L_r} \tag{10.31}$$

$$ar_{12} = \frac{R_s L_m}{\sigma L_s L_r} \tag{10.32}$$

$$ai_{22} = \omega_m \tag{10.33}$$

$$c_{12} = \frac{-L_m}{\sigma L_s L_r} \tag{10.34}$$

Hence, the stator and rotor fluxes are estimated within the $\hat{x}$ vector. Note that the resulting matrix G is reasonably simple and could be implemented without expensive computational cost. On the other hand, depending on which magnitudes are chosen for the x and y vectors, we can find many different observer combinations, but almost all of them with the G matrix expression longer and, subsequently, more computationally costly.

In addition, with this observer structure, the torque can also be estimated from the expression

$$T_{em} = \frac{3}{2}\frac{L_m}{\sigma L_r L_s}pIm\{\vec{\psi}_r^* \cdot \vec{\psi}_s\} \tag{10.35}$$

10.3 START-UP OF THE DOUBLY FED INDUCTION MACHINE BASED WIND TURBINE

Doubly fed induction machine based wind turbines require more complex start-up than other typical AC machines (e.g., squirrel cage induction machines), due to the reduced size of the back-to-back converter used to supply the rotor. The added complexity arises from the intrinsic behavior of the DFIM itself since, as studied in Chapters 3 and 4, it demands more AC voltage amplitude supply at the rotor, at lower speeds near zero, while lower AC amplitude at speeds near synchronism. If a cost effective reduced size back-to-back converter is used, its limited AC voltage availability, only permits the DFIM to be controlled in a range of speeds near synchronism. Thus, it is necessary to use an alternative acceleration process for the machine at zero speed.

Under this environment, normally the solution adopted obliges one to disconnect the stator of the machine from the grid, by using a breaker located in the stator so that the machine can be accelerated externally by the wind itself. Then, once the machine reaches the speed within the controllability margin, a connection process with the grid is carried out that concludes with the stator circuit breaker closed and the machine generating energy from the wind at normal operation.

In addition, due to the presence of the three physical windings of the rotor, in general, it is necessary to calibrate the encoder's zero-reference position with the magnetic field created by the three phase windings of the rotor. This task is not necessary, for instance in squirrel cage induction machines, due to the different physical construction of their rotor. In contrast, for instance, synchronous machines with permanent magnets also require a specific alignment of both rotor and stator fluxes to be performed normally in a start-up process.

Therefore, this section examines these two special processes for start-up, together with an illustrative example of how the wind turbine is controlled sequentially by means of different states, in a typical real wind energy generation application.

From the system configuration point of view, both encoder calibration and grid voltage synchronization processes share a common electric circuit structure; that is, the stator breaker is opened, so there is no current flow through the stator. This fact is graphically represented in Figure 10.10. As can be noticed, both processes are driven by the rotor side converter, so, in general, they need the DC bus voltage of the back-to-back converter established by the grid side converter. Only when the synchronization process is finished, is the stator breaker ready to be closed safely and normal operation of the wind turbine can start. In both cases, it is necessary to establish a fixed speed for proper start-up, this task is carried out by the pitch control of the wind turbine.

On the other hand, from the control strategy point of view, both encoder calibration and grid voltage synchronization processes share an almost-common control structure. In both cases, the rotor dq current components are controlled according to the vector control principles studied in Chapter 7. By controlling the rotor currents according to criteria described in subsequent sections, together with information for the DFIM provided by several sensors, it is possible to calibrate the encoder's

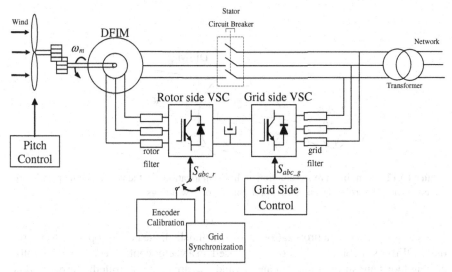

Figure 10.10   System configuration during encoder calibration and grid voltage synchronization processes.

absolute position as well as to synchronize the induced stator voltage with the grid voltage, in order to produce a proper connection to the grid.

Figure 10.11 illustrates the simplified control block diagram for both processes. The i_{dr} and i_{qr} control loops are shared by all controls; then, depending on which of the control processes is enabled, these dq rotor current references are generated according to different criteria by superior control loops. The encoder calibration operates on the estimated angle of the rotor position. While the grid voltage synchronization operates on the rotor dq currents, modifying the phase and amplitude of the induced stator voltage.

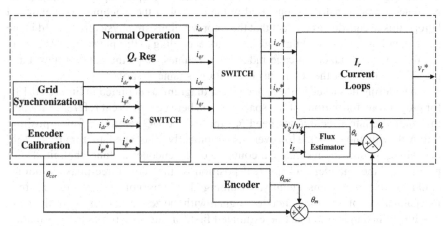

Figure 10.11   Simplified control structure considering encoder calibration and grid voltage synchronization processes.

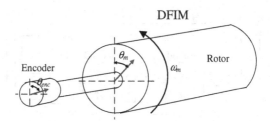

Figure 10.12   Simplified representation of different location, for the zero absolute position of the encoder and zero location of the three-phase rotor windings.

Finally, once the synchronization process is finished, the normal operation control of the DFIM is enabled. If this control is based on a vector control technique, the rotor current loops are still maintained and Q_s and T_{em} are then controlled. However, if at normal operation of the DFIM, a different control technique is adopted, such as DTC or DPC, the overall control structure of the rotor side converter is changed. In this section, for simplicity, vector control is considered as the control at normal operation.

10.3.1   Encoder Calibration

10.3.1.1   Introduction to the Problem   The problem to be solved arises when the encoder is mechanically coupled to the axis of the DFIM's rotor. In general, the absolute zero position of the encoder and the zero location of the three-phase windings do not have to be coincident, so there is an angle shift between the two zeros that must be corrected (Figure 10.12).

The alignment between the two zeros is crucial for proper implementation of the reference frame transformations used in the control of the machine. As mentioned before, this occurs due to the fact that the rotor windings are physically located in the rotor, creating the resultant flux spatial vector according to this physical distribution. In contrast, for instance, in cage induction machines, since they do not have three physical windings at the rotor, this calibration is unnecessary.

Considering this issue in the space vector diagram, as depicted in Figure 10.13a, for proper transformations of the space vectors between different reference frames, the studied control techniques need θ_m, that is, the DQ reference frame position (aligned with the rotor). However, as graphically illustrated in Figure 10.13b, depending on how the mechanical coupling of the encoder is done, the initial zero position of the encoder and the real position of the rotor three-phase windings could be unequal. Consequently, assuming that it is not realistic to ensure a mechanical coupling of the encoder aligned with the zero real position of the rotor windings, the angle θ_{cor} must be estimated first for later use in the transformation of the reference frames. The procedure for this angle shift estimation is described in the next section.

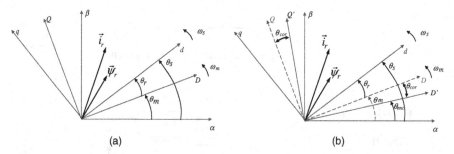

Figure 10.13 Problem representation of the encoder zero position in the space vector diagram.

10.3.1.2 Encoder Zero Position Correction

As introduced before, the encoder is calibrated by means of the rotor side converter. Figure 10.14 illustrates the system configuration during this process. The most important characteristics can be summarized as follows:

- The grid side converter operates normally, imposing the required DC bus voltage and exchanging sinusoidal currents with the grid through the grid side converter.
- The pitch control establishes a constant speed (ω_m), as the mechanical axis must be moving to detect the angle shift between the encoder and the rotor electric windings. However, it is not necessary to have one specific speed of rotation and the test can be performed at a wide range of speeds (but reasonably constant).

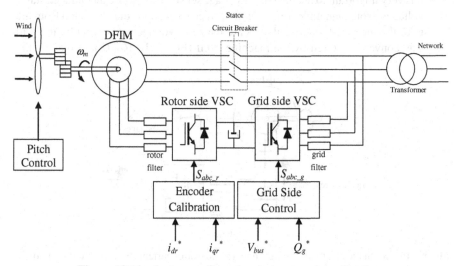

Figure 10.14 System configuration during encoder calibration.

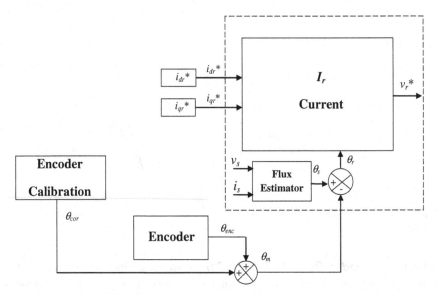

Figure 10.15 Simplified control structure of encoder calibration processes.

- The stator breaker is opened, so there is no current in the stator side.
- The rotor side converter controls the rotor currents in the dq reference frame (normally $i_{qr} = 0$) according to the vector control principles, as depicted in Figure 10.15.
- Due to the rotor currents imposed by the rotor side converter, there is a stator voltage induced in the machine. Note that, in some way, the machine operates similarly to a transformer.
- It is very important to locate a stator voltage sensor, to measure the induced stator voltages. Note that, under this situation, since the stator is disconnected from the grid, the measured stator voltages are the rotor voltages generated by the grid side converter filtered by the machine itself (Figure 10.16). Consequently, the

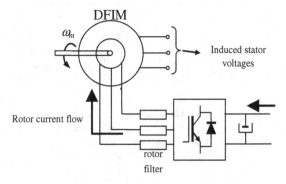

Figure 10.16 Induced stator voltage due to the rotor current flow when the stator is disconnected from the grid.

measured voltages must be free of ripples and aliasing for a proper encoder calibration. Therefore, the stator voltage sensors must be chosen accordingly for that purpose (see Chapter 5 for more details).

Hence, with the machine driven under the mentioned conditions, the angle correction is based on the estimation of the rotor flux by two different methods. The first estimation method is based on the flux expression:

$$\vec{\psi}_r^r = L_m \vec{i}_s^r + L_r \vec{i}_r^r \qquad (10.36)$$

Note that as the stator breaker is opened, there is no stator current, so the expression yields

$$\vec{\psi}_r^r = L_r \vec{i}_r^r \qquad (10.37)$$

Consequently, the rotor flux in DQ coordinates can be estimated by the block diagram in the top left-hand side of Figure 10.18.

On the other hand, the rotor flux can also be derived from the stator flux. Thus, taking into account that there is no stator current, the stator and rotor fluxes must be in phase, as deduced from Equation (10.37) and

$$\vec{\psi}_s^s = L_s \vec{i}_s^s + L_m \vec{i}_r^s \rightarrow \vec{\psi}_s^s = L_m \vec{i}_r^s \qquad (10.38)$$

This fact is graphically represented in the space vector diagram of Figure 10.17.

Consequently, an alternative way to estimate the rotor flux is to calculate the stator flux from the induced stator voltage, according to the expression

$$\vec{\psi}_s = \int (\vec{v}_s - R_s \vec{i}_s) dt \qquad (10.39)$$

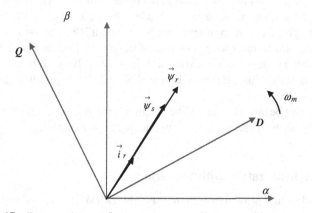

Figure 10.17 Rotor and stator flux space vectors alignment with zero stator current.

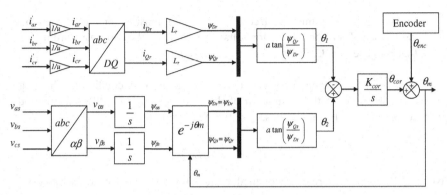

Figure 10.18   Encoder calibration algorithm.

Since the stator current is zero, this expression can be simplified to

$$\vec{\psi}_s = \int \vec{v}_s \, dt \qquad (10.40)$$

Based on this expression, the rotor flux in the DQ reference frame can be estimated according to the simplified block diagram of the bottom left-hand side of Figure 10.18. It is important to highlight that since the $\alpha\beta$ to DQ reference frame transformation is within the estimation, the error due to the encoder zero position produces an incorrect rotor flux estimation. However, this error can easily be detected by comparison to the first presented rotor flux estimation based on rotor currents. Basically, it yields a phase displacement between the rotor flux estimations of two methods. As seen in Section 10.2.3, the pure integrator used by this last estimator must not be affected by the offsets; however, in this section, this problem is not tackled for simplicity.

Finally, the encoder is calibrated according to the algorithm of the block diagram shown in Figure 10.18. In order to know whether both estimated fluxes are in phase or not, their angles are compared. When both fluxes are in phase, the subtraction of both angles is zero. K_{cor} is only a constant to accelerate the calibration. During the correction, by means of the integrator, the term θ_{cor} is added to θ_{enc} until the subtraction of both angles $(\theta_2 - \theta_1)$ gives zero, so the integrator does not modify θ_{cor}, because both fluxes are in phase, so θ_m corresponds to the position of the rotor reference frame DQ.

On the other hand, when the calibration is finished because θ_{cor} is maintained reasonably constant, the algorithm is disabled and the adjusted θ_{cor} is kept for subsequent control processes.

Finally, it must be remarked that if the machine's manufacturer ensures a calibrated encoder placement on the machine, obviously this automatic encoder calibration becomes unnecessary.

10.3.2 Synchronization with the Grid

As mentioned in the introduction to this section, DFIM based wind turbines with a reduced size of the back-to-back converter supplying the rotor require a special

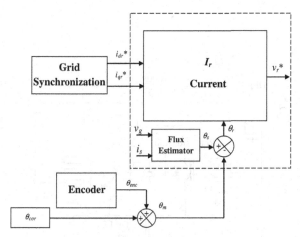

Figure 10.19 Simplified control structure of grid voltage synchronization processes.

synchronization process with the grid voltage, since control of the machine is only possible in an approximately ±30% speed range, around synchronous speed. This section studies in detail the most important characteristics of this synchronization process. However, the reader must know that the synchronization procedure presented in this section is only one representative example of the different synchronization philosophies that can be applied to solve this problem.

10.3.2.1 General Control Loops
When it comes to connect the stator of the DFIM to the grid, the process establishes a vector current control structure based synchronization. The parallelism with the encoder calibration is evident, as exhibited in the schematic control block diagram of Figure 10.19. The same current control loops of the encoder calibration process are maintained, while the corrected angle θ_{cor} must be kept as well. In this case, the main difference with the previous process is that the dq current references are generated by superior control loops. Thanks to the proper generation of these references, it is possible to synchronize the DFIM with the grid and safely close the stator breaker. It is not necessary to create new references for this task, since the schematically represented "grid synchronization" block is in charge of doing it automatically. In addition, the angle for the vector control, θ_s, is derived from the grid voltage as shown in Figure 10.19.

As mentioned before, the main objective of this control process consists of establishing the proper conditions, to proceed to the stator breaker closure and thus connection of the DFIM with the grid. For that purpose the following is carried out:

- By controlling the rotor dq currents with the stator breaker opened, a voltage in the stator side of the DFIM is induced.
- This induced stator voltage (three phases) must be as equal as possible in phase and amplitude to the grid voltage, as represented in Figure 10.20.

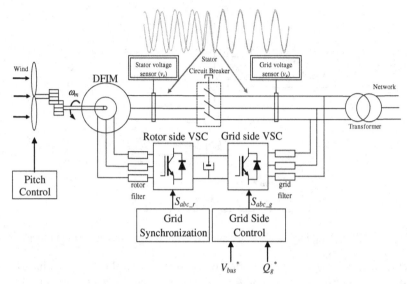

Figure 10.20   System configuration during grid voltage synchronization.

- Thus, once the induced stator voltage and the grid voltage are similar enough, the stator breaker can be closed and the normal operation regimen can start.
- Note that, as occurred in the encoder calibration process, while the stator breaker is opened, the DFIM operates like in some way as a transformer, inducing an image of the rotor chopped voltage into the stator. This induced stator voltage is far from being as sinusoidal as the grid voltage; consequently, the sensor used for the stator voltage must be prepared to filter the hormonics of this voltage in order to be compared with the grid voltage (see Chapter 5 for more details).

Hence, the grid synchronization process is based on two independent cascaded control loops. One loop is dedicated to equalizing the amplitudes of both the stator and grid voltages (*abc* phases), while the other loop is dedicated to equalizing the phases of both the stator and grid voltages (*abc* phases). The proposed control block diagram is represented in Figure 10.21. In subsequent sections the basis of this control structure is studied in detail.

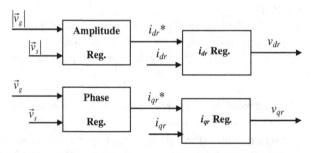

Figure 10.21   Grid voltage synchronization control loops.

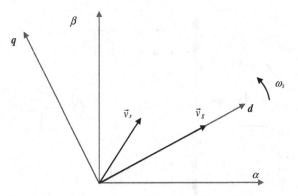

Figure 10.22   Space vector representations of the grid voltage and the induced stator voltage.

10.3.2.2 Amplitude Control Loop   We first study the amplitude control. As illustrated in Figure 10.22, if specific rotor dq currents are established by means of the rotor side converter, the space vector representations of both the induced stator and grid voltages, in general, would have different amplitudes and phases.

Thus, Figure 10.23 shows how the amplitude control can be automatically achieved by modifying the i_{dr} current component. Note that this process must be carried out with the machine rotating at constant speed, preferably not "very low" speed. As seen in Chapter 3, the relation between the stator and rotor voltage amplitudes depends on the slip of the machine as follows:

$$|\vec{v}_s| = \frac{|\vec{v}_r|}{s}, \quad s = \frac{\omega_s - \omega_m}{\omega_s} \quad (10.41)$$

Consequently, for a specific required grid voltage, at speeds not far from the synchronous speed, the needed rotor voltage will be smaller than at speeds far from the synchronous speed. Note that since θ_r is calculated from θ_m and θ_s (grid abc voltage is used), as shown in Figure 10.19, the induced stator voltage is of equal

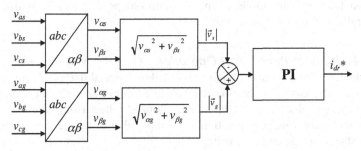

Figure 10.23   Amplitude control loop.

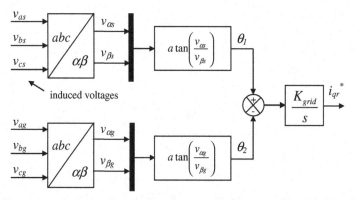

Figure 10.24   Phase control loop.

angular frequency (ω_s) as the grid voltage, at any speed (ω_m) that the machine rotates.

Finally, note that a PI controller could be a suitable regulator to correct the error between both stator and grid voltages, since both controlled variables ($\vec{v}_s$ and $\vec{v}_g$) have constant magnitudes at steady state.

10.3.2.3 Phase Control Loop

What comes next is the phase synchronization of the stator and grid voltages. As occurred with the amplitude, phase can be controlled by modifying one component of the rotor currents, that is, the q component. Hence, with this philosophy, it is possible to create a cascaded control loop as shown in Figure 10.21, modifying the q component of the rotor current whenever the phase shift of the stator voltage must be equal to the grid voltage. For that purpose, the control structure of Figure 10.24 can be employed. In order to detect a phase shift error between the induced stator and grid voltage space vectors, the angle calculation of the $\alpha\beta$ components can be used. Once the phase shift error is detected, the integrator modifies the q rotor current reference, modifying next the q component of the induced stator voltage, and thus finally correcting the phase angle shift. Once both voltages are synchronized, the angle difference ($\theta_1 - \theta_2$) provides a zero input to the integrator, stopping the phase correction. A constant, K_{grid}, is used to accelerate or decelerate the process.

To conclude, both amplitude and phase control loops are coupled, so, in general, both regulators work simultaneously.

10.3.2.4 Jump to the Normal Operation

Once the synchronization process has finished, resulting in equal (as much as possible) induced stator and grid voltages, the stator breaker can safely be closed. Its closure brings a smooth transient in stator and rotor currents, due to the fact that since a chopped voltage (rotor side converter) is used for the stator voltage generation, it is impossible to ensure an exactly equal induced stator voltage and grid voltage.

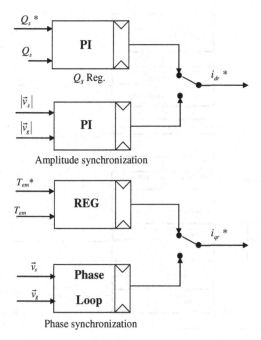

Figure 10.25   Control strategy change to normal operation.

Nevertheless, this is not a problem. What it is more critical is the change of control strategy—from synchronization to normal operation. As depicted in Figure 10.25, the most relevant change is that the two superior loops of the amplitude and phase control must be disabled, thus enabling T_{em} and Q_s control. Note that these two incoming controls (T_{em} and Q_s) must be perfectly coordinated with the outgoing controls (amplitude and phase), so they do not provoke a strong undesired transient due to a sudden change in the i_{dr} and i_{qr} references. For that purpose, T_{em} and Q_s references should be properly chosen as well as the initial conditions of the integral part of the PI regulators.

Finally, for simplicity, only the vector control technique has been considered at normal operation of the DFIM.

10.3.3 Sequential Start-up of the DFIM Based Wind Turbine

This section gives an example for the start-up of a 2 MW DFIM based wind turbine up to normal operation. For simplicity in the exposition, we do not include the LVRT process during disturbances. We only want to give a general idea of how the wind turbine can be commanded when it is connected to the grid. There are many improvements or alternative methods that can be used; however, the presented start-up procedure only seeks to provide the reader with some basic knowledge about the further work that accompanies the control strategies studied in previous sections, to integrate them in a realistic wind energy generation scenario.

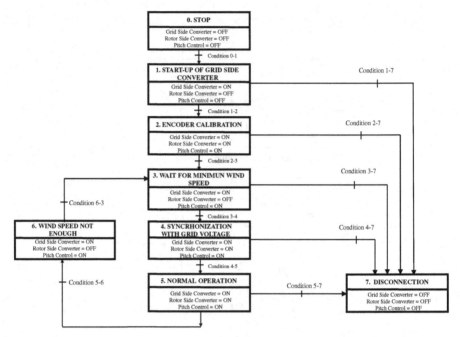

Figure 10.26 General sequential start-up.

10.3.3.1 General Sequential Procedure

Figure 10.26 illustrates the schematic state-flow for the sequential start-up of the wind turbine. As can be noticed, it is composed of eight different states, in which the wind turbine operates at different modes. Due to its complexity, different converters, reduced DC bus voltage, pitch control, stator breaker, and so on, the DFIM must go through different previous states, until it finally can operate at normal conditions, generating energy from the wind. In subsequent subsections, all the states are analyzed.

10.3.3.2 "Stop" State

This is the initial state of the start-up procedure. The system is waiting until the start is commanded.

10.3.3.3 "Start-up of Grid Side Converter" State

Once the general start-up is enabled, the DC link of the back-to-back converter is precharged. After closing the main grid breaker, the DC link is charged through the diodes of the grid side converter and the charging resistors. Figure 10.27 shows graphically the charge power circuit. After some milliseconds, the charge resistors are short-circuited and the DC bus voltage is stabilized (Figure 10.28).

Once the DC link is charged, the operation of the grid side converter is enabled, charging the DC-link voltage to normal operating conditions. Thus, the V^*_{bus} and Q^*_g references are set to the desired values, as illustrated in Figure 10.29. Since the rotor side converter is not operating, a minimum consumption of grid side currents is

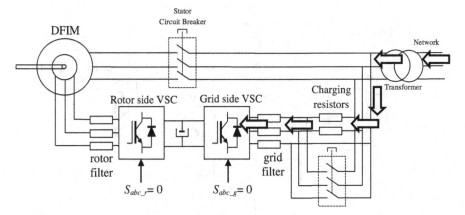

Figure 10.27 DC-link charge.

established in order to maintain the DC bus voltage. Figure 10.30 shows the most important magnitude variations during the start-up of the grid side converter. Note that at steady state, once the DC bus voltage has reached the required value, there is a small current consumption for the losses of the grid side converter and the DC bus.

10.3.3.4 "Encoder Calibration" State   During this state, the encoder is calibrated as presented in the previous section. First, the pitch control sets the rotation speed of the turbine to a given constant value. During the entire state, the speed is maintained constant and the rotor side converter is enabled. Figure 10.31 shows the system configuration during this state. Note that the grid side converter controls the power flow through the grid, ensuring sinusoidal current exchange and the desired Q_g.

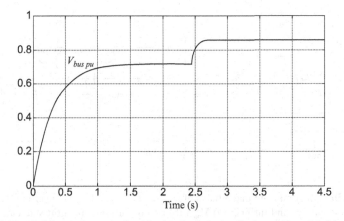

Figure 10.28   DC bus voltage during charge through the diodes of the grid side converter.

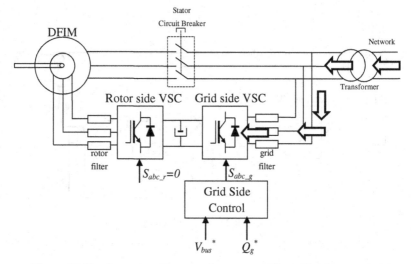

Figure 10.29   System configuration in start-up of the grid side converter.

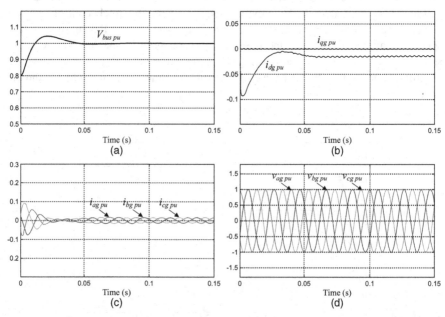

Figure 10.30   Most relevant magnitudes of start-up of the grid side converter ($Q_g = 0$) of a 2 MW DFIM based wind turbine: (a) V_{bus}, (b) and of currents, (c) grid side currents, and (d) grid voltage.

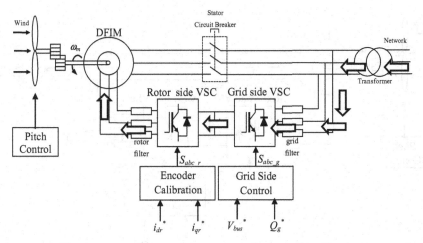

Figure 10.31   System configuration during encoder calibration.

Hence, after some milliseconds, the encoder is calibrated by synchronizing the stator and rotor flux estimations and achieving the required final angle of correction, θ_{cor}. The required power for this process is not very high, so the rotor dq current values are normally established relatively small. Figure 10.32 illustrates the most relevant

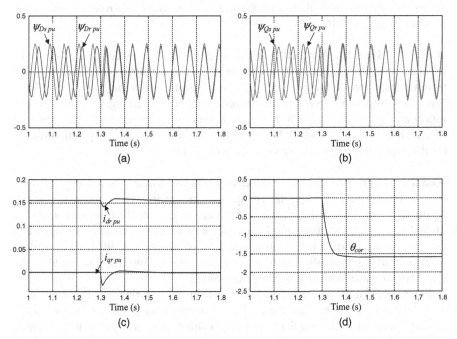

Figure 10.32   Most relevant magnitudes of encoder calibration process of a 2 MW DFIM based wind turbine: (a) D components of fluxes, (b) Q components of fluxes, (c) $i_{dr\ pu}$ and $i_{qr\ pu}$, and (d) θ_{cor} (rad).

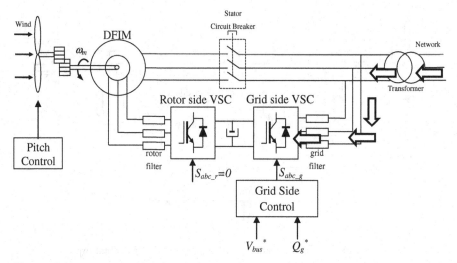

Figure 10.33   System configuration while waiting for the required speed for grid connection.

magnitudes during the calibration of the encoder. Note that once the calibration is finished, the estimated θ_{cor} value is maintained until the sequential procedure makes the encoder calibration start again. In this example, the rotor currents have been set to $i_{dr\ pu} = 0.16$ and $i_{qr\ pu} = 0$.

10.3.3.5 "Wait for Minimum Wind Speed" State

Once the calibration of the encoder is finished, the system waits until the pitch control sets the speed to the required value for the subsequent task: synchronization with the grid voltage. Note that, during this task, the rotor side converter can be disabled again since it is very simple and quick to reenable it. However, in order to avoid repeating the same task several times, it is preferable to keep the grid side converter operating while waiting for the required speed to be achieved (Figure 10.33).

10.3.3.6 "Synchronization with Grid Voltage" State

Once the speed of the wind turbine is high enough, the voltage synchronization process can be started. Again, the rotor side control must be reenabled, controlling the rotor dq currents as studied in previous sections, until each phase of the induced stator voltages and the grid voltages are approximately equal in phase and amplitude. The system configuration is graphically illustrated in Figure 10.34. During the entire task, the speed is maintained constant by the pitch control.

Figure 10.35 illustrates the most relevant magnitudes during the synchronization with the grid. Note that, since the back-to-back converter generates induced stator currents polluted by harmonics, the stator voltages that use the synchronization algorithm must be filtered.

When the voltage synchronization is finished because the stator and grid voltages are similar enough (in phase and amplitude), the stator breaker is closed minimizing

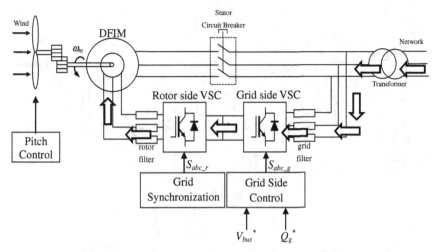

Figure 10.34 System configuration during grid synchronization.

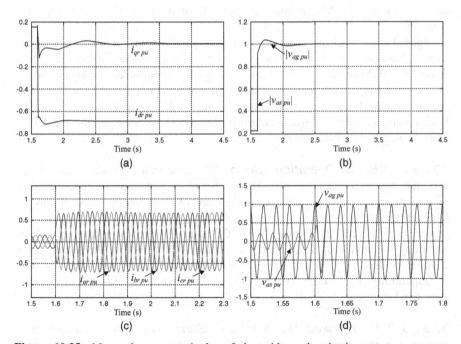

Figure 10.35 Most relevant magnitudes of the grid synchronization process at $\omega_m =$ 1000 rpm, of a 2 MW DFIM based wind turbine: (a) i_{dr} and i_{qr}, (b) modules of induced stator voltage and grid voltage, (c) rotor currents, and (d) induced a phase stator voltage and grid voltage (both filtered).

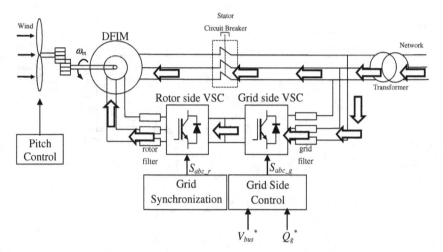

Figure 10.36 System configuration during stator breaker closure.

the overcurrents due to the transient. Thus, since the breaker takes some milliseconds in closing, once the stator and grid voltages are measured to be equal, the procedure steps to the next state. Also, in this case, the pitch control maintains the speed roughly constant.

Figure 10.36 graphically shows this state configuration and Figure 10.37 shows the stator voltage and currents during the stator voltage closure. Once the stator breaker is closed, the DFIM operates at zero torque ($i_{qr} = 0$). It is strongly recommended that the stator reactive power reference is near zero, in order to minimize the reactive stator current increase during the stator breaker closure. For that purpose, the corresponding i_{dr} is commanded by the grid synchronization algorithm, magnetizing the DFIM almost completely by the rotor.

10.3.3.7 "Normal Operation" State

When the stator breaker is closed, the wind turbine can operate normally. The change of the state must be done carefully, so that the new controllers entering do not provoke any strong transient and the DFIM begins to be controlled in torque (stator active power) and stator reactive power, as illustrated in Figure 10.38. Of course, to reach normal operation, according to the speed of the wind, the system should modify the torque stator reactive power ($T_{em} \neq 0$ and $Q_s \neq 0$) and speed references smoothly to their corresponding values. Table 10.1 and Figure 10.39 show an example of a steady state wind turbine.

10.3.3.8 "Wind Speed Not Enough" State

When the wind turbine is operating in the normal state and the speed of the wind goes down, the speed reference is decreased accordingly. If the speed goes down to its predefined minimum value there is no torque capability, so the wind turbine is no able to generate energy and the stator of the machine must be disconnected from the grid. This means that the

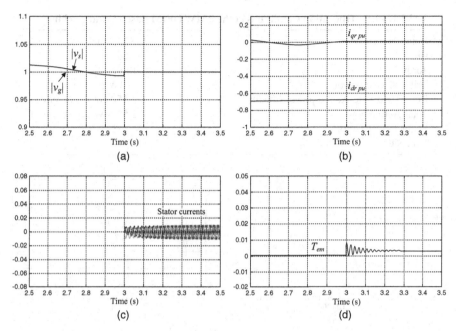

Figure 10.37 Most relevant magnitudes of the stator breaker closure, of a 2 MW DFIM based wind turbine: (a) modules of the stator and grid voltages, (b) i_{dr} and i_{qr}, (c) stator currents, and (d) torque.

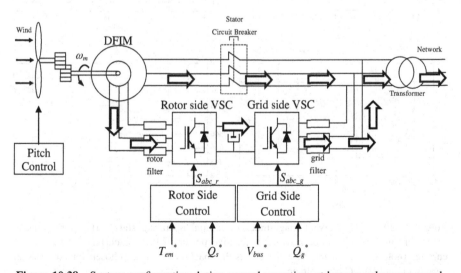

Figure 10.38 System configuration during normal operation, at hypersynchronous speed.

TABLE 10.1   Operation Conditions of the 2 MW DFIM Based Wind Turbine

Rotational speed	1875 rpm
slip	-0.25
Torque	-12.9 kN
Stator Active power	-2 MW (Generating)
Rotor/Grid Side Active power (ideally)	-0.55 MW (Generating)
Total Active power	-2.55 MW (Generating)
Stator Power factor	1
Grid Side Power factor	1

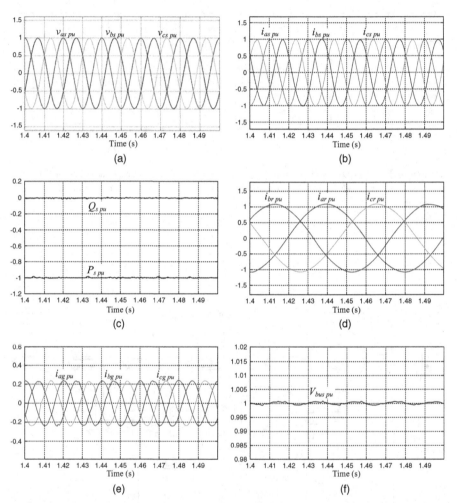

Figure 10.39 Most relevant magnitudes of normal operating state: (a) stator voltages, (b) stator currents, (c) stator active and reactive powers, (d) rotor currents, (e) grid side currents (note that the same base current as the machine has been chosen for comparison purposes), and (f) DC bus voltage.

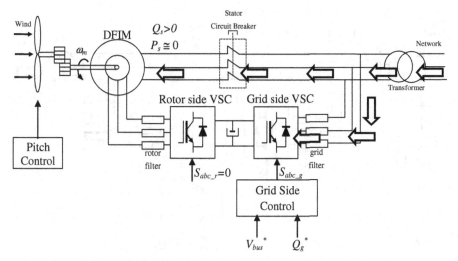

Figure 10.40 Rotor side converter inhibition.

sequential procedure jumps to the "wind speed not enough" state, where the stator breaker is opened.

However, before proceeding to open the stator breaker, it is recommended that the rotor side converter be disabled first, setting the T_{em} (P_s) and Q_s references to make i_{dr} and i_{qr} zero or near zero (magnetizing through the stator and zero torque). Thus, the rotor side converter is inhibited when almost no current flows through the rotor. This disconnection is totally safe, since it does not produce overcurrents either in the stator or in the rotor (Figure 10.40).

In that way, once the rotor is disconnected from the rotor side converter, only the stator of the machine is connected to the grid, consuming mainly reactive stator current for the magnetizing inductance (L_m). After that, the stator breaker can be opened (not at zero current).

If the process is done the other way around, disconnection of the stator first provokes such a strong perturbation (similar situation to a voltage dip) that control can be lost as seen in Chapter 9. However, alternatively, it could also be possible to shut-down the DFIM, by setting a zero stator current ($T_{em} = 0$ and $Q_s = 0$) by control, then opening the stator breaker at zero stator current and then immediately disabling the rotor side converter (Figure 10.41).

10.3.3.9 "Disconnection" State Disconnection of the wind turbine can be performed for different reasons. This section shows some of them.

1. When the user wants to stop the wind turbine, the system is disconnected in general by following several steps. The speed control is disabled, while the rotor side converter sets a breaking torque, making the speed decrease. Once the speed goes below a minimum threshold, the procedure disables both the rotor and grid side

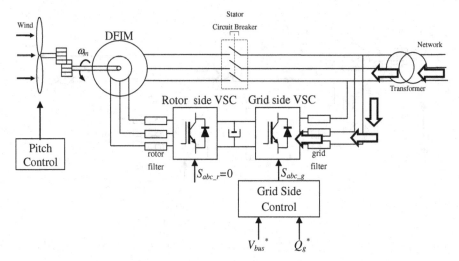

Figure 10.41 Opening the stator breaker.

converters, while the stator and main grid breakers are also opened. After that moment, the mechanical brake stops the wind turbine.

2. When a sudden disconnection is required, the mechanical brake operates until the speed is zero. The rotor and grid side converters are disabled, and the stator and main grid breakers are also opened, so the DFIM is totally disconnected from the grid. It is possible to reach this state, due to user demand, due to a failure (stator current out of margins, overspeed, etc.).

10.3.3.10 *Final Discussion*   It is important to highlight again that this is just one example of how the DFIM based wind turbine can be automated in a real context. There are many alternative improvements and modifications to the presented procedure. However, this level of exposition is enough for the reader, to achieve a general perspective and basic knowledge about the requirements to make these kinds of turbines operable.

Finally, it is also important to note that there are many other aspects of wind turbine operation that have not been considered, such as error handling, commissioning, and details about the changes between different controls when jumping into new states.

10.4 SUMMARY

This chapter has examined the basic structure and characteristics of several estimation techniques for immeasurable magnitudes of the DFIM. The reader has been introduced to the wide area associated with control, which deals with many different estimation techniques, in this case applied to the DFIM.

On the other hand, the characteristics of the start-up of DFIM based wind turbines with a reduced size back-to-back converter are studied and illustrated, with the help of several graphical examples, providing the reader with control knowledge about grid connected wind turbines. The most important aspects of the start-up are identified as the encoder calibration and synchronization with the grid voltage.

Finally, the last two chapters of this book exhibit and analyze several aspects of the stand alone operation of DFIM based wind turbines; while the last chapter presents and discusses future trends and challenges of wind turbine technology in general.

REFERENCES

1. http://www.gamesacorp.com.
2. B. K. Bose, *Modern Power Electronics and AC Drives*. Prenctice Hall, 2002.
3. http://www.gamesacorp.com.
4. http://www.acciona-energia.es.
5. http://www.ingeteam.com.
6. http://www.abb.com/.
7. ACS Catalogs for wind turbines from ABB.
8. http://www.ecotecnia.com/.
9. S. A. Gomez and J. L. R. Amenedo, "Grid Synchronisation of Doubly-Fed Induction Generators Using Direct Torque Control," in *IECON Conference Proceeding*, Vol. 4, pp. 3338–3343, 2002.
10. Y. Fang, L. Qi-hui, and J. Zhang, "Flexible Grid-Connection Technique and Novel Maximum Wind Power Tracking Algorithm for Doubly-Fed Wind Power Generator," Industrial Electronics Society, 2007. In *IECON 2007, 33rd Annual Conference of the IEEE*, 5–8 November 2007, pp. 2098–2103.
11. Y. Lang, X. Zhang, D. Xu, S. R. Hadianamrei, and H. Ma, "Stagewise Control of Connecting DFIG to the Grid," *IEEE ISIE*, 2006.
12. J. Park, K. Lee, D. K. Kerim, K. Lee, and J. P. Kordi, "An Encoder-Free Grid Synchronization Method for a Doubly-fed Induction Generator," *EPE 2007*, Aalborg.
13. G. Yuan, J. Chai, and Y. Li, "Vector Control and Synchronization of Doubly Fed Induction Wind Generator System," in *4th International PEMC Conference Proceedings* Vol. 2, pp. 886–890, 2004.
14. J.-B. G. Manel, A. Jihen, and S.-B. Ilhem, "A Novel Approach of Direct Active and Reactive Power Control Allowing the Connection of the DFIG to the Grid," in *Conference EPE'09*, Barcelona, Spain.
15. A. G. A. Khalil, D. C. Lee, and S. H. Lee, "Grid Connection of Doubly-Fed Induction Generators in Wind Energy Conversion System," in *Conference IPEMC*, 2006.
16. J. L. Da Silva, R. G. de Oliveira, S. R Silva, B. Rabelo, and W. Hofmann, "A Discussion About a Start-up Procedure of a Doubly-Fed Induction Generator System," in Nordic Workshop on Power and Industrial Electronics, June 2008.
17. S. A. Gomez and J. L. R. Amenedo, "Grid Synchronization of Doubly Fed Induction Generators Using Direct Torque Control," *IEEE IECON Conference Proceedings*, Vol. 4, pp. 3338–3343, 2002.

18. X. Zhang, D. Xu, Y. Lang, and H. Ma, "Study on Stagewise Control of Connecting DFIG to the Grid," in *IEEE 5th International Power Electronics and Motion Control Conference*, pp. 1–5, 2006.

19. A. G. Abo-Khalil, D.-C. Lee, and S.-P. Ryu, "Synchronization of DFIG Output Voltage to Utility Grid," *European Power and Energy Systems*, pp. 1–6, 2006.

CHAPTER 11

Stand-Alone DFIM Based Generation Systems

11.1 INTRODUCTION

The previous chapters have analyzed the most relevant aspects of DFIM based wind turbines connected to the grid. Under this scenario, wind turbines deliver electric energy captured from wind to a main electric grid. However, a DFIM can also be employed as a generator that operates unconnected to the grid, that is, in stand-alone mode. In addition, it can also transform energy from energy sources different from wind.

In this way, this chapter analyses the stand-alone operation of the doubly fed induction machine, showing how it is possible to increase the application possibilities of this machine. Thus, the mathematical model of the DFIM as a stand-alone is studied accompanied by its control.

Therefore, from the study carried out in preceding chapters, this chapter describes how several modeling and control differences must be introduced, to allow the DFIM to operate in stand-alone mode.

11.1.1 Requirements of Stand-alone DFIM Based System

A doubly fed induction machine is an interesting solution for primary movers other than wind turbines, with limited range of speed, like internal combustion engines [1,2] and water turbines [3–6]. These sources are more controllable than wind turbines and can assure required power availability. In the case of such power sources, it is advisable to obtain controlled stand-alone operation with a fixed amplitude and frequency of the generated voltage. This operation mode increases power availability during grid failures in grid connected systems and allows load supply in remote and isolated areas. For a rotating energy storage the DFIM can also be used [7], and operation as a rotary uninterruptible power supply (UPS) system can be applied.

Doubly Fed Induction Machine: Modeling and Control for Wind Energy Generation,
First Edition. By G. Abad, J. López, M. A. Rodríguez, L. Marroyo, and G. Iwanski.
© 2011 the Institute of Electrical and Electronic Engineers, Inc. Published 2011 by John Wiley & Sons, Inc.

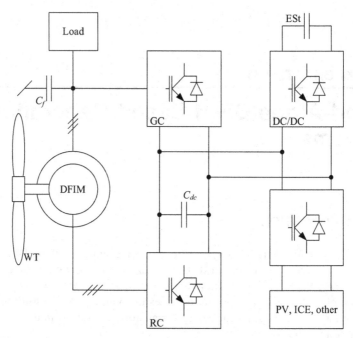

Figure 11.1. Stand-alone DFIM based wind generation system supported by DC coupled additional power source and energy storage.

Recently, the main applications of the DFIM are wind driven power generation systems connected to the grid. Single operated wind turbines cannot be applied as high-reliability stand-alone power sources, regardless of a generation system topology. However, stand-alone operation of a wind power system can be applied successfully, if the system is supported with additional energy source like an internal combustion engine (ICE), PV panels, or long-term energy storage connected to the DC or AC side.

In the case of a DFIM power system, both connections are also possible. However, connection of a supporting storage or source to the DC bus (Figure 11.1) can only partially help in reduction of an energy deficit, due to the limited power of the rotor side RC and grid side GC converters.

Apart from the support of additional storage or a source, the stator side has to be able to supply the load with high-quality generated voltage. High-frequency harmonics can be reduced by stator connected filtering capacitors C_f, similar to the VSI equipped with LC filters applied in uninterruptible power supply (UPS) systems. Symmetrical voltage in the case of an unbalanced load requires mainly adequate system topology with load access to the neutral wire. It can be realized by the use of a neutral wire connected with the neutral point of a DFIM's star connected stator (Figure 11.2a). Another solution is the use of a transformer in the case of a stator voltage that is higher (e.g., standard in wind turbine, 690 V) than the voltage

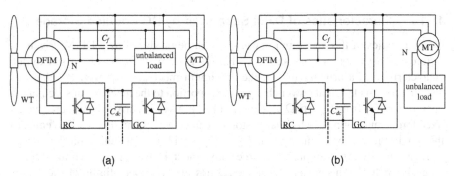

Figure 11.2. Possible system connections for unbalanced load supply.

required by the load. A transformer is used for simultaneous change of the voltage and connection type from three wire system to four wire system (Figure 11.2b). The supporting power source and energy storage with converters are neglected in Figure 11.2, as they have no influence on the connection type of the stator side.

In the case of primary movers other than wind turbines, which do not have to be supported by an additional energy source or storage (ICE, water turbine), it is necessary to obtain preliminary charging of the DC link to start the rotor connected converter RC and obtain excitation. The simplest way to charge the DC link is use of the same battery that is used for start-up in the case of an internal combustion engine, or application of a fractional power permanent magnet generator coupled on the same shaft (Figure 11.3) in the case of a water turbine.

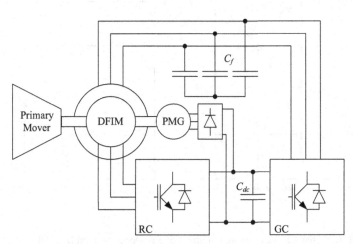

Figure 11.3. Possible scheme allowing preliminary DC link charging with fractional power PMG.

11.1.2 Characteristics of DFIM Supported by DC Coupled Storage

Characteristics of the maximum power that can be produced by a DFIM driven by different energy sources are shown in Figure 11.4. In the case of a fixed torque characteristic of a primary mover, the maximum mechanical power, P_m, that can be converted to electrical power changes in proportion to the mechanical speed in the operation range from Ω_{min} to Ω_{max} (Figure 11.4a). A system driven by a source with a fixed torque and supported by energy storage can generate a total power P_{tot} equal to the maximum power of the system P_{max} (Figure 11.4a) depending on the energy storage capacity. An example of a source with almost fixed torque and a limited speed range is an ICE. The presented characteristics are the upper limits of the power system; it is not possible to generate more stator power and other curves are a consequence of this fact.

In the case of a wind driven DFIM, mechanical power P_m in the speed range from Ω_{min} to Ω_{max} changes with the cube of mechanical speed; at minimum speed Ω_{min} equal to 66% of Ω_s the DFIM can generate at most 12.5% of P_{max}. Assuming that a positive slip power occurs when power is taken from the rotor of the DFIM and delivered to the load, the mechanical power P_m is the sum of rotor power and stator power if power losses are neglected:

$$P_m = P_s + P_r = (1 - s)P_s \qquad (11.1)$$

Hence,

$$P_s(\Omega_{min}) = \frac{P_m(\Omega_{min})}{(1 - s)} \qquad (11.2)$$

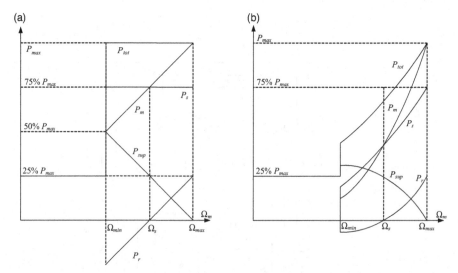

Figure 11.4.   Characteristics of maximum active power generated by a DFIM versus rotor speed for (a) a fixed torque primary mover and (b) a wind turbine, supported by DC coupled energy source.

and is equal to 19% at 0.66 of Ω_s, whereas the slip power is equal to -6.5% of P_{max}.

A wind turbine supported by a DC coupled energy source can generate a total power P_{tot} equal to the sum of mechanical power P_m and the power of the supporting source P_{sup} or is equal to the sum of the stator power P_s and the power delivered to the load by the GC equal to 25% of P_{max} (Figure 11.4b).

Below minimum speed Ω_{min} in both cases (e.g., during start-up or shut-down in the case of an ICE or for low wind speed in the case of a wind turbine), the stator of the machine has to be disconnected from the load by the switch placed between the stator and filtering capacitors. Then, the power is generated only in a limited range through the GC converter. In both cases, maximum possible generated power with a disconnected stator side of the DFIM is equal to the rated power of the GC converter (25% of P_{max}).

11.1.3 Selection of Filtering Capacitors

Application of filtering capacitors on the stator side eliminates switching frequency harmonics produced by both RC and GC converters and additionally compensates part of the reactive power needed for magnetization. In a simplified model of the machine, it can be assumed that the rotor is supplied from the VSI with current feedback; therefore, it can be treated as a machine fed from a current source on the rotor side. In such a case, two equations of the machine connected with the stator side and a third equation connected with the stator capacitors can be taken into consideration:

$$\vec{v}_s = R_s \vec{i}_s + \frac{d\vec{\psi}_s}{dt} + j\omega_s \vec{\psi}_s \tag{11.3}$$

$$\vec{\psi}_s = L_s \vec{i}_s + L_m \vec{i}_r \tag{11.4}$$

$$\vec{i}_s = -C_f \frac{d\vec{v}_s}{dt} + \vec{i}_{ld} - j\omega_s C_f \vec{v}_s \tag{11.5}$$

where v_s and v_r are stator and rotor voltage, ψ_s is the stator flux, i_s and i_r are stator and rotor currents, R_s is the stator resistance, L_s and L_m are stator and magnetizing inductances, p_b is the number of poles pairs, ω_s is the synchronous speed, C_f is the filtering capacitance, and i_{ld} is the load current.

Based on equations (11.3)–(11.5) for a filtering capacitance C_f equal to zero and a resistive load, a model of a stand-alone DFIM supplied from a current source can be derived:

$$\vec{v}_s = \frac{R_o L_m}{Z_s} \frac{d\vec{i}_r}{dt} + j\left(\frac{\omega_s R_o L_m}{Z_s}\right)\vec{i}_r - \frac{L_s}{Z_s}\frac{d\vec{v}_s}{dt} \tag{11.6}$$

where R_o is the load resistance, stator resistance is neglected, and Z_s is the stator side impedance:

$$Z_s = R_o + j\omega_s L_s \tag{11.7}$$

The differential of the rotor current in Equation (11.6) indicates that the generated voltage v_s is distorted by rotor current ripples produced by the converters. The negative sign of the stator voltage derivative is responsible for partial damping of the voltage distortions. In the limiting case—no-load operation—R_o and Z_s are infinite and Equation (11.6) is reduced to

$$\vec{v}_s = L_m \frac{d\vec{i}_r}{dt} + j\omega_s L_m \vec{i}_r \tag{11.8}$$

This indicates that the stator voltage is distorted by rotor current ripples of PWM frequency.

Power electronics supply systems generating high-quality AC voltage are equipped with *LC* filters. Inductors used for current formation are not necessary in a stand-alone DFIM, due to the high inductance of the machine itself. To obtain high-quality generated voltage, only filtering capacitors are needed on the stator side. The worst case for the voltage quality is no-load operation, due to the lowest damping ratio of the output circuit. For a high-power DFIM, the stator resistance can be neglected and the output voltage can be described by

$$\vec{v}_s = \frac{1}{1 - \omega_s^2 L_s C_f} \left(L_m \frac{d\vec{i}_r}{dt} + j\omega_s L_m \vec{i}_r - j2\omega_s L_s C_f \frac{d\vec{v}_s}{dt} - L_s C_f \frac{d^2\vec{v}_s}{dt^2} \right) \tag{11.9}$$

The negative sign of the second-order derivative of the stator voltage in Equation (11.9) is the component responsible for effective damping of the voltage distortions caused by rotor current ripples.

The first criterion for capacitor selection is that the capacitance C_f must not fully compensate the magnetizing reactive power. However, the frequency related to mechanical speed has to be taken into consideration instead of the stator voltage frequency (50 or 60 Hz). It has to meet the requirement

$$C_f < \frac{1}{4\pi^2 f_m^2 L_m} \tag{11.10}$$

At the same time, to obtain a given resonant frequency f_r of the output filter $L_{rs\sigma} C_f$, which has to be significantly smaller than the switching frequency, the capacitance must be equal to

$$C_f = \frac{1}{4\pi^2 f_r^2 \sigma L_m} \tag{11.11}$$

For high-frequency harmonics, the model of an unloaded DFIM can be simplified to an *LC* filter, which consists of equivalent machine leakage inductance $L_{rs\sigma}$ close to

σL_m, where σ is the total leakage factor, and filtering capacitor C_f. On the logarithmic scale, the resonant frequency f_r has to be obtained more or less in the middle between the operational frequency (50 or 60 Hz) and the switching frequency to avoid resonances for these frequencies.

Comparison of generated voltage without and with filtering capacitors is presented in Figure 11.5. Figures 11.5a and 11.5b show the stator voltage and rotor current of a simulated model of a 250 kW slip ring induction machine operated as a generator. The use of filtering capacitors completely eliminates switching frequency harmonics in the stator voltage (Figure 11.5a), whereas the system not equipped with capacitors on the stator side has significantly distorted stator voltage (Figure 11.5b).

Results of laboratory tests of a DFIM model (2.2 kW) are shown in Figures 11.5c and (11.5)d. Similar to the simulated model, filtering capacitors eliminate switching harmonics (Figure 11.5d), whereas the system without filtering capacitors generates strongly distorted voltage.

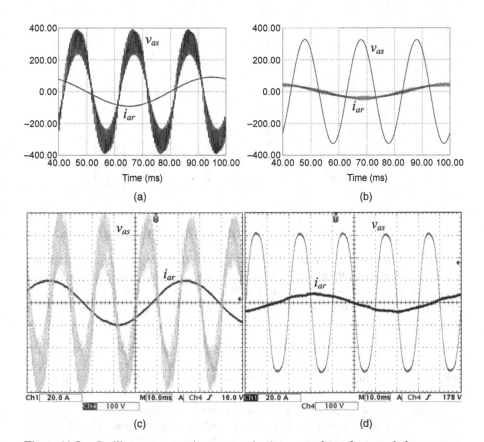

Figure 11.5. Oscillograms presenting generated voltage waveforms for a stand-alone system (a) without and (b) with filtering capacitors connected to the stator of the DFIM.

11.2 MATHEMATICAL DESCRIPTION OF THE STAND-ALONE DFIM SYSTEM

11.2.1 Model of Stand-alone DFIM

The mathematical model of a stand-alone operated DFIM is based on the same equations as the grid connected machine. Model of the grid connected DFIM with dq components of the stator and rotor currents as the state variables is desribed by Equation (11.12).

$$
\begin{bmatrix} \dfrac{di_{ds}}{dt} \\[2mm] \dfrac{di_{qs}}{dt} \\[2mm] \dfrac{di_{dr}}{dt} \\[2mm] \dfrac{di_{qr}}{dt} \end{bmatrix} =
\begin{bmatrix}
-\dfrac{R_s L_r}{L_s L_r - L_m^2} & \left(\omega_s + \dfrac{\omega_m L_m^2}{L_s L_r - L_m^2}\right) & \dfrac{R_r L_m}{L_s L_r - L_m^2} & \dfrac{\omega_m L_r L_m}{L_s L_r - L_m^2} \\[3mm]
-\left(\omega_s + \dfrac{\omega_m L_m^2}{L_s L_r - L_m^2}\right) & -\dfrac{R_s L_r}{L_s L_r - L_m^2} & -\dfrac{\omega_m L_r L_m}{L_s L_r - L_m^2} & \dfrac{R_r L_m}{L_s L_r - L_m^2} \\[3mm]
\dfrac{R_s L_m}{L_s L_r - L_m^2} & -\dfrac{\omega_m L_s L_m}{L_s L_r - L_m^2} & -\dfrac{R_r L_s}{L_s L_r - L_m^2} & \left(\omega_s - \dfrac{\omega_m L_r L_s}{L_s L_r - L_m^2}\right) \\[3mm]
\dfrac{\omega_m L_s L_m}{L_s L_r - L_m^2} & \dfrac{R_s L_m}{L_s L_r - L_m^2} & -\left(\omega_s - \dfrac{\omega_m L_r L_s}{L_s L_r - L_m^2}\right) & -\dfrac{R_r L_s}{L_s L_r - L_m^2}
\end{bmatrix}
\begin{bmatrix} i_{ds} \\[2mm] i_{qs} \\[2mm] i_{dr} \\[2mm] i_{qr} \end{bmatrix}
$$

$$
+
\begin{bmatrix}
\dfrac{L_r}{L_s L_r - L_m^2} & 0 & -\dfrac{L_m}{L_s L_r - L_m^2} & 0 \\[3mm]
0 & \dfrac{L_r}{L_s L_r - L_m^2} & 0 & -\dfrac{L_m}{L_s L_r - L_m^2} \\[3mm]
-\dfrac{L_m}{L_s L_r - L_m^2} & 0 & \dfrac{L_s}{L_s L_r - L_m^2} & 0 \\[3mm]
0 & -\dfrac{L_m}{L_s L_r - L_m^2} & 0 & \dfrac{L_s}{L_s L_r - L_m^2}
\end{bmatrix}
\begin{bmatrix} v_{ds} \\[2mm] v_{qs} \\[2mm] v_{dr} \\[2mm] v_{qr} \end{bmatrix}
\qquad (11.12)
$$

The only difference in relation to the grid connected system is that the stator voltage is not given by the power grid, but is obtained as a result of the excited machine loaded on the stator side. Assuming that the machine is not equipped with stator connected filtering capacitors, the stator voltage during a resistive load supply is calculated as

$$
\vec{v}_s = -R_o \vec{i}_s \qquad (11.13)
$$

where R_o is the load resistance, and the stator current also represents the load current.

Replacement of the stator voltage components in Equation (11.12) with components taken from Equation (11.13) gives a model of a stand-alone DFIM without filtering capacitors, loaded with resistive load, Equation (11.14), and with stator and rotor current components as state variables. Other models with rotor current and stator voltage components can be described by Equation (11.15); as for the resistive load, stator voltage and current are linearly dependent.

$$
\begin{bmatrix} \dfrac{di_{ds}}{dt} \\[2ex] \dfrac{di_{qs}}{dt} \\[2ex] \dfrac{di_{dr}}{dt} \\[2ex] \dfrac{di_{qr}}{dt} \end{bmatrix} = \begin{bmatrix} -\dfrac{(R_s+R_o)L_r}{L_sL_r-L_m^2} & \left(\omega_s+\dfrac{\omega_m L_m^2}{L_sL_r-L_m^2}\right) & \dfrac{R_r L_m}{L_sL_r-L_m^2} & \dfrac{\omega_m L_r L_m}{L_sL_r-L_m^2} \\[2ex] -\left(\omega_s+\dfrac{\omega_m L_m^2}{L_sL_r-L_m^2}\right) & -\dfrac{(R_s+R_o)L_r}{L_sL_r-L_m^2} & -\dfrac{\omega_m L_r L_m}{L_sL_r-L_m^2} & \dfrac{R_r L_m}{L_sL_r-L_m^2} \\[2ex] \dfrac{(R_s+R_o)L_m}{L_sL_r-L_m^2} & -\dfrac{\omega_m L_s L_m}{L_sL_r-L_m^2} & -\dfrac{R_r L_s}{L_sL_r-L_m^2} & \left(\omega_s-\dfrac{\omega_m L_r L_s}{L_sL_r-L_m^2}\right) \\[2ex] \dfrac{\omega_m L_s L_m}{L_sL_r-L_m^2} & \dfrac{(R_s+R_o)L_m}{L_sL_r-L_m^2} & -\left(\omega_s-\dfrac{\omega_m L_r L_s}{L_sL_r-L_m^2}\right) & -\dfrac{R_r L_s}{L_sL_r-L_m^2} \end{bmatrix} \begin{bmatrix} i_{ds} \\[2ex] i_{qs} \\[2ex] i_{dr} \\[2ex] i_{qr} \end{bmatrix}
$$

$$
+ \begin{bmatrix} -\dfrac{L_m}{L_sL_r-L_m^2} & 0 \\[2ex] 0 & -\dfrac{L_m}{L_sL_r-L_m^2} \\[2ex] \dfrac{L_s}{L_sL_r-L_m^2} & 0 \\[2ex] 0 & \dfrac{L_s}{L_sL_r-L_m^2} \end{bmatrix} \begin{bmatrix} v_{dr} \\[2ex] v_{qr} \end{bmatrix}
\tag{11.14}
$$

$$
\begin{bmatrix} \dfrac{dv_{ds}}{dt} \\[2ex] \dfrac{dv_{qs}}{dt} \\[2ex] \dfrac{di_{dr}}{dt} \\[2ex] \dfrac{di_{qr}}{dt} \end{bmatrix} = \begin{bmatrix} -\dfrac{(R_s+R_o)L_r}{L_sL_r-L_m^2} & \left(\omega_s+\dfrac{\omega_m L_m^2}{L_sL_r-L_m^2}\right) & -\dfrac{R_o R_r L_m}{L_sL_r-L_m^2} & -\dfrac{\omega_m R_o L_r L_m}{L_sL_r-L_m^2} \\[2ex] -\left(\omega_s+\dfrac{\omega_m L_m^2}{L_sL_r-L_m^2}\right) & -\dfrac{(R_s+R_o)L_r}{L_sL_r-L_m^2} & \dfrac{\omega_m R_o L_r L_m}{L_sL_r-L_m^2} & -\dfrac{R_o R_r L_m}{L_sL_r-L_m^2} \\[2ex] -\dfrac{(R_s+R_o)L_m}{R_o(L_sL_r-L_m^2)} & \dfrac{\omega_m L_s L_m}{R_o(L_sL_r-L_m^2)} & -\dfrac{R_r L_s}{L_sL_r-L_m^2} & \left(\omega_s-\dfrac{\omega_m L_r L_s}{L_sL_r-L_m^2}\right) \\[2ex] -\dfrac{\omega_m L_s L_m}{R_o(L_sL_r-L_m^2)} & -\dfrac{(R_s+R_o)L_m}{R_o(L_sL_r-L_m^2)} & -\left(\omega_s-\dfrac{\omega_m L_r L_s}{L_sL_r-L_m^2}\right) & -\dfrac{R_r L_s}{L_sL_r-L_m^2} \end{bmatrix} \begin{bmatrix} v_{ds} \\[2ex] v_{qs} \\[2ex] i_{dr} \\[2ex] i_{qr} \end{bmatrix}
$$

$$
+ \begin{bmatrix} -\dfrac{L_m}{L_sL_r-L_m^2} & 0 \\[2ex] 0 & -\dfrac{L_m}{L_sL_r-L_m^2} \\[2ex] \dfrac{L_s}{L_sL_r-L_m^2} & 0 \\[2ex] 0 & \dfrac{L_s}{L_sL_r-L_m^2} \end{bmatrix} \begin{bmatrix} v_{dr} \\[2ex] v_{qr} \end{bmatrix}
\tag{11.15}
$$

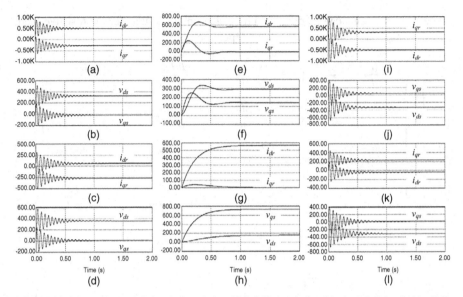

Figure 11.6. DFIM model simulation: (a,b,e,f,i,j) fully loaded and (c,d,g,h,k,l) loaded in 10% for (a–d) 66% of synchronous speed, (e–h) synchronous speed, and (i–l) 133% of synchronous speed.

Simulation results of the models are shown in Figure 11.6. The parameters taken from the data of 250 kW DFIM ($L_m = 4.2$ mH, $L_s = L_r = 4.4$ mH, $R_s = R_r = 0.02 \, \Omega$) are referred to the stator side.

For any speed, the d component of the rotor voltage is selected to achieve a stator voltage equal to 230 V for a fully loaded stator, while the q component equals zero. Thus, for no-load operation, especially at synchronous speed, the stator voltage for the same rotor voltage value is higher than for the fully loaded stator. Two waveforms from the top in each column of Figure 11.6 show the rotor current and stator voltage vector components of a fully loaded 250 kW machine, whereas the next two show the same variables for a 10% loaded machine. The left-hand column is related to the case with mechanical speed ω_m equal to 66% of ω_s; the middle column to synchronous speed ω_s, and the right column is related to the case of 133% of ω_s.

A power system equipped with filtering capacitors is able to generate voltage with significantly higher quality with reduced switching frequency harmonics. A model of a stand-alone DFIM equipped with capacitors is complemented by Equation (11.5). State equations of the machine can be created based on components of rotor and stator currents and stator voltage, Equation (11.16). In an autonomous DFIM system, rotor voltage is only one electromotive force and for given speed and load all responses of currents and voltages in the system depend only on rotor voltage vector components.

$$
\begin{bmatrix}
\dfrac{di_{ds}}{dt} \\[4pt]
\dfrac{di_{qs}}{dt} \\[4pt]
\dfrac{di_{dr}}{dt} \\[4pt]
\dfrac{di_{qr}}{dt} \\[4pt]
\dfrac{dv_{ds}}{dt} \\[4pt]
\dfrac{dv_{qs}}{dt}
\end{bmatrix}
=
\begin{bmatrix}
-\dfrac{R_s L_r}{L_s L_r - L_m^2} & \left(\omega_s + \dfrac{\omega_m L_m^2}{L_s L_r - L_m^2}\right) & \dfrac{R_r L_m}{L_s L_r - L_m^2} & \dfrac{\omega_m L_r L_m}{L_s L_r - L_m^2} & \dfrac{L_r}{L_s L_r - L_m^2} & 0 \\[10pt]
-\left(\omega_s + \dfrac{\omega_m L_m^2}{L_s L_r - L_m^2}\right) & -\dfrac{R_s L_r}{L_s L_r - L_m^2} & \dfrac{\omega_m L_r L_m}{L_s L_r - L_m^2} & \dfrac{R_r L_m}{L_s L_r - L_m^2} & 0 & \dfrac{L_r}{L_s L_r - L_m^2} \\[10pt]
\dfrac{R_s L_m}{L_s L_r - L_m^2} & -\dfrac{\omega_m L_s L_m}{L_s L_r - L_m^2} & -\dfrac{R_r L_s}{L_s L_r - L_m^2} & \left(\omega_s - \dfrac{\omega_m L_r L_s}{L_s L_r - L_m^2}\right) & -\dfrac{L_m}{L_s L_r - L_m^2} & 0 \\[10pt]
\dfrac{\omega_m L_s L_m}{L_s L_r - L_m^2} & \dfrac{R_s L_m}{L_s L_r - L_m^2} & -\left(\omega_s - \dfrac{\omega_m L_r L_s}{L_s L_r - L_m^2}\right) & -\dfrac{R_r L_s}{L_s L_r - L_m^2} & 0 & -\dfrac{L_m}{L_s L_r - L_m^2} \\[10pt]
-\dfrac{1}{C_f} & 0 & 0 & 0 & -\dfrac{1}{R_o C_f} & \omega_s \\[10pt]
0 & -\dfrac{1}{C_f} & 0 & 0 & -\omega_s & -\dfrac{1}{R_o C_f}
\end{bmatrix}
\begin{bmatrix}
i_{ds} \\[4pt]
i_{qs} \\[4pt]
i_{dr} \\[4pt]
i_{qr} \\[4pt]
v_{ds} \\[4pt]
v_{qs}
\end{bmatrix}
$$

$$
+
\begin{bmatrix}
-\dfrac{L_m}{L_s L_r - L_m^2} & 0 \\[10pt]
0 & -\dfrac{L_m}{L_s L_r - L_m^2} \\[10pt]
\dfrac{L_s}{L_s L_r - L_m^2} & 0 \\[10pt]
0 & \dfrac{L_s}{L_s L_r - L_m^2} \\[10pt]
0 & 0 \\[10pt]
0 & 0
\end{bmatrix}
\begin{bmatrix}
v_{dr} \\[4pt]
v_{qr}
\end{bmatrix}
\tag{11.16}
$$

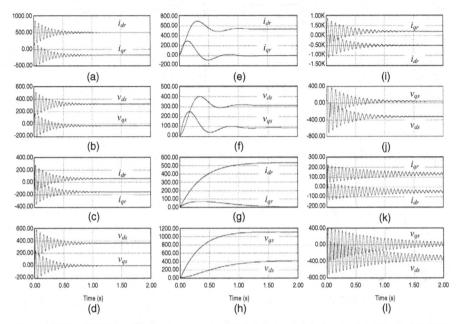

Figure 11.7. DFIM model simulation: (a,b,e,f,i,j) fully loaded and (c,d,g,h,k,l) loaded in 10% for (a–d) 66% of synchronous speed, (e–h) synchronous speed, and (i–l) 133% of synchronous speed with 1 mF of stator connected filtering capacitors.

The stator connected filtering capacitor C_f, which fully compensates the inductive character of the stator at the frequency related to the highest possible speed (1.33 of ω_s), is equal to 1.3 mF. To achieve partial compensation with some margin, a capacitor is selected to compensate around 75% of the reactive power calculated for a pulsation equal to 418 rad/s, that is, 1.33 of ω_s.

Similar to Figure 11.6, in Figure 11.7 the rotor current and stator voltage vector components are presented. There are for cases of a fully loaded (Figures 11.7a,b, (11.7)e,f, (11.7)i,j) and a 10% loaded (Figures 11.7c,d, (11.7)g,h, (11.7)k,l) stator of the machine, for mechanical speed equal to 66% of ω_s (Figures 11.7a–d), synchronous speed ω_s (Figures 11.7e–h), and 133% of ω_s (Figures 11.7i–l). It can be seen that, for a DFIM equipped with stator connected filtering capacitors, the response dynamics strongly depend on the mechanical speed, especially in the case of low load operation.

Figure 11.8 presents waveforms of the DFIM stator voltage components in the dq frame for fully loaded and 10% loaded for different values of filtering capacitors. For 0.65 mF, which is half of the capacitance compensating the reactive power during no-load operation, the system is stable for both load cases. For 1.3 mF and no-load operation, there are sustained oscillations; while for a fully loaded DFIM, the same capacitance value allows stable operation. Sustained oscillations for the fully loaded DFIM occur for 1.95 mF of filtering capacitance. It means that, for a loaded system,

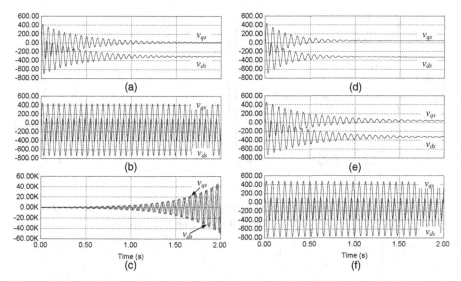

Figure 11.8. DFIM model simulation: (a–c) 10% loaded and (d–f) fully loaded with (a,d) 0.65 mF, (b,e) 1.3 mF, and (c,f) 1.95 mF filtering capacitor at 133% of synchronous speed.

especially for a nonlinear load, a larger capacitance can be used for better filtering of higher harmonics in the generated voltage.

11.2.2 Model of Stand-alone DFIM Fed from Current Source

The rotor of the DFIM is supplied from the current controlled voltage source inverter (VSI). With good approximation, the inverter can be treated as a current source. For such a case, the standard rotor side equations can be eliminated. Instead, the rotor current is given arbitrarily, and its dynamics is limited with a T_e time constant. Limitation of the rotor current rise with $1/T_e$ is necessary for elimination of the numerical calculation errors caused by a step change of the rotor current. The model is described with the following equations:

$$\frac{d\vec{i}_r}{dt} = \frac{1}{T_e}\left(\vec{i}_r^* - \vec{i}_r\right) \tag{11.17}$$

$$\vec{v}_s = R_s\vec{i}_s + \frac{d\vec{\psi}_s}{dt} + j\omega_s\vec{\psi}_s \tag{11.18}$$

$$\vec{\psi}_s = L_s\vec{i}_s + L_m\vec{i}_r \tag{11.19}$$

where $\vec{i}_r^*$ is the reference rotor current vector. Assuming reference rotor current components as input signals, the following equations for a DFIM supplied from a current source can be described:

$$
\begin{bmatrix} \dfrac{di_{dr}}{dt} \\[2ex] \dfrac{di_{qr}}{dt} \\[2ex] \dfrac{di_{ds}}{dt} \\[2ex] \dfrac{di_{qs}}{dt} \end{bmatrix}
=
\begin{bmatrix}
-\dfrac{1}{T_e} & 0 & 0 & 0 \\[2ex]
0 & -\dfrac{1}{T_e} & 0 & 0 \\[2ex]
\dfrac{L_m}{L_s T_e} & \dfrac{\omega_s L_m}{L_s} & -\dfrac{R_s}{L_s} & \omega_s \\[2ex]
-\dfrac{\omega_s L_m}{L_s} & \dfrac{L_m}{L_s T_e} & -\omega_s & -\dfrac{R_s}{L_s}
\end{bmatrix}
\begin{bmatrix} i_{dr} \\[2ex] i_{qr} \\[2ex] i_{ds} \\[2ex] i_{qs} \end{bmatrix}
+
\begin{bmatrix}
\dfrac{1}{T_e} & 0 & 0 & 0 \\[2ex]
0 & \dfrac{1}{T_e} & 0 & 0 \\[2ex]
-\dfrac{L_m}{L_s T_e} & 0 & \dfrac{1}{L_s} & 0 \\[2ex]
0 & -\dfrac{L_m}{L_s T_e} & 0 & \dfrac{1}{L_s}
\end{bmatrix}
\begin{bmatrix} i_{dr}^{*} \\[2ex] i_{qr}^{*} \\[2ex] v_{ds} \\[2ex] v_{qs} \end{bmatrix}
\tag{11.20}
$$

A model of a stand-alone machine equipped with filtering capacitors loaded with resistive load and supplied from a current source on the rotor side is described by

$$
\begin{bmatrix} \dfrac{di_{dr}}{dt} \\[2ex] \dfrac{di_{qr}}{dt} \\[2ex] \dfrac{di_{ds}}{dt} \\[2ex] \dfrac{di_{qs}}{dt} \\[2ex] \dfrac{dv_{ds}}{dt} \\[2ex] \dfrac{dv_{qs}}{dt} \end{bmatrix}
=
\begin{bmatrix}
-\dfrac{1}{T_e} & 0 & 0 & 0 & 0 & 0 \\[2ex]
0 & -\dfrac{1}{T_e} & 0 & 0 & 0 & 0 \\[2ex]
\dfrac{L_m}{L_s T_e} & \dfrac{\omega_s L_m}{L_s} & -\dfrac{R_s}{L_s} & \omega_s & \dfrac{1}{L_s} & 0 \\[2ex]
-\dfrac{\omega_s L_m}{L_s} & \dfrac{L_m}{L_s T_e} & -\omega_s & -\dfrac{R_s}{L_s} & 0 & \dfrac{1}{L_s} \\[2ex]
0 & 0 & -\dfrac{1}{C_f} & 0 & -\dfrac{1}{R_o C_f} & \omega_s \\[2ex]
0 & 0 & 0 & -\dfrac{1}{C_f} & -\omega_s & -\dfrac{1}{R_o C_f}
\end{bmatrix}
\begin{bmatrix} i_{dr} \\[2ex] i_{qr} \\[2ex] i_{ds} \\[2ex] i_{qs} \\[2ex] v_{ds} \\[2ex] v_{qs} \end{bmatrix}
$$

$$
+
\begin{bmatrix}
\dfrac{1}{T_e} & 0 \\[2ex]
0 & \dfrac{1}{T_e} \\[2ex]
-\dfrac{L_m}{L_s T_e} & 0 \\[2ex]
0 & -\dfrac{L_m}{L_s T_e} \\[2ex]
0 & 0 \\[2ex]
0 & 0
\end{bmatrix}
\begin{bmatrix} i_{dr}^{*} \\[2ex] i_{qr}^{*} \end{bmatrix}
\tag{11.21}
$$

Simulation results of the model are presented in Figure 11.9. For different filtering capacitors, the rotor current components are selected to obtain a stator voltage

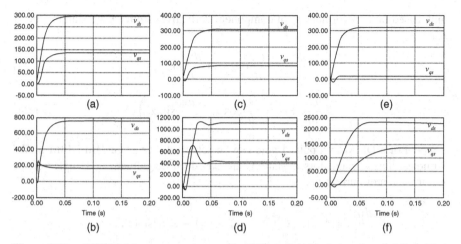

Figure 11.9. DFIM fed from a current controlled VSI model simulation: (a,c,e) fully loaded and (b,d,f) 10% loaded for (a,b) 0.1 mF (c,d), 1 mF and (e,f) 2 mF of filtering capacitors.

amplitude equal to 325 V for a fully loaded generator. Thus, for a 10% loaded DFIM, the stator voltage amplitude is higher for the same rotor current. The results are the same for any mechanical speed.

11.2.3 Polar Frame Model of Stand-alone DFIM

In grid connected power systems, the p and q power components are controlled in orthogonal frames. In a stand-alone system, the amplitude and frequency of the produced voltage have to be fixed in spite of the load and rotor speed. The fixed amplitude of the generated voltage corresponds to a fixed magnitude of its vector, while the fixed frequency corresponds to a fixed position of the vector in the rotating reference frame connected with the vector. Moreover, a fixed angle of this vector means also a fixed phase of the generated voltage. Representation of all vectors by magnitude and angle give a model of the DFIM in a polar frame.

Consider that the model in Equation (11.16) takes the form in Equation (11.22):

$$
\begin{bmatrix}
\dfrac{di_{ds}}{dt} \\[2mm]
\dfrac{di_{qs}}{dt} \\[2mm]
\dfrac{di_{ds}}{dt} \\[2mm]
\dfrac{di_{qr}}{dt} \\[2mm]
\dfrac{dv_{ds}}{dt} \\[2mm]
\dfrac{dv_{qs}}{dt}
\end{bmatrix}
=
\begin{bmatrix}
A & B & C & D & E & 0 \\[2mm]
-B & A & -D & C & 0 & E \\[2mm]
G & H & I & J & K & 0 \\[2mm]
-H & G & -J & I & 0 & K \\[2mm]
-\dfrac{1}{C_f} & 0 & 0 & 0 & -\dfrac{1}{R_oC_f} & \omega_s \\[2mm]
0 & -\dfrac{1}{C_f} & 0 & 0 & -\omega_s & -\dfrac{1}{R_oC_f}
\end{bmatrix}
\begin{bmatrix}
i_{ds} \\[2mm]
i_{qs} \\[2mm]
i_{dr} \\[2mm]
i_{qr} \\[2mm]
v_{ds} \\[2mm]
v_{qs}
\end{bmatrix}
+
\begin{bmatrix}
F & 0 \\[2mm]
0 & F \\[2mm]
L & 0 \\[2mm]
0 & L \\[2mm]
0 & 0 \\[2mm]
0 & 0
\end{bmatrix}
\begin{bmatrix}
v_{dr} \\[2mm]
v_{qr}
\end{bmatrix}
$$

$$(11.22)$$

and, based on transformations of component derivatives from a Cartesian to a polar frame,

$$\frac{d|\vec{v}|}{dt} = \frac{dv_x}{dt}\cos\alpha_v + \frac{dv_y}{dt}\sin\alpha_v \tag{11.23}$$

$$\frac{d\alpha_v}{dt} = \frac{1}{|\vec{v}|}\left(\frac{dv_y}{dt}\cos\alpha_v - \frac{dv_x}{dt}\sin\alpha_v\right) \tag{11.24}$$

a model of a DFIM in a polar frame can be achieved. For a standard DFIM model without filtering capacitors, equations are based on magnitudes and angles of the stator and rotor current vectors:

$$\frac{d|\vec{i}_s|}{dt} = A|\vec{i}_s| + C|\vec{i}_r|\cos(\alpha_{ir} - \alpha_{is}) + D|\vec{i}_r|\sin(\alpha_{ir} - \alpha_{is})$$
$$+E|\vec{v}_s|\cos(\alpha_{vs} - \alpha_{is}) + F|\vec{v}_r|\cos(\alpha_{vr} - \alpha_{is}) \tag{11.25}$$

$$\frac{d\alpha_{is}}{dt} = -B - \frac{|\vec{i}_r|}{|\vec{i}_s|}(D\cos(\alpha_{ir} - \alpha_{is}) - C\sin(\alpha_{ir} - \alpha_{is}))$$
$$+\frac{|\vec{v}_s|}{|\vec{i}_s|}E\sin(\alpha_{vs} - \alpha_{is}) + \frac{|\vec{v}_r|}{|\vec{i}_s|}F\sin(\alpha_{vr} - \alpha_{is}) \tag{11.26}$$

$$\frac{d|\vec{i}_r|}{dt} = G|\vec{i}_s|\cos(\alpha_{is} - \alpha_{ir}) + H|\vec{i}_s|\sin(\alpha_{is} - \alpha_{ir}) + I|\vec{i}_r|$$
$$+K|\vec{v}_s|\cos(\alpha_{vs} - \alpha_{ir}) + L|\vec{v}_r|\cos(\alpha_{vr} - \alpha_{ir}) \tag{11.27}$$

$$\frac{d\alpha_{ir}}{dt} = -J - \frac{|\vec{i}_s|}{|\vec{i}_r|}(H\cos(\alpha_{is} - \alpha_{ir}) - G\sin(\alpha_{is} - \alpha_{ir}))$$
$$+\frac{|\vec{v}_s|}{|\vec{i}_r|}K\sin(\alpha_{vs} - \alpha_{ir}) + \frac{|\vec{v}_r|}{|\vec{i}_r|}L\sin(\alpha_{vr} - \alpha_{ir}) \tag{11.28}$$

For a stand-alone DFIM without filtering capacitors and a supplied resistive load given in Equation (11.13), a polar frame model is described with Equations (11.29)–(11.32):

$$\frac{d|\vec{i}_s|}{dt} = (A - ER_o)|\vec{i}_s| + C|\vec{i}_r|\cos(\alpha_{ir} - \alpha_{is})$$
$$+D|\vec{i}_r|\sin(\alpha_{ir} - \alpha_{is}) + F|\vec{v}_r|\cos(\alpha_{vr} - \alpha_{is}) \tag{11.29}$$

$$\frac{d\alpha_{is}}{dt} = -B - \frac{|\vec{i}_r|}{|\vec{i}_s|}(D\cos(\alpha_{ir} - \alpha_{is}) - C\sin(\alpha_{ir} - \alpha_{is})) + \frac{|\vec{v}_r|}{|\vec{i}_s|}F\sin(\alpha_{vr} - \alpha_{is})$$
$$\tag{11.30}$$

$$\frac{d|\vec{i}_r|}{dt} = (G - KR_o)|\vec{i}_s|\cos(\alpha_{is} - \alpha_{ir}) + H|\vec{i}_s|\sin(\alpha_{is} - \alpha_{ir}) + I|\vec{i}_r| + L|\vec{v}_r|\cos(\alpha_{vr} - \alpha_{ir})$$

$$(11.31)$$

$$\frac{d\alpha_{ir}}{dt} = -J - \frac{|\vec{i}_s|}{|\vec{i}_r|}(H\cos(\alpha_{is} - \alpha_{ir}) - (G + KR_o)\sin(\alpha_{is} - \alpha_{ir})) + \frac{|\vec{v}_r|}{|\vec{i}_r|}L\sin(\alpha_{vr} - \alpha_{ir})$$

$$(11.32)$$

where the angle of the rotor voltage vector α_{ur}, as an angle of single electromotive force in the model, can be assumed arbitrarily zero. Then, all angles are related to the rotor voltage vector.

For a stand-alone DFIM equipped with filtering capacitors and supplied with a resistive load, a full polar frame model is described by Equations (11.33)–(11.37), where the coefficients are taken from Equation (11.16).

$$\frac{d|\vec{i}_s|}{dt} = -\frac{R_s L_r}{L_s L_r - L_m^2}|\vec{i}_s| + \frac{L_m}{L_s L_r - L_m^2}|\vec{i}_r|(R_r\cos(\alpha_{ir} - \alpha_{is}) + \omega_m L_r\sin(\alpha_{ir} - \alpha_{is}))$$

$$+ \frac{L_r}{L_s L_r - L_m^2}|\vec{v}_s|\cos(\alpha_{vs} - \alpha_{is}) - \frac{L_m}{L_s L_r - L_m^2}|\vec{v}_r|\cos(\alpha_{vr} - \alpha_{is}) \qquad (11.33)$$

$$\frac{d\alpha_{is}}{dt} = -\left(\omega_s + \frac{\omega_m L_m^2}{L_s L_r - L_m^2}\right) - \frac{|\vec{i}_r|}{|\vec{i}_s|}\frac{L_m}{L_s L_r - L_m^2}(\omega_m L_r\cos(\alpha_{ir} - \alpha_{is}) - R_r\sin(\alpha_{ir} - \alpha_{is}))$$

$$+ \frac{|\vec{v}_s|}{|\vec{i}_s|}\frac{L_r}{L_s L_r - L_m^2}\sin(\alpha_{vs} - \alpha_{is}) - \frac{|\vec{v}_r|}{|\vec{i}_s|}\frac{L_m}{L_s L_r - L_m^2}\sin(\alpha_{vr} - \alpha_{is}) \qquad (11.34)$$

$$\frac{d|\vec{i}_r|}{dt} = \frac{|\vec{i}_s|L_m}{L_s L_r - L_m^2}(R_s\cos(\alpha_{is} - \alpha_{ir}) - \omega_m L_s\sin(\alpha_{is} - \alpha_{ir})) - \frac{R_r L_s}{L_s L_r - L_m^2}|\vec{i}_r|$$

$$- \frac{L_m}{L_s L_r - L_m^2}|\vec{v}_s|\cos(\alpha_{vs} - \alpha_{ir}) + \frac{L_s}{L_s L_r - L_m^2}|\vec{v}_r|\cos(\alpha_{vr} - \alpha_{ir}) \qquad (11.35)$$

$$\frac{d\alpha_{ir}}{dt} = -\left(\omega_s - \frac{\omega_m L_r L_s}{L_s L_r - L_m^2}\right) + \frac{|\vec{i}_s|}{|\vec{i}_r|}\frac{L_m}{L_s L_r - L_m^2}(\omega_m L_s\cos(\alpha_{is} - \alpha_{ir}) + R_s\sin(\alpha_{is} - \alpha_{ir}))$$

$$- \frac{|\vec{v}_s|}{|\vec{i}_r|}\frac{L_m}{L_s L_r - L_m^2}\sin(\alpha_{vs} - \alpha_{ir}) + \frac{|\vec{v}_r|}{|\vec{i}_r|}\frac{L_s}{L_s L_r - L_m^2}\sin(\alpha_{vr} - \alpha_{ir}) \qquad (11.36)$$

$$\frac{d|\vec{v}_s|}{dt} = -\frac{|\vec{i}_s|}{C_f}\cos(\alpha_{vs} - \alpha_{is}) - \frac{|\vec{v}_s|}{R_o C_f} \qquad (11.37)$$

$$\frac{d\alpha_{vs}}{dt} = -\omega_s + \frac{|\vec{i}_s|}{C_f|\vec{v}_s|}\sin(\alpha_{vs} - \alpha_{is}) \qquad (11.38)$$

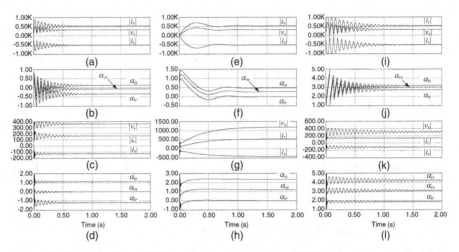

Figure 11.10. Polar frame DFIM model simulation: (a,b,e,f,i,j) fully loaded and (c,d,g,h,k,l) 10% loaded for (a–d) 66% of synchronous speed, (e–h) synchronous speed, and (i–l) 133% of synchronous speed with 1 mF of stator connected filtering capacitors.

For a synchronously rotated frame connected with the rotor voltage vector, angle α_{vr} can be assumed zero. For the system not equipped with filtering capacitors, Equations (11.37) and (11.38) are removed and every R_s in Equations (11.33)–(11.36) is replaced with the sum of stator R_s and load R_o resistance.

Figure 11.10 shows the responses of the simulated polar frame model of a fully loaded DFIM system (Figures 11.10a,b, (11.10)e,f, and (11.10)i,j) and the system loaded at 10% (Figures 11.10c,d, (11.10)g,h, and (11.10)k,l) for a rotor speed equal to $0.66\omega_s$ (Figures 11.10a–d), synchronous speed (Figures 11.10e–h), and $1.33\omega_s$ (Figures 11.10i–l). Waveforms represent vector magnitudes and angles related to the rotor voltage vector. The negative value of the stator current magnitude means that this current has negative direction corresponded to generation of the power and supply of the load. The linear dependence of all vector magnitudes on the rotor voltage vector for a given speed is shown in Figure 11.11.

Figure 11.11c shows the vector magnitudes for twice the reduced rotor voltage vector magnitude in relation to the case presented in Figure 11.11a. It can be observed that, for given load and mechanical speed, vector angle responses are independent of the rotor voltage magnitude (Figures 11.11b and 11.11d). Magnitude response is independent of the selected frame, as shown in Figure 11.12, where results of the model simulation are presented in stationary (Figures 11.12a and 11.12b) and synchronously rotating (Figures 11.12c and 11.12d) coordinate systems.

11.2.4 Polar Frame Model of Stand-alone DFIM Fed from Current Source

Similar to the Cartesian frame in polar coordinates, we can show a model of a DFIM supplied from a current controlled voltage source or current source with limited

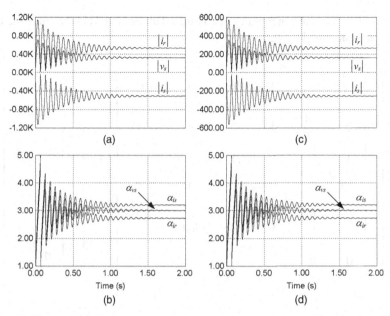

Figure 11.11.   Polar frame DFIM model simulation: (a,c) vectors magnitude and (b,d) angle responses of fully loaded system for different voltage vector magnitudes at 133% of synchronous speed.

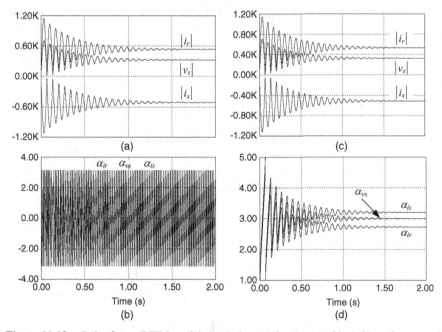

Figure 11.12.   Polar frame DFIM model simulation results presented in (a,b) stationary and (c,d) synchronously rotating coordinates.

change of the rotor current rapidity. We assume that the model (11.21) takes the form shown in Equation (11.39),

$$
\begin{bmatrix}
\dfrac{di_{dr}}{dt} \\[2ex]
\dfrac{di_{qr}}{dt} \\[2ex]
\dfrac{di_{ds}}{dt} \\[2ex]
\dfrac{di_{qs}}{dt} \\[2ex]
\dfrac{dv_{ds}}{dt} \\[2ex]
\dfrac{dv_{qs}}{dt}
\end{bmatrix}
=
\begin{bmatrix}
-\dfrac{1}{T_e} & 0 & 0 & 0 & 0 & 0 \\[2ex]
0 & -\dfrac{1}{T_e} & 0 & 0 & 0 & 0 \\[2ex]
A & B & C & D & E & 0 \\[2ex]
-B & A & -D & C & 0 & E \\[2ex]
0 & 0 & -\dfrac{1}{C_f} & 0 & -\dfrac{1}{R_o C_f} & \omega_s \\[2ex]
0 & 0 & 0 & -\dfrac{1}{C_f} & -\omega_s & -\dfrac{1}{R_o C_f}
\end{bmatrix}
\begin{bmatrix}
i_{dr} \\[2ex]
i_{qr} \\[2ex]
i_{ds} \\[2ex]
i_{qs} \\[2ex]
v_{ds} \\[2ex]
v_{qs}
\end{bmatrix}
+
\begin{bmatrix}
\dfrac{1}{T_e} & 0 \\[2ex]
0 & \dfrac{1}{T_e} \\[2ex]
-A & 0 \\[2ex]
0 & -A \\[2ex]
0 & 0 \\[2ex]
0 & 0
\end{bmatrix}
\begin{bmatrix}
i_{dr}^* \\[2ex]
i_{qr}^*
\end{bmatrix}
$$

$$\tag{11.39}$$

The Polar frame model of a DFIM with rotor connected current source is given by Equations (11.40)–(11.45).

$$\frac{d|\vec{i}_r|}{dt} = \frac{1}{T_e}\left(|\vec{i}_r|^* - |\vec{i}_r|\right) \tag{11.40}$$

$$\frac{d\alpha_{ir}}{dt} = \frac{1}{T_e}\frac{|\vec{i}_r|^*}{|\vec{i}_r|}\sin(\alpha_{ir}^* - \alpha_{ir}) \tag{11.41}$$

$$\frac{d|\vec{i}_s|}{dt} = A|\vec{i}_r|\cos(\alpha_{ir} - \alpha_{is}) + B|\vec{i}_r|\sin(\alpha_{ir} - \alpha_{is}) + C|\vec{i}_s|$$
$$+ E|\vec{v}_s|\cos(\alpha_{vs} - \alpha_{is}) - A|\vec{i}_r|^*\cos(\alpha_{ir}^* - \alpha_{is}) \tag{11.42}$$

$$\frac{d\alpha_{is}}{dt} = -D + \frac{|\vec{i}_r|}{|\vec{i}_s|}\left(A\sin(\alpha_{ir} - \alpha_{is}) - B\cos(\alpha_{ir} - \alpha_{is})\right)$$
$$+ \frac{|\vec{v}_s|}{|\vec{i}_s|}E\sin(\alpha_{vs} - \alpha_{is}) - \frac{|\vec{i}_r|^*}{|\vec{i}_s|}A\sin(\alpha_{ir}^* - \alpha_{is}) \tag{11.43}$$

$$\frac{d|\vec{v}_s|}{dt} = -\frac{|\vec{i}_s|}{C_f}\cos(\alpha_{vs} - \alpha_{is}) - \frac{|\vec{v}_s|}{R_o C_f} \tag{11.44}$$

$$\frac{d\alpha_{vs}}{dt} = -\omega_s + \frac{|\vec{i}_s|}{C_f|\vec{v}_s|}\sin(\alpha_{vs} - \alpha_{is}) \tag{11.45}$$

where the equation describing the angle α_{ir} of the rotor current vector can be replaced with a simpler version,

$$\frac{d\alpha_{ir}}{dt} = \frac{1}{T_e}\left(\alpha_{ir}^* - \alpha_{ir}\right) \tag{11.46}$$

if a quasi-step change of the reference angle is needed in the analysis or if the reference angle is changed according to a known function.

Linear change of the reference angle can be given with the equation

$$\alpha_{ir}^* = \int \omega_{ir}^s d\tau \tag{11.47}$$

and this change is defined by a reference angular speed ω_{ir}^s

The speed of the rotor current vector is related to the rotating reference frame. In the case of synchronous rotation of the rotor current vector, the differential equation with the rotor current vector angle derivative can be neglected, and only the reference amplitude of the rotor current is given. For the reference frame rotated synchronously with the rotor current vector, Equation (11.41) is eliminated and it can be assumed that

$$\alpha_{ir}^* = 0 \tag{11.48}$$

For such an assumption and with the coefficients taken from Equation (11.21), a full polar coordinates model in a synchronously rotated frame connected with the rotor current vector can be obtained as in Equations (11.49)–(11.53).

$$\frac{d|\vec{i}_r|}{dt} = \frac{1}{T_e}\left(|\vec{i}_r|^* - |\vec{i}_r|\right) \tag{11.49}$$

$$\frac{d|\vec{i}_s|}{dt} = -\frac{L_m}{L_s T_e}\left(|\vec{i}_r|^* - |\vec{i}_r|\right)\cos\alpha_{is} - \frac{\omega_s L_m}{L_s}|\vec{i}_r|\sin\alpha_{is} - \frac{R_s}{L_s}|\vec{i}_s| + \frac{1}{L_s}|\vec{v}_s|\cos(\alpha_{vs} - \alpha_{is}) \tag{11.50}$$

$$\frac{d\alpha_{is}}{dt} = -\omega_s + \frac{L_m}{L_s T_e|\vec{i}_s|}\left(|\vec{i}_r|^* - |\vec{i}_r|\right)\sin\alpha_{is} - \frac{|\vec{i}_r|}{|\vec{i}_s|}\frac{\omega_s L_m}{L_s}\cos\alpha_{is} + \frac{|\vec{v}_s|}{|\vec{i}_s|}\frac{1}{L_s}\sin(\alpha_{vs} - \alpha_{is}) \tag{11.51}$$

$$\frac{d|\vec{v}_s|}{dt} = -\frac{|\vec{i}_s|}{C_f}\cos(\alpha_{vs} - \alpha_{is}) - \frac{|\vec{v}_s|}{R_o C_f} \tag{11.52}$$

$$\frac{d\alpha_{vs}}{dt} = -\omega_s + \frac{|\vec{i}_s|}{C_f|\vec{v}_s|}\sin(\alpha_{vs} - \alpha_{is}) \tag{11.53}$$

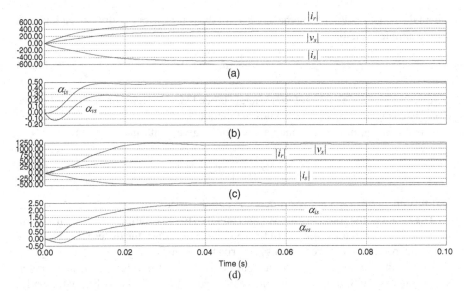

Figure 11.13. Polar frame model of a DFIM fed from a current controlled VSI: (a,c) vectors magnitudes and (b,d) angles for (a,b) fully loaded and (c,d) 10% loaded DFIM system.

Simulation results of the polar frame model of a DFIM fed from the current controlled VSI are presented in Figure 11.13. Figures 11.13a and 11.13b show the magnitudes and angles of the rotor current, stator current, and stator voltage vectors for a fully loaded DFIM system, while Figures 11.13c and 11.13d show the same variables for a 10% loaded DFIM system, respectively. Responses of the system are independent of the mechanical speed. Vector magnitudes are proportional to the rotor current vector magnitude and are the same for any rotational speed of the reference frame, whereas vector angles are independent of the rotor current vector magnitude.

11.3 STATOR VOLTAGE CONTROL

11.3.1 Amplitude and Frequency Control by the Use of PLL

Frequency of the stator voltage results from the sum of the frequency related to the mechanical speed and rotor current frequency Equation (11.54):

$$f_s = \frac{\omega_m}{2\pi} + f_{ir} \tag{11.54}$$

Amplitude of the stator voltage for a given speed and load is proportional to the rotor current amplitude and frequency. However, for the rotor current frequency equal to the slip frequency, the only way to obtain a reference amplitude of the generated voltage is to maintain adequate rotor current amplitude, as can be observed in the polar frame model of a DFIM supplied from a current controlled VSI.

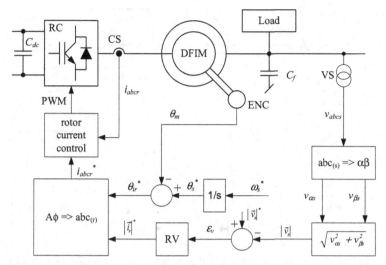

Figure 11.14. Simple stator voltage and frequency control by the use of a rotor position encoder.

A basic control diagram with rotor position sensor is presented in Figure 11.14. The reference angle of the rotor current θ_{ir}^* is obtained from Equation (11.55):

$$\theta_{ir}^* = \int \omega_s^* d\tau - p\theta_m \tag{11.55}$$

where ω_s^* is the reference speed of the frame related to the reference frequency (e.g., 50 Hz).

Control of the stator voltage amplitude is based on the voltage vector magnitude calculated from the orthogonal components in the $\alpha\beta$ frame. The output signal of the voltage vector magnitude controller RV is responsible for the reference magnitude of the rotor current vector.

Fast rotor current controllers can be realized as two-level hysteresis controllers in a three-phase system, three-state hysteresis controllers in an $\alpha\beta$ frame, or proportional controllers in a three-phase or $\alpha\beta$ frame. Proportional controllers applied in the inner control loop have no negative influence on elimination of the steady state error in the stator voltage amplitude control loop. Two PI controllers of the rotor current vector components can be used in synchronously rotated frame xy connected with the rotor current vector, as discussed later.

Control of the grid side converter is similar to the case of a grid connected system, as a superior control loop for this converter is used for stabilization of the DC-link voltage. An additional role for the grid side converter can be reduction of the stator current harmonics [8] or asymmetry [9] during supply of a nonlinear and an unbalanced load, respectively, and reactive power compensation by the stator side of the machine, if not enough reactive power is delivered through the rotor converter. Those functions are independent of the voltage stabilization by the rotor converter control.

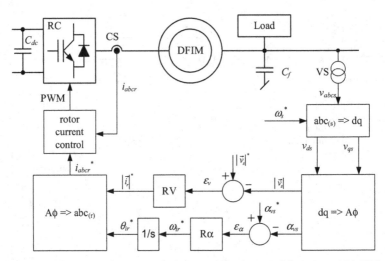

Figure 11.15. Sensorless control of the stator voltage and frequency with PLL.

Reference rotor current vector angle θ_{ir}^* can be obtained with no need of a rotor position encoder. One of the known methods is determination of rotor position by the use of a phase locked loop (PLL). In the case of a grid connected power electronics converter, a PLL used for synchronization of the generated current with first harmonics of the grid voltage produces an angle for transformation of variables [10]. In the case of DFIM sensorless control with a PLL, the produced signal is treated as a rotational speed not an angle. The angle is obtained after integration of achieved rotational speed. In the case of a grid connected DFIM, the use of PLL synchronizing stator current with grid voltage is not a reliable method, as the grid connected DFIM is nonlinear. However, in the case of a stand-alone DFIM system, a PLL can be used for synchronization of the generated stator voltage vector with arbitrary given reference voltage vector. This is presented in Figure 11.15.

A PLL is realized as a PI controller of the stator voltage vector angle α_{vs} calculated from dq orthogonal components in a synchronously rotated frame, Equation (11.56), whereas reference angle α_{vs}^* equals zero; this means that the reference voltage vector overlaps the d axis of the frame.

$$\alpha_{vs} = \tan^{-1}\left(\frac{v_{qs}}{v_{ds}}\right) \tag{11.56}$$

Other signals indicating voltage vector displacement can be derived from the calculation of the cross product of the reference and actual voltage vectors represented in, for example, $\alpha\beta$ coordinates.

$$\vec{v}_s^* \times \vec{v}_s = |\vec{v}_s|^* |\vec{v}_s|\sin\alpha_{vs} = v_{\alpha s}^* v_{\beta s} - v_{\beta s}^* v_{\alpha s} \tag{11.57}$$

so

$$\alpha_{vs} = \sin^{-1}\left(\frac{v_{\alpha s}^* v_{\beta s} - v_{\beta s}^* v_{\alpha s}}{|\vec{v}_s|^* |\vec{v}_s|}\right) \tag{11.58}$$

In the *dq* frame, Equation (11.58) is reduced to

$$\alpha_{vs} = \sin^{-1}\left(\frac{v_{qs}}{|\vec{v}_s|}\right) = \sin^{-1}\left(\frac{v_{qs}}{\sqrt{v_{ds}^2 + v_{qs}^2}}\right) \qquad (11.59)$$

as the reference vector $\vec{v}_s^*$ is placed along the *d* axis, so v_{qs}^* equals zero and v_{ds}^* equals $|\vec{v}_s|^*$.

Equivalently, v_{qs} can be used instead of α_{us} as an indication of displacement of the actual voltage vector in relation to the reference one. All mentioned input signals are correct for a PLL to obtain the reference angular speed ω_{ir}^* of the rotor current vector and consequently the rotor current vector angle θ_{ir}^* in the frame connected with rotor.

A vector diagram presenting the dynamic state of the rotor current and stator voltage vector is shown in Figure 11.16. For given speed and load, there is a base position and length of the rotor current vector $\vec{i}_{rbase}$ responsible for generation of the stator voltage related to the reference vector $\vec{v}_s^*$. Displacement of the rotor current vector $\vec{i}_r$ from the base position $\vec{i}_{rbase}$ caused by mechanical speed or load change or during start-up of the system causes also phase displacement α_{us} of the stator voltage vector $\vec{v}_s$ from reference position $\vec{v}_s^*$.

To obtain the reference position by the stator voltage vector, it is necessary to move the rotor current vector by at least an instantaneous change of its rotational speed. Phase movement of the rotor current vector is realized by the change of rotational speed of the reference rotor current vector, which rotates with the speed

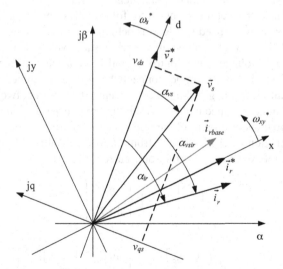

Figure 11.16.   Space vector diagram of the rotor current and stator voltage in a transient state caused by mechanical speed or load change.

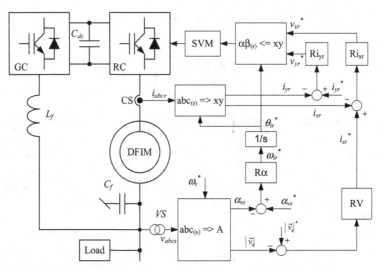

Figure 11.17. Full scheme of sensorless stator voltage control with a PLL based on the stator voltage angle controller and PI controllers of the rotor current.

ω^*_{xy} equal to

$$\omega^*_{xy} = \omega_m + \omega^*_{ir} \tag{11.60}$$

where ω^*_{ir} is a result of a voltage vector synchronization loop PLL.

Change of the reference angular speed ω^*_{ir}, forced by the PI controller, moves this vector as long as the voltage vector displacement α_{vs}, related to the reference position, is eliminated. The actual rotor current vector follows the reference position until the base position for a given speed and load is achieved.

Figure 11.17 shows a full scheme of sensorless stator voltage control with a PLL based on the stator voltage angle controller and PI controllers of the rotor current vector components in an xy frame connected with the reference rotor current. Reference magnitude and angle of the rotor current achieved from two stator voltage controllers, RV and Rα, respectively, represent the reference rotor current vector in a polar frame. Two transformations of the reference rotor current vector from the polar to $\alpha\beta$ frame, Equation (11.16), and next from $\alpha\beta$ to xy, Equation (11.62), have been applied using a reference angle of the rotor current θ^*_{ir}.

$$\begin{bmatrix} i^*_{\alpha r} \\ i^*_{\beta r} \end{bmatrix} = |\vec{i}_r|^* \begin{bmatrix} \cos\theta^*_{ir} \\ \sin\theta^*_{ir} \end{bmatrix} \tag{11.61}$$

$$\begin{bmatrix} i^*_{xr} \\ i^*_{yr} \end{bmatrix} = i^*_{\alpha r} \begin{bmatrix} \cos\theta^*_{ir} \\ \sin\theta^*_{ir} \end{bmatrix} + i^*_{\beta r} \begin{bmatrix} \sin\theta^*_{ir} \\ -\cos\theta^*_{ir} \end{bmatrix} = \begin{bmatrix} |\vec{i}_r|^* \\ 0 \end{bmatrix} \tag{11.62}$$

As a result, two components of the reference rotor current vector are obtained, in which the x component is equal to its amplitude and the y component is always zero, so in fact these transformations can be neglected. The same angle is used for transformation of the actual rotor current vector to $\alpha\beta$ and from $\alpha\beta$ to xy. Thus, PI controllers of the rotor current components can be applied, as xy components of the rotor current vector are fixed during the steady state. Two components of the rotor voltage vector are calculated by the rotor current PI controllers, Ri_{xr} and Ri_{yr}. Transformation from the xy to the $\alpha\beta$ frame connected with the rotor allow one to apply the SVM method.

The worst case for the system stability is no-load operation, because of the lowest damping ratio. For an unloaded model, rotor current and stator voltage controllers should be designed. This feature concerns every power electronics conversion system. Simplified analysis of the rotor current dynamics can be done for the simplified RL model, in which the inductance seen from the rotor side is treated as some equivalent value, L_{eq}. The equivalent inductance is determined by the calculation of stator and capacitor impedances for frequency f_m related to rotor speed ω_m. For a large power doubly fed machine, the stator winding resistance R_s can be neglected.

The equivalent inductance L_{eq} is

$$L_{eq} = L_{r\sigma} + \frac{L_s}{1 - 4\pi^2 f_m^2 L_s C_f} \tag{11.63}$$

where the stator inductance L_s is a sum of magnetizing L_m and stator leakage $L_{s\sigma}$ inductance.

Dynamics of the current responses in the full model and simplified RL model are very similar, as can be seen in a result of simulation in Figure 11.18. The unloaded generator is considered, because it is the worst case from the system stability point of view, and for an equivalent $R_r L_{eq}$ model the rotor current controllers can be prototyped.

Comparison of the open-loop RL model response and response of the RL model equipped with a PI current controller is shown in Figure 11.19a. Input signals have been selected to achieve the same current in the steady state. The same PI controller has been used in the full model in both axes of the xy frame to control the rotor

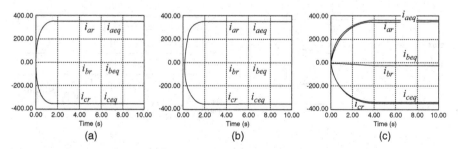

Figure 11.18. Current responses of a full model i_{ar}, i_{br}, i_{cr} and equivalent RL model i_{aeq}, i_{beq}, i_{ceq} for DC rotor current at different rotor speeds: (a) 1000 rpm, (b) 1500 rpm, and (c) 2000 rpm.

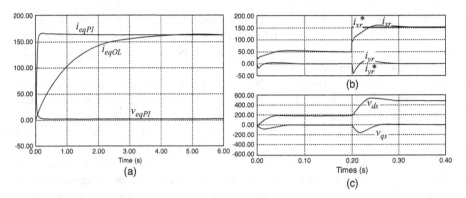

Figure 11.19. Simulation results of (a) simplified and (b,c) full models, with current PI controllers.

current components. Simulation results are also shown in Figure 11.19. The reference rotor current vector components i_{xr}^*, i_{yr}^* and current responses i_{xr}, i_{yr} are shown in Figure 11.19b, whereas voltage responses in the dq frame are shown in Figure 11.19c.

Full model responses to the step change of stator voltage vector magnitude and phase are shown in Figure 11.20. Figure 11.20a shows the reference and actual stator voltage vector magnitudes during a step change of reference value, while Figure 11.20b shows the responses of the reference and actual rotor current vector components that correspond to Figure 11.20a. Stator voltage vector dq components (Figure 11.20c), reference angular speed of the rotor current vector (Figure 11.20d), and stator phase voltage and rotor phase current (Figure 11.20e) have been presented as the responses to the step change of reference voltage vector angle from zero to π.

The function of the stator voltage vector angle, Equation (11.56), is nonlinear and returns values in the range from $-\pi$ to π, as shown in Figure 11.21, similar to the

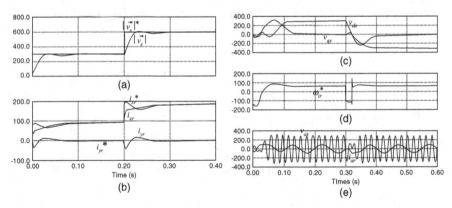

Figure 11.20. Responses to the step change of the stator voltage vector (a,b) magnitude and (c–e) phase.

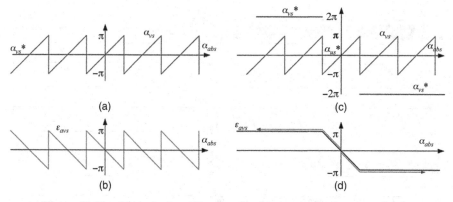

Figure 11.21. Method of modification of reference voltage vector angle α_{us}^*.

position encoder overflow. In some cases, error ε_α of the Rα controller has a periodical character and the average value of this error is zero. As a result, if the Rα controller cannot deliver an adequate signal of reference rotor current angular speed ω_{ir}^* before the first overflow, it may not be possible during the next periods. Modification of the reference angle α_{vs}^*, based on the function overflow (Figure 11.21c) results in monotonic dependence of the error on the actual value of the voltage vector angle (Figure 11.21d). Practical realization of the method is shown in Figure 11.22. Initial value of α_{vs}^* equals zero. Every change of the angle higher than π causes an α_{vs}^* increase by 2π and an opposite change causes a decrease by -2π with simultaneous limitation of α_{vs}^* to 2π and -2π and limitation of the error ε_α to $-\pi$ and π.

Simulation results of an unloaded system, starting up at 1000 rpm without and with modified reference signal of the stator voltage angle, is shown in Figure 11.23. Comparisons of unsuccessful start-up (Figure 11.23a–e) and successful start-up in the case of modification of the reference stator voltage angle (Figures 11.23f–j) are presented.

Transient states caused by mechanical speed change during supply of linear (resistive) and nonlinear (three-phase diode rectifier loaded by resistor) load are presented in Figures 11.24a,b and 11.24c,d, respectively. Implementation of stator connected capacitors allow one to obtain satisfactory quality of the stator voltage v_{as} at nonlinear load current i_{ald}.

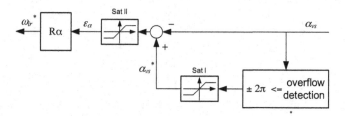

Figure 11.22. Practical realization of modification of reference voltage vector angle α_{us}^*.

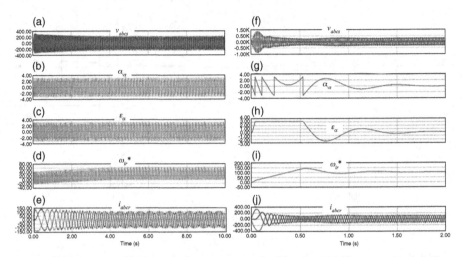

Figure 11.23. Stator three-phase voltage v_{abcs}, stator voltage vector angle α_{us}, angle controller error ε_α, reference angular speed of the rotor current vector ω_{ir}^*, and three-phase rotor current i_{abcr} during (a–e) unsuccessful and (f–j) successful (fghij) start-up of stand-alone DFIM.

Currently, the main applications of the DFIM are in wind turbines, which generally are not reliable as an isolated power source. Therefore, stand-alone operation mode of a DFIM has been reported in only a few publications, whereas the grid connected DFIM systems are described widely. From the small number of publications on the stand-alone DFIM, the most attention has been addressed in References [8,11,12], where in opposition to the direct voltage control method presented in the current chapter, stator flux oriented control is described in different versions. Those are the only proposals, where in a different manner, rotor speed or position sensors are eliminated. However, there is no complex solution to different problems arising from use of a stand-alone system. Even if there are complementary control methods for the grid side converter proposed for reduction of stator voltage harmonics [8] or

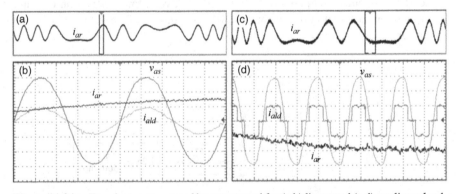

Figure 11.24. Transient states caused by rotor speed for (a,b) linear and (c,d) nonlinear load.

imbalance [9], it may not be enough and an adequate control for the rotor side converter during a nonlinear and unbalanced load supply may be necessary.

11.3.2 Voltage Asymmetry Correction During Unbalanced Load Supply

Similar to the case of a grid connected DFIM working with an unbalanced grid, a stand-alone power system should be able to operate with an unbalanced load. Independent of the DFIM system topology dedicated to unbalanced load supply, it is not possible to compensate the zero sequence component by three wire converters. The main control part presented in the previous section, responsible for amplitude and frequency stabilization, is connected with a positive sequence, whereas elimination of the negative sequence component requires additional controllers [13,14].

It can be realized by control of the orthogonal components of the stator voltage vector in the $dq_{(2)}$ frame, rotating with negative synchronous speed $-\omega_s$, opposite to the main polar frame (Figure 11.25). Transformation of the stator voltage vector from a three-phase system $abc_{(s)}$ to $dq_{(2)}$ gives two signals, which contain a DC component representing the negative sequence and a 100 Hz component representing a positive sequence. A lowpass filter (LPF) in each negative sequence component control loop reduces significantly the content of the 100 Hz harmonics (corresponding to the positive sequence). It allows one to use PI controllers for elimination of negative sequence components, for which reference signals v^*_{ds2}, v^*_{qs2} equal zero. Output signals of negative sequence controllers, Rv_{d2} and Rv_{q2}, are responsible for the reference

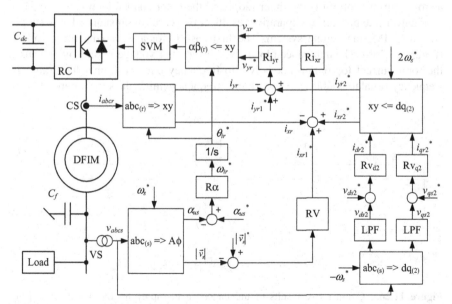

Figure 11.25. Scheme of the control method based on positive and negative sequence components for voltage asymmetry reduction during unbalanced load supply.

rotor current negative sequence dq components i^*_{dr2}, i^*_{qr2}, and they are represented in the negatively rotated frame $dq_{(2)}$. It is necessary to transform this signal to the rotating frame xy connected to the positive sequence reference rotor current, to obtain an equivalent reference current. Negative sequence reference rotor current components after transformation to the xy frame are represented by a signal of frequency equal to 100 Hz.

Instead of the transformation from $abc_{(s)}$ to $dq_{(2)}$, calculation of the reference rotor current negative sequence components, and transformation from $dq_{(2)}$ to xy, another realization of stator voltage asymmetry correction is possible. Calculated dq components v_{ds} and v_{qs} (Figure 11.15), needed to obtain polar coordinates of the stator voltage vector, contain DC components corresponding to the positive sequence and 100 Hz component related to the negative sequence. Implementation of resonant PI controllers designed for the 100 Hz frequency allows elimination of the negative sequence component in the stator voltage. Output signals from resonant converters have a 100 Hz frequency and are responsible for reference rotor current negative sequence components. From the control point of view, it is only an other way of calculating the same reference rotor current signals and there is no difference in results.

In both cases, the total reference current is a sum of positive (DC component) and negative (100 Hz) sequence components and is compared with actual xy components of the rotor current vector. The PI rotor current controllers are designed for the DC component. However, a proportional part can successfully control the negative sequence current. It causes the steady state error in the rotor current control loop but has no negative influence on elimination of negative sequence in the stator voltage, as rotor current control is the inner loop, and the outer control loop is a PI type.

Figure 11.26 presents comparative results of two cases of unbalanced load supply with a DFIM stand-alone system. The first case is for a sinusoidal rotor voltage (Figure 11.26b). The unbalanced load, even if the rotor voltage is sinusoidal, causes the rotor current to contain not only slip frequency but also higher harmonics of frequency equal to double the synchronous frequency minus the slip (Figure 11.26c).

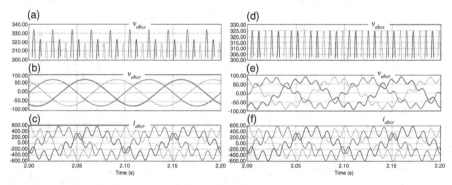

Figure 11.26. Comparative results of unbalanced load supply for (a–c) sinusoidal rotor voltage, and for (d–f) rotor voltage with negative sequence component for full compensation of asymmetry.

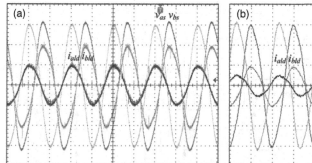
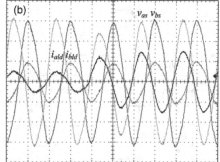

Figure 11.27. Laboratory tests results of 2.2 kW slip ring induction machine for unbalanced load in (a) steady state and in (b) transient caused by step change of the load in single phase.

For an unbalanced load, in each phase there are different voltage drops in machine resistances and leakage inductances. Even if the negative sequence of the current is delivered from the rotor converter through the machine to the stator side, asymmetry of the stator voltage is not fully eliminated (Figure 11.26a). Full compensation of the stator voltage asymmetry (Figure 11.26d) requires additional content of negative sequence in the rotor voltage (Figure 11.26e), which causes more negative sequence in the rotor current (Figure 11.26f) than in the case of a sinusoidal rotor voltage.

Figure 11.27 presents laboratory tests results of 2.2 kW slip ring induction machine. Stator voltage asymmetry is significantly reduced. Moreover, the control method is fast enough and allows one to obtain short transient states during the step change of the load in the single phase (Figure 11.27b).

11.3.3 Voltage Harmonics Reduction During Nonlinear Load Supply

A similar control method, used for voltage asymmetry correction with an unbalanced load, can be used for stator voltage harmonics compensation at nonlinear load supply. Each harmonic can be represented as a vector rotating with angular speed related to the harmonics frequency. By assuming that odd harmonics are negative sequence components, even harmonics are positive sequence components, and there are no multiples of third harmonics in the three wire system, a control method for elimination of stator voltage harmonics can be synthesized (Figure 11.28).

By having dq components of the stator voltage vector in a synchronously rotated frame (Figure 11.15), harmonic vector dq components in every respective frame can be obtained by transformation from dq to $dq_{(h)}$, where h is the number of harmonics. The transformation requires fewer calculations; for example, for 5th and 7th harmonics, transformation is done by the use of $-6\omega_s$ and $6\omega_s$, respectively,

After calculation of the reference rotor current harmonic components, the required transformations also use $6\omega_s$ and $-6\omega_s$ for 5th and 7th harmonics, respectively. It means, that $\sin(6\omega_s t)$ and $\cos(6\omega_s t)$ can be calculated only once per calculating period and can be used many times in both transformations. Similarly, for 11th and 13th harmonics, $12\omega_s$ and $-12\omega_s$ can be used, respectively, for transformation of stator

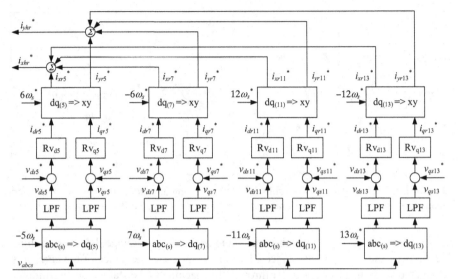

Figure 11.28. Control part responsible for stator voltage harmonics compensation at nonlinear load.

voltage from the *dq* synchronous frame to the *dq*$_{(h)}$ frame for each harmonic (Figure 11.29). The *x* and *y* components of the reference rotor current vector, responsible for harmonics compensation, are, respectively, the sum of all *x* and *y* components of each harmonic considered in the control part.

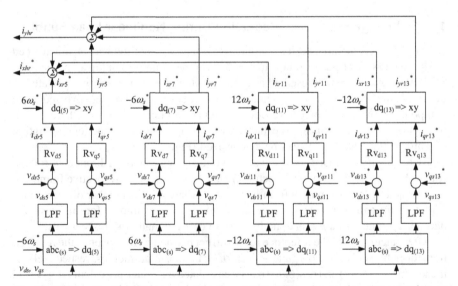

Figure 11.29. Control part responsible for stator voltage harmonics compensation at nonlinear load.

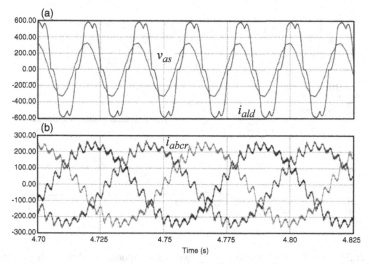

Figure 11.30. (a) Stator phase voltage v_{as} and load current i_{ald} and (b) rotor three-phase current with harmonics i_{abcr} during nonlinear load supply by DFIM.

Another way to calculate the reference rotor current components for stator voltage harmonics compensation is the use of resonant controllers for dq components of each harmonic in a synchronously rotated dq frame. The d components of 5th and 7th harmonics are represented in this frame, by a common 300 Hz component of the v_{ds} signal. Thus, a common resonant controller designed for the 300 Hz frequency can be used. The same is true for q components of both harmonics. For 11th and 13th harmonics, a 600 Hz PI resonant controller frequency is correct.

In practice, controllers for 11th and 13th harmonics can be neglected in all presented methods, as filtration by a stator connected filtering capacitor is enough. Moreover, for a high power DFIM system, 11th and 13th harmonics are too high to be effectively reduced by the rotor converter, as the converter's switching frequency is not much higher.

Steady state simulation results of a DFIM with a nonlinear load are presented in Figure 11.30. This control part, responsible for stator voltage harmonics compensation, is taken from Figure 11.29. Current harmonics needed by the nonlinear load (Figure 11.30a) are provided by the rotor converter. Rotor current shown in Figure 11.30b contains a fundamental harmonic equal to the slip frequency and higher harmonics for compensation of the voltage distortions.

Figure 11.31 shows results of a simulation during start-up of a stand-alone DFIM system with nonlinear load. The initial transient lasts about 1 second and originates from the control of the stator voltage vector angle. After this period, harmonic components are reduced to zero by harmonic components PI controllers, as can be observed in Figure 11.31b for 5th harmonic components and in Figure 11.31c for 7th harmonic components. Simulation test results of transient states caused by mechanical speed change of a DFIM supplying a nonlinear load are shown in Figure 11.32.

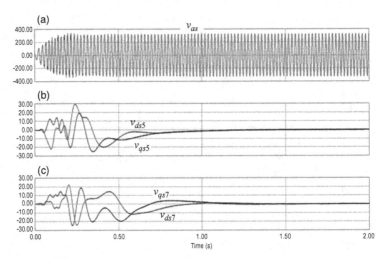

Figure 11.31. (a) Stator phase voltage v_{as} and (b,c) responses of stator voltage harmonics controllers.

For the control part, from Figure 11.29, it is assumed that, the nonlinear load is symmetrical and is supplied in a three wire system. For a four wire system and simultaneous supply of an unbalanced and nonlinear load, a combination of the control parts from Figures 11.25 and 11.29 is necessary. Moreover, it may be

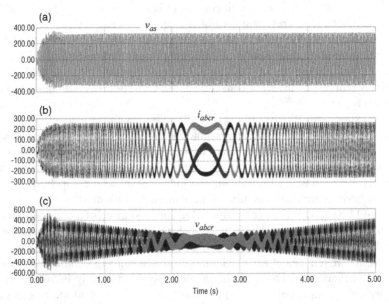

Figure 11.32. (a) Stator phase voltage v_{sa}, (b) three-phase rotor current i_{abcr}, and (c) three-phase rotor voltage v_{abcr} during nonlinear load supply at variable speed.

necessary to take into consideration the odd harmonics positive sequence and even harmonics negative sequence. Compensation of any zero sequence component is not possible, unless the rotor side is equipped with four slip rings and four wire converter topology. However, in a small number of such loads, application of a fourth ring in the rotor side to obtain high quality voltage of the four wire stand-alone DFIM is unnecessary.

Stator voltage asymmetry correction and compensation of harmonics can be supported by an active filter function applied in control of the grid side converter. Active filter operation of GSC can provide reduction of shaft torque pulsations caused by rotor and stator current harmonics. However, for high amounts of harmonics or asymmetry, supporting operation of GSC may not be enough, especially during operation near the bottom and top limits of mechanical speed. Then, adequate control in the rotor side converter can significantly increase the stator voltage quality.

11.4 SYNCHRONIZATION BEFORE GRID CONNECTION BY SUPERIOR PLL

Direct voltage control can easily be adopted to obtain synchronization of the stator voltage with the grid [15], before controlled connection by a grid connection switch (GCS) (Figure 11.33). Controlled connection of a DFIM based wind turbine to the power network eliminates the impact of magnetizing current appearing in a directly connected stator of a DFIM without synchronization.

The DFIM's advantages can be utilized in a sources other then wind turbines. Adjustable speed hydro power generation plants and a range of power generation systems driven by internal combustion engines can successfully apply the DFIM as a

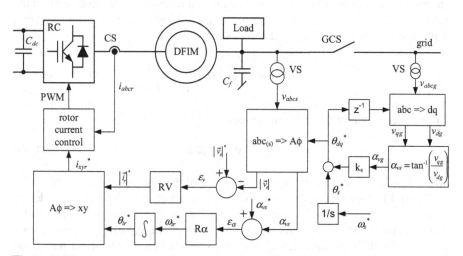

Figure 11.33. Direct voltage control method with superior PLL used for synchronization of the DFIM with the power network.

power generation unit. For such type of energy sources, stand-alone mode is recommended, as during a grid failure they can still operate and supply a selected part of the grid connected load. This is one of the goals of the so-called distributed generation concept. In the case of any power system designed for both operation modes, depending on the grid voltage conditions, synchronization is also needed to protect the load supplied in stand-alone mode against the rapid change of the load supplied voltage phase. Such a phase change, in the case of connection without prior synchronization, may negatively influence loads like electrical motors and may cause disturbances of the voltage in a weak power grid.

Another issue is controlled disconnection after the loss of mains. In the case of any grid connected power system, during grid failures, uncontrolled voltage in the grid is maintained by the local power system (e.g., DFIM) unless grid failure is detected [16]. Even if the methods of islanding detection are developed for a specific topology and control method of the grid connected power source, they can easily be adopted for different types of power systems including the DFIM. Controlled stand-alone operation after mains outage detection has to be applied immediately, without a change of the stator voltage phase in relation to the phase of grid voltage before failure.

Synchronization is realized by reduction of the grid voltage vector angle α_{vg} related to the dq reference frame rotating with speed

$$\omega_{dq}^* = \omega_s^* - \frac{d\alpha_{vg}}{dt} \tag{11.64}$$

where positive α_{vg} means that the grid voltage vector is delayed in reference to the d axis of the dq frame.

The angle α_{vg} can be achieved in the same way as for the stator voltage angle α_{vs} in the main control loop, so based on $\tan^{-1}$ or $\sin^{-1}$ functions with dq components of the grid voltage vector as function arguments, or based on cross product of stator and grid voltage vectors in, for example, the $\alpha\beta$ frame:

$$\alpha_{ug} = \sin^{-1}\left(\frac{v_{\alpha s}v_{\beta s} - v_{\beta s}v_{\alpha s}}{|\vec{v}_s||\vec{v}_g|}\right) \tag{11.65}$$

The rotation angle of the dq frame equals

$$\theta_{dq}^* = \theta_s^* - k_s\alpha_{vg} \tag{11.66}$$

where k_s is a factor allowing iterative decrease of the grid voltage vector angle.

Figure 11.34 shows a vector diagram of the stator voltage and grid voltage during synchronization. In the initial state after grid voltage recovery, the grid voltage vector overtakes the d axis and the stator voltage by angle α_{vg}. The angle θ_{dq}^* in the next step is calculated with Equation (11.66); this causes the movement of the dq frame and reference stator voltage vector. Basic control synchronizes the stator voltage vector with the new reference position, so the actual stator voltage vector achieves a new

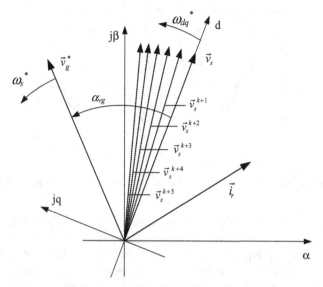

Figure 11.34. Vector diagram during synchronization of the stator voltage with the grid.

position. Successive positions of the stator voltage vector in the next steps are shown in Figure 11.34.

An oscillogram showing synchronization of the stator phase voltage v_{as} with the power network phase voltage v_{ag} is shown in Figure 11.35a; the final part, with closing of the grid connection switch (GCS) at t_{con} and change of operation mode, is shown in Figure 11.35b. A 2.2 kW DFIM has been loaded at 50% with a resistive load. Very fast synchronization is achieved, but for other than a resistive type of load, a longer time process may be required to avoid rapid change of the load supply voltage phase.

Change of the operation mode from grid connection to stand-alone operation after grid failure detection has to be realized without any change of the supply voltage seen by the load. This is an uninterruptible load supply [17] for electromechanical variable

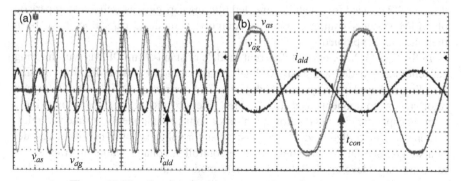

Figure 11.35. Synchronization process of stator voltage with grid.

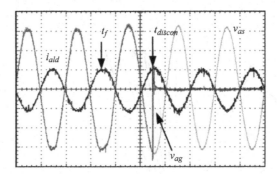

Figure 11.36. Controlled disconnection of DFIM after grid failure detection.

speed power units. Between failure instant t_f and disconnection instant t_{discon}, voltage in the part of the power grid between the stator and grid failure point is maintained by the DFIM power unit (Figure 11.36), but this voltage is out of control, due to the fact that in this period the control method for grid connection mode is applied. After DFIM disconnection by the GCS, operating mode has been changed to stand-alone and the voltage parameters have been controlled. Stator voltage phase remains unchanged in relation to the grid voltage phase before loss of main. The mains outage detection methods are independent of the power generation system topology and the methods developed for other generation systems can be adapted to the DFIM.

11.5 SUMMARY

In the introduction of this chapter, basic features and requirements of stand-alone power systems with DFIMs have been discussed. Basic topologies allowing black start and stand-alone operation of the DFIM and topologies of a four wire system with neutral point access on the stator side for unbalanced and single phase load supply have been shown. Steady state characteristics of a stand-alone DFIM supported by an additional energy source and storage connected to the DC side, which is quite important in the case of a wind turbine as the primary mover of the DFIM, are discussed. The need for a stator connected filtering capacitor as well as a comparison of DFIM systems equipped and not equipped with those capacitors have been presented and explained.

In Section 11.2, we present different models of stand-alone DFIMs with filtering capacitors and loaded with resistive load. An initial model is described in Cartesian coordinates based on a typical model of a grid connected DFIM with stator and rotor current components as state variables. The next model is extended by using a filtering capacitor circuit equation, with stator voltage components as state variables. To achieve a simplified model for analysis of stator voltage control methods, a model of a stand-alone DFIM fed from a current controlled voltage source inverter on the rotor side was developed. Polar coordinate models have been derived. All models have been implemented in C language and simulated.

Section 11.3 is devoted to the sensorless stator voltage control metod for a standalone DFIM, based on PLL synchronization of actual and reference stator voltage vector and PI controllers of the rotor current. Additional control parts for stator voltage asymmetry correction during unbalanced load supply and stator voltage harmonics compensation during nonlinear load supply have been discussed. The results of the control methods, simulated and verified in a small laboratory unit, are shown.

In Section 11.4, the method of stator voltage synchronization with the grid voltage is analyzed. A PLL, superior in relation to the stator voltage control, is shown and laboratory results of DFIM synchronization with the grid are presented.

REFERENCES

1. W. Koczara and G. Iwanski, "Fuel Saving Variable Speed Generating Set," *International Conference on Clean Electrical Power–ICCEP'09*, 9–11 June 2009, Capri, Italy, pp. 22–28.
2. R. Pena, R. Cardenas, G. M. Asher, J. C. Clare, J. Rodriguez, and P. Cortes, "Vector Control of a Diesel-Driven Doubly Fed Induction Machine for a Stand-Alone Variable Speed Energy System," *28th Annual Conference of the Industrial Electronics Society–IECON 02*, 5–8 November 2002, Vol. 2, pp. 985–990.
3. S. Furuya, T. Taguchi, K. Kusunoki, T. Yanagisawa, T. Kageyama, and T. Kanai, "Successful Achievement in a Variable Speed Pumped Storage Power System at Yagisawa Power Plant," *Power Conversion Conference–PCC'93, Yokohama*, 19–21 April 1993, pp. 603–608.
4. T. Kuwabara, A. Shibuya, H. Furuta, E. Kita, and K. Mitsuhashi, "Design and Dynamic Response Characteristics of 400 MW Adjustable Speed Pumped Storage Unit for Ohkawachi Power Station," *IEEE Trans. Energy Conversion*, Vol. 11, No. 2, pp. 376–384, June 1996.
5. K. Grotenburg, F. Koch, I. Erlich, and U. Bachmann, "Modeling and Dynamic Simulation of Variable Speed Pump Storage Units Incorporated into the German Electric Power System," *9th European Conference on Power Electronics and Applications–EPE'09*, Graz, Austria, p.10.
6. S. Breban et al., "Variable Speed Small Hydro Power Plant Connected to AC Grid or Isolated Loads," *Eur. Power Electron. Drives Assoc. J.*, Vol. 17, No. 4, January, 2008.
7. H. Akagi and H. Sato, "Control and Performance of a Doubly-Fed Induction Machine Intended for a Flywheel Energy Storage System," *IEEE Trans. Power Electron.*, Vol. 17, No. 1, pp. 109–116, 2002.
8. A. K. Jain and V. T. Ranganathan, "Wound Rotor Induction Generator With Sensorless Control and Integrated Active Filter for Feeding Nonlinear Loads in a Stand-Alone Grid," *IEEE Trans. Ind. Electron.*, Vol. 55, No. 1, pp. 218–228, January 2008.
9. R. Pena, R. Cardenas, E. Escobar, J. Clare, and P. Wheeler, "Control System for Unbalanced Operation of Stand-Alone Doubly Fed Induction Generators," *IEEE Trans. Energy Conversion*, Vol. 22, No. 2, pp. 544–555, June 2007.
10. L. Guilherme, B. Rolim, D. R. Costa, and M. Aredes, "Analysis and Software Implementation of a Robust Synchronizing PLL Circuit Based on the pq Theory," *IEEE Trans. Ind. Electron.*, Vol. 53, pp. 1919–1926, December 2006.

11. R. Cardenas, R. Pena, J. Proboste, G. Asher, and J. Clare, "MRAS Observer for Sensorless Control of Standalone Doubly Fed Induction Generators," *IEEE Trans. Energy Conversion*, Vol. 20, pp. 710–718, December 2005.

12. D. G. Forchetti, G. O. Garcia, and M. I. Valla, "Adaptive Observer for Sensorless Control of Stand-Alone Doubly Fed Induction Generator," *IEEE Trans. Ind. Electron.*, Vol. 56, No. 10, pp. 4174–4180, 2009.

13. G. Iwanski and W. Koczara, "Sensorless Direct Voltage Control of the Stand-Alone Slip-Ring Induction Generator," *IEEE Trans. Ind. Electron.*, Vol. 54, No. 2. pp. 1237–1239, April 2007.

14. M. Chomat, L. Schreier, and J. Bendl, "Control Method for Doubly Fed Machine Supplying Unbalanced Load" *9th European Conference on Power Electronics and Applications–EPE*, Toulouse, France, 2003 (CD Proc.).

15. G. Iwanski and W. Koczara, "DFIM Based Power Generation System with UPS Function for Variable Speed Applications," *IEEE Trans. Ind. Electron.*, Vol. 55, No. 8, pp. 3047–3054, August 2008.

16. R. Teodorescu and F. Blaabjerg, "Flexible Control of Small Wind Turbines with Grid Failure Detection Operating in Stand-alone and Grid-Connected Mode," *IEEE Trans. Power Electron.*, Vol. 19, pp. 1323–1332, September 2004.

17. J. M. Guerrero, L. G. de Vicuna, and J. Uceda, "Uninterruptible Power Supply Systems Provide Protection," *IEEE Ind. Electron. Mag.*, Vol. 1, No. 1, pp. 28–38, 2007.

CHAPTER 12

New Trends on Wind Energy Generation

12.1 INTRODUCTION

The preceding chapters have concentrated on the technical aspects of the DFIM based wind energy generation systems. Deep and numerous theoretical analyses accompanied by a wide range of illustrative examples have provided the reader with the technical fundamentals and basis of the DFIM, especially the power electronic converter and the wind energy generation application.

In this last chapter, the new trends in wind turbine technology are identified and discussed, focusing not only on the DFIM based wind energy generation but also considering alternative wind turbine concepts and solutions, as well as emerging technologies. Novel wind energy generation issues are treated, such as the wind farm location and electric grid integration; as a result of an increasing interest in this field, more people from different scientific and social disciplines are becoming involved.

However, the material presented in this chapter is not intended to be definitive, since the rapid advances in technology all around the world, together with the different technological and particular interests of manufacturers, researchers, and industry in general, make it almost impossible to predict the scenario for wind turbines in the long term.

Instead, the approach taken is to present and discuss the potential major areas for development of wind energy generation technology, taking into account also that it cannot evolve totally independently, since it is closely influenced by the technological advances in other scientific areas, such as the power electronics and drives and mechanics.

In order to identify and summarize the new trends in wind energy generation, this chapter is organized as follows. The first section of this chapter answers the question of *what* must be improved or innovated in wind power generation in the future; that is, it identifies the general future challenges, considering basically the present state of wind turbine technology and its margin for improvement.

The second section of this chapter identifies the impact on wind turbine technology to achieve these future challenges, that is, *how* these challenges can be achieved.

Doubly Fed Induction Machine: Modeling and Control for Wind Energy Generation,
First Edition. By G. Abad, J. López, M. A. Rodríguez, L. Marroyo, and G. Iwanski.
© 2011 the Institute of Electrical and Electronic Engineers, Inc. Published 2011 by John Wiley & Sons, Inc.

Therefore, technical details and tendencies of wind turbine technology are revealed that can contribute to a more sustainable and efficient wind energy generation.

12.2 FUTURE CHALLENGES FOR WIND ENERGY GENERATION: WHAT MUST BE INNOVATED

This section presents, sets priorities, and summarizes the future challenges for wind energy generation. The technological advances and enhanced efforts of wind turbine manufacturers and researchers should tend to improve the following four main aspects of wind power generation:

1. Wind farm location
2. Power, efficiency, and reliability increase
3. Electric grid integration
4. Environmental concerns

These four distinct challenges are not so different from each other, since they are closely interrelated. However, for a more simple exposition the division into four main challenges or objectives has been followed. The subsequent sections specify and justify each proposed future challenge.

12.2.1 Wind Farm Location

An important issue is to provide the possibility of increasing the potential locations of wind farms, in new sites with different characteristics than present locations and of course, maintaining the wind power generation as a competitive energy resource.

12.2.1.1 Onshore–Offshore An increase in offshore wind farms is expected in the future, especially for those countries whose wind power generation is already significant and who are finding limits in land for the location of wind farms. Thus, the present onshore wind turbine technology is serving as the basis for new offshore wind turbines. However, new efforts are needed basically in two issues:

- The offshore wind farms require innovative technology for wind turbine suspention for deep water. In addition, the reliability and robustness of these turbines is also crucial if offshore locations are used.
- On the other hand, the generated energy of the offshore sites must be transmitted though submarine cables, so technology for energy evacuation to onshore must be introduced as well. There are mainly two possible solutions for energy transmission through submarine cables: AC transmission and converter based transmission in DC (HVDC technology), as schematically represented in Figure 12.1.

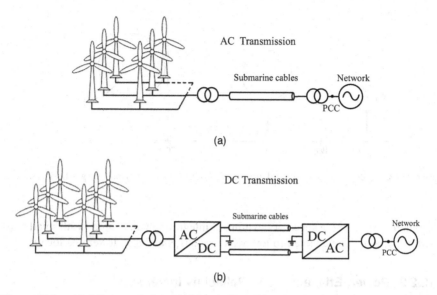

Figure 12.1   (a) AC transmission wind farm. (b) DC transmission wind farm with HVDC.

On the other hand, the offshore wind turbine technology itself is being altered to newer generator-gear-converter concepts, as presented in subsequent sections, compared to onshore wind turbine technology.

Thus, thanks to this new generation of offshore wind turbines, it is possible to facilitate the increase of wind energy generation while reducing the visual impact on land.

12.2.1.2 Wind Maps and Prediction

It is important to generate more knowledge related to wind forecasts and predictions. The wind expectations and wind behavior of specific terrains, determines the best locations for wind farms, layout configurations, and specific needs that are fundamental for the efficiency of the wind farm.

12.2.1.3 Aerodynamic Behavior of Wind Turbines

Very closely related to the previous issue is the understanding of the aerodynamic behavior of different configurations of wind turbines. This fact contributes directly to wind turbine optimization, in terms of efficient wind power generation. Thus, it is expected that research oriented to better understand wind flux and distribution around the wind turbine and its blades, under several external conditions, will enable optimization of the wind turbine mechanical design and the measurement techniques of wind conditions.

12.2.1.4 Low Wind Scenarios–Low Wind Turbines

Despite the fact that wind turbine sizes have been increasing since their first usage, it is also expected that the demand for small wind turbines will increase, for instance, in small village energy production or stand-alone applications.

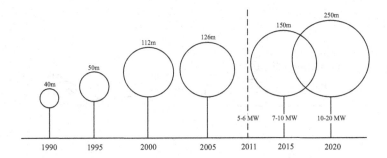

Figure 12.2 Growth in size of wind turbine. (*Source*: http://www.wind-energy-the-facts.com/.)

Therefore, it is also necessary to think of integrating wind energy with small wind turbines that can be adapted to low wind scenarios and also to remote locations.

12.2.2 Power, Efficiency, and Reliability Increase

Since the beginning of wind energy generation, the sizes of the individual wind turbines have been progressively increasing. Wind turbine manufacturers started offering products of less than 1 MW, while expectations in the near future tend to sizes greater than 5 MW (Figure 12.2).

The basic reason for the increase of power is that the efficiency of the individual wind turbines is improved as well. Not only does the efficiency increase impact on the generated energy, but it also impacts manufacturing, maintenance, and costs of wind turbines.

Thus, the following challenges result from the size increase of wind turbines.

12.2.2.1 Manufacturing and Logistics Issues related to manufacturing process improvements and reduction of civil work costs must be aligned and in accordance with the increase in the size of wind turbines.

12.2.2.2 Reliability and Maintenance A very important aspect of the wind turbine is the reliability and, associated with it, the maintenance. The reliability of each separated element of the wind turbine, as well as the whole wind turbine itself, is a continuously researched issue. As the size increases, reliability and availability of the wind generated energy become a major issue, since a failure in a single turbine would make a large capacity unavailable.

Hence, newer concepts of "preventive maintenance" (e.g., applied during low wind periods) or "predictive maintenance" (by means of analysis and supervision during normal operation) are expected to be introduced in wind turbine technology.

In addition, well-known concepts such as element redundancy and reduced number of elements are also introduced in wind turbine technology in order to improve reliability. Thus, risky or weak elements are identified and substituted by alternative or more reliable solutions.

12.2.2.3 Reduced Costs Finally, in addition to all these aspects of increasing the efficiency and reliability of the wind turbine, obviously we must pay special attention to the costs of wind energy generation in general. The new wind turbine technology must still be a competitive energy generation resource in terms of costs.

12.2.3 Electric Grid Integration

The increase of wind energy penetration affects the electric grid and influences its behavior too. Mainly due to the variability of the wind, wind energy generation presents some differences from more traditional or conventional energy resources. Thus, the increase of wind energy penetration on the grid requires additional efforts to ensure the stability and sustainability of the electric system.

12.2.3.1 Wind Variability In general, if wind energy penetration is high, it means that dependency on the wind is high to cover energy demands. This disadvantage of the wind energy generation provokes mainly two solutions to mitigate this:

- Disposition of additional energy generation (operating reserves) to cover the demand for energy when the wind is low. This additional energy can be supplied by a generation system based on a different energy resource or can also be based on a storage facility system. By using any of these solutions, the energy cost impact must be reduced, so optimized design plays an important role.
- The wind forecast accuracy is improving so it is possible to avoid problems derived from low wind periods, activating in an optimized way the operating reserves.

12.2.3.2 Grid Planning Operation The integration of wind energy also requires coordination efforts with the grid. Wind farms can also accomplish complementary functionalities such as voltage and frequency regulation, operation during disturbances (LVRT—low voltage ride through), energy management, reactive power management, and power quality improvements. These functionalities together with the energy generation itself can contribute to a more optimized and reliable electric grid system.

12.2.3.3 Grid Codes and Standards In order to regularize wind energy generation and integration with the grid, countries are continuously developing new grid codes and standards that adapt the wind energy generation to the specific characteristics of their electric grids. These grid codes are similar in essence from one country to another and impose characteristics such as generated power quality and behavior during grid failures. Hence, wind farms must be adapted to different grid codes, for proper integration in the grid.

12.2.4 Environmental Concerns

The concern about environmental implications of wind energy generation has gained significant relevance during the last few years, and still an increase in production is expected due to the spreading population of wind turbines around the world.

Although many environmental benefits can be derived from wind energy generation compared to other energy resources, it is still possible to emphasize several environmental challenges that must be solved.

12.2.4.1 *Greenhouse Gases Reduction* Probably the most obvious and popular benefit of the employment of wind energy as an alternative and clean energy resource from fossil fuels is the reduction of CO_2 emissions. There are well supported theories justifying a global climate change due to the global increase of CO_2 emissions.

Hence, wind energy generation permits a reduction in our dependence on fossil fuels for energy generation, together with a reduction in the CO_2 global emissions.

12.2.4.2 *Visual Effect* An environmental challenge related to wind energy generation is the visual impact of the wind turbines located in landscapes near inhabited areas. Alternative locations could be exclusion areas like the sea. Placing wind turbines some kilometers out from the coast may significantly reduce their visual impact.

However, as mentioned before, this solution is also accompanied by new challenges related to wind turbine technology.

12.2.4.3 *Noise Reduction* Sound emissions of recent wind turbine models have been reduced significantly from initial designs. In today's designs, the main source of noise emission is the blade turning noise. Special focus on blade design research can contribute to reduce this issue in the future. Nevertheless, several studies have found noise from modern wind turbines much lower than other common noise sources such as road traffic and construction.

12.2.4.4 *Bird Fatalities* Animal or bird fatalities provoked by wind turbines are mainly significant when large migratory groups of birds cross wind parks. Bigger wind turbines with lower rotating speeds of the blades are more advantageous for birds than smaller turbines, with faster rotation speeds.

Therefore, new bigger wind turbine models together with careful location of wind parks out of the way of migratory routes can contribute to mitigate this important environmental concern.

12.3 TECHNOLOGICAL TRENDS: HOW THEY CAN BE ACHIEVED

The previous section has identified the main innovations and improvements that wind energy generation requires to be more competitive and sustainable in the future. Therefore, this section tries to provide specific details about how the previously mentioned improvements can be achieved from a technological perspective of wind energy generation. In other words, this section summarizes the technological advances and trends in wind turbines that can contribute to achieve the aforementioned general objectives:

1. Wind farm location
2. Power, efficiency, and reliability increase
3. Electric grid integration
4. Environmental concerns

For that purpose and seeking to provide a simple exposition, the technological challenges have been divided into two groups:

1. Mechanical structure of the wind turbine
2. Power train technology

Thus, for instance, by improving the mechanical structure of the wind turbine, developing newer blade concepts, and enabling the introduction of better materials, we can develop the next generation of bigger, more efficient, and more reliable wind turbines. Moreover, these mechanical advances inevitably must be aligned and coordinated with advances in the power train technology, allowing, for instance, the introduction of newer generator and drive train designs, together with an adequate offshore energy evacuation technology. Finally, this may allow us to locate the new more efficient and reliable wind turbines offshore providing a sustainable electric grid integration of wind energy that is environmentally clean.

12.3.1 Mechanical Structure of the Wind Turbine

The mechanical configuration of the wind turbine is a continuously advancing area. Obviously, it plays a very important role in wind energy generation, and technological innovations in the structure of wind turbines necessarily impacts on the overall wind turbine development.

In this section, five main research issues within the mechanical structure of wind turbines have been distinguished: blades, materials, fatigue, mechanical tests, and manufacturing processes.

12.3.1.1 Blade Concepts Larger blade scaling is necessary for multimegawatt wind turbines. It thus becomes necessary to take into account design innovations such as:

- Improved wind inflow—optimized aerodynamic shapes (to improve the wind energy capture efficiency)
- Load mitigation control
- Larger, lighter, and more flexible blades (but also stronger)
- Acoustic noise mitigation

12.3.1.2 Materials An other important area where efforts must be focused is the existing and new materials for blade construction of wind turbines.

- *Existing Materials.* Based mainly on fiber reinforced composites, the most important challenge is to develop fatigue life prediction methods.
- *New Materials.* They can contribute to stronger and stiffer blade designs. Thus, for instance, reinforcements based on materials such as carbon fiber and polymer resins, may allow smarter blade concepts. On the other hand, environmental considerations such as the recyclability and production processes of blades using new materials must also be taken into account.

12.3.1.3 Fatigue and Reliability Designs that prevent damage from fatigue can also improve the reliability and life of the blades. Hence, design processes can be improved by iterative methods considering load histories obtained from tracking data (structural heath monitoring), developing fatigue life predictions based on simulation tools and drawing conclusions from the obtained performances.

12.3.1.4 Mechanical Tests Design advances require design validations and verifications. For that purpose, special mechanical tests must be carried out when the blades are not operating normally in the wind turbine. Therefore, concepts such as modal testing, nondestructive tests, and full-scale tests on prototype blades can detect problems before they occur and contribute to preventive designs yielding resolution initiatives.

12.3.1.5 Production Processes and Manufacturing The main problems derived from the production processes of blades are the defects and imperfections. In general, these defects reduce the strength and lifetime of the blades. In structural designs, these imperfections are considered by using safety margins and oversizing, but unfortunately, this reduces the economic competitiveness of the product.

Consequently, better understanding of what can influence production defects is an important matter of innovation for a more sustainable blade design.

12.3.2 Power Train Technology

With the same philosophy as in the previous sections, there are many advances in power electronics and AC machines that can contribute to more efficient and sustainable wind energy generation. This section tries to summarize the most remarkable technological aspects of power electronics and drives, dividing the exposition into six concepts:

- Gear–generator–converter
- Offshore energy evacuation
- Power quality, grid support, and energy storage concepts
- Innovative wind turbine

- New generator designs
- New converter topologies

12.3.2.1 Gear–Generator–Converter Concept From the electronic and electric point of view, the most important element of the wind turbine generation system is the gear–generator–converter system. These three elements are closely related to each other and, in general, the design of each of these three elements is carried out in a coordinated fashion, considering individual and overall characteristics. Thus, for instance, if a design with a specific generator is going to be used, the converter and the gearbox must be accordingly designed and adapted to the voltage, current, speeds, and so on of the whole system. Of course, in the same way, the mechanical and aerodynamic structure must also be coordinately designed.

On the contrary, for instance, if a gearless system (direct drive) is wanted, the generator must be specially designed to operate at the low speeds of the wind turbine's rotor (in close relation with the blades), and also accordingly the converter must meet the supply necessities of the generator and the connection to the grid.

Hence, this section tries to summarize the emerging most representative gear–generator–converter concepts. Some of them are prototypes of different wind turbine manufacturers, which in the future can supplant today's mostly used DFIM based wind turbines. The new wind turbine concepts summarized in this section are also accompanied by some other more classic concepts already examined in Chapter 1.

It is difficult to know if, in the future, there will be a leading wind turbine concept that be used for all power range, onshore location, and offshore location wind turbines, or, on the contrary, if there will be a diversification of the wind turbine concepts as seems to be occurring right now.

Direct Drive Multipole Wound Rotor Synchronous Generator One example of a direct drive wind turbine is the Enercon concept. The necessity of the gear is avoided with a special generator design based on a wound rotor synchronous generator (annular generator) as shown in Figure 12.3. The main characteristics of this configuration can be summarized as follows:

- The generator is a low speed synchronous generator that must be fed through a rotor and a stator (excited rotor).

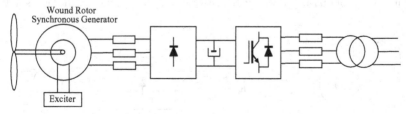

Figure 12.3 Direct drive concept based on multipole wound rotor synchronous generator (a DC-DC converter may be added in the DC-link for lower speeds).

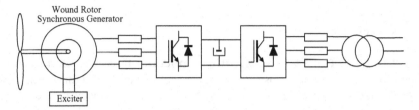

Figure 12.4   Direct drive concept based on multipole wound rotor synchronous generator with back-to-back converter.

- Full power generated by the converter is evacuated by a full scale converter, composed of a diode rectifier, a DC link, and an active front end.
- Low voltage ride through and grid code fulfillment are achieved.

In a similar way, with very similar features, Mtorres has developed a very close concept based on a multipole wound rotor synchronous generator but with a slightly different converter topology (Figure 12.4). In this case, a back-to-back converter is employed to evacuate the generated energy, requiring a different control strategy for the machine.

Direct Drive Permanent Magnet Synchronous Generator   Also with a direct drive concept based on a different generator design, we have the permanent magnet synchronous generator solution. Figure 12.5 shows the gearless turbine concept. The generator operates at low speeds and is fed by a full scale back-to-back converter. There are some manufacturers that have adopted this concept, for instance, GoldWind and General Electric.

Permanent Magnet Synchronous Generator with Different Stage Gearboxes   Another common wind turbine concept allows the generator to operate at medium speed, using a multipole generator and different stage gearboxes. There are several manufacturers that have adopted this configuration (Figure 12.6), resulting in a competitive solution (e.g., Multibrid, Winwind, Vestas, GE Wind, Clipper).

Alternatively, the Gamesa G128–4.5 MW wind turbine concept is also a permanent magnet multipole based generator but incorporates a modular technology solution

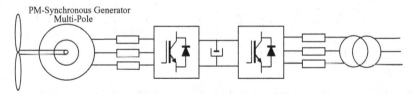

Figure 12.5   Direct drive permanent magnet synchronous generator based wind turbine.

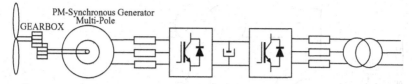

Figure 12.6 Wind turbine based on permanent magnet synchronous generator with different stage gearboxes.

and uses a two-stage gear (Figure 12.7). With different full scale converters, the different stators of the generator are fed. Thanks to its modular design, mounting and transportation are easier, and the reliability is increased.

Medium Voltage Doubly Fed Induction Generator   Another wind turbine is the Acciona medium voltage concept. The generator is specially designed so it can be connected to a medium voltage (12 kV), reducing considerably the size of the transformer (Figure 12.8). In this way, the wind turbine itself presents very similar features to the traditional DFIM based wind turbines, but it needs an specially designed generator.

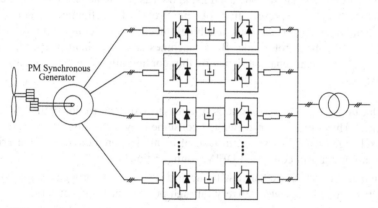

Figure 12.7 Modular permanent magnet synchronous generator based wind turbine. Gamesa concept.

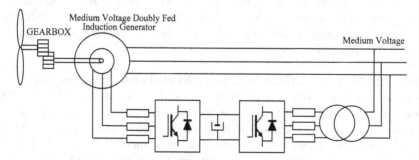

Figure 12.8 Medium voltage doubly fed generator, Acciona concept.

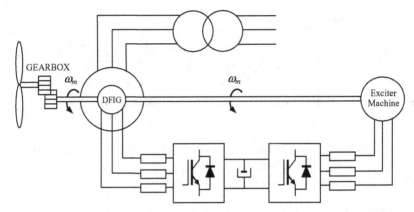

Figure 12.9 Exciter mechanically coupled to the DFIM based wind turbine. Ingeteam concept.

Exciter Mechanically Coupled to the DFIM Several authors have also reported a different configuration based on the DFIM known as the "Kramer drive." Figure 12.9 shows a schematic representation of the proposed configuration. An exciter (permanent magnet synchronous machine) mechanically coupled to the DFIM feeds the VSC of the rotor side. Both machines share mechanical speed, torque, and power. The main characteristics of this configuration can be summarized as follows:

- One of the main objectives of this system is to avoid connection of the VSC to the grid. Thus, the power generated/consumed by the rotor of the DFIM is exchanged with the exciter, instead of being delivered directly to the grid, as in a normal grid connected DFIM configuration.
- In that way, power exchange with the grid is performed through the stator of the DFIM, avoiding power delivery through power electronic converters.
- Thanks to this, the low voltage ride through capability of the wind turbine (reliability against voltage dips) can be improved.
- On the other hand, this system compared to a normal DFIM configuration system, increases the complexity of the machine concept, since two machines are needed instead of only one, increasing also the volume of the system that must be placed in the wind turbine.

DFIM with Grid Side Converter Connected in Series An alternative DFIM based wind turbine system is depicted in Figure 12.10. It is a special DFIM configuration, with the stator of the machine connected to the grid and a transformer connected in series. The transformer is fed by the grid side converter of the DFIM itself. Thanks to this configuration, the stator of the machine is supplied by the addition (vector) of two voltages—the grid voltage and the voltage created by

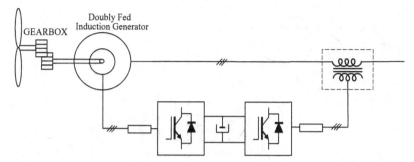

Figure 12.10 DFIM with grid side converter connected in series configuration (w2pS concept).

the grid side converter. The main characteristics of this configuration can be summarized as follows:

- It provides stator voltage control during faulty operation of the grid voltage (transient and steady state), without the need of any additional hardware protection. Thus, if the grid voltage suffers a voltage drop, the grid side converter is in charge of increasing the voltage seen by the stator, still allowing exchange of active and reactive powers according to grid code requirements.
- Equivalently, if the grid voltage suffers any other type of contingency (e.g., imbalance), the contribution created by the grid side voltage tries to mitigate the effect on the stator voltage of the DFIM.
- Thanks to this configuration, it is also possible to maintain the wind turbine connected even with low wind speed, reducing the stator voltage by means of the grid side converter and operating at very low mechanical speed.
- On the other hand, the special transformer requirement slightly complicates the system configuration compared to the normal DFIM connected to the grid.

Reconfigurable Wind Turbine Based on DFIM   Especially for offshore wind farms, where system failure results in a very expensive repair, concepts such as availability, redundancy, and derating become major issues. Thus, the system shown in Figure 12.11, developed by Ingeteam, based on a DFIM, can operate in three different modes:

1. Full converter with asynchronous machine short-circuiting the rotor of the DFIM
2. As a typical DFIM based wind turbine
3. Asynchronous machine directly connected to the grid, without the back-to-back converter and at fixed speed

Therefore, this versatile system, which will also use the parallel connection of several back-to-back converters, is able to operate at different derated modes when a

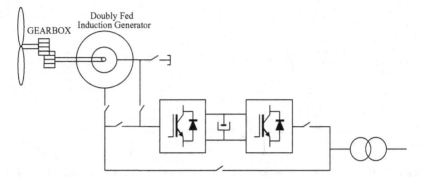

Figure 12.11   Reconfigurable wind turbine based on DFIM (several converters can be placed in parallel).

failure occurs (e.g., one of the converters fails), increasing the availability of the system and reducing the necessity of a total disconnection of the wind turbine.

Double Inverter Fed DFIM   Several authors have reported the possibility to operate a DFIM fed by converters for both the stator and rotor. As illustrated in Figure 12.12, it is possible to use different configurations depending on how the converters are connected. But the basic idea is to be able to control the imposed

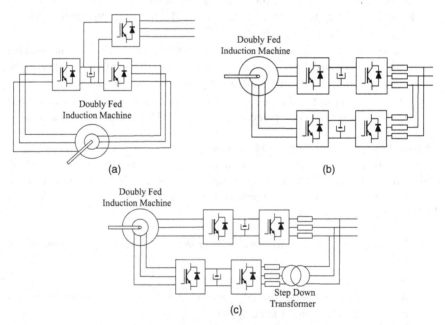

Figure 12.12   Double inverter fed DFIM drive: (a) common front end, (b) two independent converters with the same characteristics, and (c) two independent converters with the same characteristics, feeding a machine with a stator–rotor turn ratio different from 1.

voltage on both the rotor and stator sides (amplitude and frequency), with reversible capacity of power exchange. The main characteristics of these configurations are summarized as follows:

- The relation between the DFIM stator and rotor voltage ratio and the available voltage source converters (voltage, power, etc.), implies different supplying combinations, requiring transformers of different characteristics as shown in Figure 12.12.
- Thanks to these configurations, it is possible to operate the machine in all four quadrants of torque–speed range. Thus, an increase of the speed range can be achieved compared to the grid connected configuration, accompanied with speed reversibility, making this system suitable to work in a wide range of applications.
- In addition, the start-up of the machine can inherently be done by the control system of the normal operation, making this drive more versatile than the grid connected configuration.
- Obviously, these advantages are obtained by increasing the complexity of the supply system by introducing more power electronics.
- Probably, this is not a competitive solution for wind energy generation applications but it could be useful in some drive applications.

12.3.2.2 Offshore Energy Evacuation
One of the particularities of the offshore wind farms is that the generated energy must be transported by submarine cables to onshore. These cables (Figure 12.13) present a parasitic

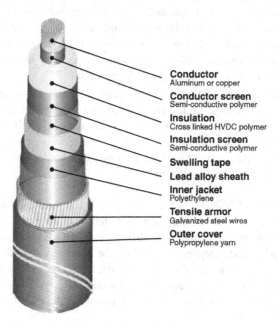

Conductor
Aluminum or copper

Conductor screen
Semi-conductive polymer

Insulation
Cross linked HVDC polymer

Insulation screen
Semi-conductive polymer

Swelling tape

Lead alloy sheath

Inner jacket
Polyethylene

Tensile armor
Galvanized steel wires

Outer cover
Polypropylene yarn

Figure 12.13   Schematic of 220 kV submarine cables. (*Source*: ABB.)

capacitance much more significant than land cables. For a given active power (active current) that must be transmitted through the cable, there is an additional reactive current circulating through the cable (reactive current through the capacitance of the cable), yielding a significant reduction of the active power that can be effectively transmitted through this cable, without reaching its maximum current limit.

Therefore, as mentioned in Section 12.2.1.1, there are two possibilities for transporting the generated energy onshore through submarine cables: AC transmission (assuming the strong capacitive character of the cables) or DC transmission (assuming two stages of conversion AC/DC and DC/AC, by means of power electronics).

AC Transmission (HVAC) When this solution is adopted, the AC energy generated by the offshore wind turbines is transmitted onshore by AC cables. Under this situation, the most challenging characteristics can be distinguished as follows:

- Design and evaluation of the transmission layout. The choice of the submarine cables (voltage, distance from coast, number of clusters, characteristics of the offshore and onshore substations, etc.), together with the selection of an efficient compensation of the capacitance in the cables through passive elements (inductances in general), is an important issue that must be adapted to each wind farm.

- On the other hand, the capacitive character of the submarine cables not only produces difficulties with the static energy transmission, but also is susceptible to resonant problems due to the combination of inductive elements (grid side filters of wind turbines, transformers, and compensation inductances) and capacitive elements (submarine cables). Hence, a combination of passive and active filters with optimized size and location is necessary for proper and safe energy transport to onshore. In addition, a detailed frequency model predicting the behavior of the designed layout is very important.

- Finally, grid code compliance of AC offshore wind farms, especially their behavior under voltage faults, is an issue that can require additional power electronics (reactive power compensators, crowbars, damping elements, etc.) located either in the offshore and onshore substations or even at each wind turbine itself. Thus, to meet the LVRT requirements, adequate designs of these compensators must be introduced.

DC Transmission (HVDC) When this solution is adopted, the AC energy generated by offshore wind turbines is transmitted onshore by DC cables. For that purpose, power electronic converters allowing AC/DC conversion and DC/AC conversion are necessary. Depending on how the energy is generated by the wind turbines, it is possible that the offshore converter could be passive (based only on diodes). However, in a realistic general case, the onshore converter must be an active front end, so the active and reactive powers delivered to the PCC (point of common coupling) can be controlled.

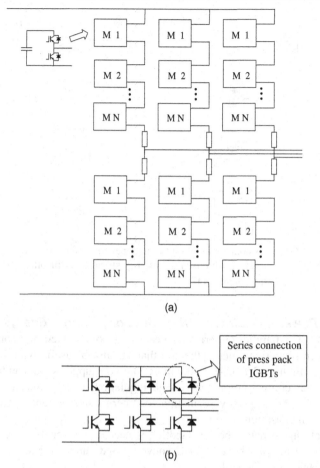

(a)

(b)

Figure 12.14   (a) MMC VSC topology. (b) Two-level VSC based on a series connection of IGBTs.

Thus, although the converter topology used in the past for similar applications (HVDC) was mainly based on thyristor valves, today, the tendency has moved to VSC, which provides more control capacity for the transmission, based mostly on IGBTs. In this way, the high voltage transmission levels (hundreds of kilovolts) require that the converter topology must be prepared to operate with a high number of semiconductors to reach those voltage levels. Today, two leading technologies are very well positioned, the MMC (modular multilevel VSC) shown in Figure 12.14a and the two-level VSC based on a series connection of IGBTs (Figure 12.14b). In both cases, there are still efforts required to adapt this topology to the grid codes for grid integration, as well as to achieve a more efficient, cost effective, and reliable transmission option. Nevertheless, it seems that the most important manufacturers are tending to the modular multilevel converter solution rather than the series connections of IGBTs, based on the classic two-level converter.

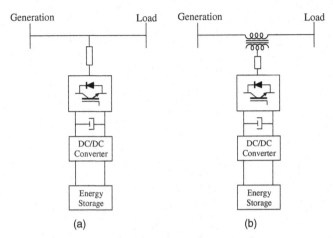

Figure 12.15 (a) Series connection of multifunctional converter. (b) Parallel connection of multifunctional converter (In both cases the energy storage is optional and depends on the functionality of the converter).

12.3.2.3 Power Quality, Grid Support, and Energy Storage Concepts
The parallel or series connection of power electronic converters can provide several functionalities to the grid that can also be useful with the growth of wind energy (Figure 12.15). Thus, the most common supports that can be found are reactive power compensation for voltage regulation, energy management active power storage for frequency regulation, or energy support and active filter for harmonics compensation.

Note that only if active power handling is needed does energy storage become necessary, with its associated DC/DC converter and battery (or alternative energy storage) arrangement technology.

Therefore, in scenarios with close interaction of wind energy generation combined with some other energy resources and loads of different characteristics (harmonic consumption, variability, etc.), we can conclude that multifunctional converters can provide the necessary support for grid stability and efficient operation. Consequently, the design of the converters and their control and operation represent a big challenge for increasing development of wind farms and integrating them into the grid.

12.3.2.4 New Generator Designs
As seen in this chapter and in Chapter 1, several generator concepts can be found in wind turbine technology. In general, the most generalized ones can be grouped as:

- *Asynchronous Generators.* Doubly fed at different voltages and with a cage induction rotor (currently less used).
- *Synchronous Generators.* Both permanent magnet at medium and high speeds, and multipole externally excited and with permanent magnet at low speed, for direct drive.

In general, strong efforts must be made to improve the designs and to introduce newer generator concepts. This section covers possible future generator improvements and designs valid for wind turbine technology. Nevertheless, for already existing generator concepts, the margin for improvement in designs can simply be summarized into the following aspects (all of them interrelated):

- Improve the power and torque densities (torque and power capacity for given longitude and diameter of the machine).
- Reduce weights, diameters, longitudes, and dimensions in general.
- Reduce the costs.
- Improve the magnet designs (deterioration, sensitivity to high temperature, etc.).
- Improve the efficiency and energy yield.

Permanent Magnet Synchronous Generators   Probably, the most important drawbacks of direct drive (or low speed and high torque) generators can be identified as their cost (together with an uncertain availability of magnets now a days) and their large diameter. In low speed (multipole) permanent magnet synchronous generators, the most common topology is the radial flux concept.

However, there also exist some alternative generator designs known as the transverse flux and axial flux concepts. In a generic classification of permanent magnet generators, it is possible to find some other concepts as well. Figure 12.16 illustrates the most promising permanent magnet generator topologies for wind turbines. From these three types, radial flux generators have mostly been used for high power applications; however, in the future, improved designs based on these three topologies can be expected—reducing losses, increasing torque–power densities, and simplifying construction aspects.

Brushless Doubly Fed Generator   Some authors have also reported the possibility of using other machine concepts, but not looking at higher power wind turbines. Figure 12.17 shows the brushless doubly fed generator. This wind turbine concept is very close to the classical doubly fed induction machine based wind

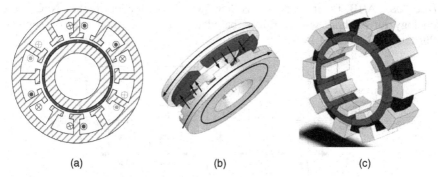

(a)   (b)   (c)

Figure 12.16   Permanent magnet synchronous generators: (a) radial flux generator with concentrated windings, (b) axial flux generator, and (c) transversal flux generator.

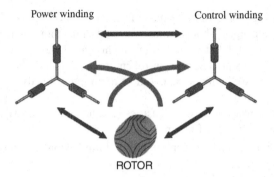

Figure 12.17 Brushless doubly fed generator concept.

turbine, but instead of a wound rotor, this machine has a control three phase winding of lower power in the stator (double stator). The major advantage of this machine compared to the classic DFIM is that it avoids the necessity of the brushes; however, a main disadvantage is that its complexity leads to a more complex control than in other type of machines.

12.3.2.5 New Converter Topologies   Although in the past, the most commonly used converter topology was the two-level VSC, higher power wind turbines are demanding greater voltage operation, other than the 690 V from the past. This enabled the introduction of different converter topologies, such as the multilevel converters, which are more suitable for medium voltage. Thus, topologies that have commonly been used in medium voltage drives for high power applications are now a realistic solution also for higher power wind turbines.

Consequently, there exists the possibility not only to adapt known converter topologies to wind power generation applications, but also to use emerging topologies in order to achieve a more efficient and reliable power conversion and control of the wind turbines.

Medium Voltage IGCT Based 3L-NPC Full Scale Converter for Offshore Wind Turbines   ABB, for instance seems that, has started to adopt the 3L-NPC VSC topology for offshore wind turbines, based on a full scale converter concept as illustrated in Figure 12.18.

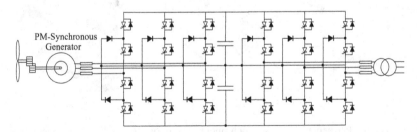

Figure 12.18 Medium voltage IGCT based 3L-NPC wind turbine.

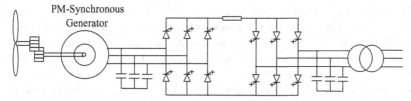

Figure 12.19   CSC based wind turbine.

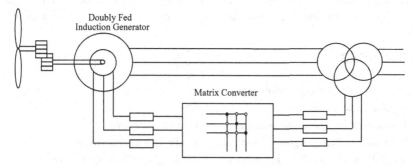

Figure 12.20   Matrix converter based wind turbine.

Current Source Converter Based PMSG Wind Turbine   There have been scientific efforts to test the current source converter topology in wind energy generation applications. This solution, based on SGCTs, is also commonly employed in medium voltage drives (Figure 12.19).

Matrix Converter Based DFIM Wind Turbine   Leading companies such as Vestas have proposed innovative converter designs such as matrix topologies. See Figure 12.20.

12.4 SUMMARY

This chapter has outlined the most relevant future trends in wind energy generation, involving a wide perspective of concerns. It has been structured: identifying first the general improvements and objectives that wind energy generation must address in the future. Then, many technological advances and possibilities that can contribute to achieve those objectives have been presented and discussed.

   It is anticipated that, as a result of the wide range of emerging technologies, there will not be a unique gear–generator–converter technology for wind turbines in the near future, and manufacturers will tend to diversify the wind turbine solutions, establishing their own particular solutions, in many cases very close to one another, but probably still combining the direct drive solution with a full scale converter, the geared full scale converter solution, and the geared doubly fed based solution with reduced scale converter.

However, future wind turbine technology must also innovate the mechanical structure of the wind turbine itself, guarantee increasing penetration into the market with efficient and reliable grid integration, widen the location of wind turbines by developing their capabilities (low wind, offshore, etc.), and increase diversification while paying attention to environmental concerns.

Consequently, in order to optimize wind turbine technology in a more efficient and reliable way, it is necessary to coordinate the innovation efforts, yielding a more integrated design. At the same time the application itself must be considered, together with all the elements of the wind turbine: the electric generator, the converter, the gear, the mechanical structure, the aerodynamics, the location, grid integration, and so on.

REFERENCES

1. http://www.wind-energy-the-facts.com/.
2. http://www.windplatform.eu/.
3. European Wind Energy Association, available at www.ewea.org.
4. International Energy Agency Wind, available at http://www.ieawind.org/.
5. World Wind Energy Association (WWEA), available at www.wwindea.org.
6. Global Wind Energy Council, available at http://www.gwec.net.
7. Spanish Wind Energy Association (AEE), available at www.aeeolica.org.
8. American Wind Energy Association, available at http://www.awea.org/.
9. EPE Wind Energy Chapter (EPE-WE), available at http://www.epe-association.org.
10. Sandia National Laboratory Wind Energy Technology, available at http://www.sandia.gov/wind/links.htm.
11. Renewable Energy database, available at http://re-database.com/.
12. Danish Wind Industry Association, available at http://www.talentfactory.dk.
13. http://www.thewindpower.net.
14. R. Saikki, "Technology Trends of Wind Power Generators," Nordic Conference, Stockholm, September 2009.
15. J. Moya, "Future Trends and Needs in the Wind Power Generation," Gamesa Corporation, Pennsylvania Wind Energy Symposium 2008.
16. S. Butterfield, "Technology Overview and Challenges," NREL, Pennsylvania Wind Energy Symposium 2008.
17. B. Hayman, J. Wedel-Heinen, and P. Brondsted, "Materials Challenges in Present and Future Wind Energy," *MRS Bull.*, Vol. 33, pp. 343–353, April 2008.
18. "Wind Power and the Environment—Benefits and Challenges," available at www.ewea.org.
19. www.wilkipedia.com.
20. http://www.ingeteam.com.
21. http://www.ingeconcleanpower.com/.
22. http://w2ps.es/.
23. B. K. Bose, *Modern Power Electronics and AC Drives*. Prenctice Hall, 2002.
24. Y. Kawabata, E. Ejiogu, and T. Kawabata, "Vector-Controlled Double-Inverter-Fed Wound-Rotor Induction Motor Suitable for High-Power Drives," *IEEE Trans. Ind. Appl.*, Vol. 35, pp. 1058–1066, September/October 1999.

25. F. Bonnet, P. E. Vidal, and M. Pietrzak-David, "Dual Direct Torque Control of Doubly Fed Induction Machine," *IEEE Trans. Ind. Electron.*, Vol. 54, No. 5, pp. 2482–2490, October 2007.

26. G. Poddar and V. T. Ranganathan, "Direct Torque and Frequency Control of Double-Inverter-Fed Slip-Ring Induction Motor Drive," *IEEE Trans. Ind. Electron.*, Vol. 51, No. 6, pp. 1329–1337, December 2004.

27. G. Poddar and V. T. Ranganathan, "Sensorless Double-Inverter-Fed Wound-Rotor Induction-Machine Drive," *IEEE Trans. Ind. Electron.*, Vol. 53, No. 1, pp. 86–95, February 2006.

28. J. M. Corcelles, S. Arnaltes, D. Santos, and J. L. R. Amenedo, "Asynchronous Generator with Double Supply," International Patent WO2008/077974.

29. http://www.abb.com/.

30. http://www.siemens.com.

31. http://www.nrel.gov/.

32. G. Ugalde, "Study on Concentrated Windings Permanent Magnet Machines for Direct Drive Applications," Ph.D. thesis, University of Mondragon, 2009.

33. D. Bang, H. Polinder, G. Shrestha, and J. A. Ferreira, "Review of Generator Systems for Direct-Drive Wind Turbines," in *EWEC 2008*.

34. M. R. Dubois, H. Polinder, and J. A. Ferreira, "Comparison of Generator Topologies for Direct-Drive Wind Turbines," *Proc. Nordic Countries Power and Industrial Electronic Conference Dalborg*, 2000, pp. 22–26.

35. G. Shrestha, H. Polinder, D. J. Bang, and J. A. Ferreira, "Review of Energy Conversion System for Large Wind Turbines," in *EWEC 2008*.

36. I. Sarasola, J. Poza, M. A. Rodriguez, and G. Abad, "Direct Torque Control Design and Experimental Evaluation for the Brushless Doubly Fed Machine," *Energy Conversion and Management*, Vol. 52, No. 2, pp. 1226–1234, February 2011.

37. J. Dai, D. Xu, and B. Wu, "A Novel Control Scheme for Current-Source-Converter-Based PMSG Wind Energy Conversion Systems," *IEEE Trans. Power Electron.*, Vol. 24, No. 4, pp. 963–972, April 2009.

38. A. V. Rebsdorf and L. Helle, "Variable Wind Speed Turbine Having a Matrix Converter," U.S. Patent, 6566764. May 2003.

39. J. Mayor et al., "Wind Turbine Method Operation and System". US Patent, 0057446. March 2011.

List of Wind Turbine Manufacturers

Manufacturer	Web page
AAER (Canada)	http://www.pioneerwindenergy.com/
Acciona (Spain)	http://www.acciona-energia.es
ACSA (Spain)	http://www.acsaeolica.com
Alstom-Ecotecnia (France, Spain)	http://www.alstom.com/power
Areva Wind (France, Germany)	http://www.areva-wind.com/
AMSC (USA)	http://www.amsc.com/products/applications/enewable/
Aeronautica Windpower (USA)	http://aeronauticawind.com
Alizeo (France)	http://www.groupe-alizeo.com
Avantis (China, Germany, Austria)	http://www.avantis-energy.com

(Continued)

(*Continued*)

Manufacturer	Web page
AWE (Canada)	http://www.awe-wind.com/
Bard (Germany)	http://www.bard-offshore.de/
Blaster (Norway)	http://www.blaaster.no
China Creative Wind Energy (China)	http://ccwewind.com
Clipper (UK)	http://www.clipperwind.com
Cwel (India)	http://www.cwel.in
Darwind (Netherland, China)	http://www.xemc-darwind.com/
Ddis (France)	http://www.ddiswt.com
Doosan (South Korea)	http://www.doosanheavy.com/
Dong Fang (China)	http://www.dongfang.com.cn
Elecon Engineering (India)	http://www.elecon.com
Elsewedy (Egipt)	http://www.elsewedyelectric.com/
Enercon (Germany)	http://www.enercon.de
Envision (China)	http://www.envisioncn.com
Eviag (Germany)	http://www.eviag.com/
EWT (Netherland)	http://www.ewtinternational.com/
Fuhrlander (Germany)	http://www.fuhrlaender.de/
Gamesa (Spain)	http://www.gamesacorp.com
GC China Turbine Corporation (China)	http://www.gcchinaturbine.com
Ghodwat (India)	http://www.ghodawatenergy.com
Hewind (China)	http://www.hewind.com
IMPSA (Argentina)	http://www.impsa.com.ar/
Mitsubishi (Japan)	http://www.mhi.co.jp/
Mingyang (China)	http://www.mywind.com.cn
GE Energy (USA)	http://www.ge-energy.com/
Hyundai (South Korea)	http://www.hyundai-elec.com
Hyosung (South Korea)	http://www.hyosung.com
Jeumont (France)	http://www.jeumontelectric.com
Leitwind (Italia)	http://www.leitwind.com
Mervento (Findland)	http://www.mervento.fi
Moncada Energy Group (Italy)	http://www.moncadaenergy.com/
Mtorres	http://www.mtorres.es/
Multibrid	http://www.multibrid.com
Nordex (Germany)	http://www.nordex-online.com/
Northern Power (USA)	http://www.northernpower.com/
Norwin (Denmark)	http://www.norwin.dk/
RePower (Germany)	http://www.repower.de/
Samsung (South Korea)	http://www.shi.samsung.co.kr/
Sany (China)	http://www.sanyse.com.cn
Shanghai Electric Company (China)	http://www.shanghai-electric.com
Schuler (Germany)	http://www.schulergroup.com
Siemens (Germany)	http://www.energy.siemens.com/
Sinovel (China)	http://www.sinovel.com
Statoil (Norway)	http://www.statoil.com
Subaru (Japan)	http://www.fhi.co.jp
Suzlon (India)	http://www.suzlon.com
Vensys (Germany)	http://www.vensys.de/
Vestas (Denmark)	http://www.vestas.com
Unison (South Korea)	http://www.unison.co.kr/
W2E (Germany)	http://www.w2e-rostock.de/
Wikow (Czech republic)	http://www.wikov.com
Windflow (New Zeland)	http://www.windflow.co.nz/
WinWinD (Findland)	http://www.winwind.fi/
Wind Direct (Austria)	http://www.windtec.at
Zephyros (Netherlands)	http://www.peeraer.com/zephyros/

Appendix

A.1 SPACE VECTOR REPRESENTATION

Space vector notation is a commonly extended tool that can be applied in AC machines to represent the flux, voltage, and current magnitudes in a compact manner. By using the space vector representation, it is possible to derive models and obtain the differential equations representing their behavior, in a simpler way than using the classic three-phase representation. Space vector notation is used in many analyses in this book.

A.1.1 Space Vector Notation

The three phase magnitudes representing the system ideally can be written

$$x_a = \hat{X}\cos(\omega t + \phi)$$
$$x_b = \hat{X}\cos(\omega t + \phi - 2\pi/3) \qquad (A.1)$$
$$x_c = \hat{X}\cos(\omega t + \phi + 2\pi/3)$$

with constant angular frequency ω, amplitude $\hat{X}$, and constant phase shift ϕ. This balanced three-phase system can be represented in a plane, as a space vector $\vec{x}$, that rotates at ω angular speed across the origin of the three axes $\vec{a}, \vec{b}, \vec{c}$ spatially shifted 120°, as shown in Figure A.1. The projection of the rotating space vector $\vec{x}$ on each of the axes provides the instantaneous magnitudes x_a, x_b, x_c. The axes are defined as follows:

$$\vec{a} = 1 \qquad (A.2)$$

$$\vec{b} = e^{j(2\pi/3)} \qquad (A.3)$$

$$\vec{c} = e^{j(4\pi/3)} \qquad (A.4)$$

Doubly Fed Induction Machine: Modeling and Control for Wind Energy Generation,
First Edition. By G. Abad, J. López, M. A. Rodríguez, L. Marroyo, and G. Iwanski.
© 2011 the Institute of Electrical and Electronic Engineers, Inc. Published 2011 by John Wiley & Sons, Inc.

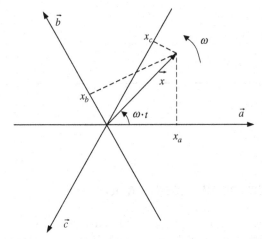

Figure A.1 Space vector representation of axes $\vec{a}, \vec{b}, \vec{c}$.

Thus, mathematically, the space vector can be expressed as

$$\vec{x} = |\vec{x}| e^{j(\omega t + \phi)} \tag{A.5}$$

with the amplitude of the space vector equal to the amplitude of the three-phase magnitudes:

$$|\vec{x}| = \hat{X} \tag{A.6}$$

On the basis of the space vector notation, the three phase magnitudes may be alternatively represented by the same rotating space vector, by two phase magnitudes (x_α and x_β) in the real–imaginary complex plane, as illustrated in Figure A.2. In this case, the projections of the rotating space vector on the $\alpha\beta$ axes provide the two phase magnitudes x_α and x_β. This fact can be represented

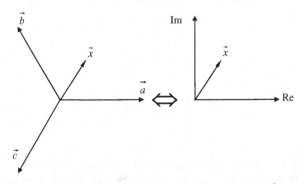

Figure A.2 Space vector representation in stationary $\alpha\beta$ axes.

mathematically as

$$\vec{x} = x_\alpha + jx_\beta = \tfrac{2}{3}(x_a + ax_b + a^2x_c) \tag{A.7}$$

where

$$a = e^{j(2\pi/3)} \tag{A.8}$$

The constant 2/3 of expression (A.7) is chosen to scale the space vectors according to the peak amplitude of the three phase magnitudes, that is, to give relation (A.6).

The $\alpha\beta$ components of the space vector can be calculated from the abc magnitudes as follows:

$$x_\alpha = \text{Re}\{\vec{x}\} = \tfrac{2}{3}\left(x_a - \tfrac{1}{2}x_b - \tfrac{1}{2}x_c\right) \tag{A.9}$$

$$x_\beta = \text{Im}\{\vec{x}\} = \frac{1}{\sqrt{3}}(x_b - x_c) \tag{A.10}$$

These last two expressions are commonly represented in matrix form:

$$\begin{bmatrix} x_\alpha \\ x_\beta \end{bmatrix} = \frac{2}{3} \begin{bmatrix} 1 & -\dfrac{1}{2} & -\dfrac{1}{2} \\ 0 & \dfrac{\sqrt{3}}{2} & -\dfrac{\sqrt{3}}{2} \end{bmatrix} \cdot \begin{bmatrix} x_a \\ x_b \\ x_c \end{bmatrix} \tag{A.11}$$

with the T matrix often called the Clarke direct transformation:

$$T = \frac{2}{3} \begin{bmatrix} 1 & -\dfrac{1}{2} & -\dfrac{1}{2} \\ 0 & \dfrac{\sqrt{3}}{2} & -\dfrac{\sqrt{3}}{2} \end{bmatrix} \tag{A.12}$$

This fact can be represented schematically as in Figure A.3.

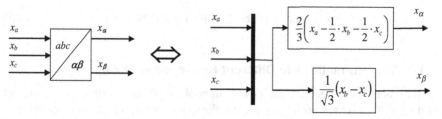

Figure A.3 $\alpha\beta$ Components calculation from abc components.

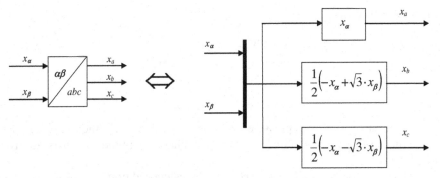

Figure A.4 *abc* Components calculation from $\alpha\beta$ components.

In addition, the inverse Clarke transformation gives the inverse relation:

$$\begin{bmatrix} x_a \\ x_b \\ x_c \end{bmatrix} = \begin{bmatrix} 1 & 0 \\ -\dfrac{1}{2} & \dfrac{\sqrt{3}}{2} \\ -\dfrac{1}{2} & -\dfrac{\sqrt{3}}{2} \end{bmatrix} \cdot \begin{bmatrix} x_\alpha \\ x_\beta \end{bmatrix} \qquad (A.13)$$

graphically represented in Figure A.4.

It is necessary to highlight the following:

- x_α and x_β are sinusoidal varying magnitudes.
- The amplitudes of x_α and x_β are equal to the amplitudes of x_a, x_b, and x_c.
- x_α and x_β are 90° phase shifted.
- x_α is equal to x_a.

θ is the angular position that can be calculated from the angular frequency ω as

$$\theta = \int \omega \, dt = \omega t \qquad (A.14)$$

Note that the following can be written:

$$\vec{x} = x_\alpha + jx_\beta = |\vec{x}| \cos(\omega t + \phi) + j|\vec{x}| \sin(\omega t + \phi) \qquad (A.15)$$

A.1.2 Transformations to Different Reference Frames

On the other hand, as seen in Chapter 4 of this book, for developing the dynamic model of the DFIM, it is very useful to represent the space vectors in different rotatory and stationary reference frames. Hence, three phase magnitudes of the DFIM (fluxes, current, and voltages) are represented with the space vector notation, but in different reference frames.

In order to denote that one space vector is referred to one specific reference frame, the superscript notation is introduced. Three different reference frames can be distinguished:

1. *The Stator Reference Frame (α-β).* Aligned with the stator, the rotating speed of the frame is zero (stationary), and the space vector referred to it rotates at the synchronous speed ω_s.

$$\vec{x}^s = x_\alpha + jx_\beta \qquad (A.16)$$

   The *s* superscript denotes space vectors referred to the stator reference frame.

2. *The Rotor Reference Frame (D-Q).* Aligned with the rotor, the rotating speed of the frame is the electric angular speed of the rotor ω_m, and the space vector referred to it rotates at the slip speed ω_r.

$$\vec{x}^r = x_D + jx_Q \qquad (A.17)$$

   The *r* superscript denotes space vectors referred to the rotor reference frame.

3. *The Synchronous Reference Frame (d-q).* The rotating speed of the frame is the synchronous speed ω_s, and the space vector referred to it does not rotate, that is, it presents constant real and imaginary parts.

$$\vec{x}^a = x_d + jx_q \qquad (A.18)$$

   The *a* superscript denotes space vectors referred to the synchronous reference frame.

Figure A.5 shows the three different reference frame representations.

Note that the angular frequency of the reference frame can be different from the angular frequency of the magnitudes being represented in space vector notation.

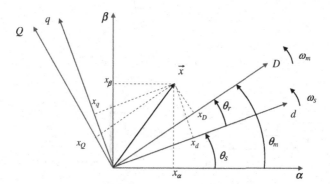

Figure A.5 Space vector representation in different reference frames.

For referencing space vectors into different reference frames, the rotational transformation is used. So, for instance, to transform from $\alpha\beta$ coordinates to DQ coordinates, the following operation is used:

$$\vec{x}^r = \begin{bmatrix} x_D \\ x_Q \end{bmatrix} = \begin{bmatrix} \cos\theta_m & -\sin\theta_m \\ \sin\theta_m & \cos\theta_m \end{bmatrix} \cdot \begin{bmatrix} x_\alpha \\ x_\beta \end{bmatrix} \qquad (A.19)$$

where θ_m is the electric angular position of the shaft,

$$\theta_m = \int \omega_m dt = \omega_m \cdot t \qquad (A.20)$$

we use M as the direct rotational transformation:

$$M = \begin{bmatrix} \cos\theta_m & -\sin\theta_m \\ \sin\theta_m & \cos\theta_m \end{bmatrix} \qquad (A.21)$$

Hence, the inverse rotational transformation is defined with the matrix

$$M^{-1} = \begin{bmatrix} \cos\theta_m & \sin\theta_m \\ -\sin\theta_m & \cos\theta_m \end{bmatrix} \qquad (A.22)$$

In addition, note also that the next relation holds:

$$\vec{x}^r = e^{-j\theta_m}\vec{x}^s \qquad (A.23)$$

or

$$\vec{x}^s = e^{j\theta_m}\vec{x}^r \qquad (A.24)$$

Graphically, this transformation of coordinates can be represented as illustrated in Figures A.6 and A.7.

Finally, the relations with the synchronous reference frame are

$$\vec{x}^a = e^{-j\theta_s}\vec{x}^s \qquad (A.25)$$

or

$$\vec{x}^a = e^{-j\theta_r}\vec{x}^r \qquad (A.26)$$

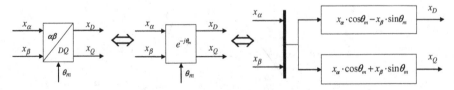

Figure A.6 DQ Components calculation from $\alpha\beta$ components.

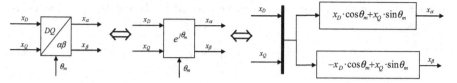

Figure A.7 $\alpha\beta$ components calculation from DQ components.

with

$$\theta_r = \int \omega_r dt = \omega_r t, \qquad \theta_s = \int \omega_s dt = \omega_s t \tag{A.27}$$

θ_r, θ_s, ω_r, and ω_s are as described in Chapter 4.

A.1.3 Power Expressions

When considering voltage and currents as magnitudes represented in space vector notation, the active and reactive powers can be calculated according to established expressions. Therefore, the apparent power (S) is defined as the complex power formed by the reactive power (Q) and active power (P):

$$S = P + jQ \tag{A.28}$$

Thus, by using the space vector notation, the apparent power is calculated as follows:

$$S = \tfrac{3}{2}(\vec{v} \cdot \vec{i}^*) \tag{A.29}$$

where the superscript $*$ represents the conjugate of a space vector found as

$$\vec{x} = x_\alpha - jx_\beta \tag{A.30}$$

Consequently, the apparent power is

$$\begin{aligned} S &= \tfrac{3}{2}\left[(v_\alpha + jv_\beta)(i_\alpha - ji_\beta)\right] \\ &= \tfrac{3}{2}\left[(v_\alpha i_\alpha + i_\beta v_\beta) + j(v_\beta i_\alpha - v_\alpha i_\beta)\right] \end{aligned} \tag{A.31}$$

Accordingly, the active and reactive powers are calculated as

$$P = \tfrac{3}{2}(v_\alpha i_\alpha + i_\beta v_\beta) \tag{A.32}$$

$$Q = \tfrac{3}{2}(v_\beta i_\alpha - v_\alpha i_\beta) \tag{A.33}$$

Note that the term $\tfrac{3}{2}$ is necessary in the power expressions in order to maintain the correspondence of powers from the *abc* coordinates and the $\alpha\beta$ coordinates. On the other hand, equivalent expressions can be derived in dq and DQ coordinates as follows:

$$P = \tfrac{3}{2}(v_D i_D + i_Q v_Q) \tag{A.34}$$

$$Q = \tfrac{3}{2}(v_Q i_D - v_D i_Q) \tag{A.35}$$

$$P = \tfrac{3}{2}(v_d i_d + i_q v_q) \tag{A.36}$$

$$Q = \tfrac{3}{2}(v_q i_d - v_d i_q) \tag{A.37}$$

A.2 DYNAMIC MODELING OF THE DFIM CONSIDERING THE IRON LOSSES

The dynamic models developed in Chapter 4 of this book do not take into account the core loss that may be present in the DFIM. In this section, the $\alpha\beta$ and the dq model differential equations of the DFIM are developed considering the iron losses. Traditionally, most of the models have ignored this phenomenon causing in certain cases inaccuracies in the control strategies based on these models. Consequently, in order to develop a model as close as possible to the real machine, the iron loss will be introduced in the machine's model equations. In general, this analysis is oriented to low power machines, since they present more significant power loss than higher power machines.

In general, when a DFIM presents significant iron losses and the control strategy that is driving the machine does not consider these losses, the following errors are committed:

- There is an error in the estimated flux (rotor or stator), since part of the stator and rotor currents are delivered through the iron losses but do not create flux, while the estimator used by the control considers that all the current is creating the flux. This error is revealed as phase shift and amplitude variations between the real and estimated flux.
- If there is an error in the estimated flux, there is an error in the orientation based on this flux, necessary for vector control (dq reference frame alignment) or direct control strategies (sector calculation for vector injection in DTC or DPC).
- This orientation error provokes one more undesired effect:
  o In vector control based techniques, there is an additional decoupling between the d and q axes, producing an accuracy error in the controlled torque.

○ In direct control techniques, in addition to the problems affecting vector control techniques, it can provoke time intervals where the injected vectors are not appropriate, because the flux is supposed to be in the wrong sector. This normally leads to torque and flux oscillations. Note that this effect only would occur in the neighborhood of a sector change.

• These facts can be more significantly felt under certain operating conditions of the machine, especially at torques/powers far from the rated (low) ones.

• Finally, it can be concluded that the iron losses present in the machine but not considered in control can lead to slight deteriorations and inaccuracies that in general do not produce instability, but they are responsible for control performance degradations.

Therefore, this section gives the basis for considering the losses of the DFIM at the modeling stage. In that way, based on this DFIM model and testing it with the studied control strategies by means of a simulation based evaluation, it could be possible to assess if there are going to be problematic situations due to the iron losses. This gives the designer the opportunity to provide corrective actions. However, at this stage, no solutions of control considering iron losses are provided.

A.2.1 $\alpha\beta$ Model

In this section, the model of the DFIM is developed using the space vector representation in the stator reference frame and considering the iron loss. The voltage equations of the typical DFIM model are still valid:

$$\vec{v}_s^s = R_s \vec{i}_s^s + \frac{d\vec{\psi}_s^s}{dt} \tag{A.38}$$

$$\vec{v}_r^s = R_r \vec{i}_r^r + \frac{d\vec{\psi}_r^s}{dt} - j\omega_m \vec{\psi}_r^s \tag{A.39}$$

The iron loss of the machine is modeled as a resistance in parallel to the mutual magnetizing inductance of each phase, as shown in Figure A.8.

This resistance R_{fe} provokes an active current consumption (i_{fe}) but does not create flux. The flux is created by the current flow through the mutual magnetizing and leakage inductances. Hence, the flux expressions in the stator reference frame are given by

$$\vec{\psi}_s^s = L_{\sigma s} \vec{i}_s^s + L_m \vec{i}_m^s \tag{A.40}$$

$$\vec{\psi}_r^s = L_m \vec{i}_m^s + L_{\sigma r} \vec{i}_r^s \tag{A.41}$$

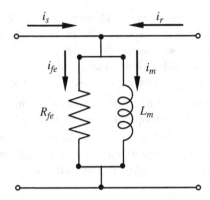

Figure A.8 Iron loss modeling at one of the *abc* phases.

Additionally, from the current and voltage relations, two new expressions are added to the model:

$$\vec{i}_s^s + \vec{i}_r^s = \vec{i}_{fe}^s + \vec{i}_m^s \tag{A.42}$$

$$R_{fe}\vec{i}_{fe}^s = L_m\frac{d\vec{i}_m^s}{dt} \tag{A.43}$$

Therefore, by means of all derived equations, Figure A.9 shows the electrical model of the DFIM in stator coordinates considering the iron loss.

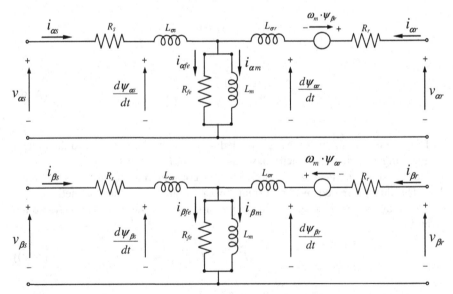

Figure A.9 $\alpha\beta$ Model of the DFIM in stator coordinates considering the iron loss.

On the other hand, the instantaneous power transmitted through the rotor and the stator is calculated as follows:

$$p_s(t) = v_{as}(t) \cdot i_{as}(t) + v_{bs}(t) \cdot i_{bs}(t) + v_{cs}(t) \cdot i_{cs}(t)$$
$$= \tfrac{3}{2}(v_{\alpha s} i_{\alpha s} + v_{\beta s} i_{\beta s}) = P_s \qquad (A.44)$$

$$p_r(t) = v_{ar}(t) i_{ar}(t) + v_{br}(t) i_{br}(t) + v_{cr}(t) i_{cr}(t)$$
$$= \tfrac{3}{2}(v_{\alpha r} i_{\alpha r} + v_{\beta r} i_{\beta r}) = P_r \qquad (A.45)$$

Therefore, the total instantaneous power of the machine, $p(t)$, is the sum of the stator and the rotor instantaneous powers. Note that in a balanced three-phase system (with no zero sequence components), the active power (P) is equivalent to the instantaneous power $(p(t))$.

$$p(t) = \tfrac{3}{2}(v_{\alpha s} i_{\alpha s} + v_{\beta s} i_{\beta s}) + \tfrac{3}{2}(v_{\alpha r} i_{\alpha r} + v_{\beta r} i_{\beta r}) \qquad (A.46)$$

By substituting expressions (A.38) and (A.39) into expression (A.46) and rearranging the terms, the total power transmitted to the machine is given by

$$p(t) = \tfrac{3}{2}R_s|\vec{i}_s|^2 + \tfrac{3}{2}R_r|\vec{i}_r|^2$$
$$+\tfrac{3}{2}\mathrm{Re}\left\{ \frac{d\vec{\psi}_s}{dt} \cdot \vec{i}_s^* + \frac{d\vec{\psi}_r}{dt} \cdot \vec{i}_r^* \right\} - \tfrac{3}{2}\mathrm{Re}\left\{ j\omega_m \vec{\psi}_r \cdot \vec{i}_r^* \right\} \qquad (A.47)$$

where, for simplicity, the superscript s of the space vectors has been omitted. Now substituting Equations (A.40) and (A.41) into expression (A.47), in order to remove the flux derivates, it is possible to obtain

$$p(t) = \tfrac{3}{2}R_s|\vec{i}_s|^2 + \tfrac{3}{2}R_r|\vec{i}_r|^2 + \tfrac{3}{2}R_{fe}|\vec{i}_{fe}|^2$$
$$+\tfrac{3}{2}\mathrm{Re}\left\{ L_{\sigma s}\frac{d\vec{i}_s}{dt} \cdot \vec{i}_s^* + L_{\sigma r}\frac{d\vec{i}_r}{dt} \cdot \vec{i}_r^* + L_m\frac{d\vec{i}_m}{dt} \cdot \vec{i}_m^* \right\}$$
$$-\tfrac{3}{2}\mathrm{Re}\left\{ j\omega_m, \vec{\psi}_r \cdot \vec{i}_r^* \right\} \qquad (A.48)$$

Hence, three different terms are distinguished:

$$P_{loss} = \tfrac{3}{2}R_s|\vec{i}_s|^2 + \tfrac{3}{2}R_r|\vec{i}_r|^2 + \tfrac{3}{2}R_{fe}|\vec{i}_{fe}|^2 \qquad (A.49)$$

where P_{loss} represents the resistive active power losses.

$$P_{mag} = \tfrac{3}{2}\mathrm{Re}\left\{L_{\sigma s}\frac{d\vec{i}_s}{dt}\cdot\vec{i}_s^* + L_{\sigma r}\frac{d\vec{i}_r}{dt}\cdot\vec{i}_r^* + L_m\frac{d\vec{i}_m}{dt}\cdot\vec{i}_m^*\right\} \tag{A.50}$$

P_{mag} is the stored magnetic power, and is equal to zero.

$$P_{mec} = -\tfrac{3}{2}\,\mathrm{Re}\left\{j\omega_m\vec{\psi}_r\cdot\vec{i}_r^*\right\} \tag{A.51}$$

P_{mec} is the mechanical power produced by the DFIM, that is, the power stored in the equivalent voltage source $j\omega_m\vec{\psi}_r$, that will be transmitted to the mechanical system by means of the electromagnetic torque. So using this last result, the torque produced by the machine can be calculated from the mechanical power as follows:

$$\begin{aligned}T_{em} = \frac{P_{mec}}{\omega_m/p} &= -\tfrac{3}{2}p\,\mathrm{Re}\left\{j\vec{\psi}_r\vec{i}_r^*\right\}\\ &= -\tfrac{3}{2}p\,\mathrm{Im}\left\{\vec{\psi}_r^*\cdot\vec{i}_r\right\} = \tfrac{3}{2}p(\psi_{\alpha r}i_{\beta r} - \psi_{\beta r}i_{\alpha r})\end{aligned} \tag{A.52}$$

Note that the equivalent torque expressions derived in the model without considering the iron losses (Chapter 4), due to the presence of the term R_{fe}, in this case do not remain.

A.2.2 *dq* Model

In this section, the model differential equations of the DFIM are developed using the space vector representation in the synchronous reference frame and considering core loss. Again, the voltage equations are equivalent to the voltage equations derived from the "classic" model of the DFIM (Chapter 4).

$$\vec{v}_s^a = R_s\vec{i}_s^a + \frac{d\vec{\psi}_s^a}{dt} + j\omega_s\vec{\psi}_s^a \tag{A.53}$$

$$\vec{v}_r^a = R_r\vec{i}_r^a + \frac{d\vec{\psi}_r^a}{dt} + j(\omega_s - \omega_m)\vec{\psi}_r^a \tag{A.54}$$

The rest of the flux and current equations are obtained from expressions (A.40) to (A.43), transforming them to the synchronous reference frame:

$$\vec{\psi}_s^a = L_{\sigma s}\vec{i}_s^a + L_m\vec{i}_m^a \tag{A.55}$$

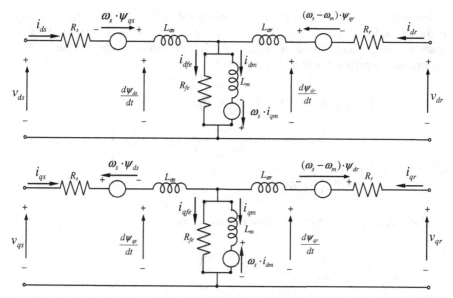

Figure A.10 *dq* Model of the DFIM in synchronous coordinates, considering the iron loss.

$$\vec{\psi}_r^a = L_m \vec{i}_m^a + L_{\sigma r} \vec{i}_r^a \qquad (A.56)$$

$$\vec{i}_s^a + \vec{i}_r^a = \vec{i}_{fe}^a + \vec{i}_m^a \qquad (A.57)$$

$$R_{fe} \vec{i}_{fe}^a = L_m \frac{d\vec{i}_m^a}{dt} + j\omega_s \vec{i}_m^a \qquad (A.58)$$

Figure A.10 shows the electric model of the DFIM in synchronous coordinates, considering the iron loss.

Added to this, the instantaneous power transmitted through the rotor and the stator is calculated as follows, using the expression (A.46) in *dq* coordinates:

$$p(t) = \tfrac{3}{2}(v_{ds}i_{ds} + v_{qs}i_{qs}) + \tfrac{3}{2}(v_{dr}i_{dr} + v_{qr}i_{qr}) \qquad (A.59)$$

Substituting Equations (A.53) and (A.54) in the last expression:

$$p(t) = \tfrac{3}{2}R_s|\vec{i}_s|^2 + \tfrac{3}{2}R_r|\vec{i}_r|^2 + \tfrac{3}{2}\mathrm{Re}\left\{\frac{d\vec{\psi}_s}{dt}\cdot\vec{i}_s^* + \frac{d\vec{\psi}_r}{dt}\cdot\vec{i}_r^*\right\}$$
$$-\tfrac{3}{2}\mathrm{Re}\left\{j\omega_s\vec{\psi}_s\cdot\vec{i}_s^*\right\} - \tfrac{3}{2}\mathrm{Re}\left\{j(\omega_s-\omega_m)\vec{\psi}_r\cdot\vec{i}_r^*\right\} \qquad (A.60)$$

For simplicity, the superscripts have been omitted in this case again. Then, as done in the $\alpha\beta$ model, by substituting in this last equation expressions (A.55) to (A.58), it leads to an equivalent expression of the power in dq coordinates (despite the fact that the original expression is slightly different):

$$p(t) = \tfrac{3}{2}R_s|\vec{i}_s|^2 + \tfrac{3}{2}R_r|\vec{i}_r|^2 + \tfrac{3}{2}R_{fe}|\vec{i}_{fe}|^2$$

$$+\tfrac{3}{2}\mathrm{Re}\left\{ L_{\sigma s}\frac{d\vec{i}_s}{dt}\cdot\vec{i}_s^* + L_{\sigma r}\frac{d\vec{i}_r}{dt}\cdot\vec{i}_r^* + L_m\frac{d\vec{i}_m}{dt}\cdot\vec{i}_m^* \right\}$$

$$-\tfrac{3}{2}\mathrm{Re}\left\{ j\omega_m\vec{\psi}_r\cdot\vec{i}_r^* \right\} \tag{A.61}$$

Consequently, the torque expression is calculated in the same manner:

$$T_{em} = \frac{P_{mec}}{\omega_m/p} = -\tfrac{3}{2}p\ \mathrm{Re}\left\{ j\vec{\psi}_r\cdot\vec{i}_r^* \right\}$$

$$= -\tfrac{3}{2}p\ \mathrm{Im}\left\{ \vec{\psi}_r^*\cdot\vec{i}_r \right\} = \tfrac{3}{2}p(\psi_{dr}i_{qr} - \psi_{qr}i_{dr}) \tag{A.62}$$

A.2.3 State-Space Representation of $\alpha\beta$ Model

As stated in Chapter 4, a representation of the $\alpha\beta$ model in state-space equations is very useful for simulation purposes. Next, one of the different state-space representation combinations will be shown. It employs the state-space magnitudes $\vec{i}_s^s$, $\vec{i}_r^s$, and $\vec{i}_{fe}^s$.

$$\frac{d}{dt}\begin{bmatrix} \vec{i}_s^s \\ \vec{i}_r^s \\ \vec{i}_{fe}^s \end{bmatrix} = \begin{bmatrix} \left[-\dfrac{R_s}{L_{\sigma s}}\right] & 0 & \left[-\dfrac{R_{fe}}{L_{\sigma s}}\right] \\[2ex] \left[j\omega_m\dfrac{L_m}{L_{\sigma r}}\right] & \left[-\dfrac{R_r}{L_{\sigma r}}+j\omega_m\dfrac{L_r}{L_{\sigma r}}\right] & \left[-\dfrac{R_{fe}}{L_{\sigma r}}-j\omega_m\dfrac{L_m}{L_{\sigma r}}\right] \\[2ex] \left[-\dfrac{R_s}{L_{\sigma s}}+j\omega_m\dfrac{L_m}{L_{\sigma r}}\right] & \left[-\dfrac{R_r}{L_{\sigma r}}+j\omega_m\dfrac{L_r}{L_{\sigma r}}\right] & \left[-R_{fe}\left(\dfrac{1}{L_{\sigma s}}+\dfrac{1}{L_{\sigma r}}+\dfrac{1}{L_m}\right)-j\omega_m\dfrac{L_m}{L_{\sigma r}}\right] \end{bmatrix}\cdot\begin{bmatrix} \vec{i}_s^s \\ \vec{i}_r^s \\ \vec{i}_{fe}^s \end{bmatrix}$$

$$+\begin{bmatrix} \dfrac{1}{L_{\sigma s}} & 0 \\[2ex] 0 & \dfrac{1}{L_{\sigma r}} \\[2ex] \dfrac{1}{L_{\sigma s}} & \dfrac{1}{L_{\sigma r}} \end{bmatrix}\cdot\begin{bmatrix} \vec{v}_s^s \\ \vec{v}_r^s \end{bmatrix} \tag{A.63}$$

If each space vector is replaced by its $\alpha\beta$ components, the state-space representation yields

$$
\frac{d}{dt}
\begin{bmatrix}
i_{\alpha s}\\
i_{\beta s}\\
i_{\alpha r}\\
i_{\beta r}\\
i_{\alpha fe}\\
i_{\beta fe}
\end{bmatrix}
=
\begin{bmatrix}
-\dfrac{R_s}{L_{\sigma s}} & 0 & 0 & 0 & -\dfrac{R_{fe}}{L_{\sigma s}} & 0\\[10pt]
0 & -\dfrac{R_s}{L_{\sigma s}} & 0 & 0 & 0 & -\dfrac{R_{fe}}{L_{\sigma s}}\\[10pt]
0 & 0 & -\dfrac{R_r}{L_{\sigma r}} & -\omega_m\dfrac{L_r}{L_{\sigma r}} & \dfrac{R_{fe}}{L_{\sigma r}} & \omega_m\dfrac{L_m}{L_{\sigma r}}\\[10pt]
0 & 0 & \omega_m\dfrac{L_r}{L_{\sigma r}} & -\dfrac{R_r}{L_{\sigma r}} & -\omega_m\dfrac{L_m}{L_{\sigma r}} & \dfrac{R_{fe}}{L_{\sigma r}}\\[10pt]
\dfrac{R_s}{L_{\sigma s}} & \omega_m\dfrac{L_m}{L_{\sigma r}} & \dfrac{R_r}{L_{\sigma r}} & \omega_m\dfrac{L_m}{L_{\sigma r}} & -R_{fe}\!\left(\dfrac{1}{L_{\sigma s}}+\dfrac{1}{L_{\sigma r}}+\dfrac{1}{L_m}\right) & -\omega_m\dfrac{L_m}{L_{\sigma r}}\\[10pt]
-\omega_m\dfrac{L_m}{L_{\sigma r}} & \dfrac{R_s}{L_{\sigma s}} & -\omega_m\dfrac{L_m}{L_{\sigma r}} & \dfrac{R_r}{L_{\sigma r}} & \omega_m\dfrac{L_m}{L_{\sigma r}} & -R_{fe}\!\left(\dfrac{1}{L_{\sigma s}}+\dfrac{1}{L_{\sigma r}}+\dfrac{1}{L_m}\right)
\end{bmatrix}
\begin{bmatrix}
i_{\alpha s}\\
i_{\beta s}\\
i_{\alpha r}\\
i_{\beta r}\\
i_{\alpha fe}\\
i_{\beta fe}
\end{bmatrix}
$$

$$
+
\begin{bmatrix}
\dfrac{1}{L_{\sigma s}} & 0 & 0 & 0\\[8pt]
0 & \dfrac{1}{L_{\sigma s}} & 0 & 0\\[8pt]
0 & 0 & \dfrac{1}{L_{\sigma r}} & 0\\[8pt]
0 & 0 & 0 & \dfrac{1}{L_{\sigma r}}\\[8pt]
-\dfrac{1}{L_{\sigma s}} & 0 & -\dfrac{1}{L_{\sigma r}} & 0\\[8pt]
0 & -\dfrac{1}{L_{\sigma s}} & 0 & -\dfrac{1}{L_{\sigma r}}
\end{bmatrix}
\begin{bmatrix}
v_{\alpha s}\\
v_{\beta s}\\
v_{\alpha r}\\
v_{\beta r}
\end{bmatrix}
$$

(A.64)

REFERENCES

1. W. Leonhard, *Control of Electrical Drives.* Springer-Verlag, 1985.
2. E. Levi, "Impact of Iron Loss on Behavior of Vector Controlled Induction Machines," *IEEE Trans. Ind. Appl.*, Vol. 31, No. 6, pp. 1287–1296, December, 1995.

INDEX

3L-NPC, *See* Three-level neutral point clamped multilevel topology,
Active and reactive power representations under unbalanced conditions, 454
Active neutral point clamped multilevel topology, 116
Active power calculation, 610
Active power control, 37
Active power oscillations cancellation strategy (APOC), 457
Active vector selection
   of P-DPC, 420, 422, 439, 441
   of P-DTC, 404, 405, 449
Air density, 7
Amplitude
   and frequency control by the use of PLL of stand-alone systems, 558
   of the space vector, 603
   of the stator voltage, 558
Analysis of the unbalanced voltage distortion based on the amplitude of space vectors, 461
Angle δ between the flux space vectors, 367, 388
Apparent power calculation, 609
$\alpha\beta$ components, 605
$\alpha\beta$ model of the DFIM, 212
   considering iron losses, 611

Back to back power electronic converter, 26, 87
Balanced voltage, 267
Balanced voltage dips, 268
Balancing of DC bus capacitors voltages, 128, 444
Base values, 175

Behavior
   under asymmetrical dips with crowbar protection, 488
   under three-phase dips with crowbar protection, 486
Betz limit, 8
Block diagram of vector control, 146, 313
Braking chopper, 492

Cascaded H bridge multilevel topology, 115
Characteristics of DFIM supported by DC coupled storage, 540
Circle trajectory of fluxes, 370, 390
Clarke transformation, 94, 605
Classic DPC, *See* Direct power control (DPC) of the DFIM,
Classic DTC, *See* Direct torque control (DTC) of the DFIM,
Classification of AC machine's controls, 366
Combination of Crowbar
   and braking chopper, 493
   and software solutions during voltage dips, 490
Complex conjugate, 214
Constant switching frequency, 399, 416
Constant switching period, h, 400, 418, 431
Continental outage, 480
Control block diagram
   of P-DPC, 420, 434
   of P-DTC, 403, 449
Control for unbalanced voltage
   based on DPC, 452
   based on DTC, 461

Doubly Fed Induction Machine: Modeling and Control for Wind Energy Generation,
First Edition. By G. Abad, J. López, M. A. Rodríguez, L. Marroyo, and G. Iwanski.
© 2011 the Institute of Electrical and Electronic Engineers, Inc. Published 2011 by John Wiley & Sons, Inc.

619

Printed in the United States
By Bookmasters